Cyclodextrin Glucanotransferase

Foundations and Frontiers in Enzymology

Cyclodextrin Glucanotransferase

A Comprehensive Guide from Gene to Applications

Edited by

Kiransinh N. Rajput

Department of Microbiology & Biotechnology, University School of Sciences, Gujarat University, Ahmedabad, Gujarat, India

Series Editor

Munishwar Nath Gupta

ELSEVIER

ACADEMIC PRESS

An imprint of Elsevier

Academic Press is an imprint of Elsevier
125 London Wall, London EC2Y 5AS, United Kingdom
50 Hampshire Street, 5th Floor, Cambridge, MA 02139, United States

ISBN: 978-0-443-41694-1

For Information on all Academic Press publications
visit our website at https://www.elsevier.com/books-and-journals

Publisher: Megan Ball
Acquisitions Editor: Ali Afzal-Khan
Editorial Project Manager: Harshita Bathla
Production Project Manager: Rajeswari R.
Cover Designer: Gopalakrishnan Venkatraman

Typeset by MPS Limited, Chennai, India

Contents

List of contributors

Ténor Dias-Mendel Allode
Laboratory of Natural Sciences and Applications (LSNA), Higher Normal School of Natitingou, National University of Sciences, Technologies, Engineering, and Mathematics of Abomey, Natitingou, Benin

Sulaiman Al-Zuhair
United Arab Emirates University, Al Ain, United Arab Emirates

Irene Andressa
Department of Food Technology, Federal University of Viçosa, University Campus, Viçosa, MG, Brazil

Nurul Aini Mohd Azman
Faculty of Chemical and Process Engineering Technology, Universiti Malaysia Pahang Al-Sultan Abdullah Lebuhraya Tun Khalil Yaakob, Gambang, Pahang, Malaysia

Nurul Nabila Huda Baharudin
Faculty of Chemical and Process Engineering Technology, Universiti Malaysia Pahang Al-Sultan Abdullah Lebuhraya Tun Khalil Yaakob, Gambang, Pahang, Malaysia

Tushar Banerjee
School of Life Sciences, Devi Ahilya Vishwavidyalaya, Indore, Madhya Pradesh, India

Grégorio Crini
Université Marie et Louis Pasteur, Laboratoire Chrono-environnement, Besançon, France

Jeferson Silva Cunha
Department of Food Technology, Federal University of Viçosa, University Campus, Viçosa, MG, Brazil

Manoela Maciel dos Santos Dias
Department of Food Technology, Federal University of Viçosa, University Campus, Viçosa, MG, Brazil

Marc Fourmentin
Université Littoral Côte d'Opale, UR 4493, LPCA, Laboratoire de Physico-Chimie de l'Atmosphère, Dunkerque, France

Sophie Fourmentin
Université Littoral Côte d'Opale, UR 4492, UCEIV, Unité de Chimie Environnementale et Interactions sur le Vivant, Dunkerque, France

Isha Jain
School of Life Sciences, Devi Ahilya Vishwavidyalaya, Indore, Madhya Pradesh, India

Arijit Jana
Department of Microbiology, Vidyasagar University, Midnapore, West Bengal, India

Rushikesh Joshi
Department of Biochemistry & Forensic Science, University School of Sciences, Gujarat University, Ahmedabad, Gujarat, India

Vodounon Cyrille Junior Alode
Laboratory of Natural Sciences and Applications (LSNA), Higher Normal School of Natitingou, National University of Sciences, Technologies, Engineering, and Mathematics of Abomey, Natitingou, Benin

Miriana Kfoury
Université Littoral Côte d'Opale, UR 4492, UCEIV, Unité de Chimie Environnementale et Interactions sur le Vivant, Dunkerque, France

Niha Mohan Kulshreshtha
Department of Biotechnology, B. Lal Institute of Biotechnology, Jaipur, Rajasthan, India

Dario Lacalamita
Università degli Studi di Bari Aldo Moro, Dipartimento di Chimica, Bari, Italy

Bruno Ricardo de Castro Leite Júnior
Department of Food Technology, Federal University of Viçosa, University Campus, Viçosa, MG, Brazil

Chong Liu
Department of Chemical & Materials Engineering, University of Auckland, New Zealand

Ajay Malviya
Department of Microbiology, Barkatullah University, Bhopal, Madhya Pradesh, India

Rohaida Che Man
Faculty of Chemical and Process Engineering Technology, Universiti Malaysia Pahang Al-Sultan Abdullah Lebuhraya Tun Khalil Yaakob, Gambang, Pahang, Malaysia

Paras C. Mevada
Department of Microbiology & Biotechnology, University School of Sciences, Gujarat University, Ahmedabad, Gujarat, India

Keshab Chandra Mondal
Department of Microbiology, Vidyasagar University, Midnapore, West Bengal, India

Krishnendu Mondal
Department of Microbiology, Vidyasagar University, Midnapore, West Bengal, India; Department of Microbiology, Kalinga University, Raipur, Chhattisgarh, India

Chiara Mongioví
Istituto per i Processi Chimico-Fisici, CNR, Bari, Italy

Nadia Morin-Crini
Université Marie et Louis Pasteur, Laboratoire Chrono-environnement, Besançon, France

Siti Kholijah Abdul Mudalip
Faculty of Chemical and Process Engineering Technology, Universiti Malaysia Pahang Al-Sultan Abdullah Lebuhraya Tun Khalil Yaakob, Gambang, Pahang, Malaysia

Ishfaq Nabi Najar
Fermentation and Microbial Biotechnology Division, CSIR-IIIM, Jammu and Kashmir, India

Babatunde A. Ogunbadejo
United Arab Emirates University, Al Ain, United Arab Emirates

Ana Flávia Coelho Pacheco
Cândido Tostes Dairy Institute, Agricultural Research Corporation of Minas Gerais (EPAMIG), Tenente Luiz de Freitas, Juiz de Fora, MG, Brazil

Flaviana Coelho Pacheco
Department of Food Technology, Federal University of Viçosa, University Campus, Viçosa, MG, Brazil

Rakeshkumar Panchal
Department of Microbiology & Biotechnology, University School of Sciences, Gujarat University, Ahmedabad, Gujarat, India

Dimple S. Pardhi
Department of Microbiology & Biotechnology, University School of Sciences, Gujarat University, Ahmedabad, Gujarat, India

Drashti Patel
Department of Microbiology & Biotechnology, University School of Sciences, Gujarat University, Ahmedabad, Gujarat, India; Department of Biochemistry & Forensic Science, University School of Sciences, Gujarat University, Ahmedabad, Gujarat, India

Radhika Patel
P. G. Department of Biosciences, Sardar Patel University, Satellite Campus, Vadtal Road, Bakrol, Gujarat, India

Sakshi Patel
Department of Biotechnology, B. Lal Institute of Biotechnology, Jaipur, Rajasthan, India

Vimal Prajapati
Aspee Shakilam Agri Biotechnology Institute, Navsari Agricultural University, Surat, Gujarat, India

Khushbu Rabadiya
Department of Microbiology & Biotechnology, University School of Sciences, Gujarat University, Ahmedabad, Gujarat, India

Kiransinh N. Rajput
Department of Microbiology & Biotechnology, University School of Sciences, Gujarat University, Ahmedabad, Gujarat, India

Tejalba R. Rathod
Department of Microbiology & Biotechnology, University School of Sciences, Gujarat University, Ahmedabad, Gujarat, India

Vikram H. Raval
Department of Microbiology & Biotechnology, University School of Sciences, Gujarat University, Ahmedabad, Gujarat, India

Fábio Ribeiro dos Santos
Department of Food Technology, Federal University of Viçosa, University Campus, Viçosa, MG, Brazil

Prayatna Sharma
Department of Microbiology, Sikkim University, Gangtok, Sikkim, India

Anjali D. Solanki
Department of Microbiology & Biotechnology, University School of Sciences, Gujarat University, Ahmedabad, Gujarat, India

Preetibala Solanki
Department of Microbiology, Barkatullah University, Bhopal, Madhya Pradesh, India

Siti Zubaidah Sulaiman
Faculty of Chemical and Process Engineering Technology, Universiti Malaysia Pahang Al-Sultan Abdullah Lebuhraya Tun Khalil Yaakob, Gambang, Pahang, Malaysia

Nagendra Thakur
Department of Microbiology, Sikkim University, Gangtok, Sikkim, India

Ujjval Trivedi
P. G. Department of Biosciences, Sardar Patel University, Satellite Campus, Vadtal Road, Bakrol, Gujarat, India

Lucia Dhiantika Witasari
Universitas Gadjah Mada, Faculty of Agricultural Technology, Department of Food and Agricultural Product Technology, Yogyakarta, Indonesia

Preface

The starch processing industry has undergone a radical transformation over the last few decades, moving from simple hydrolysis to the sophisticated production of high-value functional carbohydrates. At the vanguard of this biotechnological evolution stands cyclodextrin glucanotransferase (CGTase), a distinguished member of the α-amylase family that converts starch into cyclic oligosaccharides known as cyclodextrins. This edited book, *Cyclodextrin Glucanotransferase: A Comprehensive Guide from Gene to Applications*, serves as a definitive resource, tracing the journey of this enzyme from its genetic foundations to its vast industrial utility. The versatility of CGTase lies in its complex catalytic repertoire, including cyclization, coupling, and disproportionation reactions. Such multifaceted capabilities have rendered the enzyme indispensable for the synthesis of molecular "host" architectures. Cyclodextrins encapsulate diverse "guest" molecules, thereby significantly enhancing their solubility, chemical stability, and biological efficacy. As global demand for natural and functional bioactive ingredients continues to surge, a nuanced understanding of CGTase becomes an imperative for both rigorous academic exploration and industrial innovation.

This book is organized into four thematic sections, comprising 14 comprehensive chapters that provide a holistic view of the CGTase and cyclodextrins. The Introduction section lays the groundwork by introducing the α-amylase family. It provides an exhaustive overview of the enzyme's structural architecture and functional significance. The Mechanism and cyclodextrin glucanotransferase production section delves into the "how" and "where" of the enzyme. It explores the catalytic mechanisms that drive starch conversion and examines major microbial sources, with a dedicated focus on *Bacillus* spp. and other emerging bacterial genera. The Purification and engineering of cyclodextrin glucanotransferases section focuses on the biotechnological toolkit. This section traverses the complexities of gene cloning, heterologous overexpression, and the rigorous protocols of extraction, purification, and biochemical characterization. Furthermore, it addresses the frontiers of protein engineering and immobilization strategies aimed at optimizing enzymatic resilience and throughput. The last section, Cyclodextrins and their applications, bridges the gap between biocatalysis and the consumer end product. It delineates the physicochemical properties of standard and large-ring cyclodextrins, assesses the current global production landscape, and details the pervasive applications of these molecules across the food, pharmaceutical, and chemical sectors.

This book is meticulously curated to serve multidisciplinary professionals. It offers a robust theoretical and practical foundation for research scholars embarking upon their explorations in enzymology and carbohydrate chemistry. The detailed chapters on protein engineering and catalytic mechanisms offer deep insights into the current frontiers of CGTase research for scientists and teachers. Simultaneously, for industry professionals within the starch and pharmaceutical domains, this guide serves as a practical manual for enzyme production and large-scale application. I hope that this comprehensive guide inspires new research and facilitates the development of more efficient, sustainable processes in the ever-expanding world of enzyme technology.

The culmination of this academic endeavor would not have been possible without the intellectual contributions of our esteemed authors and subject matter experts, whose scholarly rigor is reflected in every chapter. I am deeply grateful to the technical support of my dedicated research scholars, especially Dimple S. Pardhi and Drashti Patel, along with Tejal Rathod, Anjali Solanki, and Paras Mevada, whose tireless efforts have made the timely completion of this book project possible. Special gratitude is extended to my organization, Gujarat University, for providing the institutional support and intellectual environment conducive to such an undertaking. Finally, I owe an immeasurable commitment of gratitude to my family; their unwavering patience and encouragement served as the quiet strength behind this project.

I sincerely hope that this comprehensive guide inspires novel explorations and facilitates the development of sustainable, efficient processes in the expanding realm of enzyme technology.

Kiransinh N. Rajput

SECTION I

Introduction

CHAPTER

1 Introduction to α-amylase family

Radhika Patel[1], Ujjval Trivedi[1] and Vimal Prajapati[2]
[1]*P. G. Department of Biosciences, Sardar Patel University, Satellite Campus, Vadtal Road, Bakrol, Gujarat, India,*
[2]*Aspee Shakilam Agri Biotechnology Institute, Navsari Agricultural University, Surat, Gujarat, India*

Introduction

Enzymes are biocatalysts that are important for every living organism. They catalyze a variety of hydrolysis and synthesis reactions. Unlike other chemical catalysts, enzymes speed up the reactions by lowering the activation energy, bringing the catalyzed reaction to equilibrium speedily (Singhania et al., 2015). Enzymes have been classified into six major groups according to their mode of action: oxidoreductase, transferase, hydrolase, lyase, isomerase, and ligase. Table 1.1 shows the list of classes and some industrial enzymes (Boyce 2001; Singhania et al., 2015). Among these is a large and biologically significant class of hydrolases that catalyze the cleavage of bonds through the addition of water. A prominent group of hydrolases includes glycoside hydrolases (GHs), which are involved in the synthesis, degradation, and rearrangement of α-1,4- and α-1,6-linked glucans. These enzymes are crucial to energy metabolism, structural polysaccharide synthesis, and industrial processes. As starch is the major storage carbohydrate in plants and a central feedstock in industrial bioprocessing, research on GH13 α-amylases connects core concepts of enzymology with the structural organization and physicochemical characteristics of starch (Lombard et al., 2014).

Starch

Starch is a polymer that is synthesized by plants as a result of photosynthesis. It is synthesized in plastids present in leaves, used as a storage material during night periods, and also in amyloplastids found in seeds, roots, and tubers, as long-term storage. In plants, starch granules are stored in different organs depending on the species. For example, they accumulate in tubers such as potatoes and cassava, in grains such as maize, wheat, rice, and barley, in whole fruits, such as apples and bananas, and in stems such as sago (Apriyanto et al., 2022). Starch is considered to be

Cyclodextrin Glucanotransferase. DOI: https://doi.org/10.1016/B978-0-443-41694-1.00005-8

Table 1.1 Classification of enzymes used in industrial processes.

Sr. No.	Class	Industrial enzymes
1.	Oxidasereductase	Catalase, glucose oxidase, laccase
2.	Transferase	Glucosyltransferases, fructosyltransferases
3.	Hydrolase	Amylase, cellulases, lipases
4.	Lyase	Pectate lyases, α-acetolactate decarboxylases
5.	Isomerase	Mutases, topoisomerases
6.	Ligase	Aminoacyl t-RNA synthetases

utilized most abundantly from the available carbohydrates (40%–75%) of modern dietary energy intake. Starch in its raw crystalline form is digested slowly than cooked starch (Hardy et al., 2015; Van Der Maarel et al., 2002). In addition to using starch-containing plant parts directly as food, starch is extracted and used either as such or after chemical or enzymatic processing to produce diverse products, including starch hydrolysates, glucose sirups, fructose, starch or maltodextrin derivatives, and cyclodextrins (Fig. 1.1). Although many plants produce starch, only a few are significant for industrial starch processing (Van Der Maarel et al., 2002). Corn starch is widely used industrially as a thickening and gelling agent. Wheat starch is used in the baking industry and the production of adhesives. Rice starch is used as laundry starch and in cosmetic dusting powders. Potato starch is used in the food, textile, paper, and adhesive industries (Guzmán-Maldonado et al., 1995).

Starch is composed of two polysaccharide fractions: amylose and amylopectin (Fig. 1.2A). Amylose is traditionally described as a linear polymer of α-D-glucopyranosyl units linked primarily by (1→4) glycosidic bonds; however, it is now recognized that a small proportion of amylose molecules contain occasional α-(1→6) linkages, introducing slight branching. The average amylose content of starches is 20%–25%. In contrast, amylopectin is a highly branched polysaccharide consisting of hundreds of short α-(1→4)-glucan chains interconnected through α-(1→6) bonds at branch points, which account for approximately 5%–6% of its linkages. Amylopectin is also considered one of the largest molecules in nature, as a complete amylopectin consists of, on average, about 2,000,000 glucose units. While amylose contributes to the formation of helical structures and compact granule packing, the extensive branching of amylopectin imparts a more open, branched architecture, a structural feature shared with glycogen. Amylopectin is water-soluble, while amylose and starch are insoluble in cold water, thereby aiding their extraction from natural sources. If a starch–water mixture is heated, gelatinization occurs, wherein starch absorbs water, resulting in irreversibly swollen particles. In the course of gelatinization, amylose is leached from the starch molecules, resulting in a thick solution. If a thick starch solution is cooled, a gel is formed. The gel formation process is known as retrogradation, in which amylose is primarily

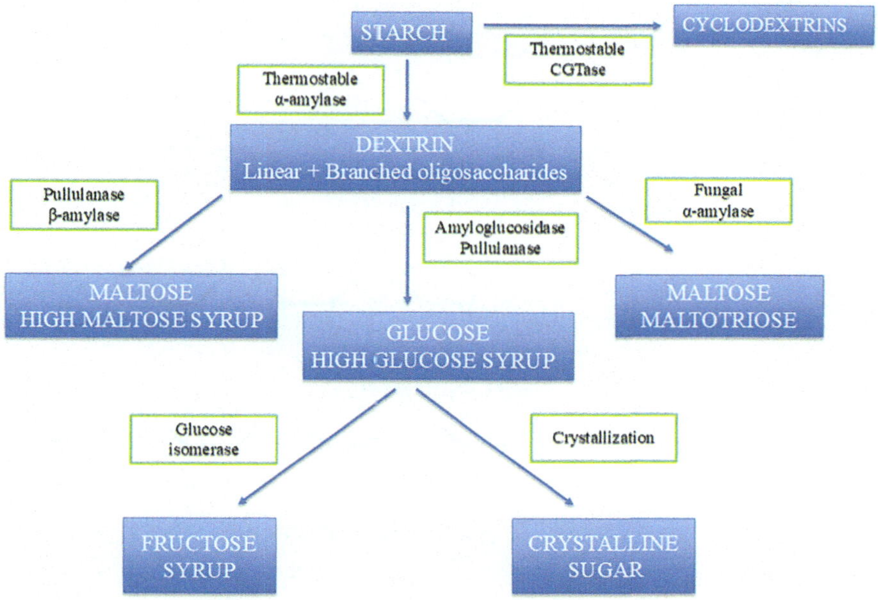

FIGURE 1.1

Industrial processing of starch into different products.

involved, while amylopectin is refractory to retrogradation due to its highly branched structure. Together, the linearity of amylose and the branched nature of amylopectin define the physicochemical properties of starch, influencing its digestibility, gelatinization, and functional applications (Pérez & Bertoft, 2010; Buleon et al., 1998; Xu et al., 2023; Myers et al., 2000).

Starch converting enzyme: amylase

The enzymatic hydrolysis of starch is carried out by amylolytic enzymes. The term amylase is derived from the Greek word "amylon," which means starch. Payen and Persoz first precipitated a product by starch splitting from malt in 1833, and they named it "diastase," derived from the Greek word "diastasis," meaning separation (Salt & Schenker, 1976). Amylases are broadly categorized as endo and exoamylases: α-amylase, β-amylase, and γ-amylase, with α-amylase being a particularly important industrial enzyme (Miłek & Lamkiewicz, 2022).

α-Amylase (EC 3.2.1.1) is an endo-GHs that randomly cleaves internal α-(1→4) glycosidic bonds, producing maltooligosaccharides and α-limit dextrins. These enzymes are ubiquitous across archaea, bacteria, and fungi, reflecting their essential role in carbohydrate metabolism and energy mobilization. As the earliest discovered

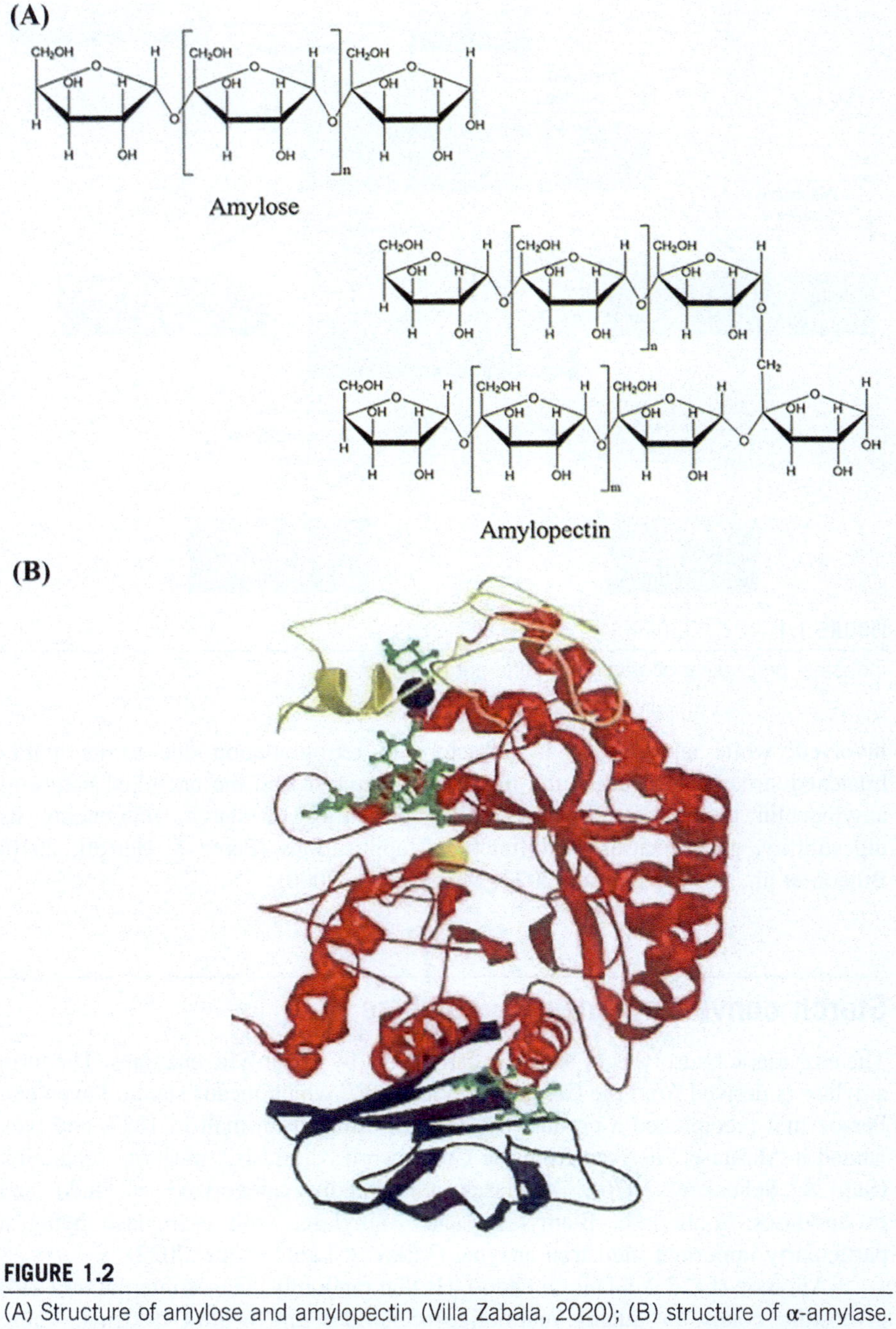

FIGURE 1.2

(A) Structure of amylose and amylopectin (Villa Zabala, 2020); (B) structure of α-amylase. Domain A (*red*), which is the largest and represents a typical $((\beta/\alpha)_8)$ TIM-barrel structure, domain B (*yellow*), and domain C (*purple*). In the catalytic center, the Ca^{+} ion (*blue*) and the chloride ion (*yellow*) spheres are present along with the green structures bound to active sites (Souza & Magalhães, 2010).

biological catalyst, α-amylase has become a foundational model for investigating protein structure, catalytic mechanisms, and the complex interplay between sequence and function. Salivary and pancreatic amylases are α-amylases, while plant-origin amylases are β-amylases (Far et al., 2020; Salt & Schenker 1976).

β-Amylase (EC 3.2.1.2), also called glycogenase or saccharogen amylase, is an exoamylase produced by bacteria, fungi, and plants. It hydrolyzes every second α-(1→4) bond from the nonreducing ends of starch to release maltose. In seeds, β-amylase remains dormant prior to germination, whereas α-amylase and proteases are active. During ripening, β-amylase contributes to maltose production, imparting sweetness, and is a key enzyme in malt preparation from cereal grains. While absent from animal tissues, β-amylase can occur in microbes of the digestive system. γ-Amylase (EC 3.2.1.3), also called glucoamylase or amyloglucosidase, hydrolyzes both α-(1→4) and α-(1→6) glycosidic bonds at the nonreducing ends, releasing glucose. Unlike other amylases, it preferentially targets α-(1→6) linkages and functions optimally at acidic pH (~3) (Paul et al., 2021; Ali et al., 2025). Table 1.2 shows different families of GH with their action type and bonds cleaved, along with the products formed.

Interest in the evolutionary relationships of α-amylase intensified approximately 25 years ago following the recognition that it belongs to a broader family of starch-active hydrolases and related enzymes that share homologous amino acid sequences and mechanistic features, despite catalyzing distinct reactions. A representative member of this family is cyclodextrin glucanotransferase (CGTase; EC 2.4.1.19), which shares the initial catalytic step with α-amylase but, unlike α-amylase, utilizes a saccharide acceptor instead of water, thereby promoting transglycosylation rather than hydrolysis. Consequently, it became evident that CGTases and α-amylases, despite their distinct catalytic specificities, exhibit substantial sequence similarity. This resemblance has historically complicated the accurate functional assignment of α-amylase versus CGTase activity and has posed challenges for resolving their evolutionary relationships (Janeček & Gabriško, 2016).

α-Amylase family

Glycosyl hydrolases (EC 3.2.1.x) are the key enzymes involved in carbohydrate metabolism found in the major kingdoms archaea, eubacteria, and eukaryotes. Initially, in the late 1980's and early 1990's, the work on systematic comparison of the primary sequences of glycosyl hydrolases was carried out, revealing (1) families containing multiple EC entries and (2) enzymes with similar substrate specificities that belonged to unrelated families. A classification that more accurately reflected sequence—and consequently structural—similarity was therefore considered valuable, particularly in view of the rapidly increasing number of glycosyl hydrolase genes being sequenced and the comparatively slower increase in the number of three-dimensional structures being solved (Svensson, 1988; Henrissat 1991).

Table 1.2 Classification and action of starch-modifying enzymes.

Enzyme	EC number	Action type	Bonds cleaved	Main products
α-Amylase glycoside hydrolase 13 family	EC 3.2.1.1	Endo-hydrolase	α-1,4	Maltose,maltotriose, α-dextrins
β-Amylase GH14	EC 3.2.1.2	Exo-hydrolase	α-1,4	Maltose, β-limit dextrins
Glucoamylase GH15	EC 3.2.1.3	Exo-hydrolase	α-1,4 & α-1,6	Glucose
Isoamylase GH13	EC 3.2.1.33	Debranching	α-1,6	Linear amylosic chains
Cyclodextrin glucanotransferase GH13	EC 2.4.1.19	Cyclization	α-1,4	α, β, γ-cyclodextrins
GBE GH13	EC 2.4.1.18	Branching	α-1,4 (cleave)	Branched starch/glycogen-like polymers

Structural characteristics

In the sequence-based classification of carbohydrate-active enzymes, α-amylases are among the most prevalent GHs. According to the sequence-based classification of GHs, α-amylase belongs to the GH13 or α-amylase family. This family includes enzymes that possess several major characteristics: (1) they act on reactions that involve α-glycosidic bonds, hydrolyzing these bonds to form α-anomeric monosaccharides and/or β-anomeric monosaccharides or participate in transglycosylations that form new α-(1–4) or α-(1–6) glycosidic bonds; (2) they possess a $(\beta/\alpha)_8$ TIM-barrel fold that has a conserved topology with a core containing the catalytic residues; and (3) they possess four highly conserved regions within their amino acid sequence that comprise residues participating in catalysis as well as residues that are fundamental in ensuring that the TIM-barrel topology is properly maintained (Henrisstat 1991; Kuriki & Imanaka, 1999). Fig. 1.2B shows the structure of α-amylase, where the polypeptide chain is folded in three domains: A, B and C. α-Amylases display a conserved domain organization consisting of three domains, designated A, B, and C. The largest domain, domain A, has a characteristic barrel-like $(\beta/\alpha)_8$ architecture, which houses the catalytic core. Domain B, which is located between domains A and C and is held by a disulfide bridge to domain A, also assists in the formation of the substrate-binding cleft. Domain C, which is located at the C-terminus, has a β-sheet-rich structure and is a structurally independent domain. The active site of α-amylases is located in a long cleft between domains A and B, where a bound Ca^{2+} ion is responsible for the stabilization of the three-dimensional structure of the enzyme. The catalytic mechanism involves conserved acidic residues located in a multisubsite binding pocket, which enables the cleavage of internal α-1,4-glycosidic bonds.

GHs are a vast group of enzymes that cleave glycosidic bonds in glycosides, glycans, and glycoconjugates. GHs are classified into different families in the carbohydrate-active enzymes or CaZy (www.cazy.org), on the basis of sequence similarities and predicted structures (Vuong & Wilson 2010). The database was originally created in 1991, for classification of GHs and till 2023, this section represents the largest count of 186 GH families. Among these 186 families, four, that is, GH13—the largest and main, GH57—the second and smaller, GH119—a very small family related to GH57—and in the family GH126, the presence of pure amylase specificity remains, but it may be considered among the α-amylase GH families. In particular, the two major α-amylase families, GH13 and GH57, encompass not only α-amylases but also the majority of other amylolytic enzymes, including α-glucosidases, pullulanases, amylopullulanases, isoamylases, CGTases, 4-α-glucanotransferases, α-glucan branching enzymes, α-glucan debranching enzymes, and sucrose isomerases. In contrast, the other two classical "amylases," β-amylase and glucoamylase, each form distinct GHs families, GH14 and GH15, respectively (Janeček & Svensson 2022; Janeček 2023).

Advances in sequence and structural analysis techniques to interpret the functions of GHs revealed that there is no direct association between catalysis, substrate specificity, and EC classification systems. Several GH families, including GH13, GH57, GH119, and GH126, include enzymes with different α-bond cleavage activities with unique evolutionary and genetic backgrounds. Detailed characterization of the enzymatic functions of all family-13 enzymes further revealed that these enzymes exhibit a retaining mechanism with a covalently linked glycosyl-enzyme intermediate (Janeček et al., 2014; CAZypedia Consortium, no.date). Family GH57 comprises a group of enzymes that adopt an incomplete (β/α) $_7$-barrel fold, in contrast to the canonical $(\beta/\alpha)_8$-barrel of GH13. The catalytic domain is followed by a helical bundle and lacks a classical transition-state stabilizer, with members functioning as α-amylases, branching enzymes, and 4-α-glucanotransferases (Janeček et al., 2014). Analysis of over 5000 GH57 sequences resulted in the division of the family into 10 subfamilies that are functionally consistent, eight of which are linked to particular enzymatic functions. Two subfamilies, GH57_9 and GH57_10, have uncharacterized members despite the presence of conserved catalytic regions, which mark them as targets for future research (Poláček et al., 2025). Family GH119 is a small group that is evolutionarily and structurally related to GH57, sharing the incomplete $(\beta/\alpha)_7$-barrel architecture and two-residue catalytic machinery. Functionally, GH119 enzymes act as α-amylases, as exemplified by the *Bacillus circulans* IgtZ protein, while retaining distinct conserved sequence features that differentiate them from GH57 (Vuillemin et al., 2024; Janeček & Kuchtová, 2012). Another Family, GH126, is atypical among α-amylase-related families, as its members show closer structural homology to inverting β-glucan-active hydrolases and adopt an unusual $(\alpha/\alpha)_6$-barrel fold. Although capable of degrading amylose and glycogen, the catalytic mechanism of GH126 enzymes remains unresolved, with evidence suggesting either inverting or retaining behavior (Janeček et al., 2014). The catalytic mechanism and catalytic fold of different families of glycosyl hydrolases are shown in Table 1.3

Table 1.3 Catalytic mechanism and catalytic fold of different family of glycosyl hydrolase.

Family	Principal specificities	Catalytic fold	Catalytic mechanism
Glycoside hydrolase 13 (GH13) family	α-Amylase, pullulanase, cyclodextrin glucanotransferase, branching enzyme	$(\beta/\alpha)_8$-barrel	Retaining
GH70	Glucansucrases, transglycosylases	$(\beta/\alpha)_8$-barrel	Retaining
GH77	4-α-Glucanotransferase	$(\beta/\alpha)_8$-barrel	Retaining
GH57	α-Amylase, amylopullulanase	$(\beta/\alpha)\backslash_7$-barrel	Retaining
GH119	α-Amylase (e.g., IgtZ)	Incomplete $(\beta/\alpha)_8$-barrel	Retaining

Taxonomical distribution and hierarchical classification of glycoside hydrolase 13 α-amylase family

There are three main enzyme classes: (1) hydrolases (EC 3); (2) transferases (EC 2); and (3) isomerases (EC 5) in the α-amylase family GH13. At a higher hierarchical level, the CAZy database has introduced clans that group together multiple glycosyl hydrolase families based on shared structural features. Within this framework, GH13 is classified under the GH-H clan, together with the GH70 and GH77 families. Clan members share a common evolutionary ancestor, a $(\beta/\alpha)_8$-barrel fold, and a retaining catalytic mechanism (Janeček & Svensson 2022).

The α-amylase subfamilies include GH13_1, GH13_5, GH13_6, GH13_7, GH13_15, GH13_19, GH13_24, GH13_27, GH13_28, GH13_32, GH13_36, GH13_37, and GH13_42. The non-α-amylase subfamilies in the GH13 includes GH_2, GH13_3, GH13_4, GH13_8, GH13_9, GH13_10, GH13_11, GH13_12, GH13_13, GH13_14, GH13_16, GH13_17, GH13_18, GH13_20, GH13_21, GH13_22, GH13_23, GH13_25, GH13_26, GH13_29, GH13_30, GH13_31, GH13_33, GH13_34, GH13_35, GH13_38, GH13_39, GH13_40, and GH13_41, which consist of enzymes having different specificities and catalytic behaviors (Paul et al., 2021). At present, in the CaZy Database, there are 49 GH13 subfamilies, from GH13_1 to GH13_49. Initially, there were just 10 subfamilies, GH13_1 to GH13_10, in 1997. In 2006, a major expansion occurred by the addition of 22 new subfamilies, GH13_11 to GH_32, followed by the continuous addition of 17 subfamilies (GH13_33 to GH13_49) during the years 2010 to 2024.

Hierarchical clustering of GH13 subfamily profiles across major taxonomic groups reveals both deep evolutionary conservation and pronounced lineage-specific diversification within the α-amylase family (Lombard et al., 2014). The phylogenetic heatmap highlights extensive microbial diversity across all three domains of life, with bacterial sequences accounting for approximately 77.5% of the 225,899 GH13 entries currently cataloged in the CAZy database. Clustering based on subfamily composition broadly reflects phylogenetic relationships, with Firmicutes (~52,000 sequences) and Proteobacteria (~48,000 sequences) exhibiting the greatest diversification of GH13 subfamilies while also revealing instances of functional convergence among distantly related taxa occupying similar ecological niches.

Variation in heatmap (Fig. 1.3) intensity across the 49 recognized GH13 subfamilies (GH13_1–GH13_49) illustrates differential patterns of retention and expansion within the family. Core α-amylase subfamilies, including GH13_1, GH13_11, and GH13_36, display pan-domain distribution consistent with their central roles in α-glucan metabolism, whereas more specialized functional groups, such as pullulanases (GH13_9, GH13_14, and GH13_15) and maltogenic α-amylases (GH13_19 and GH13_20), show pronounced lineage-specific enrichment, reflecting adaptation to distinct carbohydrate environments (Janeček et al., 2014). Recent phylogenomic analyzes have further expanded the family by establishing 11 novel subfamilies between 2019 and 2024 (GH13_41–GH13_49), with the most recent addition, GH13_49, representing a neopullulanase-like clade. Notably, the identification of GH13_43 in 2020, comprising

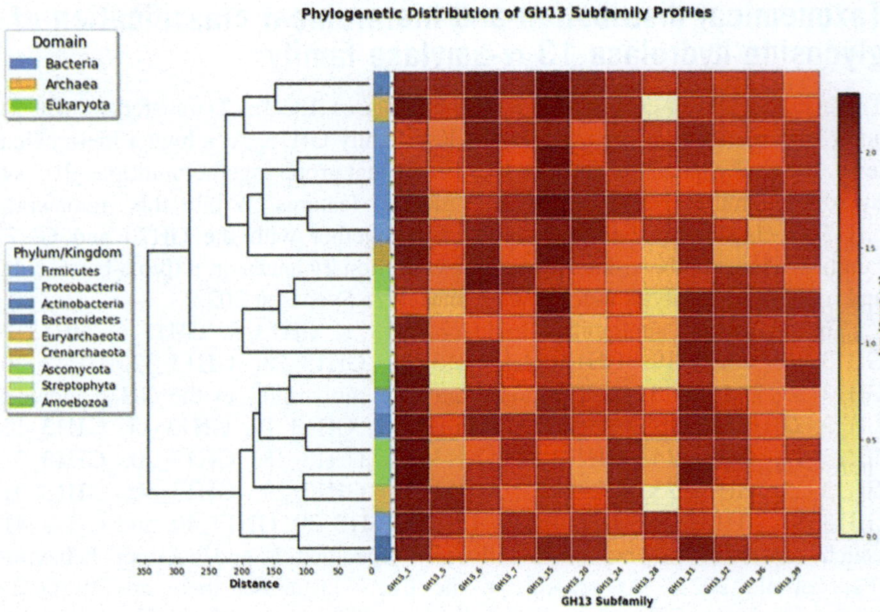

FIGURE 1.3

Phylogenetic landscape of glycoside hydrolase 13 (*GH13*) family subfamily distribution. Hierarchical clustering of major taxonomic groups based on their GH13 subfamily composition is shown. The heatmap shows log-transformed abundances for 12 representative GH13 subfamilies.

α-1,4-glucan lyases, marked a significant biochemical advance by introducing a previously unrecognized eliminative catalytic mechanism within the GH13 family. Collectively, these evolutionary patterns, together with more than 730 solved three-dimensional structures and 884 biochemically characterized enzymes, underscore GH13 as the most structurally and functionally diverse clan within the GH superfamily, with ongoing metagenomic and synthetic biology efforts continuing to expand its catalytic and biotechnological potential (CAZy Database).

Within the GH13 family, there is a critical subgroup, the CGTase (EC 2.4.1.19), classified in subfamily GH13_2. While α-amylases primarily perform hydrolysis, CGTases are enigmatic enzymes that prioritize intramolecular transglycosylation (cyclization) to produce nonreducing cyclic dextrins from starch. The conventional classification of CGTases within the GH13_2 subfamily was based on four- or five-domain structures, which were mainly associated with cyclodextrin producers. However, recent research has revealed a new set of CGTases that possess a reduced three-domain (ABC) structure, which represents the catalytic domain. These enzymes have been shown to possess improved hydrolysis capabilities and to produce fermentable sugars and linear dextrins. This finding marks the fifth monophyletic lineage within the GH13_2 subfamily and indicates that the utilization of cyclodextrins in carbohydrate metabolism is a major adaptive strategy for extremely thermophilic bacteria (Centeno-Leija et al., 2022).

As the GH13 family unites a broad array of α-amylase-related enzymes that retain a conserved $(\beta/\alpha)_8$ TIM-barrel catalytic core and the canonical retaining catalytic triad (Asp–Glu–Asp), yet diverge extensively in substrate specificity, accessory modules, and surface topology, these subfamily-level specializations are principally encoded by variations in domain B loop length and sequence, differential acquisition of carbohydrate-binding modules or surface binding sites, and insertions/deletions in peripheral regions (notably domain C) that modulate substrate access and product profile (Stam et al., 2006; Kuriki & Imanaka, 1999). Such structural permutations underlie distinct functional outcomes—cyclodextrin formation by CGTases, extracellular raw-starch degradation by bacterial and fungal raw-starch amylases, thermostable catalysis in archaeal enzymes, chloride-activated digestion in animal amylases, and transglycosylation or α-1,6-branch processing in neopullulanase-type enzymes—each of which correlates with characteristic sequence fingerprints and domain architectures (Klein & Schulz, 1991; Liu et al., 2017; Dong et al., 1997; Kuriki & Imanaka, 1999). Representative structural examples and PDB entries that illustrate these principles are summarized in Table 1.3, which maps subfamily scope to defining structural features, representative enzymes, and primary structural references (Kim et al., 1999; Robert et al., 2003; Janeček et al., 2014; Stam et al., 2006).

The GH13 family is polyspecific in a unique manner, requiring a subcategorization into 49 subfamilies to more accurately link amino acid sequence with enzymatic function. The subfamilies can be divided into five main functional groups: endo-acting hydrolases, debranching enzymes, transglycosylases (including CGTases), exo-acting/saccharifying enzymes, and dual-specificity /intermediary enzymes.

The endo-acting hydrolases correspond to the classical α-amylase activity, which catalyzes random hydrolysis of internal α-1,4-glycosidic bonds to quickly degrade the viscosity of starch polymers. This family can be classified by taxonomic origin, including fungal amylases in GH13_1, bacterial liquefying amylases in GH13_5, and animal-specific amylases in GH13_15 and GH13_24. The debranching enzymes, on the other hand, are specialized for the hydrolysis of α-1,6-glycosidic bonds at the branch points of amylopectin and glycogen. The most important members of this family include isoamylases and glycogen debranching enzymes in GH13_11, as well as several pullulanases classified into subfamilies.

A very specialized functional group is that of the transglycosylases, which favor the formation of new glycosidic bonds over simple hydrolysis. This group is dominated by the GH13_2 subfamily (CGTases), which are specialized in the intramolecular transglycosylation (cyclization) reaction to yield nonreducing cyclic dextrins from starch. Other enzymes in this group are the starch-branching enzymes of the GH13_8 and GH13_9 subfamilies, which catalyze the formation of α-1,6-branches, and the amylosucrases of the GH13_4 subfamily, which produce α-glucans from sucrose.

The exo-acting/saccharifying enzymes support the final steps of starch hydrolysis by removing terminal units or short oligosaccharides from the nonreducing ends of the substrate. This category of enzymes includes bacterial saccharifying amylases (GH13_28) and several α-glucosidases in subfamilies GH13_17, GH13_21, and GH13_30. Finally, the family also includes intermediary enzymes that display dual specificity for α-1,4 and α-1,6 linkages. Subfamilies GH13_20 and GH13_36 include enzymes such as neopullulanases and cyclomaltodextrinases, which act as an evolutionary and functional intermediates between purely hydrolytic and more specialized debranching or transglycosidic enzymes.

Environmental adaptations of glycoside hydrolase 13 family enzymes

Extremophiles are microorganisms that are able to survive and reproduce in extreme environments that are unfavorable for life. They are able to survive in extreme physicochemical conditions, such as high or low temperatures, extreme pH, high salinity, and high pressure. Based on their adaptive traits, extremophiles are classified as thermophiles (high temperature), psychrophiles (low temperature), halophiles (high salt), acidophiles (acidic pH), alkaliphiles (alkaline pH), and barophiles or piezophiles (high pressure). The majority of the extremophiles belong to the domains archaea and eubacteria (Rampelotto, 2013).

Industrial bioprocessing often requires enzymes to operate under conditions of high temperature, extreme pH, or high salinity. GH13 enzymes from extremophiles exhibit specific structural configurations to survive these environments. Based on the sources, thermophilic enzymes exhibit distinct differences in their amino acid composition compared to mesophilic counterparts to maintain structural integrity at high temperatures. These differences are primarily seen in the preference for specific residues that enhance stability, rigidity, and resistance to chemical degradation. Some structural variations that form a strong basis of thermostability in these thermophilic proteins, as compared to mesophilic protiens, are: (1) their changes in the amino acid compositions. Thermophilic proteins generally possess a higher ratio of charged and polar amino acids—specifically glutamate (Glu), arginine (Arg), tyrosine (Tyr), aspartate (Asp), and lysine (Lys)—as compared to residues, such as alanine (Ala), asparagine (Asn), threonine (Thr), serine (Ser), and valine (Val), (2) higher GC content in their coding sequence, (3) higher number of disulfide bridges, (4) additional hydrogen bonds and greater ionic interactions, (5) higher hydrophobicity in the core, (6) stabilization by heat-stable chaperones, and (7) a greater number of ionic interactions (Prakash & Jaiswal, 2010; Szilágyi & Závodszky, 2000).

Most of the GH13 enzymes display an optimal pH range of 4.0–7.0. The pH of the medium is an important factor that influences the protonation and deprotonation of amino acids. Proteins tend to drop their primary structure and start unfolding when exposed to exceptionally low or high pH. This is because of the repulsion caused by similar charges of the functional groups of amino acid residues. The molten globule state that is formed during protein unfolding causes a decrease in tertiary structure and an increase in secondary structure (Yadav & Prakash, 2011). The acid stability of α-amylases is due to (1) decreasedpermeability of the cell membrane to protons, (2) the presence of a secondary transport system for the efflux of protons, (3) the well-developed repair machinery of both DNA and proteins, and (4) the presence of tetraether lipids in the archaeal membrane, which have a low probability of being degraded by acids, and (5) the presence of acidic amino acid residues such as Asp and Glu, which confer a high negative charge to the protein and stabilize the protein structure at a tremendously low pH (Sharma et al., 2016; Gupta et al., 2021). Those organisms that are capable of surviving and continuing their metabolism at pH > 9 are known as alkaliphiles and are even active in the pH range 10–13. Alkaliphilic microorganisms play a significant role in the laundry and detergent industry through the production of alkali-stable enzymes such as α-amylases. Their ability to remain active at high pH is supported by specific physiological adaptations, most notably modifications in cell wall composition. These organisms increase the acidic character of their cell walls by synthesizing polymers such as teichuronic acids and negatively charged amino acid polymers (e.g., poly-γ-D-glutamic acid), which help maintain cellular stability and functionality under alkaline conditions (Kulkarni et al., 2019; Gupta et al., 2021) (Table 1.4).

Halophiles are microorganisms that grow optimally in environments with high-salt concentrations, although their dependence on salt varies. Some species are merely salt-tolerant and can survive moderate salinity, whereas extreme halophiles require very high-salt levels for growth. These organisms are commonly found in habitats such as deep marine environments, salt pans, and salt marshes. Their proteins are structurally adapted to remain stable and functional under high ionic conditions that would typically denature nonhalophilic proteins (Tadeo et al., 2009). These microorganisms adapt to extreme salinity through specialized molecular and physiological strategies. Comparative analyzes of halophilic proteins reveal a characteristic enrichment of acidic amino acid residues and a concurrent reduction in basic residues, particularly positively charged lysine. This compositional bias enhances protein solvation and stability in high-salt environments by promoting favorable interactions with surrounding ions and maintaining structural integrity. Halophilic proteins also display a distinctive surface architecture dominated by aspartate and glutamate residues and an overall low hydrophobic character. Mutational studies have demonstrated that altering these surface features can significantly modify salt dependence, allowing the conversion of halophilic proteins toward mesophilic behavior and, conversely, the engineering of mesophilic proteins to function efficiently under obligate high-salt conditions (Tadeo et al., 2009; Tan et al., 2008; Ahmad et al., 2022).

Industrial and biotechnological application of glycoside hydrolase 13 family

α-Amylase, pullulanase, cyclodextrin glycosyltransferase, and maltogenic amylase are generally used in industrial starch processing, with α-amylase being the most widely used (Table 1.5). α-Amylases are used not only for starch liquefaction and

Table 1.4 Comparative study of a few Glycoside hydrolase 13 (*GH13*) subfamilies, showing representative enzymes, PDB IDs, structural features, and functions.

Subfamily	Representative enzymes	PDB ID	Key structural features	Function/substrate	References
GH13_1	Taka-amylase A (*Aspergillus oryzae*); fungal α-amylases	6TAA 2TAA	$(\beta/\alpha)_8$-barrel, TIM-barrel; CBM20/21 SBDs; raw-starch binding sites	Extracellular starch degradation; raw-starch binding	Swift et al. (1991); Stam et al. (2006); Matsuura et al. (1984)
GH13_2	Secreted fungal α-amylases; yeast α-amylases; cyclodextrin glucanotransferase (CGTase) (*Bacillus circulans*)	1CDG	TIM-barrel with extended loops; secretion signals; glycosylation sites; CGTase subsites for cyclodextrin formation	Secreted starch digestion; cyclodextrin production	Klein & Schulz, (1991); Lawson et al. (1994)
GH13_5	Liquefying bacterial α-amylases (*Bacillus stearothermophilus*); maltogenic amylases	3BH4 5A2A	Distinct domain B motifs; thermostable scaffolds	Liquefaction of starch; maltooligosaccharide production	Kim et al. (1999); Chai et al. (2016); Alikhajeh et al. (2010)
GH13_6	Plant α-amylases (barley AMY1, rice AMY3); green-algal amylases	3WN6	Short polypeptides; "sugar-tongs" in domain C; Ca^{2+} binding	Seed/endosperm starch mobilization	Robert et al. (2003); Ochiai et al. (2014)
GH13_7	Hyperthermophilic archaeal α-amylases (*Pyrococcus furiosus*, Thermococcus spp.)	4AEF	Thermostability adaptations; archaeal motifs	Active > 80°C; starch hydrolysis in extreme environments	Dong et al. (1997); Park et al. (2013)
GH13_15 / GH13_24	Mammalian pancreatic α-amylase; human salivary α-amylase; insect α-amylases	5U3A	Chloride-activation sites; glycosylation; SBS	Chloride-dependent starch digestion	Goldbach et al. (2019)
GH13_28	Bacterial saccharifying α-amylases (*Bacillus subtilis*)	1BAG	Domain B distinct from GH13_5	Saccharifying activity (maltose/maltodextrin production)	Fujimoto et al. (1998)
GH13_32	Actinobacterial "animal-type" α-amylases; maltotriohydrolases	1AQH 1 AQM	C-terminal CBM20/25; animal-type motifs	Maltotriohydrolase activity; saccharifying	Janeček et al. (2014); Aghajari et al. (1998)
GH13_36	Neopullulanase; oligo-1,6-glucosidase relatives	6BS6	MPDLN fingerprint in region V	α-1,6-branched glucan degradation; transglycosylation	Kuriki & Imanaka, (1999); Arnal et al. (2018); Oslancova & Janeček (2002)
GH13_37	Marine bacterial raw-starch amylases (AmyP)	5H06	Very short loop replacing domain B; catalytic + C-terminal SBD	Rapid raw-starch degradation (e.g., rice starch)	Liu et al. (2017)

Table 1.5 Applications of glycoside hydrolase 13 family enzymes in different fields.

Application	Process/use	Enzyme(s)
Glucose/ fructose syrup production	Starch liquefaction	α-Amylase
Glucose/ fructose syrup production	Starch saccharification	Amyloglucosidase, pullulanase, maltogenic α-amylase, α-amylase, isoamylase
Detergents and industrial uses	Laundry detergents and cleaners; reduction of haze formation in juices, baking, brewing; digestibility of animal feed; fiber and cotton desizing; sanitary waste treatment	α-Amylase
Cyclodextrin production	Cyclodextrin production	Cyclodextrin glycosyltransferase
Specialty starch products	Thermoreversible starch gels	Amylomaltase
Specialty starch products	Cycloamylose	Amylomaltase, branching enzyme, cyclodextrin glycosyltransferase

saccharification but also for producing stable, viscous starch solutions for textile warp sizing, haze reduction in beer and juices, and improving the digestibility of animal feed. α-Amylases are also being increasingly used in laundry and dishwashing detergents, where enzymes adapted for moderate temperatures and alkaline pH conditions enhance starch removal during low-temperature washing. On the other hand, other GH13 enzymes, such as amylomaltase and branching enzyme, have fewer industrial applications.

Starch is one of the most abundant renewable biological resources and a primary dietary carbohydrate worldwide. Its hydrolysis underpins numerous industrial processes, where earlier acid-based methods have largely been replaced by enzymatic approaches due to their higher specificity, milder reaction conditions, and the availability of efficient starch-degrading microorganisms. Enzymatic starch conversion is especially important in the food industry, producing a wide range of value-added products, including glucose and maltose sirups, fructose sweeteners, cyclodextrins, and fat-mimetic ingredients. In the baking industry, fungal α-amylases belonging to the GH13_1 subfamily, including enzymes from *Aspergillus oryzae*, play a key role in enhancing bread quality by increasing loaf volume, improving crust coloration, and contributing to flavor development. These enzymes hydrolyze starch in the dough into fermentable sugars and low-molecular-weight dextrins, thereby supplying baker's yeast with substrates needed for efficient fermentation. As a result, dough aeration is improved, and the final product exhibits a softer crumb and reduced firmness (Dekker, 1994).

For starch liquefaction, highly thermostable bacterial α-amylases from the GH13_5 subfamily, particularly those derived from *Bacillus licheniformis*, are favored because they remain active at elevated temperatures (70°C–100°C). These enzymes partially hydrolyze gelatinized starch, rapidly reducing slurry viscosity and preparing the substrate for efficient downstream saccharification. Owing to their thermal stability and catalytic efficiency, they are widely employed in industry for the large-scale production of glucose and fructose sirups used in processed foods and beverages (Mobini-Dehkordi & Javan, 2012).

In brewing, α-amylases belonging to the GH13_28 subfamily, including enzymes from *Bacillus subtilis*, are added during mashing to enhance the conversion of grain starch into fermentable sugars such as maltose. These enzymes randomly cleave α-1,4-glycosidic bonds, promoting starch liquefaction—an effect that is particularly valuable when adjunct grains are used alongside barley. Efficient enzymatic starch breakdown supports optimal yeast fermentation, thereby contributing to consistent alcohol production and overall beer quality (Dekker, 1994).

For flavor encapsulation applications, CGTases of the GH13_2 subfamily are employed to produce cyclodextrins—cyclic oligosaccharides that function as protective carriers for volatile aromas and sensitive flavor compounds. By forming inclusion complexes, these modified starch derivatives improve the stability and retention of flavors during processing steps such as spray drying. Enzymatic modification can also generate highly porous starch structures with enhanced surface area, further improving their capacity to adsorb and retain aroma compounds in functional encapsulation matrices (Madene et al., 2006).

At present, more environmentally safe enzyme-based liquid detergents are preferred and used. α-Amylase, together with protease, lipase, and cellulase, is employed in the production of enzyme-based high-quality detergents. In the detergent market, α-amylases are major components in contemporary laundry and dishwashing detergents, in which they catalyze the efficient degradation of starch-based food residues such as chocolate, sauces, and cereal products, even at moderate temperatures. Their addition to detergent compositions facilitates efficient cleaning under energy-saving, low-temperature conditions without compromising fabric quality. In the textile industry, α-amylases are commonly used for enzymatic desizing, which is a process that selectively removes starch-based sizing agents from woven fabrics without damaging the underlying fibers. They are also used in bio-washing processes to create softness and controlled fading effects, especially in denim processing (Samanta, 2022; Maarel et al., 2002).

The ongoing depletion of fossil fuel resources due to rising global energy demand and the threat of climate change has accelerated the development of sustainable, renewable biofuels such as ethanol. Starch is a major raw material for the production of first-generation (1G) bioethanol due to its abundance and high carbohydrate content. After liquefaction, starch is hydrolyzed by α-amylases into smaller fermentable sugars, which are then fermented by microorganisms such as *Saccharomyces cerevisiae* to produce ethanol (Ashok et al., 2024). More integrated process designs, such as simultaneous saccharification and fermentation and consolidated bioprocessing, have further combined enzymatic hydrolysis and fermentation to improve conversion efficiency and lower costs. In recent years, artificial intelligence (AI) and machine learning technologies have been increasingly used to optimize α-amylase-based bioethanol production. This allows for the modeling of complex nonlinear kinetics, the prediction of optimal temperature, pH, enzyme, and fermentation conditions, and real-time process control. As a result, AI-assisted optimization can improve sugar extraction, reduce energy consumption, prevent contamination, and make bioethanol production more economical (Owusu & Marfo, 2023).

Conclusion

GH13 family is one of the most well-characterized and versatile enzyme families that participate in carbohydrate metabolism. GH13 enzymes share a common structural and functional template, and they have evolved subfamily-specific solutions to facilitate the degradation of a broad spectrum of α-glucans in all domains of life. The structural and functional versatility of GH13 enzymes has made them a paradigm for studying enzyme evolution, substrate specificity, and the relationship between protein structure and catalytic function.

However, recent progress in the fields of genomics, structural biology, and phylogenetic taxonomy has significantly expanded our knowledge of GH13 diversity, and new breakthroughs in the application of AI tools for structure prediction, optimization, and protein design are revolutionizing the field of GH13 research. In addition, the combination of sustainable production methods and circular bioeconomy concepts will further accelerate the development of efficient GH13 biocatalysts for specific applications. Nevertheless, future research and rational design of the GH13 enzyme family will continue to play a crucial role in advancing basic glycobiology research and overcoming various challenges in food, biotechnology, and biomedicine.

References

Aghajari, N., Haser, R., Feller, G., & Gerday, C. (1998). Crystal structures of the psychrophilic α-amylase from *Alteromonas haloplanctis* in its native form and complexed with an inhibitor. *Protein Science, 7*(3), 564–572.

Ahmad, A., Rahamtullah, & Mishra, R. (2022). Structural and functional adaptation in extremophilic microbial α-amylases. *Biophysical Reviews, 14*(2), 499–515.

Ali, Z., Abdullah, M., Yasin, M. T., Amanat, K., Sultan, M., Rahim, A., & Sarwar, F. (2025). Recent trends in production and potential applications of microbial amylases: A comprehensive review. *Protein Expression and Purification, 227*, 106640.

Alikhajeh, J., Khajeh, K., Ranjbar, B., Naderi-Manesh, H., Lin, Y. H., Liu, E., & Chen, C. J. (2010). Structure of *Bacillus amyloliquefaciens* α-amylase at high resolution: Implications for thermal stability. *Structural Biology and Crystallization Communications, 66*(2), 121–129.

Apriyanto, A., Compart, J., & Fettke, J. (2022). A review of starch, a unique biopolymer–Structure, metabolism and in planta modifications. *Plant science, 318*, 111223.

Arnal, G., Cockburn, D. W., Brumer, H., & Koropatkin, N. M. (2018). Structural basis for the flexible recognition of α-glucan substrates by *Bacteroides thetaiotaomicron* SusG. *Protein Science, 27*(6), 1093–1101.

Ashok, P. P., Dasgupta, D., Ray, A., & Suman, S. K. (2024). Challenges and prospects of microbial α-amylases for industrial application: A review. *World Journal of Microbiology and Biotechnology, 40*(2), 44.

Boyce, S. (2001). Enzyme classification and nomenclature. *Encyclopedia of life sciences.*

Buleon, A., Colonna, P., Planchot, V., & Ball, S. (1998). Starch granules: Structure and biosynthesis. *International Journal of Biological Macromolecules, 23*(2), 85–112.

CAZypedia Consortium. (n.d.). *Glycoside hydrolase family 13.*

Centeno-Leija, S., Espinosa-Barrera, L., Velazquez-Cruz, B., Cardenas-Conejo, Y., Virgen-Ortíz, R., Valencia-Cruz, G., & Serrano-Posada, H. (2022). Mining for novel cyclomaltodextrin glucanotransferases unravels the carbohydrate metabolism pathway via cyclodextrins in *Thermoanaerobacterales*. *Scientific Reports, 12*(1), 730.

Chai, K. P., Othman, N. F. B., Teh, A. H., Ho, K. L., Chan, K. G., Shamsir, M. S., & Ng, C. L. (2016). Crystal structure of *Anoxybacillus* α-amylase provides insights into maltose binding of a new glycosyl hydrolase subclass. *Scientific reports, 6*(1), 23126.

Dekker, R. F. H. (1994). Enzymes in food and beverage processing. 1. *Food Australia, 46*(3), 136–139.

Dong, G., Vieille, C., Savchenko, A., & Zeikus, J. G. (1997). Cloning, sequencing, and expression of the gene encoding extracellular alpha-amylase from *Pyrococcus furiosus* and biochemical characterization of the recombinant enzyme. *Applied and Environmental Microbiology, 63*(9), 3569–3576.

Far, B. E., Ahmadi, Y., Khosroshahi, A. Y., & Dilmaghani, A. (2020). Microbial alpha-amylase production: Progress, challenges and perspectives. *Advanced Pharmaceutical Bulletin, 10*(3), 350.

Fujimoto, Z., Takase, K., Doui, N., Momma, M., Matsumoto, T., & Mizuno, H. (1998). Crystal structure of a catalytic-site mutant α-amylase from *Bacillus subtilis* complexed with maltopentaose. *Journal of Molecular Biology, 277*(2), 393–407.

Goldbach, L., Vermeulen, B. J., Caner, S., Liu, M., Tysoe, C., van Gijzel, L., Yoshisada, R., Trellet, M., Ingen, H., Brayer, G. D., Bonvin, A. M. J. J., & Jongkees, S. A. (2019). Folding then binding vs folding through binding in macrocyclic peptide inhibitors of human pancreatic α-amylase. *ACS Chemical Biology, 14*(8), 1751–1759.

Gupta, N., Beliya, E., Paul, J. S., Tiwari, S., Kunjam, S., & Jadhav, S. K. (2021). Molecular strategies to enhance stability and catalysis of extremophile-derived α-amylase using computational biology. *Extremophiles: Life Under Extreme Conditions, 25*(3), 221–233.

Guzmán-Maldonado, H., Paredes-López, O., & Biliaderis, C. G. (1995). Amylolytic enzymes and products derived from starch: A review. *Critical Reviews in Food Science & Nutrition, 35*(5), 373–403.

Hardy, K., Brand-Miller, J., Brown, K. D., Thomas, M. G., & Copeland, L. (2015). The importance of dietary carbohydrate in human evolution. *The Quarterly Review of Biology, 90*(3), 251–268.

Henrissat, B. (1991). A classification of glycosyl hydrolases based on amino acid sequence similarities. *Biochemical journal, 280*(2), 309–316.

Janeček, Š. (2023). Advances in amylases—What's going on? *Molecules (Basel, Switzerland), 28*(21), 7268.

Janeček, Š., & Gabriško, M. (2016). Remarkable evolutionary relatedness among the enzymes and proteins from the α-amylase family. *Cellular and Molecular Life Sciences, 73*(14), 2707–2725.

Janeček, Š., & Kuchtová, A. (2012). In silico identification of catalytic residues and domain fold of the family GH119 sharing the catalytic machinery with the α-amylase family GH57. *FEBS Letters, 586*(19), 3360–3366.

Janeček, Š., & Svensson, B. (2022). How many α-amylase GH families are there in the CAZy database? *Amylase, 6*(1), 1–10.

Janeček, Š., Svensson, B., & MacGregor, E. A. (2014). α-Amylase: An enzyme specificity found in various families of glycoside hydrolases. *Cellular and Molecular Life Sciences, 71*(7), 1149–1170.

Kim, J. S., Cha, S. S., Kim, H. J., Kim, T. J., Ha, N. C., Oh, S. T., & Oh, B. H. (1999). Crystal structure of a maltogenic amylase provides insights into a catalytic versatility. *Journal of Biological Chemistry, 274*(37), 26279–26286.

Klein, C., & Schulz, G. E. (1991). Structure of cyclodextrin glycosyltransferase refined at 2.0 Å resolution. *Journal of Molecular Biology, 217*(4), 737–750.

Kulkarni, S., Dhakar, K., & Joshi, A. (2019). Alkaliphiles: Diversity and bioprospection. In *Microbial diversity in the genomic era*. Academic Press, 239–263.

Kuriki, T., & Imanaka, T. (1999). The concept of the α-amylase family: Structural similarity and common catalytic mechanism. *Journal of Bioscience and Bioengineering, 87*(5), 557–565.

Lawson, C. L., van Montfort, R., Strokopytov, B., Rozeboom, H. J., Kalk, K. H., de Vries, G. E., Penninga, D., Dijkhuizen, L., & Dijkstra, B. W. (1994). Nucleotide sequence and X-ray structure of cyclodextrin glycosyltransferase from *Bacillus circulans* strain 251 in a maltose-dependent crystal form. *Journal of Molecular Biology, 236*(2), 590–600.

Liu, Y., Yu, J., Li, F., Peng, H., Zhang, X., Xiao, Y., & He, C. (2017). Crystal structure of a raw-starch-degrading bacterial α-amylase belonging to subfamily 37 of the glycoside hydrolase family GH13. *Scientific Reports, 7*(1), 44067.

Lombard, V., Golaconda Ramulu, H., Drula, E., Coutinho, P. M., & Henrissat, B. (2014). The carbohydrate-active enzymes database (CAZy) in 2013. *Nucleic Acids Research, 42*(D1), D490–D495.

Madene, A., Jacquot, M., Scher, J., & Desobry, S. (2006). Flavour encapsulation and controlled release–a review. *International journal of food science and technology, 41*(1), 1–21.

Matsuura, Y., Kusunoki, M., Harada, W., & Kakudo, M. (1984). Structure and possible catalytic residues of taka-amylase A. *The Journal of Biochemistry, 95*(3), 697–702.

Miłek, J., & Lamkiewicz, J. (2022). The starch hydrolysis by α-amylase Bacillus spp.: An estimation of the optimum temperatures, the activation and deactivation energies. *Journal of Thermal Analysis and Calorimetry, 147*(24), 14459–14466.

Mobini-Dehkordi, M., & Javan, F. A. (2012). Application of alpha-amylase in biotechnology. *Journal of Biology and Today World, 1*(1), 39–50.

Myers, A. M., Morell, M. K., James, M. G., & Ball, S. G. (2000). Recent progress toward understanding biosynthesis of the amylopectin crystal. *Plant Physiology, 122*(4), 989–998.

Ochiai, A., Sugai, H., Harada, K., Tanaka, S., Ishiyama, Y., Ito, K., & Mitsui, T. (2014). Crystal structure of α-amylase from *Oryza sativa*: Molecular insights into enzyme activity and thermostability. *Bioscience, Biotechnology, and Biochemistry, 78*(6), 989–997.

Oslancova, A., & Janeček, Š. (2002). Oligo-1, 6-glucosidase and neopullulanase enzyme subfamilies from the α-amylase family defined by the fifth conserved sequence region. *Cellular and Molecular Life Sciences CMLS, 59*(11), 1945–1959.

Owusu, W. A., & Marfo, S. A. (2023). Artificial intelligence application in bioethanol production. *International Journal of Energy Research, 2023*(1), 7844835.

Park, J. T., Song, H. N., Jung, T. Y., Lee, M. H., Park, S. G., Woo, E. J., & Park, K. H. (2013). A novel domain arrangement in a monomeric cyclodextrin-hydrolyzing enzyme from the hyperthermophile *Pyrococcus furiosus*. *Biochimica et Biophysica Acta (BBA)-Proteins and Proteomics, 1834*(1), 380–386.

Paul, J. S., Gupta, N., Beliya, E., Tiwari, S., & Jadhav, S. K. (2021). Aspects and recent trends in microbial α-amylase: A review. *Applied Biochemistry and Biotechnology, 193*(8), 2649–2698.

Pérez, S., & Bertoft, E. (2010). The molecular structures of starch components and their contribution to the architecture of starch granules: A comprehensive review. *Starch-Stärke, 62*(8), 389–420.

Poláček, A., Lombard, V., Coutinho, P. M., Terrapon, N., & Janeček, Š. (2025). Dividing the α-amylase family GH57 of starch hydrolases and related enzymes into subfamilies using evolutionary, clustering and functional criteria. *International Journal of Biological Macromolecules, 309*, 142823.

Prakash, O., & Jaiswal, N. (2010). α-Amylase: An ideal representative of thermostable enzymes. *Applied Biochemistry and Biotechnology, 160*(8), 2401–2414.

Rampelotto, P. H. (2013). Extremophiles and extreme environments. *Life (Chicago, Ill: 1978), 3*(3), 482–485.

Robert, X., Haser, R., Gottschalk, T. E., Ratajczak, F., Driguez, H., Svensson, B., & Aghajari, N. (2003). The structure of barley α-amylase isozyme 1 reveals a novel role of domain C in substrate recognition and binding: A pair of sugar tongs. *Structure (London, England: 1993), 11*(8), 973–984.

Samanta, S. (2022). Structural and catalytical features of different amylases and their potential applications. *Jordan Journal of Biological Sciences, 15*(2).

Salt, W. B., & Schenker, S. (1976). Amylase—its clinical significance: a review of the literature. *Medicine, 55*(4), 269–289.

Sharma, A., Parashar, D., & Satyanarayana, T. (2016). Acidophilic microbes: Biology and applications. *Biotechnology of Extremophiles: Advances and Challenges*, 215–241.

Singhania, R.R., Patel, A.K., Thomas, L., Goswami, M., Giri, B.S., & Pandey, A. (2015). Industrial enzymes. In Industrial biorefineries & white biotechnology (pp. 473-497). Elsevier.

Souza, P. M. D., & Magalhães, P. D. O. E. (2010). Application of microbial α-amylase in industry-A review. *Brazilian Journal of Microbiology, 41*(4), 850–861.

Stam, M. R., Danchin, E. G., Rancurel, C., Coutinho, P. M., & Henrissat, B. (2006). Dividing the large glycoside hydrolase family 13 into subfamilies: Towards improved functional annotations of α-amylase-related proteins. *Protein Engineering, Design and Selection, 19*(12), 555–562.

Svensson, B. (1988). Regional distant sequence homology between amylases, α-glucosidases and transglucanosylases. *FEBS Letters, 230*(1-2), 72–76.

Swift, H. J., Brady, L., Derewenda, Z. S., Dodson, E. J., Dodson, G. G., Turkenburg, J. P., & Wilkinson, A. J. (1991). Structure and molecular model refinement of *Aspergillus oryzae* (TAKA) α-amylase: An application of the simulated-annealing method. *Structural Science, 47*(4), 535–544.

Szilágyi, A., & Závodszky, P. (2000). Structural differences between mesophilic, moderately thermophilic and extremely thermophilic protein subunits: Results of a comprehensive survey. *Structure (London, England: 1993), 8*(5), 493–504.

Tadeo, X., López-Méndez, B., Trigueros, T., Laín, A., Castaño, D., & Millet, O. (2009). Structural basis for the aminoacid composition of proteins from halophilic archea. *PLoS Biology, 7*(12), e1000257.

Tan, T. C., Mijts, B. N., Swaminathan, K., Patel, B. K., & Divne, C. (2008). Crystal structure of the polyextremophilic α-amylase AmyB from *Halothermothrix orenii*: Details of a productive enzyme–substrate complex and an N domain with a role in binding raw starch. *Journal of Molecular Biology, 378*(4), 852–870.

Villa Zabala, C. C. (2020). An overview on starch structure and chemical nature. *Starch-based Nanomaterials,* 3–9.

Vuillemin, M., Prieto, E. S. M., Pilgaard, B., Siebenhaar, S., Holck, J., Henrissat, B., & Alghamdi, K. M. (2024). Biochemical exploration of family GH119 reveals a single α-amylase specificity and confirms shared catalytic machinery with GH57 enzymes. *International Journal of Biological Macromolecules, 262*, 129783.

Vuong, T. V., & Wilson, D. B. (2010). Glycoside hydrolases: Catalytic base/nucleophile diversity. *Biotechnology and Bioengineering, 107*(2), 195–205.

Xu, H., Zhou, J., Liu, X., Yu, J., Copeland, L., & Wang, S. (2023). Methods for characterizing the structure of starch in relation to its applications: A comprehensive review. *Critical Reviews in Food Science and Nutrition, 63*(20), 4799–4816.

Yadav, J. K., & Prakash, V. (2011). Stabilization of α-amylase, the key enzyme in carbohydrates properties alterations, at low pH. *International journal of food properties, 14*(6), 1182–1196.

CHAPTER

Cyclodextrin gluconotransferase: an overview 2

Ténor Dias-Mendel Allode and Vodounon Cyrille Junior Alode

Laboratory of Natural Sciences and Applications (LSNA), Higher Normal School of Natitingou, National University of Sciences, Technologies, Engineering, and Mathematics of Abomey, Natitingou, Benin

History of cyclodextrin gluconotransferase

Cyclodextrin glucanotransferases (CGTases) are enzymes that produce cyclodextrins (CDs), cyclic oligosaccharides with numerous industrial applications. The discovery of CGTases is the result of decades of collaborative work by researchers from diverse backgrounds worldwide, reflecting successive advances in microbiology, enzymatic biochemistry, and molecular biotechnology. These enzymes, primarily produced by bacteria of the genus *Bacillus*, are capable of converting starch into CDs. The scientific context of the late 19th century largely explains the delayed understanding of CDs: industrial microbiology and controlled fermentations were still in their infancy, driven by the work of Pasteur, Koch, and Fischer, who had laid the foundations of microbial fermentation and sugar stereochemistry. However, the analytical tools available at that time remained very limited (Szejtli, 1998). Researchers lacked modern chromatography, NMR spectroscopy, and high-resolution X-ray crystallography, which would not truly develop until the mid-20th century. The absence of these precise analytical tools made the accurate identification of complex oligomeric structures difficult (Davis & Higson, 2011; Villiers, 1903). The identification of products resulting from the enzymatic hydrolysis of starch relied primarily on empirical criteria such as solubility, crystallization, and elemental analysis, which explains the numerous misinterpretations concerning the true nature of the cyclic dextrins observed by Villiers and his colleagues (Loftsson & Duchêne, 2007). This technological context significantly delayed the understanding of the cyclic structure of CDs and their inclusion properties. It was only with the gradual development of structural methods (X-ray diffraction, followed by NMR and mass spectrometry) that rigorous confirmation of cyclic structures and the distinction between α-, β-, and γ-CDs could be established (Hofmann et al., 1989; Ochiai et al., 1935).

This combination of an emerging industrial context and analytical limitations explains why the true nature of CDs was only revealed several decades after their

Cyclodextrin Glucanotransferase. DOI: https://doi.org/10.1016/B978-0-443-41694-1.00014-9

first observations (Crini, 2020a; Loftsson & Duchêne, 2007; Saenger, 1980). The history of CGTase discovery is linked to the discovery of CDs. This discovery unfolded in several stages. First, we have the pre-discovery period, which covers the discovery of CDs before CGTases and runs from 1891 to 1903. The first publication on CDs by the Frenchman Villiers in 1891 marked the start of the work of several researchers, including Schardinger, Pringsheim, Karrer, Freudenberg, French, Cramer, Casu, Bender, Saenger, Nagai, Szejtli, and Pitha (Crini, 2020b). Indeed, the first appearance of what would become known as CDs dates back to 1891, when Villiers observed the formation of a crystalline product during the hydrolysis of starch by a culture of *Bacillus amylobacter*, now called *Bacillus circulans* (Davis & Higson, 2011). He determined its composition to be $(C_6H_{10}O_5)_2 \cdot 3H_2O$, which he standardized as "cellulosin" (Allode et al., 2024; Loftsson & Duchêne, 2007). Villiers described this product as a "soluble amylose," without identifying its exact nature, in his article titled "On an Intermediate Product of Amylic Fermentation," published in the Comptes Rendus de l'Académie des Sciences (Villiers, 1903). Following Villiers' work, the European scientific community greeted these results with a mixture of curiosity and skepticism, giving rise to lively debates about the true nature of "cellulosin" and its origin (French, 1957; Villiers, 1891). At that time, several chemists interpreted this compound as a linear dextrin or as a partially hydrolyzed derivative of starch, due to the lack of analytical methods to reveal its cyclic structure (Crini, 2020a; Del Valle, 2004; French, 1957). This discussion dominated for several years and delayed the recognition of the true cyclic structure of these oligomers (French, 1957; Morin-Crini et al., 2021). Thus for several decades, "crystalline dextrins" were considered atypical products of amylose degradation rather than true cyclic macromolecules, which considerably delayed understanding of their physicochemical properties and enzymatic formation mechanisms (Crini, 2020a; Saenger et al., 1976). It was only with the systematic isolation of α-, β-, and γ-CDs by Cramer and collaborators that the erroneous hypothesis of a strictly linear structure was definitively abandoned in favor of the model of cyclized glycosidic chains (Crini, 2020b; Loftsson & Duchêne, 2007). In 1903, the Austrian microbiologist Franz Schardinger published a paper describing two crystalline compounds, A and B, that he had isolated from a bacterial digestion product of potato starch. He was only able to isolate very small amounts of compound A but significantly more of compound B, which he identified as Villiers' cellulosin (Davis & Higson, 2011).

In his paper, Schardinger suggested that "crystalline dextrin" would be a more appropriate name than "cellulosin" for these compounds. Later, he renamed them α-dextrin and β-dextrin (Davis & Higson, 2011; Loftsson & Duchêne, 2007). Following Franz Schardinger's groundbreaking work on crystallized dextrins in 1903, Austrian and German laboratories played a significant role in disseminating, exploring, and interpreting the concept of cyclic dextrins. In particular, the teams led by Hans Pringsheim in Germany continued to study these isolated products, which they still called *Schardinger dextrins*, and to explore their physicochemical properties, although a clear structure remained elusive for several decades (Crini, 2020b). Their work fueled the scientific debate that gradually contributed to the development of isolation and identification methods (Saenger et al., 1976; Schardinger, 1903). This intellectual ferment paved the way for the systematic isolation of α-, β-, and γ-CDs by Cramer and his collaborators (Gerhard, 2014; Jicsinszky et al., 2025). However, this period of knowledge dissemination was hampered by the historical upheavals of the early 20th century. In particular, the First World War (1914–18) diverted the attention of the scientific community toward the war effort and led to a significant slowdown in fundamental research in carbohydrate chemistry, including on CDs, as human and material resources were massively mobilized for military applications and urgent industrial needs. It was only after the 1930s that advances in chemical and crystallographic analysis allowed chemists like Freudenberg and his collaborators to propose a cyclic structure for these molecules and to go beyond Schardinger's simple empirical observations (Crini, 2020b; Loftsson & Duchêne, 2007). This discovery was the starting point for understanding CDs, even though their precise structure was not known. It wasn't until 1935 that Friedrich Cramer succeeded in isolating three distinct types of CDs: α-, β-, and γ-CDs. These molecules, composed of six, seven, and eight glucose units respectively, linked by α-(1→4) bonds, were the first to be properly characterized (Loftsson & Duchêne, 2007; Ochiai et al., 1935). After the discovery of α-, β-, and γ-CDs by Cramer (1935), the scientific community gradually adopted a nomenclature to designate these compounds. After initial names such as "cellulosin" and "crystalline dextrin," various chemists proposed alternative terms like "Schardinger dextrins," "cycloamyloses," or "cycloglucans" before the simpler term "cyclodextrin" gradually became established in the following years (Allode et al., 2024; Loftsson & Duchêne, 2007; Saenger et al., 1976; Schardinger, 1903; Villiers, 1891). This terminological evolution reflects the growing understanding that these molecules are not linear starch degradation products but rather cyclic oligosaccharides with well-defined ring structures (Loftsson & Duchêne, 2007).

Alongside this conceptual maturation, the Second World War, which lasted from 1939 to 1945, had a negative impact on research in carbohydrate and starch chemistry: scientific and industrial efforts were largely redirected toward war-related applications, temporarily slowing fundamental investigations into starch derivatives and their structure (Crini, 2020b; Loftsson & Duchêne, 2007). It was only after the end of the conflict that human and material resources could once again be mobilized toward carbohydrate chemistry, enabling the development of analytical techniques and the revival of structural and applied studies in the following decades (Crini, 2020b). It was not until the 1950s that microbial production of carbohydrates was attributed to a specific enzyme. It should be recalled that during the exploratory period of research into microbial species capable of producing the enzyme, Schardinger had shown that *Bacillus macerans* produced CDs, thereby demonstrating CGTase activity (Loftsson & Duchêne, 2007). The enzymatic activity was described as "cyclodextrin-producing amylase" in several publications from the 1960s. Kuriki et al. (1961) at the Tokyo University of Science identified the enzyme responsible for CD synthesis from starch and showed that the activity of this enzyme differed from that of classical amylases. This was later named CGTase (Kuriki et al., 1961). This was published in an article entitled "*Formation of cyclohexaamylose by an enzymatic reaction*" in the journal *Agricultural and Biological Chemistry*. Following the identification of CGTase, Japan established itself as a key player in amylolytic enzyme research during postwar scientific and industrial reconstruction. Japanese researchers, notably (Kuriki et al., 1961), made significant contributions to the characterization of CGTases and the understanding of their catalytic properties, in a scientific environment supported by the state and the agri-food industry. In parallel, several Japanese companies, such as Hayashibara Biochemical Laboratories, the National Food Research Institute, and Kirin, launched industrial programs for the enzymatic valorization of starch in the late 1960s and throughout the 1970s, aiming for the large-scale production of CDs for the food and pharmaceutical sectors (Loftsson & Duchêne, 2007; Szejtli, 1998). This approach gradually replaced traditional acidic methods in the food and fermentation industries. This dynamic marked the shift from primarily academic research to the structured industrial exploitation of CGTases, firmly establishing Japan as a world leader in amylolytic enzymes. As Kobayashi (1995) points out, "the combination of fine microbiological expertise and a coherent industrial strategy enabled Japan to dominate the global market for amylolytic enzymes as early as the 1970s.“

The work of researchers such as Lubbert Dijkhuizen, Jan van der Veen, and Brian Dijkstra in the 1970s and 1980s led to a detailed understanding of the catalytic mechanism of CGTase (Van Der Veen et al., 2000b; van der Veen, Uitdehaag et al., 2000). This work showed that the enzyme acts via a double-displacement mechanism, typical of group 13 glycoside hydrolases (GH) (Kelly et al., 2009; van der Veen et al., 2000). From the late 1970s through the 1980s, research on CGTases gradually came to be structured around international collaborations among laboratories in the Netherlands, Germany, and Japan, which have fostered the exchange of expertise in enzymology and carbohydrate biochemistry. In particular, Dutch teams at the University of Groningen, led by Dijkhuizen and van der Veen, worked closely with German laboratories to characterize in detail the three transglycosylase reactions catalyzed by CGTases and to establish correlations between structure and product specificity (van der Veen, Uitdehaag et al., 2000). These collaborations were complemented by regular exchanges with Japanese researchers specializing in the study of amylolytic enzymes, contributing to a more integrated view of CGTases as model enzymes within the glycoside hydrolase family. At the same time, the rise of the first enzyme databases and molecular classifications, such as the carbohydrate-active enzymes (CAZy) database and the BRENDA database, had a considerable impact on the description and classification of CGTases, allowing the integration of sequence and structural information within a coherent evolutionary framework (Henrissat, 1991; Schomburg, 2004). The information in the resources enabled placement of CGTases within the GH13 family of GH based on sequence and catalytic motif similarities, facilitating the identification of new enzyme variants and their comparison on a global scale (Crini, 2020b; Henrissat & Bairoch, 1993).

The late 1980s and early 1990s marked a major turning point in the study of CGTases and, more broadly, industrial enzymes, with the rise of genetic engineering and molecular biotechnology. The ability to clone and express genes was demonstrated by cloning the *cgt gene* in *B. macerans* (Takano et al., 1986). They made it possible to approach enzymes not only through their observed activity, but as modifiable molecular objects focused on targeted gene manipulation, sequence analysis, and understanding gene-protein-function relationships (Hofmann et al., 1989; van der Veen, Uitdehaag et al., 2000). This shift moved enzymology from a descriptive approach to molecular enzymology, where site-directed mutagenesis, heterologous expression, and combined structural analysis made it possible to identify and modify the determinants of stability and selectivity (Zhang et al., 2017). This transition paved the way for the production of CGTases in recombinant hosts such as *Escherichia coli* and *Bacillus subtilis*, facilitating the generation of homogeneous and reproducible enzymes for research and industrialization (Takano et al., 1986; van der Veen, Uitdehaag et al., 2000). Simultaneously, the integration of molecular biology approaches, including DNA sequencing and recombinant expression, profoundly altered enzyme research strategies, enabling the systematic linking of genetic structure to the functional properties of CGTases for the first time (Lindley, 1994; Strokopytov et al., 1995).

In concrete terms, these advances have accelerated the generation of CGTase variants better suited to industrial conditions (thermostability, solvent tolerance, and selectivity toward α/β/γ CDs), thereby reducing separation costs and facilitating the transition to pilot and commercial scales (Ibrahim et al., 2013; Pardhi et al., 2023). Hofmann et al. (1989) first demonstrated the three-dimensional structure of CGTases at 3.4 Å resolution by X-ray crystallography using CGTase from *B. circulans*. It was not until the work of other researchers between 1990 and 2000 that higher resolution structures and enzyme-substrate complexes were developed (Schmidt et al., 1998; Uitdehaag et al., 1999, 2000). This work allowed for the precise visualization of the enzyme's active site and its interaction with various substrates. These results significantly enhanced our understanding of the structure-function relationship of CGTase. Since then, studies on various aspects of CGTases have proliferated. Today, CGTases are at the heart of research aimed at producing high-purity CDs, encapsulating pharmaceutical molecules, and valorizing starchy waste through enzymatic bioconversion. The story of the discovery of CGTases is therefore collective, spanning nearly a century and connecting researchers from different nationalities and disciplines. From Antoine Villiers's first observation in 1903 to the resolution of the three-dimensional structure in 2000, each step has contributed to making CGTases indispensable enzymes in the modern biotechnology industry. Thanks to these discoveries, it is now possible to produce CDs on a large scale for use in a variety of applications, ranging from drug formulation to the stabilization of food flavorings.

Significance of cyclodextrin gluconotransferase

Industrial enzymes and the global ecological transition

Faced with the climate emergency, accelerating environmental degradation, and mass extinctions of species, the transition to greener, more sustainable production of liquid fuels and basic chemicals has become imperative (Cui et al., 2022; Wu et al., 2021). In this context, conventional chemical processes, based on high-temperature reactions, toxic solvents, and heavy metals, pose major environmental limitations, particularly in terms of greenhouse gas emissions, energy consumption, and the generation of polluting waste (Farhan et al., 2025). Industrial enzymes are a promising alternative to these traditional processes (Farhan et al., 2025). As biocatalysts, they catalyze low-energy chemical reactions, with exceptional selectivity and high substrate specificity, which significantly reduce purification steps, the formation of undesirable by-products, and energy consumption compared to conventional chemical reactions (Farhan et al., 2025). A recent review of advances in industrial biocatalysis highlights that integrating enzymes into production chains not only reduces overall environmental impact but also optimizes process efficiency in sectors as diverse as pharmaceuticals, food and beverage, textiles, and biofuels (Farhan et al., 2025; Mesbah, 2022). Indeed, enzymes often operate under mild temperature and pH conditions, eliminating the need for aggressive metal catalysts or volatile organic solvents commonly used in conventional chemical processes (Cui et al., 2022). These environmental benefits of enzymes directly align with the principles of green chemistry, which aim to minimize the use of hazardous substances and promote resource conservation (Sheldon et al., 2021). For example, the ability of biocatalysts to operate in aqueous reactions and use renewable raw materials often eliminates dependence on fossil resources for chemical synthesis (Mesbah, 2022; Sheldon et al., 2021). In this context, enzymes are a key technology for reducing the chemical industry's carbon footprint and supporting international carbon-neutrality goals (Radley et al., 2023). Beyond energy savings, enzymes offer additional ecological advantages: they are biodegradable and nontoxic, and can be produced on a large scale by cultivable microorganisms, thereby reducing the environmental impact of the supply chain (Cui et al., 2022). In this context, enzyme-associated biocatalysis in protein engineering enables the design of more robust, thermostable, and tolerant enzymes, better suited to harsh industrial conditions, thus expanding their applications in processes where conventional chemical catalysts previously held a monopoly (Radley et al., 2023).

Biocatalysts also contribute directly to circular economy models by allowing the valorization of renewable biomass for the production of higher value-added molecules, instead of using nonrenewable fossil raw materials (Sheldon & Woodley, 2018; Sheldon et al., 2021). This transition is particularly important in a world facing climate change, where reducing the carbon intensity of industrial processes has become imperative for achieving the sustainable development goals (SDGs) set by the international community. Finally, in the spectrum of amylolytic enzymes, such as CGTase, their role goes beyond simple transglycosylation: they are part of a high-functional-yield biocatalytic platform, capable of efficiently transforming starch into functional products while meeting the ecological expectations of modern industries.

Cyclodextrin gluconotransferase in the functional hierarchy of GH13 enzymes

Among the GH listed in the CAZy server, the GH13 α-amylase family is one of the most important. It was initially divided into 35 subfamilies by Janeček & Svensson (2022) and Stam et al. (2006). The CAZy database administrators subsequently subdivided them into 36 subfamilies. Initially, the oligo-1,6-glucosidase and neopullulanase subfamilies were defined based on the sequence of the fifth conserved region as a selection marker. This subfamily is located outside the α-amylase $(\beta/\alpha)_8$ catalytic barrel, in the B domain, specifically at a longer loop connecting the β3 strand to the α3 helix of the barrel (Janeček & Zámocká, 2020; Majzlová et al., 2013). This family includes enzymes with diverse functions, such as α-amylases, pullulanases, branching enzymes, maltogenic amylases, and CGTases. Despite a generally conserved structural architecture, these enzymes are strongly distinguished by their catalytic profiles and biological roles. (Janeček & Zámocká, 2020). Among GH13 enzymes, CGTase occupies a unique functional position due to its catalytic multifunctionality. Unlike classical α-amylases, whose main activity relies on the endoglycosidic hydrolysis of α-(1,4) linkages, CGTase is capable of catalyzing four distinct types of reactions: cyclization, coupling, dismutation, and hydrolysis (Allode et al., 2024; Kelly et al., 2009). This enzymatic versatility confers rare functional flexibility to CGTase, allowing it to direct the reaction flow toward the formation of cyclic, linear, or mixed products depending on the reaction conditions (Allode et al., 2024; Pardhi et al., 2023). This ability to generate multiple catalytic fluxes from the same active site justifies the classification of CGTase as a multifluid enzyme (Saini et al., 2022). In a biological context, this property suggests an evolutionary role intermediate between strictly hydrolytic enzymes and those specialized in transglycosylation.

Industrially, it allows for fine-tuning of product profiles without major modifications to the biocatalyst, simply by adjusting the substrate, pH, temperature, or the presence of additives (Leemhuis et al., 2010; van der Veen, Uitdehaag et al., 2000). Compared to conventional α-amylases, CGTase therefore offers significant added functional value. Where hydrolysis primarily yields dextrins or fermentable sugars, CGTase enables the formation of CDs, molecules with high functional and economic value (Esteso & Romero, 2024; Hedges, 2009). This difference in catalytic purpose explains why CGTase, although belonging to the same GH13 family, is considered a strategic enzyme rather than a simple functional variant of amylases (Biwer et al., 2002). From an evolutionary perspective, several phylogenetic studies suggest that CGTases constitute an evolutionary pivot within the

GH13 family, having diverged from ancestral amylolytic enzymes through progressive specialization of intramolecular transglycosylation (Kelly et al., 2009; Stam et al., 2006). This functional evolution was accompanied by a remarkable conservation of catalytic motifs, notably the Asp-Glu-Asp triad, while incorporating structural adjustments that favor cyclisation over hydrolysis (Janeček & Zámocká, 2020). From an industrial and biotechnological perspective, this particular hierarchical position confers upon CGTase the status of a platform enzyme within the GH13 family. It lies at the interface between starch degradation and the synthesis of complex functional molecules, which distinguishes it from strictly catabolic enzymes. This duality explains the continued interest in CGTase in the fields of bioeconomy, green chemistry, and high-value-added enzymatic processes (Crini, 2020a; Spiridon & Anghel, 2025). Thus CGTase cannot be considered solely a starch-transforming enzyme, but rather a central catalytic tool within the GH13 family, linking microbial enzymatic biodiversity to modern industrial needs. This hierarchical position fully justifies its scientific and economic importance and explains why it is the subject of sustained interest in contemporary research in enzymology and industrial biotechnology.

Cyclodextrin gluconotransferase as a platform enzyme in modern bioprocesses

In the current context of the transition to more sustainable industrial processes, certain enzymes occupy a particularly strategic position due to their ability to structure complete transformation chains. The concept of a platform enzyme aligns with the logic described by Sheldon & Woodley (2018), according to which certain enzymes constitute central catalytic nodes, capable of fueling multiple transformation pathways and generating a diversity of value-added molecules from renewable substrates (Bornscheuer et al., 2012; Sheldon & Woodley, 2018).

Their value lies as much in their catalytic versatility as in their ability to integrate into various biorefining schemes (Bozkurt et al., 2024; Paul et al., 2024). CGTase could be fully classified in this category. Unlike specialized enzymes that lead to a limited number of products, CGTase allows, from an abundant and renewable substrate such as starch, the formation of an entire family of cyclic and linear carbohydrate structures, the diversity of which depends closely on reaction conditions (Alhussein et al., 2025; Muniz et al., 2024; Pardhi et al., 2023). This catalytic versatility, resulting from its ability to catalyze cyclization, coupling, and disproportionation reactions, gives CGTase a structuring role in modern bioprocesses, particularly those based on the valorization of plant biomass (Allode et al., 2024; Fouda-Mbanga et al., 2025; Kelly et al., 2009; Saini et al., 2022). In the absence of a formalized and unified definition in the literature, the concept of a platform enzyme can be understood through three fundamental criteria: (1) the ability to transform widely available raw materials, (2) the production of several intermediates or molecular families, and (3) adaptability to industrial environments. As such, CGTase meets all of these requirements. Starch, derived from cereals, tubers, or agro-industrial residues, is a renewable global resource, and its enzymatic transformation is a central challenge for the circular bioeconomy (Crini et al., 2020; Oehlenschläger et al., 2024; Sheldon & Woodley, 2018). Compared with other major industrial enzymes such as cellulases, lipases, and proteases, CGTase is distinguished by the structure of its intermediate products. Cellulases primarily produce fermentable sugars, whereas lipases and proteases act on more specific matrices (Astray et al., 2009; Sheldon & Woodley, 2018). CGTase, on the other hand, generates molecular architectures capable of interacting with a wide variety of compounds, thus conferring a particularly broad functional potential (Astray et al., 2009; Biwer et al., 2002; EM Martin Del Valle, 2004; Rajput et al., 2016). This unique characteristic stems from CGTase's ability to catalyze intramolecular transglycosylation reactions, giving rise to stable yet tunable cyclic structures. Depending on the pH, temperature, enzyme source, or nature of the starch substrate, product distribution can be directed, providing the process with a flexibility rarely found in industrial enzyme systems (Allode et al., 2024; van der Veen et al., 2000). Another key element of CGTase's platform nature lies in its substrate tolerance. Many microbial CGTases are capable of acting on native, modified, or partially hydrolyzed starches, including those derived from agricultural by-products, illustrating the catalytic plasticity of this enzyme (Benavent-Gil et al., 2021; Ibrahim et al., 2012; Pardhi et al., 2023).

Operationally, CGTase also exhibits properties favorable to industrialization. Many CGTases from *Bacillus spp.* display good thermal stability, tolerance to pH variations, and high catalytic activity, characteristics frequently observed among enzymes of the α-amylase family (van der Maarel et al., 2002). Several studies have shown that thermostable CGTases retain significant activity at temperatures above 60°C, reducing the risk of contamination and improving process efficiency (Ibrahim et al., 2012; Rajput et al., 2016). The adaptability of CGTase is also demonstrated by its compatibility with various bioprocess strategies, including enzyme immobilization, catalytic reuse, and process engineering. While the technical aspects of these approaches are developed in other chapters, their mere feasibility illustrates the potential of CGTase as a key enzyme within integrated industrial systems (Rabadiya et al., 2024). Beyond its catalytic performance, CGTase fits into a broader vision of modern biotechnology, where enzymes are no longer considered simply as reaction tools, but as architectural elements of bio-based supply chains. In this sense, CGTase links the primary transformation of starch to subsequent molecular functionalization steps without resorting to heavy or polluting chemical processes. Thus CGTase can be considered a platform enzyme not only because of the diversity of structures it generates, but also because of its ability to structure complete value chains around starch biomass. This central position explains the sustained scientific interest it has received for several decades, as well as its strategic role in the evolution of bioprocesses toward more sustainable, flexible, and economically viable models.

Central role of cyclodextrin gluconotransferase in starch economy

Starch is one of the most abundant and strategic biological resources worldwide. Found primarily in cereals, tubers, and certain legumes, it is a plant polysaccharide widely distributed throughout terrestrial biomass (Carvalho et al., 2024; Global Market Insights, 2025; Tester et al., 2004). Its annual global production amounts to hundreds of millions of tons, making starch a major renewable raw material. According to the Global Market Insights report (2025), the global industrial starch market was valued at US$58.5 billion in 2024 and is projected to grow at a compound annual growth rate of over 6% between 2025 and 2034 (Global Market Insights, 2025). Historically used as a food and energy source, starch now occupies a central place in the development of the bioeconomy, due to its polysaccharide structure rich in glucosidic units easily mobilized by enzymatic means (Christophe & Steven, 2009; van der Maarel et al., 2002). In modern industrial systems, the value of starch no longer lies solely in its conversion into fermentable sugars but in its capacity to be transformed into a wide variety of high-value functional molecules.

This evolution reflects the gradual shift from an approach centered on energy production to one of integrated biomass valorization, characteristic of the biorefinery concept (Tomasik & Horton, 2012; van der Maarel et al., 2002). This transition relies heavily on biocatalysts that selectively convert starch macromolecules while preserving their chemical integrity, enabling the targeted production of highly functional compounds (Sheldon & Woodley, 2018; Straathof, 2014). In this context, CGTase occupies a unique position. Unlike conventional hydrolytic enzymes, CGTase does not simply depolymerize starch but allows for controlled restructuring of the carbohydrate chain, leading to the formation of cyclic molecules called CD (Allode et al., 2024). This nondestructive transformation capacity gives CGTase a strategic role in the starch value chain, ensuring continuity between raw plant biomass and high-quality functional intermediates (Biwer et al., 2002; Kelly et al., 2009). Comparing chemical and enzymatic starch transformation processes clearly highlights the economic and environmental advantages of the biocatalytic pathway. Conventional chemical processes generally involve harsh temperature and pH conditions or the use of solvents, resulting in high energy consumption and the generation of undesirable byproducts. Conversely, the enzymatic transformation catalyzed by CGTase occurs under moderate conditions, with high reaction specificity, which maximizes yield while minimizing material loss (Crini, 2020a). CGTase thus fits into a logic of progressive starch valorization, from upstream to downstream stages of the industrial chain. Upstream, it can be integrated into systems using native or partially processed starches, including those derived from agricultural by-products. Downstream, the generated cyclic molecules constitute versatile intermediates that can be integrated into multiple industrial processes without requiring further complex chemical transformations (Podgorbunskikh et al., 2022; Tomasik & Horton, 2012). This intermediate position gives CGTase a structuring role in the starch economy, comparable to that played by certain key enzymes in lignocellulosic processes. It acts as a biochemical node, linking primary agricultural production to industrial value chains, while maintaining consistency between economic performance and environmental sustainability. The indirect economic benefits of using CGTase are also significant. Catalytic selectivity allows for a substantial reduction in subsequent purification steps, thereby lowering operating costs. Furthermore, limiting the formation of byproducts reduces the effluent treatment load and improves the overall environmental performance of the process (Pardhi et al., 2023). From a circular bioeconomy perspective, CGTase contributes to optimizing the use of biological resources by enabling the valorization of starches of varying quality, including those derived from agro-industrial residues (Awasthi et al., 2020; Carvalho et al., 2024).

This approach is fully aligned with the principles of the circular bioeconomy, which aims to maximize the value of biological streams while reducing losses and environmental impacts (Carus & Dammer, 2018; Ubando et al., 2020). The possibility of processing starch locally also strengthens the resilience of starch supply chains to climatic and economic fluctuations, particularly in regions heavily dependent on agricultural production, where enzymatic bioprocesses represent a strategic lever for sustainable economic development. Furthermore, structuring the value chain around CGTase promotes the upgrading of starch-based products (Palmero et al., 2025). In addition, enzymatic methods that better preserve the molecular structure of starch facilitate the creation of bio-based materials and specialized products with a reduced environmental footprint compared to conventional chemical processes (Rashwan et al., 2024). Finally, the use of immobilized enzymes to efficiently convert polysaccharide-rich waste streams into high-value products adds a further economic dimension to these sustainable strategies (Andler & Goddard, 2018; Olaniyan et al., 2025). Thus CGTase should not be considered solely as a transforming enzyme, but as a central player in the modern starch economy. Its role extends beyond the simple production of cyclic compounds to encompass an integrated vision of plant biomass valorization, where technological efficiency, economic performance, and environmental responsibility converge (Pardhi et al., 2023). The CDs produced by CGTase are used in numerous industrial sectors (food, pharmaceuticals, cosmetics, chemicals), illustrating the versatility and impact of this biocatalysis across diverse value chains, far beyond the simple conversion of a substrate (Saini et al., 2022; Spiridon & Anghel, 2025). In a global context marked by the need to reduce dependence on fossil resources, CGTase illustrates how a microbial enzyme can serve as a structuring lever for the bioeconomy. It establishes a functional link between agriculture, biotechnology, and industry, making starch not just a raw material but a true pillar of sustainable industrial systems in the 21st century (Pardhi et al., 2023).

Cyclodextrins as bioeconomic intermediates

In the conceptual architecture of the bioeconomy, certain molecules are emerging as structuring functional intermediates, capable of linking high-value-added valorization steps while respecting sustainability principles. Among these, CDs occupy a privileged position. Rather than being considered as end products intended for specific uses, CDs can be conceptualized as "molecular building blocks" in a network of biochemical and industrial transformations, offering valuable functional modularity in integrated production chains (Kfoury & Fourmentin, 2023; Musuc, 2024; Spiridon & Anghel, 2025; Tan & Lamers, 2021). Native CDs are cyclic oligosaccharides composed of six (α-), seven (β-), or eight (γ-) D-glucose units linked in an α-(1,4) chain (Allode et al., 2024). This configuration gives these molecules a hydrophobic internal cavity surrounded by a hydrophilic external surface. This characteristic allows them to interact with numerous molecules through noncovalent inclusion complexes (Allode et al., 2024; Crini, 2020a; Loftsson & Duchêne, 2007). This dual surface-hydrophilic externally, hydrophobic internally-is a major advantage, not only for the solubilization or stabilization of compounds, but also as an anchor point for subsequent reactions or transformations. Within the framework of a circular bioeconomic model, CDs are no longer simply isolated entities but rather structuring intermediates that link raw material processing operations to hybrid process networks combining green chemistry and biotechnology. Their ability to form stable inclusion complexes with a wide variety of molecules relies on noncovalent host-guest interactions that enhance the solubility, stability, and reactivity of the encapsulated compounds (Esteso & Romero, 2024; Nicolaescu, Belu, et al., 2025). These multicomponent complexes can then be used as green encapsulation devices, allowing modulation of the bioavailability or reactivity of reaction intermediates or facilitating multistep processes without the use of aggressive solvents or reagents (Aiassa et al., 2021; Nicolaescu, Ionescu, et al., 2025). Furthermore, recent variants of these derivatives show significant potential for green chemistry in the remediation or selective extraction of hydrophobic substances, illustrating the successful integration of CDs into sustainable industrial processes (Fouda-Mbanga et al., 2025; Nicolaescu, Belu, et al., 2025; Nicolaescu, Ionescu, et al., 2025). This intermediate function of CDs is consistent with the principles of green chemistry and modern biocatalysis, which value modular molecules capable of acting as "bridges" between heterogeneous biotransformation steps, and which favor integrated strategies based on functional intermediates capable of linking different transformation steps while limiting material and energy losses (Sheldon & Woodley, 2018). Unlike a traditional approach based on isolated unit steps, using CDs as universal functional modules simplifies reaction flows, reduces byproduct formation, and improves overall process efficiency (Kfoury & Fourmentin, 2023; Morais & Bogel-Łukasik, 2013). Furthermore, thanks to their noncovalent host-guest interactions, CDs promote efficient symbiosis between biocatalysis and mild chemical reactions, facilitating the development of sustainable hybrid cascades adapted to the demands of the contemporary bioeconomy (Crini, 2014; Schrittwieser et al., 2018). Another dimension of the bioeconomic role of CDs is their ability to facilitate reactions that are difficult to achieve with conventional methods.

Through the formation of inclusion complexes, CDs can modulate the kinetics, selectivity, and stability of reactants or intermediates by creating favorable and selective reaction microenvironments (Breslow & Dong, 1998; Rekharsky & Inoue, 1998). This property not only improves the efficiency of bioprocesses but also reduces the need for toxic catalysts, organic solvents, or energy-intensive conditions, in accordance with green chemistry principles (Jødal et al., 2021; Pirvu et al., 2025; Tian & Liu, 2021). The integration of CDs into hybrid production chains thus enhances the sustainability, safety, and overall performance of bio-industrial processes (Del Valle, 2004; Loftsson & Brewster, 2010). From an economic perspective, CDs serve as versatile molecular interfaces that link diverse technological platforms. They allow for the insertion of additional transformation modules into value chains without systematically resorting to costly or energy-intensive transformations (Crini, 2014; Del Valle, 2004). Thanks to their ability to modulate the physicochemical properties of molecules, they enable the development of modular and adaptable approaches that promote the multistep optimization of industrial processes (Loftsson & Duchêne, 2007). This capacity of CDs to serve as functional nodes within bioeconomic networks is particularly relevant in the current context of the transition to more sustainable processes. Industrialists and researchers are increasingly exploring production architectures where functional intermediates such as CDs optimize biomass utilization while integrating strategies for waste reduction, coproduct recycling, and energy efficiency (Singh et al., 2002). Ultimately, considering CDs as bioeconomic intermediates rather than finished products opens up new methodological and conceptual avenues. This perspective positions CGTase not only as an enzyme that produces CDs but also as a catalyst that facilitates access to a class of highly modular intermediates that link successive transformation steps within integrated value chains. This vision transforms the perception of CDs from isolated entities into structuring elements of bio-integrated production, a central argument in evaluating the significance of CGTase in modern bioprocesses.

Contribution of cyclodextrin gluconotransferase to the circular bioeconomy

The transition to sustainable and resilient production models is a central priority of contemporary public policies and industrial strategies. In this context, the circular bioeconomy is emerging as an integrative framework that combines renewable biological resources with energy-, resource-, and waste-efficient processes.

Unlike the linear "extract-produce-discard" paradigm, the circular bioeconomy aims to close material and energy cycles while creating value from previously underutilized biological flows (Carus & Dammer, 2018; D'Amato et al., 2017; European Commission, 2018; Ferraz & Pyka, 2023; Stegmann et al., 2020; Tan & Lamers, 2021). The CGTase fits perfectly within this dynamic. It catalyzes the enzymatic conversion of starch, an abundant and renewable resource, into CDs and other functional products or intermediates, without resorting to aggressive or energy-intensive processes (Esteso & Romero, 2024; Sheldon & Woodley, 2018).

This enzyme enables the use of biological raw materials derived from cereals, tubers, or agricultural residues. This aligns perfectly with the principles of the circular bioeconomy, which prioritize renewable and accessible inputs to reduce dependence on fossil resources (Läufer, 2017; van der Maarel et al., 2002). One of the key advantages of using CGTase is its ability to operate in aqueous environments and at moderate temperatures, thereby reducing the need for organic solvents and metal catalysts, which are associated with significant environmental impacts (Sheldon & Brady, 2018). Enzymatic processes also limit the formation of undesirable byproducts and improve the environmental footprint of starch transformation processes (Radley et al., 2023; van der Maarel et al., 2002; Woodley, 2022). Furthermore, the absence of severe chemical degradation steps reduces the production of complex residues and facilitates the treatment of industrial effluents, thereby contributing to the eco-design of modern biotechnological processes (Woodley, 2022). Another central aspect of CGTase's contribution to the circular bioeconomy is the functional recyclability of CDs. Thanks to their unique cyclic structure, characterized by a hydrophobic internal cavity and a hydrophilic external surface, CDs can form reversible inclusion complexes with a wide variety of molecules (Loftsson & Brewster, 2010; Szejtli, 2004). This property allows their reuse in several functional cycles, notably as solubilizing agents, catalytic supports, or components of multistep systems in green chemistry processes (EM Martin Del Valle, 2004; Nicolaescu, Ionescu, et al., 2025). Furthermore, CGTase does not function in isolation but can be integrated into complex biotechnological systems. In integrated biorefineries, microorganisms are used to simultaneously produce CGTase and other complementary enzymes involved in the pre- or post-treatment of biomass. This synergistic effect between microbial biocatalysis and structured enzymology fosters the emergence of intensified processes capable of utilizing a wide range of biological substrates while maintaining high circularity (D'Amato et al., 2017; Nielsen, 2017; Tan & Lamers, 2021).

This synergistic approach also aligns with the SDGs adopted by the United Nations, particularly those related to responsible consumption and production (SDG 12), clean and affordable energy (SDG 7), and innovation for sustainable industries (SDG 9) (Arora & Mishra, 2019; Colglazier, 2015). Microbial biotechnologies and enzymatic biocatalysis are now recognized as major levers for achieving these objectives by promoting the use of renewable resources and reducing the environmental footprint of industrial processes (De Lorenzo et al., 2018; Sheldon et al., 2021). Finally, CGTase's contribution to the circular bioeconomy is not limited to the technical aspects of transformations. It also has an economic and social dimension, enabling the creation of local value chains based on renewable resources and accessible enzymatic technologies, which can support regional development and reduce economic disparities related to energy and technology supplies. Thus CGTase illustrates how a microbial enzyme can contribute to a profound reorganization of production systems, moving from linear extraction to circular valorization. By integrating principles of sustainability, low environmental impact, and functional modularity, CGTase helps to make the circular bioeconomy not just a political aspiration but a tangible technological and economic reality.

Strategic positioning of cyclodextrin gluconotransferase in contemporary research

The GH13 family is among the most studied enzyme groups due to its remarkable structural and functional diversity, reflecting ongoing research on amylolytic enzymes (Nurul Ilmi & Zulaika, 2025; Janeček & Svensson, 2022; Majzlová et al., 2013). Within this scientific landscape, CGTase occupies a strategic position due to its multifunctional potential and its role as a model organism at the interface between basic research and biotechnological applications (Leemhuis et al., 2010). Scientific interest in this enzyme has intensified alongside the rise of industrial biotechnologies, leading to a proliferation of work aimed at optimizing its catalytic properties and broadening its range of applications (Bart A. Van Der Veen et al., 2000a). A bibliometric analysis of works published since the 2000s reveals a steady increase in the number of articles devoted to CGTase, particularly in fields related to enzyme engineering, industrial microbiology, and structural biology (Chen et al., 2022; Leemhuis et al., 2010; Saini et al., 2022). This scientific growth is partly explained by the increased availability of genomic sequencing tools and advanced molecular modeling techniques, which have made it possible to explore the genetic and functional diversity of CGTases from different microorganisms (Centeno-Leija et al., 2022; Liu, Wu, et al., 2022).

Advances in comparative genomics have thus facilitated the identification of new enzyme variants with distinct catalytic or structural characteristics, contributing to enriching the overall understanding of the functional evolution of amylolytic enzymes (Centeno-Leija et al., 2022; Kelly et al., 2009; Saini et al., 2022). Beyond its scientific data, CGTase stands out as a model system for studying structure-function relationships within the α-amylase family. The availability of numerous three-dimensional structures of CGTases, obtained by X-ray crystallography or computer-aided modeling, has allowed for the precise exploration of the molecular determinants responsible for enzymatic specificity and structural stability (Leemhuis et al., 2010; van der Maarel et al., 2002). This knowledge has fostered interdisciplinary approaches that combine bioinformatics, biophysics, and enzymology to elucidate the mechanisms underlying the adaptive evolution of GH13 enzymes (Chen et al., 2022; Janeček & Svensson, 2022). In this context, CGTase illustrates how minor sequence variations can lead to significant changes in catalytic function (Ara et al., 2021; Leemhuis et al., 2010). Furthermore, scientific interest in CGTase extends far beyond classical enzymology. This enzyme now represents a conceptual bridge between basic research and biotechnological innovation, as its study requires the simultaneous mobilization of several emerging disciplines. Contemporary research notably involves metagenomics, which enables the identification of new CGTases in extreme or little-explored environments, as well as synthetic biology, which aims to design optimized enzyme systems from recombinant genetic modules (Muniz et al., 2024; Singh et al., 2002). The integration of these approaches has profoundly renewed our understanding of amylolytic enzymes, highlighting their functional plasticity and potential to adapt to diverse environments (Centeno-Leija et al., 2022; Chen et al., 2022; Pardhi et al., 2023). Furthermore, CGTase plays a structuring role in the development of emerging scientific disciplines, particularly in biocatalyst engineering. The molecular models developed for this enzyme have contributed to improved algorithms for predicting enzyme structure and function. These methodological advances extend well beyond the specific context of CGTase and contribute to the rise of rational enzyme design, a key area of modern biotechnology. In this context, CGTase is frequently used as a reference system to test new enzyme engineering or computational optimization strategies (Centeno-Leija et al., 2022; Muniz et al., 2024; Pardhi et al., 2023; Tonkova, 1998). From a forward-looking perspective, CGTase appears as a particularly promising model for anticipating future directions in enzyme research.

The rise of omics technologies, the application of artificial intelligence to structural biology, and evolutionary engineering should enable a more in-depth exploration of the natural diversity and functional plasticity of this enzyme. These advances could lead to a better understanding of enzyme evolution mechanisms and open new perspectives for designing biocatalysts tailored to emerging industrial needs. Thus CGTase is emerging not only as a major object of scientific study but also as an indicator of future trends in biotechnology enzyme research.

Industrial applications of cyclodextrin gluconotransferase

Global climate change poses a major challenge to humanity's sustainable development. This has led to an increasing reliance on enzymes to replace chemicals in various industrial sectors (Mondal et al., 2022). CGTases play a significant role in biotechnology and various industries due to their unique ability to convert starch or starch derivatives into dendritic enzymes (D-Cs) through the cyclization reaction (Kelly et al., 2009; Rajput et al., 2016). D-Cs are cyclic oligosaccharides typically composed of six, seven, or eight α-(1,4)-linked D-glucose units (Rajput et al., 2016). They have paved the way for an endless array of applications in several fields, including pharmaceuticals, food processing, analytical chemistry, cosmetics, the environment, water and air pollution control, and textiles (Allode et al., 2024; Yu et al., 2025). These applications are possible thanks to their hydrophilic outer surface, which is soluble in water, and their nonpolar hydrophobic cavity (Rajput et al., 2016). This unique property of CDs-the ability to form inclusion complexes with a wide range of solid, liquid, and gaseous compounds-is therefore the source of the various applications mentioned above (Pardhi et al., 2023). In the pharmaceutical industry, this property has been widely used to improve the solubility, stability, and bioavailability of many poorly absorbed or unstable pharmaceutical compounds

(Dai et al., 2025; Kali et al., 2024). They also have the ability to enable targeted accumulation of the active ingredient at a specific site, prolong the duration of systemic circulation, promote synergistic therapeutic outcomes, and exhibit superior biocompatibility profiles (Dai et al., 2025). Several scientific articles demonstrate the role of CGTases in the pharmaceutical industry. A publication by Mohammad et al. (2020) in *the Nanomedicine Research Journal* clearly addresses the applications of CDs in various drug delivery systems and electrochemical detection. It describes the implementation of CDs in a variety of specific drug delivery systems, such as oral, controlled or targeted, ocular, transdermal, nasal, and rectal protein and peptide delivery systems. Kim et al. (2020) compiled a list of some drugs developed using CDs.

CDs play an important role in the food industry due to their unique ability to form host-guest complexes with a wide variety of compounds (Gonzalez Pereira et al., 2021). This property of CDs to form host-guest complexes allows them to encapsulate numerous molecules such as flavorings, vitamins, colorings, and even unsaturated fats or other lipophilic molecules. They thus contribute to improving physicochemical stability, solubility, and bioavailability, while extending the shelf life of food products (Gonzalez Pereira et al., 2021; Liu et al., 2022). Furthermore, they enhance sensory qualities and prevent microbiological contamination (Liu et al., 2022). Thanks to their low toxicity, CDs are also safe for human consumption, making them versatile additives used as excipients, chelating agents, or flavor maskers. CDs stabilize bioactive enzymes such as lipases, amylases, and proteases (Karavaeva et al., 2024). Treating food with CDs is an effective way to reduce undesirable tastes, such as bitterness, and abnormal flavor properties in certain food products. Furthermore, CDs can be used to improve the stability of many groups of compounds, such as anthocyanins and volatile compounds, during food processing (Kelanne et al., 2024; Sanjoy et al., 2013). CGTases, by catalyzing the conversion of starch to CDs, are therefore key enzymes in this process. CDs constitute an exceptional type of building block in a molecular architecture (Gonzalez Pereira et al., 2021). Their activity allows for the production of CDs at a lower cost, an essential condition for supporting the growth of their new applications: prebiotics, active agents in food packaging, nanosensors, or even activators for the extraction of bioactive compounds (Matencio et al., 2020). However, the quality of the CDs produced depends closely on the enzymatic production conditions, and particular attention must be paid to solvent residues, the concentrations used, and the synergy with other stabilizing agents to ensure their effectiveness without altering the sensory properties of the food (Kelanne et al., 2024). In the field of analytical chemistry, CD plays an important role due to its ability to form inclusion complexes with various substances, including chiral compounds (Morin, 2009). This property is frequently employed in several analytical methods, including liquid chromatography, gas chromatography, and capillary electrophoresis. They can be grafted onto the stationary phase or added to a mobile phase. This modifies the retention time of analytes and allows for the separation of enantiomers (Estefania, 2019). Their inherent chirality confers exceptional selectivity against racemic mixtures, which lead to the formation of various diastereomeric complexes. For example, β-CD-based columns fixed to polydimethylsiloxane have proven effective in separating enantiomers of phyto-pharmaceutical compounds. In addition to their function in chiral separation, CDs increase the solubility of several organic compounds, thus facilitating their detection, including in biological matrices. They are also used in the development of enzyme sensors.

CDs are emerging as essential, versatile tools in analytical chemistry for the qualitative and quantitative analysis of various compounds (Allode et al., 2024). In the cosmetics industry, CDs occupy a position at the intersection of the pharmaceutical and food industries. They act as carriers by encapsulating molecules, thereby adding numerous properties to improve solubility, stability, and bioavailability in cosmetic product formulations (Khaoulani, 2015; Tarimci, 2011). They are also used to stabilize oils, waxes, and fragrances in lotions, body creams, toothpastes, and fabric softeners (Ferreira et al., 2023; Tarimci, 2011). In the perfume industry, CDs can reduce the volatility of fragrances, ensure the gradual release of fragments, and are effective in deodorants for trapping odor molecules. These properties have enabled the development of innovative and effective products, such as the Febreze® brand (Procter & Gamble®), which has used CDs to neutralize unpleasant odors since 1998 (Estefania, 2019). In his thesis published in 2015, Khaoulani demonstrated the use of CDs in wastewater treatment through adsorption onto CD polymers (Khaoulani, 2015). In recent years, CDs have proven to be valuable complexing agents in environmental protection, particularly in removing pollutants from soil, water, and air (Viglianti et al., 2007). Soil treatment using aqueous CD-based solutions has been employed to solubilize pollutants. CDs are used as additives to improve efficiency and reduce treatment time compared to using water alone or conventional surfactants. Indeed, CDs have improved the solubility of typical organic contaminants in soils (such as polycyclic aromatic hydrocarbons), thereby increasing their bioavailability for microbial degradation (Estefania, 2019; Khaoulani, 2015). The applications of CDs in the textile industry are constantly evolving, and numerous studies are underway to exploit their potential in this field (Allode et al., 2024). The permanent fixation of CDs gives new textiles interesting properties (Buschmann et al., 2001). Thanks to their ability to form inclusion complexes, CDs can be used to eliminate unpleasant odors from textiles. Thus, the formation of body odor is reduced by complexing the organic compounds in sweat (Buschmann et al., 2001). This property can also be used to encapsulate active compounds such as perfumes, antioxidants, antibacterial agents, or dyes. CDs can be used as finishing agents for textiles (Allode et al., 2024; Grigoriu et al., 2008). For example, CDs can be applied to fabrics to create a protective layer that improves water and dirt resistance. Furthermore, CDs can be used to enhance the dyeing properties of textiles. This can improve dyeing efficiency by allowing better penetration of dyes into the fabric fibers. Finally, CDs can be used for medical applications in the textile industry, such as the manufacture of dressings or protective clothing that release medications or antibacterial agents to treat wounds or prevent infections (Allode et al., 2024; Bilensoy, 2011).

In conclusion, thanks to the discovery of CGTases, several industries benefit from more effective and stable products that meet consumer demands for comfort and safety. Their discovery and development perfectly illustrate how a microbial enzyme can be transformed into a versatile biotechnological tool. Each year, thousands of tons of CDs are manufactured from starch by various producers, and demand continues to grow (Leemhuis et al., 2010). CDs are enzymatically modified starches. They are produced from starch, specifically amylose, by enzymatic conversion using CGTases. They are a family of cyclic oligosaccharides typically containing six (α-cyclodextrin), seven (β-cyclodextrin), or eight (γ-cyclodextrin) 1,4-linked d- glucose units (Biwer et al., 2002; Dodziuk, 2006). The production of native CDs from CGTase involves several key steps: (i) culturing the micro-organism producing the CGTase; (ii) separating, concentrating, and purifying the enzyme; (iii) liquefaction of the starch by raising the temperature or by using a hydrolytic enzyme. The enzyme CGTase is then added to the resulting solution. This enzyme, via two catalytic groups (an acid-base group Glu 257 and a nucleophilic group Asp 229), induced the intramolecular transglycosylation reaction responsible for the formation of CDs. and (iv) separation of the CDs from the mixture, their purification, and their crystallization (Ellouze et al., 2011; Sanjoy et al., 2013). In fact, four different transglycosylation reactions can occur: (1) the cyclization reaction, (2) the chain elongation reaction by coupling, (3) the disproportionation reaction by intermolecular transglycosylation, and (4) active site transfer by water hydrolysis. All these reactions influence the quantity and final size distribution of the CDs produced (Ellouze et al., 2011).

Description of solvent-free procedure

The solvent-free process was first developed for the production of β-CD. On an industrial scale, it is particularly used in Japan. Due to its low solubility, β-CD can be easily purified by crystallization steps. The purification of α- and γ-CD has been achieved by complex and expensive chromatography, resulting in low yields and a range of byproducts. However, more economical processes have recently been published, which may also offer potential environmental advantages. CDs produced without an organic complexing agent can be used in the food industry without restriction, unlike CDs produced by solvent-based processes. The production of β-CD begins with starch liquefaction and enzymatic conversion identical to those used in a solvent-based process, except for the absence of a complexing agent. At the end of the reaction, CGTase is inactivated, the pH is reduced, and glucoamylase is added. Glucoamylase converts unused starch and other non-cyclic dextrins, which can interfere with purification, into glucose and maltose. The solution is then clarified with activated carbon, filtered, and concentrated under reduced pressure to approximately 60% (w/v) dissolved solids. In some cases, deionization is required prior to activated carbon treatment. After crystallization and recrystallization, the precipitated β-CD is isolated, washed, centrifuged, and dried.

The remaining material, consisting of glucose, maltose, α-CD, and γ-CD, is concentrated into a syrup and can be used as a food additive (Biwer et al., 2002; Sanjoy et al., 2013).

Cyclodextrin production process

Over the past few decades, the industrial production of cyclodextrins has seen significant development due to their increased use in the pharmaceutical, food, cosmetic, and environmental sectors (Allode et al., 2025). Thanks to their ability to form inclusion complexes with numerous molecules, cyclodextrins improve the solubility, stability, and bioavailability of various compounds of interest (Allode et al., 2024). According to Del Valle (2004), improvements in enzymatic production processes have reduced manufacturing costs while maintaining excellent product quality. Furthermore, Szejtli (1998) emphasized that the synthesis of cyclodextrins now constitutes a key biotechnological pathway for exploiting starch from renewable resources. Similarly, Biwer et al. (2002) described industrial-scale optimized processes that reduced costs through the use of thermostable CGTases and simplified purifications. This same observation was made by Ben Amara et al. (2022), who optimized the production of β-cyclodextrins from potato peel starch, achieving 57.26 g/L through experimental designs, with environmentally responsible purification at a 20% yield, demonstrating a low-cost strategy to minimize manufacturing expenses. Two types of CD production processes are generally used: the solvent-based process and the solvent-free process (Biwer et al., 2002). The solvent-based process requires an organic solvent, primarily toluene, ethyl alcohol, or acetone, which acts as a complexing agent and selectively extracts a specific type of CD, thus directing the enzymatic reaction to produce the particular type of CD of interest. The solvent-free process does not require a complexing agent and produces a mixture of cyclodextrins (CDs) that can be subsequently separated by chromatography or selective precipitation (Li et al., 2014; Sanjoy et al., 2013). The performance of the enzymatic cyclodextrin production process is highly dependent on the physicochemical conditions of the reaction medium. pH, temperature, substrate concentration, and the presence of certain ions directly influence the activity and selectivity of CGTases (Allode et al., 2025; Rosso et al., 2002; Wang et al., 2018). Gawande and Patkar (2001) showed that pH variations can modify the enzymatic stability and relative distribution of the cyclodextrins produced. Similarly, Van Der Veen et al. (2000a) indicated that medium viscosity and starch concentration play a determining role in substrate accessibility, thus affecting overall production yields.

Description of procedure with solvent

The process begins with the liquefaction of starch. Liquefaction is achieved using a thermostable α-amylase, acids, mechanical disintegration, or thermostable CGTases. Additionally, some α-amylases require calcium ions. If α-amylase is used, it must be inactivated by heat before any further processing. On an industrial scale, liquefaction is typically achieved by jet cooking. After liquefaction, the starch solution is cooled to the enzymatic reaction temperature, and then the CGTase and an organic complexing agent are added. CDs are produced enzymatically, and the desired CD type forms a complex with the complexing agent and precipitates. Once the conversion is complete, the CD-agent complex is separated from the reaction solution by centrifugation or filtration. The remaining solution contains unused starch, linear dextrins, glucose, maltose, CGTase, unused organic complexing agent, other byproducts, and water. The separated complex is washed, and the filtrate is distilled to recover excess complexing agent. In the next step, the CD complex is suspended in water and cleaved by heating. The complexing agent is then separated by steam distillation. Some complexing agents cannot be isolated by distillation and must be extracted by liquid-liquid extraction. The resulting solution is concentrated by vacuum distillation; it sometimes requires treatment with activated carbon. In the crystallization step, the CD is precipitated, then filtered, washed, and dried. The downstream process separates only the CDs from the rest of the reaction solution, but not the individual CDs. Thus, the choice of an appropriate enzyme and complexing agent determines the selectivity of CD formation and recovery (Biwer et al., 2002). Despite its good selectivity, the production process for cyclodextrins using organic solvents presents certain environmental and health limitations. The use of solvents such as toluene or acetone can lead to the presence of undesirable residues in the final product, requiring additional purification steps to comply with regulatory standards. (Del Valle, 2004; Duan et al., 2023). Crini (2014) reports that these processes increase the environmental footprint due to volatile organic compound emissions and the energy requirements associated with solvent recovery. These constraints have fostered the development of more environmentally friendly alternative processes (Astray et al., 2009). From an economic perspective, the solvent-based process involves high costs related to the purchase, recovery, and recycling of solvents, as well as the necessary purification steps. Szejtli (1998) indicates that these operations can represent a significant portion of the overall cost of cyclodextrin production. Therefore, improving enzyme selectivity and optimizing reaction conditions are essential levers for reducing industrial costs and improving process profitability (Duan et al., 2023; Rojas et al., 2019).

Description of solvent-free procedure

The solvent-free process was initially developed for the production of β-CD. On an industrial scale, it is particularly prevalent in Japan. The development of the solvent-free process is closely linked to advances made in Japan in the field of enzymatic biotechnology. After World War II, significant research efforts were undertaken to valorize starch through enzymatic processes (French, 1957). Due to its low solubility, β-CD can be easily purified by crystallization steps. The purification of α- and γ-CD has been achieved through complex and costly chromatography, resulting in low yields and a range of byproducts. However, more economical processes have recently been published, which may also offer potential environmental benefits. Solvent-free cyclodextrin production processes offer significant environmental advantages. The absence of organic solvents reduces the generation of chemical waste and limits toxicological risks (Biwer et al., 2002; Nicolaescu, Belu, et al., 2025). These processes are fully aligned with the principles of green chemistry and facilitate the use of cyclodextrins in the food and pharmaceutical industries, where regulatory requirements are particularly stringent. CDs produced without organic complexing agents can be used in the food industry without restriction, unlike CDs produced by solvent-based processes. The production of β-CD begins with starch liquefaction and enzymatic conversion identical to that used in a solvent-based process, except for the absence of a complexing agent. At the end of the reaction, CGTase is inactivated, the pH is lowered, and glucoamylase is added. Glucoamylase converts unused starch and other non-cyclic dextrins, which can interfere with purification, into glucose and maltose. The solution is then clarified with activated carbon, filtered, and concentrated under reduced pressure to approximately 60% (w/v) dissolved solids. In some cases, deionization is required prior to activated carbon treatment. After crystallization and recrystallization, the precipitated β-CD is isolated, washed, centrifuged, and dried. The remainder, composed of glucose, maltose, α- and γCD, is concentrated into a syrup and can be used as a food additive (Biwer et al., 2002; Sanjoy et al., 2013). However, the separation and purification of the different cyclodextrins remain significant technological challenges. Due to their very similar chemical structures, the separation of α-, β-, and γ-cyclodextrins is complex and often requires costly purification techniques. Recent advances in enzyme biotechnology have led to the development of new strategies to improve cyclodextrin production. Immobilizing CGTases on solid supports and using continuous enzyme reactors increase enzyme stability and facilitate reuse. In conclusion, both solvent-based and solvent-free cyclodextrin production processes offer complementary advantages and limitations. The solvent-based process provides greater selectivity in cyclodextrin production, while the solvent-free process is distinguished by its environmental benefits and compliance with regulatory requirements. Advances in enzyme engineering and process optimization are helping to improve the performance, sustainability and profitability of industrial cyclodextrin production, thereby reinforcing their importance in the biotechnological valorization of starch.

Conclusion

CGTase has evolved from a historically obscure amylolytic enzyme into a strategic biocatalyst central to modern biotechnology, green chemistry, and circular bioeconomy. Its unique ability to catalyze cyclization, coupling, and transglycosylation reactions positions it at the interface between starch conversion and the generation of high-value functional molecules such as CDs. Advances in genomics, structural biology, and protein engineering have transformed CGTase into a model enzyme for understanding GH13 family evolution and tailoring catalytic properties for industrial needs. As a platform enzyme, it supports flexible bioprocess design, enabling the valorization of renewable starch resources into multifunctional intermediates for pharmaceuticals, food, materials science, and environmental technologies. The enzyme's compatibility with mild, environmentally benign reaction conditions further underscores its importance in achieving sustainability and reducing reliance on fossil-based chemical processes. With ongoing developments in omics, enzyme engineering, and integrated biorefineries, CGTase stands poised to play an even more influential role in shaping next-generation biomanufacturing frameworks. Ultimately, CGTase exemplifies how microbial enzymes can drive innovation, strengthen circular value chains, and accelerate the global transition toward sustainable industrial systems.

References

Aiassa, V., Garnero, C., Longhi, M. R., & Zoppi, A. (2021). Cyclodextrin multicomponent complexes: Pharmaceutical applications. *Pharmaceutics, 13*, 1099. https://doi.org/10.3390/pharmaceutics13071099.

Alhussein, H. A. A., Al-Manhel, A. J. A., & Abd, A.-H. H. A. (2025). A comprehensive overview of cyclodextrins in terms of production, properties, and applications. *Piŝevye Sistemy, 8*, 196–203. https://doi.org/10.21323/2618-9771-2025-8-2-196-203.

Allode, T. D.-M., Vodounon, A. C., Dossou, A. G., Honzounnon, N. C., Migan, A. D. F. M., & Hlouedje, W. H. (2024). Applications of cyclodextrin-glucanotransferase in the biosynthesis of cyclodextrins: Characteristics, sources and production. *American Journal of Microbiological Research, 12*, 63–78. https://doi.org/10.12691/ajmr-12-3-4.

Allode, T. D.-M., Vodounon, C. A., Dossou, G. A., Honzounnon, N. C., Migan, M. A. D. F., Sina, H., Senou, M. (2025). Characterization of soil microbiota for the production of cyclodextrin glucanotransferase in northwestern Benin. *Egyptian Journal of Chemistry, 0*, 0–0. https://doi.org/10.21608/ejchem.2025.393766.11907.

Andler, S. M., & Goddard, J. M. (2018). Transforming food waste: How immobilized enzymes can valorize waste streams into revenue streams. *npj Science of Food, 2*, 19. https://doi.org/10.1038/s41538-018-0028-2.

Ara, K. Z. G., Linares-Pastén, J. A., Jönsson, J., Viloria-Cols, M., Ulvenlund, S., Adlercreutz, P., & Karlsson, E. N. (2021). Engineering CGTase to improve synthesis of alkyl glycosides. *Glycobiology, 31*, 603–612. https://doi.org/10.1093/glycob/cwaa109.

Arora, N. K., & Mishra, I. (2019). United Nations sustainable development goals 2030 and environmental sustainability: Race against time. *Sustainable Environment Research, 2*, 339–342. https://doi.org/10.1007/s42398-019-00092-y.

Astray, G., Gonzalez-Barreiro, C., Mejuto, J. C., Rial-Otero, R., & Simal-Gándara, J. (2009). A review on the use of cyclodextrins in foods. *Food Hydrocolloids, 23*, 1631–1640. https://doi.org/10.1016/j.foodhyd.2009.01.001.

Awasthi, M. K., Sarsaiya, S., Patel, A., Juneja, A., Singh, R. P., Yan, B., Awasthi, S. K., Jain, A., Liu, T., Duan, Y., Pandey, A., Zhang, Z., & Taherzadeh, M. J. (2020). Refining biomass residues for sustainable energy and bio-products: An assessment of technology, its importance, and strategic applications in circular bio-economy. *Renewable and Sustainable Energy Reviews, 127*, 109876. https://doi.org/10.1016/j.rser.2020.109876.

Benavent-Gil, Y., Rosell, C. M., & Gilbert, E. P. (2021). Understanding CGTase action through the relationship between starch structure and cyclodextrin formation. *Food Hydrocolloids, 112*, 106316. https://doi.org/10.1016/j.foodhyd.2020.106316.

Ben Amara, F., Bouzid, M., Sahnoun, M., Ben Nasr, Y., Jaouadi, B., Bejar, S., Jemli, S. (2022). Valorization of Potato Peels Starch for Efficient β-Cyclodextrin Production and Purification through an Eco-Friendly Process. *Starch - Stärke, 74*, 2200037. https://doi.org/10.1002/star.202200037.

Bilensoy, E. (2011). *Cyclodextrins in pharmaceutics, cosmetics, and biomedicine: Current and future industrial applications* (1st ed.). Wiley. https://doi.org/10.1002/9780470926819.

Biwer, A., Antranikian, G., & Heinzle, E. (2002). Enzymatic production of cyclodextrins. *Applied Microbiology and Biotechnology, 59*, 609–617. https://doi.org/10.1007/s00253-002-1057-x.

Bornscheuer, U. T., Huisman, G. W., Kazlauskas, R. J., Lutz, S., Moore, J. C., & Robins, K. (2012). Engineering the third wave of biocatalysis. *Nature, 485*, 185–194. https://doi.org/10.1038/nature11117.

Bozkurt, E. U., Ørsted, E. C., Volke, D. C., & Nikel, P. I. (2024). Accelerating enzyme discovery and engineering with high-throughput screening. *Natural Product Reports*. https://doi.org/10.1039/D4NP00031E.

Breslow, R., & Dong, S. D. (1998). Biomimetic reactions catalyzed by cyclodextrins and their derivatives. *Chemical Reviews, 98*, 1997–2012. https://doi.org/10.1021/cr970011j.

Buschmann, H.-J., Knittel, D., & Schollmeyer, E. (2001). New textile applications of cyclodextrins. *Journal of Inclusion Phenomena and Macrocyclic Chemistry, 40*, 169–172. https://doi.org/10.1023/A:1011892600388.

Carus, M., & Dammer, L. (2018). The circular bioeconomy: Concepts, opportunities, and limitations. *Industrial Biotechnology, 14*, 83–91. https://doi.org/10.1089/ind.2018.29121.mca.

Carvalho, H. J. M., Barcia, M. T., & Schmiele, M. (2024). Non-conventional starches: Properties and potential applications in food and non-food products. *Macromol, 4*, 886–909. https://doi.org/10.3390/macromol4040052.

Centeno-Leija, S., Espinosa-Barrera, L., Velazquez-Cruz, B., Cárdenas-Conejo, Y., Virgen-Ortíz, R., Valencia-Cruz, G., Saenz, R. A., Marín-Tovar, Y., Gómez-Manzo, S., Hernández-Ochoa, B., Rocha-Ramirez, L. M., Zataraín-Palacios, R., Osuna-Castro, J. A., López-Munguía, A., & Serrano-Posada, H. (2022). Mining for novel cyclomaltodextrin glucanotransferases unravels the carbohydrate metabolism pathway via cyclodextrins in Thermoanaerobacterales. *Scientific Reports, 12*, 730. https://doi.org/10.1038/s41598-021-04569-x.

Chen, H., Liu, Y., Ren, X., Wang, J., Zhu, L., & Lu, Y. (2022). Title incomplete in provided data]. *Applied and Environmental Microbiology, 88*, e01027-22. https://doi.org/10.1128/aem.01027-22.

Christophe, C., & Steven, B. (2009). Starch: Its synthesis, its mobilization, its evolutionary history. *Cahiers Agricultures, 18*, 315–322. https://doi.org/10.1684/agr.2009.031.

Colglazier, W. (2015). Sustainable development agenda: 2030. *Science (New York, N.Y.), 349*, 1048–1050. https://doi.org/10.1126/science.aad2333.

Crini, G. (2014). Review: A History of Cyclodextrins. *Chemical Reviews, 114*, 10940–10975. https://doi.org/10.1021/cr500081p.

Crini, G. (2020a). The contribution of Franz Schardinger to cyclodextrins: A tribute on the occasion of the centenary of his death. *Journal of Inclusion Phenomena and Macrocyclic Chemistry, 97*, 19–28. https://doi.org/10.1007/s10847-020-00990-3.

Crini, G. (2020b). Twenty years of dextrin research: A tribute to Professor Hans Pringsheim (1876–1940). *Journal of Inclusion Phenomena and Macrocyclic Chemistry, 98*, 11–27. https://doi.org/10.1007/s10847-020-01013-x.

Crini, G., Fourmentin, S., & Lichtfouse, E. (2020). *The history of cyclodextrins*. Springer International Publishing. https://doi.org/10.1007/978-3-030-49308-0.

Cui, J., Zheng, G.-W., Black, G., Iqbal, H. M. N., & Bilal, M. (2022). Editorial: Enzyme biocatalysts: Design and application. *Frontiers in Chemistry, 10*, 851857. https://doi.org/10.3389/fchem.2022.851857.

D'Amato, D., Droste, N., Allen, B., Kettunen, M., Lähtinen, K., Korhonen, J., Leskinen, P., Matthies, B. D., & Toppinen, A. (2017). Green, circular, bio economy: A comparative analysis of sustainability avenues. *Journal of Cleaner Production, 168*, 716–734. https://doi.org/10.1016/j.jclepro.2017.09.053.

Dai, Z., Yang, H., Yin, P., Liu, X., Zhang, L., Dou, Y., & Sun, S. (2025). Applications of cyclodextrin-based drug delivery systems in inflammation-related diseases. *Pharmaceutics, 17*, 378. https://doi.org/10.3390/pharmaceutics17030378.

Davis, F., & Higson, S. (2011). *Macrocycles: Construction, chemistry and nanotechnology applications* (1st ed.). Wiley. https://doi.org/10.1002/9780470980200.

Dodziuk, H. (Ed.). (2006). Cyclodextrins and Their Complexes: Chemistry, Analytical Methods, Applications, 1st ed. Wiley. https://doi.org/10.1002/3527608982.

Duan, M., Wang, Y., Tan, D., Yang, G., Deng, Y., Ran, G., Li, J. (2023). Economical synthesis of γ-cyclodextrin catalyzed by oriented cyclodextrin glycosyltransferase displayed on bacterial polyhydroxyalkanoate nanogranules. *Microbial Cell Factories, 22*, 181. https://doi.org/10.1186/s12934-023-02191-2.

Ellouze, F., Amar, N. B., & Deratani, A. (2011). Large-ring cyclodextrins: Synthesis, purification and applications. *Comptes Rendus Chimie, 14*, 967–971. https://doi.org/10.1016/j.crci.2011.02.006.

Estefania, O. (2019). Nano-reactors based on amphiphilic cyclodextrins for catalysis and vectorization (Doctoral thesis). University of Picardy Jules Verne, France.

Esteso, M. A., & Romero, C. M. (2024). Cyclodextrins: Properties and applications. *International Journal of Molecular Sciences, 25*, 4547. https://doi.org/10.3390/ijms25084547.

European Commission. (2018). *A sustainable bioeconomy for Europe: Strengthening the connection between economy, society and the environment*. Publications Office of the European Union. https://doi.org/10.2777/792130.

Farhan, M., Hasani, I. W., Khafaga, D. S. R., Ragab, W. M., Ahmed Kazi, R. N., Aatif, M., Muteeb, G., & Fahim, Y. A. (2025). Enzymes as catalysts in industrial biocatalysis: Advances in engineering, applications, and sustainable integration. *Catalysts, 15*, 891. https://doi.org/10.3390/catal15090891.

Ferraz, D., & Pyka, A. (2023). Circular economy, bioeconomy, and sustainable development goals: A systematic literature review. *Environmental Science and Pollution Research*. https://doi.org/10.1007/s11356-023-29632-0.

Ferreira, L., Mascarenhas-Melo, F., Rabaça, S., Mathur, A., Sharma, A., Giram, P. S., Pawar, K. D., Rahdar, A., Raza, F., Veiga, F., Mazzola, P. G., & Paiva-Santos, A. C. (2023). Cyclodextrin-based dermatological formulations: Dermopharmaceutical and cosmetic applications. *Colloids and Surfaces B: Biointerfaces, 221*, 113012. https://doi.org/10.1016/j.colsurfb.2022.113012.

Fouda-Mbanga, B. G., Tywabi-Ngeva, Z., Badawy, W. M., Ebite, C., Onotu, O. P., Abogidi, C., Uzordinma, A. P., & Kaba, S. (2025). Green cyclodextrin-derivatives for sustainable remediation of pesticides and heavy metals: A review. *Journal of Molecular Structure, 1328*, 141326. https://doi.org/10.1016/j.molstruc.2025.141326.

French, D. (1957). The Schardinger dextrins. In *Advances in Carbohydrate Chemistry*. Elsevier, 189–260. https://doi.org/10.1016/S0096-5332(08)60209-X.

Gerhard, W. (2014). Germany and the cyclodextrin science and technology. *Cyclodextrin News, 28*, 15.

Global Market Insights. (2025). Industrial starch market size – Growth forecast, 2025–2034 (Report No. GMI6512).

Gonzalez Pereira, A., Carpena, M., García Oliveira, P., Mejuto, J. C., Prieto, M. A., & Simal-Gandara, J. (2021). Main applications of cyclodextrins in the food industry as the compounds of choice to form host–guest complexes. *International Journal of Molecular Sciences, 22*, 1339. https://doi.org/10.3390/ijms22031339.

Gawande, B. N., Patkar, A .Y. (2001). Purification and properties of a novel raw starch degrading-cyclodextrin glycosyltransferase from Klebsiella pneumoniae AS- 22. *Enzyme and Microbial Technology, 28*, 735–743. https://doi.org/10.1016/s0141-0229(01)00347-7.

Grigoriu, A.-M., Luca, C., Gabriela, L., & Grigoriu, A. (2008). Cyclodextrins' applications in the textile industry. *Cellulose Chemistry and Technology, 42*, 103–112.

Hedges, A. (2009). Cyclodextrins. In *Starch*. Elsevier, 833–851. https://doi.org/10.1016/B978-0-12-746275-2.00022-7.

Henrissat, B. (1991). A classification of glycosyl hydrolases based on amino acid sequence similarities. *Biochemical Journal, 280*, 309–316. https://doi.org/10.1042/bj2800309.

Henrissat, B., & Bairoch, A. (1993). New families in the classification of glycosyl hydrolases based on amino acid sequence similarities. *Biochemical Journal, 293*, 781–788. https://doi.org/10.1042/bj2930781.

Hofmann, B. E., Bender, H., & Schulz, G. E. (1989). Three-dimensional structure of cyclodextrin glycosyltransferase from Bacillus circulans at 3.4 Å resolution. *Journal of Molecular Biology, 209*, 793–800. https://doi.org/10.1016/0022-2836(89)90607-4.

Ibrahim, A. S. S., Al-Salamah, A. A., El-Tayeb, M. A., El-Badawi, Y. B., & Antranikian, G. (2012). A novel cyclodextrin glycosyltransferase from alkaliphilic *Amphibacillus* sp. NPST-10: Purification and properties. *International Journal of Molecular Sciences, 13*, 10505–10522. https://doi.org/10.3390/ijms130810505.

Ibrahim, A. S. S., Al-Salamah, A. A., El-Toni, A. M., El-Tayeb, M. A., & Elbadawi, Y. B. (2013). Immobilization of cyclodextrin glucanotransferase on aminopropyl-functionalized silica-coated superparamagnetic nanoparticles. *Electronic Journal of Biotechnology, 16*. https://doi.org/10.2225/vol16-issue6-fulltext-8.

Nurul Ilmi, A., & Zulaika, E. (2025). Exploring amylolytic enzymes from GH families in maltooligosaccharide biosynthesis: Classification, catalysis, and industrial potential. *International Journal of Scientific Research Archive, 15*, 1728–1736. https://doi.org/10.30574/ijsra.2025.15.3.1924.

Janeček, Š., & Svensson, B. (2022). How many α-amylase GH families are there in the CAZy database? *Amylase, 6*, 1–10. https://doi.org/10.1515/amylase-2022-0001.

Janeček, Š., & Zámocká, B. (2020). A new GH13 subfamily represented by the α-amylase from the halophilic archaeon Haloarcula hispanica. *Extremophiles: Life Under Extreme Conditions, 24*, 207–217. https://doi.org/10.1007/s00792-019-01147-y.

Jicsinszky, L., Martina, K., & Cravotto, G. (2025). Cyclodextrin-based artificial enzymes. *Sustainable Chemistry and Pharmacy, 44*, 101945. https://doi.org/10.1016/j.scp.2025.101945.

Jødal, A.-S. S., Czaja, T. P., Van Den Berg, F. W. J., Jespersen, B. M., & Larsen, K. L. (2021). The effect of α-, β-, and γ-cyclodextrin on wheat dough and bread properties. *Molecules (Basel, Switzerland), 26*, 2242. https://doi.org/10.3390/molecules26082242.

Kali, G., Haddadzadegan, S., & Bernkop-Schnürch, A. (2024). Cyclodextrins and derivatives in drug delivery: New developments, relevant clinical trials, and advanced products. *Carbohydrate Polymers, 324*, 121500. https://doi.org/10.1016/j.carbpol.2023.121500.

Khaoulani, S. (2015). Wastewater treatment by adsorption on cyclodextrin polymers and development of chemical sensors based on chalcogenide glass membranes for the detection of Hg 2+ ions (PhD thesis). University of the Littoral Opal Coast.

Karavaeva, L., Kucherenko, A., Rudi, R. Bulanov, M., Glazova, N. V., & Kiprushkina, E. I. (2024). The study of stabilizing food enzymes using β-cyclodextrins. *Functional Foods in Health and Disease, 14*, 638–652. https://doi.org/10.31989/ffhd.v14i9.1384.

Kelanne, N., Yang, B., & Laaksonen, O. (2024). Potential of cyclodextrins in food processing for improving sensory properties of food. *Food Innovation and Advances, 3*, 1–10. https://doi.org/10.48130/fia-0024-0001.

Kelly, R. M., Dijkhuizen, L., & Leemhuis, H. (2009). The evolution of cyclodextrin glucanotransferase product specificity. *Applied Microbiology and Biotechnology, 84*, 119–133. https://doi.org/10.1007/s00253-009-1988-6.

Kfoury, M., & Fourmentin, S. (2023). Cyclodextrins as building blocks for new materials. *Beilstein Journal of Organic Chemistry, 19*, 889–891. https://doi.org/10.3762/bjoc.19.66.

Kim, D.-H., Lee, S.-E., Pyo, Y.-C., Tran, P., & Park, J.-S. (2020). Solubility improvement and application of cyclodextrins in local drug delivery. *Journal of Pharmaceutical Investigation, 50*, 17–27. https://doi.org/10.1007/s40005-019-00434-2.

Kuriki, T., Yamamoto, T., & Tsuchiya, M. (1961). Formation of cyclohexaamylose by an enzymatic reaction. *Agricultural and Biological Chemistry, 25*, 671–678.

Leemhuis, H., Kelly, R. M., & Dijkhuizen, L. (2010). Engineering of cyclodextrin glucanotransferases and the impact for biotechnological applications. *Applied Microbiology and Biotechnology, 85*, 823–835. https://doi.org/10.1007/s00253-009-2221-3.

Li, Z., Chen, S., Gu, Z., Chen, J., & Wu, J. (2014). Alpha-cyclodextrin: Enzymatic production and food applications. *Trends in Food Science & Technology, 35*, 151–160. https://doi.org/10.1016/j.tifs.2013.11.005.

Lindley, N. D. (1994). Physiology of microbial growth at high cell densities. *FEMS Microbiology Reviews, 14*(1). https://doi.org/10.1111/j.1574-6976.1994.tb00066.x.

Liu, Y., Chen, Y., Gao, X., Fu, J., & Hu, L. (2022). Application of cyclodextrin in food industry. *Critical Reviews in Food Science and Nutrition, 62*, 2627–2640. https://doi.org/10.1080/10408398.2020.1856035.

Liu, Z., Wu, G., & Wu, H. (2022). Molecular cloning and optimized production and characterization of recombinant cyclodextrin glucanotransferase from *Bacillus* sp. T1. *3 Biotech, 12*, 58. https://doi.org/10.1007/s13205-022-03111-8.

Loftsson, T., & Brewster, M. E. (2010). Pharmaceutical applications of cyclodextrins: Basic science and product development. *Journal of Pharmacy and Pharmacology, 62*, 1607–1621. https://doi.org/10.1111/j.2042-7158.2010.01030.x.

Loftsson, T., & Duchêne, D. (2007). Cyclodextrins and their pharmaceutical applications. *International Journal of Pharmaceutics, 329*, 1–11. https://doi.org/10.1016/j.ijpharm.2006.10.044.

De Lorenzo, V., Prather, K. L., Chen, G., O'Day, E., Von Kameke, C., Oyarzún, D. A., Hosta-Rigau, L., Alsafar, H., Cao, C., Ji, W., Okano, H., Roberts, R. J., Ronaghi, M., Yeung, K., Zhang, F., & Lee, S. Y. (2018). The power of synthetic biology for bioproduction, remediation and pollution control. *EMBO Reports, 19*, e45658. https://doi.org/10.15252/embr.201745658.

Läufer, A. (2017). Starch biorefinery enzymes. In K. Wagemann, & N. Tippkötter (Eds.). *Biorefineries* (pp. 137–152). Springer. https://doi.org/10.1007/10_2016_60.

van der Maarel, M. J. E. C., van der Veen, B., Uitdehaag, J. C. M., Leemhuis, H., & Dijkhuizen, L. (2002). Properties and applications of starch-converting enzymes of the α-amylase family. *Journal of Biotechnology, 94*, 137–155. https://doi.org/10.1016/S0168-1656(01)00407-2.

Majzlová, K., Pukajová, Z., & Janeček, Š. (2013). Tracing the evolution of the α-amylase subfamily GH13_36 covering the amylolytic enzymes intermediate between oligo-1,6-glucosidases and neopullulanases. *Carbohydrate Research, 367*, 48–57. https://doi.org/10.1016/j.carres.2012.11.022.

Matencio, A., Navarro-Orcajada, S., García-Carmona, F., & López-Nicolás, J. M. (2020). Applications of cyclodextrins in food science: A review. *Trends in Food Science & Technology, 104*, 132–143. https://doi.org/10.1016/j.tifs.2020.08.009.

Mesbah, N. M. (2022). Industrial biotechnology based on enzymes from extreme environments. *Frontiers in Bioengineering and Biotechnology, 10*, 870083. https://doi.org/10.3389/fbioe.2022.870083.

Mohammad, A., Singh, S., & Swain, S. (2020). Cyclodextrins: Concept to applications, regulatory issues and challenges. *Nanomedicine Research Journal, 5*. https://doi.org/10.22034/nmrj.2020.03.001.

Mondal, S., Mondal, K., Halder, S. K., Thakur, N., & Mondal, K. C. (2022). Microbial amylase: Old but still at the forefront of all major industrial enzymes. *Biocatalysis and Agricultural Biotechnology, 45*, 102509. https://doi.org/10.1016/j.bcab.2022.102509.

Morais, A. R., & Bogel-Łukasik, R. (2013). Green chemistry and the biorefinery concept. *Sustainable Chemical Processes, 1*, 18. https://doi.org/10.1186/2043-7129-1-18.

Morin, P. (2009). Separation of chiral pharmaceutical molecules by chromatographic and electrophoretic techniques. *Annales Pharmaceutiques Françaises, 67*, 241–250. https://doi.org/10.1016/j.pharma.2009.03.008.

Morin-Crini, N., Fourmentin, S., Fenyvesi, É., Lichtfouse, E., Torri, G., Fourmentin, M., & Crini, G. (2021). 130 years of cyclodextrin discovery for health, food, agriculture, and industry: A review. *Environmental Chemistry Letters, 19*, 2581–2617. https://doi.org/10.1007/s10311-020-01156-w.

Muniz, I. D. C. B., Santos, J. B., De Oliveira, R. M., Santos, F. G., Souza Junior, E. C. D., Oyama, L., Fontan, R. D. C. I., & Bonomo, R. C. F. (2024). Current advances in obtaining novel cyclodextrin glycosyltransferases for optimizing the synthesis of cyclodextrins. *Process Biochemistry, 145*, 195–209. https://doi.org/10.1016/j.procbio.2024.07.008.

Musuc, A. M. (2024). Cyclodextrins: Advances in chemistry, toxicology, and multifaceted applications. *Molecules (Basel, Switzerland), 29*, 5319. https://doi.org/10.3390/molecules29225319.

Nicolaescu, O. E., Belu, I., Mocanu, A. G., Manda, V. C., Rău, G., Pîrvu, A. S., Ionescu, C., Ciulu-Costinescu, F., Popescu, M., & Ciocîlteu, M. V. (2025). Cyclodextrins: Enhancing drug delivery, solubility and bioavailability for modern therapeutics. *Pharmaceutics, 17*, 288. https://doi.org/10.3390/pharmaceutics17030288.

Nicolaescu, O. E., Ionescu, C., Samide, A., Tigae, C., Spînu, C. I., & Oprea, B. (2025). Advancements in cyclodextrin complexes with bioactive secondary metabolites and their pharmaceutical applications. *Pharmaceutics, 17*, 506. https://doi.org/10.3390/pharmaceutics17040506.

Nielsen, P. H. (2017). Microbial biotechnology and circular economy in wastewater treatment. *Microbial Biotechnology, 10*, 1102–1105. https://doi.org/10.1111/1751-7915.12821.

Ochiai, E., Tsuda, K., & Ikuma, S. (1935). Synthese von C-substituted pyridyl-pyrrol-derivaten. *Berichte der Deutschen Chemischen Gesellschaft, 68*, 1551–1555. https://doi.org/10.1002/cber.19350680827.

Oehlenschläger, K., Schepp, E., Stiefelmaier, J., Holtmann, D., & Ulber, R. (2024). Simultaneous fermentation and enzymatic biocatalysis: A useful process option? *Biotechnology for Biofuels and Bioproducts, 17*, 67. https://doi.org/10.1186/s13068-024-02519-z.

Olaniyan, S. A., Hussein, J. B., Oke, M. O., Akinwande, B. A., Workneh, T. S., Ayodele, M., & Adeyemi, I. A. (2025). Integrated bioprocessing of cassava residues for enzymatic starch recovery, citric acid production, and effluent detoxification. *Scientific Reports, 15*, 21433. https://doi.org/10.1038/s41598-025-06841-w.

Palmero, M. G., Carrasco, A., Sen, P., de la, Mainar-Toledo, M. D., Malo, S. A., & Herrer, F. J. R. (2025). Starch valorisation as biorefinery concept integrated by an agro-industry case study to improve sustainability. *Sustainability, 17*. https://doi.org/10.3390/su17156808.

Pardhi, D. S., Rabadiya, K. J., Panchal, R. R., Raval, V. H., Joshi, R. G., & Rajput, K. N. (2023). Cyclodextrin glucanotransferase: Fundamentals and biotechnological implications. *Applied Microbiology and Biotechnology, 107*, 5899–5907. https://doi.org/10.1007/s00253-023-12708-9.

Paul, C., Hanefeld, U., Hollmann, F., Qu, G., Yuan, B., & Sun, Z. (2024). Enzyme engineering for biocatalysis. *Molecular Catalysis, 555*, 113874. https://doi.org/10.1016/j.mcat.2024.113874.

Pirvu, A. S., Varut, R.-M., Trasca, D.-M., Stoica, G. A., Radivojevic, K., Carmen, S., Arsenie, C. C., & Popescu, C. (2025). Cyclodextrins as active therapeutic agents: Beyond their role as excipients. *Pharmaceuticals, 18*. https://doi.org/10.3390/ph18101592.

Podgorbunskikh, E., Sapozhnikov, A., Kuskov, T., Gurova, D., Kopylova, A., Bychkov, A., & Lomovsky, O. (2022). Comprehensive enzymatic conversion of starch for the food industry. *Polymers, 14*, 4575. https://doi.org/10.3390/polym14214575.

Rabadiya, K., Pardhi, D., Thaker, K., Patoliya, J., Rajput, K., & Joshi, R. (2024). A review on recent upgrades and strategies to enhance cyclodextrin glucanotransferase properties for its applications. *International Journal of Biological Macromolecules, 259*, 129315. https://doi.org/10.1016/j.ijbiomac.2024.129315.

Radley, E., Davidson, J., Foster, J., Obexer, R., Bell, E. L., & Green, A. P. (2023). Engineering enzymes for environmental sustainability. *Angewandte Chemie, 135*, e202309305. https://doi.org/10.1002/ange.202309305.

Rajput, K. N., Patel, K. C., & Trivedi, U. B. (2016). β-Cyclodextrin production by cyclodextrin glucanotransferase from an alkaliphile Microbacterium terrae KNR 9 using different starch substrates. *Biotechnology Research International, 2016*, 1–7. https://doi.org/10.1155/2016/2034359.

Rashwan, A. K., Younis, H. A., Abdelshafy, A. M., Osman, A. I., Eletmany, M. R., Hafouda, M. A., & Chen, W. (2024). Plant starch extraction, modification, and green applications:

A review. *Environmental Chemistry Letters, 22*, 2483–2530. https://doi.org/10.1007/s10311-024-01753-z.

Rekharsky, M. V., & Inoue, Y. (1998). Complexation thermodynamics of cyclodextrins. *Chemical Reviews, 98*, 1875–1918. https://doi.org/10.1021/cr970015o.

Rosso, A. M., Ferrarotti, S. A., Krymkiewicz, N., Nudel, B. (2002). Optimization of batch culture conditions for cyclodextrin glucanotransferase production from Bacillus circulans DF 9R. *Microbial Cell Factories, 1*. https://doi.org/10.1186/1475-2859-1-3.

Rojas, M. J., Amaral-Fonseca, M., Fernandez-Lafuente, R., De Lima Camargo Giordano, R., Tardioli, P. W. (2019). Recovery of starch from cassava bagasse for cyclodextrin production by sequential treatment with α-amylase and cyclodextrin glycosyltransferase. *Biocatalysis and Agricultural Biotechnology, 22*, 101411. https://doi.org/10.1016/j.bcab.2019.101411.

Saenger, W. (1980). Cyclodextrin inclusion compounds in research and industry. *Angewandte Chemie International Edition in English, 19*, 344–362. https://doi.org/10.1002/anie.198003441.

Saenger, W., Beyer, K., & Manor, P. C. (1976). Topography of cyclodextrin inclusion complexes. VI. The crystal and molecular structure of α-cyclodextrin–p-iodoaniline trihydrate. *Acta Crystallographica Section B, 32*, 120–128. https://doi.org/10.1107/S0567740876002446.

Saini, K., Pathak, V. M., Tyagi, A., & Gupta, R. (2022). Microbial cyclodextrin glycosyltransferases: Sources, production, and application in cyclodextrin synthesis. *Catalysis Research, 2*, 1–69. https://doi.org/10.21926/cr.2203029.

Sanjoy, K. D., Rajan, R., Sheba, D., Nasimul, G., Jasmina, K., & Arunabha, N. (2013). Cyclodextrins—The molecular container. *Research Journal of Pharmaceutical, Biological and Chemical Sciences, 4*, 1694–1720.

Schardinger, F. (1903). Über thermophile Bakterien aus verschiedenen Speisen und Milch sowie über einige Umsetzungsprodukte derselben in kohlenhydrathaltigen Nährlösungen, darunter krystallisierte Polysaccharide (Dextrin) aus Stärke. *Zeitschrift für Untersuchung der Nahrungs- und Genussmittel, 6*, 865–880. https://doi.org/10.1007/BF02067497.

Schmidt, A. K., Cottaz, S., Driguez, H., & Schulz, G. E. (1998). Structure of cyclodextrin glycosyltransferase complexed with a derivative of its main product β-cyclodextrin. *Biochemistry, 37*, 5909–5915. https://doi.org/10.1021/bi9729918.

Schomburg, I. (2004). BRENDA, the enzyme database: Updates and major new developments. *Nucleic Acids Research, 32*, 431D–433D. https://doi.org/10.1093/nar/gkh081.

Schrittwieser, J. H., Velikogne, S., Hall, M., & Kroutil, W. (2018). Artificial biocatalytic linear cascades for preparation of organic molecules. *Chemical Reviews, 118*, 270–348. https://doi.org/10.1021/acs.chemrev.7b00033.

Sheldon, R. A., Basso, A., & Brady, D. (2021). New frontiers in enzyme immobilization: Robust biocatalysts for a circular bio-based economy. *Chemical Society Reviews, 50*, 5850–5862. https://doi.org/10.1039/D1CS00015B.

Sheldon, R. A., & Brady, D. (2018). The limits to biocatalysis: Pushing the envelope. *Chemical Communications, 54*, 6088–6104. https://doi.org/10.1039/C8CC02463D.

Sheldon, R. A., & Woodley, J. M. (2018). Role of biocatalysis in sustainable chemistry. *Chemical Reviews, 118*, 801–838. https://doi.org/10.1021/acs.chemrev.7b00203.

Singh, M., Sharma, R., & Banerjee, U. C. (2002). Biotechnological applications of cyclodextrins. *Biotechnology Advances, 20*, 341–359. https://doi.org/10.1016/S0734-9750(02)00020-4.

Spiridon, I., & Anghel, N. (2025). Cyclodextrins as multifunctional platforms in drug delivery and beyond: Structural features, functional applications, and future trends. *Molecules (Basel, Switzerland), 30*, 3044. https://doi.org/10.3390/molecules30143044.

Stam, M. R., Danchin, E. G. J., Rancurel, C., Coutinho, P. M., & Henrissat, B. (2006). Dividing the large glycoside hydrolase family 13 into subfamilies. *Protein Engineering, Design & Selection, 19*, 555–562. https://doi.org/10.1093/protein/gzl044.

Stegmann, P., Londo, M., & Junginger, M. (2020). The circular bioeconomy: Its elements and role in European bioeconomy clusters. *Resources, Conservation and Recycling X, 6*, 100029. https://doi.org/10.1016/j.rcrx.2019.100029.

Straathof, A. J. J. (2014). Transformation of biomass into commodity chemicals using enzymes or cells. *Chemical Reviews, 114*, 1871–1908. https://doi.org/10.1021/cr400309c.

Strokopytov, B., Penninga, D., Rozeboom, H. J., Kalk, K. H., Dijkhuizen, L., & Dijkstra, B. W. (1995). X-ray structure of cyclodextrin glycosyltransferase complexed with acarbose. *Biochemistry, 34*, 2234–2240. https://doi.org/10.1021/bi00007a018.

Szejtli, J. (1998). Introduction and general overview of cyclodextrin chemistry. *Chemical Reviews, 98*, 1743–1754. https://doi.org/10.1021/cr970022c.

Szejtli, J. (2004). Past, present and future of cyclodextrin research. *Pure and Applied Chemistry, 76*, 1825–1845. https://doi.org/10.1351/pac200476101825.

Takano, T., Fukuda, M., Monma, M., Kobayashi, S., Kainuma, K., & Yamane, K. (1986). Molecular cloning, DNA nucleotide sequencing, and expression in Bacillus subtilis cells of the Bacillus macerans cyclodextrin glucanotransferase gene. *Journal of Bacteriology, 166*, 1118–1122. https://doi.org/10.1128/jb.166.3.1118-1122.1986.

Tan, E. C. D., & Lamers, P. (2021). Circular bioeconomy concepts—A perspective. *Frontiers in Sustainability, 2*, 701509. https://doi.org/10.3389/frsus.2021.701509.

Tarimci, N. (2011). Cyclodextrins in the cosmetic field. In E. Bilensoy (Ed.). *Cyclodextrins in pharmaceutics, cosmetics, and biomedicine* (pp. 131–144). Wiley. https://doi.org/10.1002/9780470926819.ch7.

Tester, R. F., Karkalas, J., & Qi, X. (2004). Starch—Composition, fine structure and architecture. *Journal of Cereal Science, 39*, 151–165. https://doi.org/10.1016/j.jcs.2003.12.001.

Tian, B. R., & Liu, Y. M. (2021). Cyclodextrin-active natural compounds in food applications: A review of antibacterial activity. *Turkish Journal of Chemistry, 45*, 1707–1724. https://doi.org/10.3906/kim-2106-51.

Tomasik, P., & Horton, D. (2012). Enzymatic conversions of starch. In *Advances in carbohydrate chemistry and biochemistry*. Elsevier, 59–436. https://doi.org/10.1016/B978-0-12-396523-3.00001-4.

Tonkova, A. (1998). Bacterial cyclodextrin glucanotransferase. *Enzyme and Microbial Technology, 22*, 678–686. https://doi.org/10.1016/S0141-0229(97)00263-9.

Ubando, A. T., Felix, C. B., & Chen, W.-H. (2020). Biorefineries in circular bioeconomy: A comprehensive review. *Bioresource Technology, 299*, 122585. https://doi.org/10.1016/j.biortech.2019.122585.

Uitdehaag, J. C. M., Van Alebeek, G.-J. W. M., Van Der Veen, B. A., Dijkhuizen, L., & Dijkstra, B. W. (2000). Structures of maltohexaose and maltoheptaose bound at the donor sites of cyclodextrin glycosyltransferase. *Biochemistry, 39*, 7772–7780. https://doi.org/10.1021/bi000340x.

Uitdehaag, J. C. M., Mosi, R., Kalk, K. H., Van Der Veen, B. A., Dijkhuizen, L., Withers, S. G., & Dijkstra, B. W. (1999). X-ray structures of cyclodextrin glycosyltransferase reaction intermediates. *Nature Structural Biolcgy, 6*, 432–436. https://doi.org/10.1038/8235.

Del Valle, E. M. M. (2004). Cyclodextrins and their uses: A review. *Process Biochemistry, 39*, 1033–1046. https://doi.org/10.1016/S0032-9592(03)00258-9.

Van Der Veen, B. A., Van Alebeek, G.-J. W. M., Uitdehaag, J. C. M., Dijkstra, B. W., & Dijkhuizen, L. (2000a). Rational design of cyclodextrin glycosyltransferase to increase α-cyclodextrin production. *Journal of Molecular Biology, 296*, 1027–1038. https://doi.org/10.1006/jmbi.2000.3528.

Van Der Veen, B. A., Van Alebeek, G.-J. W. M., Uitdehaag, J. C. M., Dijkstra, B. W., & Dijkhuizen, L. (2000b). The three transglycosylation reactions catalyzed by cyclodextrin glycosyltransferase proceed via different kinetic mechanisms. *European Journal of Biochemistry, 267*, 658–665. https://doi.org/10.1046/j.1432-1327.2000.01031.x.

van der Veen, B. A., Uitdehaag, J. C. M., Dijkstra, B. W., & Dijkhuizen, L. (2000). Engineering of cyclodextrin glycosyltransferase reaction and product specificity. *Biochimica et Biophysica Acta, 1543*, 336–360. https://doi.org/10.1016/S0167-4838(00)00233-8.

Viglianti, C., De Brauer, C., Laforest, V., Bourgois, J. (2007). Use of cyclodextrins for the remediation of soils contaminated by polycyclic aromatic hydrocarbons.

Villiers, A. (1891). On the fermentation of starch by the action of butyric acid. *Comptes Rendus de l'Académie des Sciences,* 536–538.

Villiers, A. (1903). On an intermediate product of amyl fermentation. *Comptes Rendus Hebdomadaires des Séances de l'Académie des Sciences,* 1254–1257.

Wang, H., Zhou, W., Li, H., Bu, R. (2018). Optimization of the fermentation conditions for the mutant strain of β -cyclodextrin glycosyltransferase H167C to produce cyclodextrins. 3 Biotech 8, 165. https://doi.org/10.1007/s13205-018-1182-6.

Woodley, J. M. (2022). Biocatalysis for future sustainable manufacturing. *The Biochemist, 44*, 6–8. https://doi.org/10.1042/bio_2022_112.

Wu, S., Snajdrova, R., Moore, J. C., Baldenius, K., & Bornscheuer, U. T. (2021). Biocatalysis: Enzymatic synthesis for industrial applications. *Angewandte Chemie International Edition, 60*, 88–119. https://doi.org/10.1002/anie.202006648.

Yu, P. M., Gilvanova, E. A., & Aktuganov, G. E. (2025). Improved purification technique for isolation of novel CGTase from Caldalkalibacillus mannanilyticus IB-OR17-B1. *Preparative Biochemistry & Biotechnology, 55*, 223–229. https://doi.org/10.1080/10826068.2024.2386558.

Zhang, J., Zhang, Y., & Li, M. (2017). High-level secretion and characterization of cyclodextrin glycosyltransferase in recombinant Komagataella phaffii. *Journal of Biotechnology, 259*, 126–134. https://doi.org/10.1016/j.jbiotec.2017.07.031.

CHAPTER

Structure of cyclodextrin glucanotransferases

3

Jeferson Silva Cunha[1], Flaviana Coelho Pacheco[1], Fábio Ribeiro dos Santos[1], Irene Andressa[1], Ana Flávia Coelho Pacheco[2], Manoela Maciel dos Santos Dias[1] and Bruno Ricardo de Castro Leite Júnior[1]

[1]*Department of Food Technology, Federal University of Viçosa, University Campus, Viçosa, MG, Brazil,*

[2]*Cândido Tostes Dairy Institute, Agricultural Research Corporation of Minas Gerais (EPAMIG), Tenente Luiz de Freitas, Juiz de Fora, MG, Brazil*

Introduction

Cyclodextrin glucanotransferase (CGTase) is part of the α-amylase family of glycosyl hydrolases (Chen et al., 2023), showing high structural similarity to classical α-amylases (Ogunbadejo & Al-Zuhair, 2023) due to the presence of the three typical structural domains called A, B, and C. Distinctively, CGTase also has two additional domains, labeled D and E (Lim et al., 2021). This unique structural characteristic gives the enzyme multiple catalytic functions, allowing it to act in four distinct types of reactions: cyclization, disproportionation, coupling, and hydrolysis (Han et al., 2022). For this reason, CGTase has been employed in various applications, including the production of transglycosylated compounds with bioactive potential (Han et al., 2014; Khummanee et al., 2019; Huang et al., 2019; Xiao et al., 2022), the synthesis of biosurfactants (Svensson & Adlercreutz, 2011), the reduction of bitterness in stevioside, steviol, and rebaudioside (Jung et al., 2007; Muñoz-Labrador et al., 2020), the enhancement of bread volume (Zhao et al., 2022), as well as the improvement of flavonoid solubility, such as that of genistein (Han et al., 2020) and hydroxyflavones (Wang et al., 2025).

Among the various processes mediated by CGTase, the production of cyclodextrins (CDs) is particularly noteworthy (Ji et al., 2022; Cunha et al., 2024). CDs are cyclic oligosaccharides composed of α-(1→4)-linked glucose units, characterized by a hydrophobic central cavity and a hydrophilic outer surface (Lee et al., 2025). This amphiphilic structure enables CDs to form inclusion complexes with hydrophobic molecules of industrial and pharmaceutical interest, thereby enhancing their solubility, stability, and enabling controlled release (Hatem et al., 2024). As of 2023, the global CDs market was valued at USD 289.3 million, with projections estimating growth to USD 390 million by 2027 (Global Market Insights, 2024).

Cyclodextrin Glucanotransferase. DOI: https://doi.org/10.1016/B978-0-443-41694-1.00004-6

Depending on the microbial origin and the structural characteristics of the enzyme, CGTases can produce a mixture of α-, β-, and γ-CDs (Lim et al., 2021). Based on the predominance of the type of CD formed in the early phases of the cyclization reaction, these enzymes are classified as α-CGTases, β-CGTases, or γ-CGTases (Lim et al., 2021; Schöffer et al., 2017). This catalytic specificity is directly related to the producing microorganism and reflecting variations in the composition and amino acid sequence of the secreted protein (Saini et al., 2022a). Three-dimensional (3D) crystallographic structures of CGTases from several species, such as *Bacillus circulans*, *Thermoanaerobacterium thermosulfurigenes*, and *Paenibacillus macerans*, have been deposited in the Research Collaboratory for Structural Bioinformatics (RCSB) protein data bank (Lawson et al., 1994; Wind et al., 1998; Leemhuis et al., 2002; Jiang et al., 2018). Understanding of these structures has been essential for identifying catalytic residues and enzyme regions responsible for selectivity in the formation of different types of CDs, thereby supporting protein engineering strategies aimed at enhancing enzyme specificity and efficiency (Kelly et al., 2009).

Therefore this chapter aims to present the main structural characteristics of microbial CGTases. Initially, an overview of the industrial and economic relevance of these enzymes is presented, followed by a description of their general properties. Finally, an in-depth analysis is provided of the primary and 3D structures, as well as the catalytic mechanisms involved in their reactions and their potential applications.

Importance of enzymes in industrial processes

More than four decades ago, the primary sources of industrial enzymes were plants and animals. However, in the early 1980s, a significant increase in the use of enzymes derived from microorganisms was observed. This transition was quickly adopted by the bioprocessing industry due to the superior adaptability of microorganisms to industrial conditions. Microbial enzymes offer several advantages, such as cost-effective production, scalability, and low-cost fermentation processes, which are facilitated by the use of simple nutrient media (Priyanka et al., 2022).

Moreover, the growing demand for more sustainable production of industrially relevant compounds has driven the expansion of the enzyme market across various regions of the world, spurring a continuous search for new biocatalysts. Over the past two decades, there has been substantial growth in the application of enzymes in a wide range of biotechnological processes and applications, including food and beverage production, textiles, biofuels, dyes, animal feed, biomaterials, pharmaceuticals, and nutraceuticals (Elisashvili et al., 2017; Maghraby et al., 2023; dos Santos et al., 2024; Pacheco et al., 2025).

The use of enzymes in industry is essential for the implementation of "green approaches" aligned with the principles of green chemistry and sustainable energy, which rely on environmentally friendly processes that do not generate pollutants during reactions (De et al., 2024). In addition, enzymatic biotransformation processes offer several advantages over chemical catalysis, including higher yields, faster reaction rates under mild conditions, lower energy requirements, water solubility, biodegradability, minimal occurrence of undesirable side reactions, and nontoxicity (Magalhães et al., 2022; Patel et al., 2023).

The global enzyme market is projected to expand at a compound annual growth rate (CAGR) of 4.9% from 2024 to 2030 (Large, no date). In this context, proteases hold the largest share of the global market, closely followed by amylases and lipases. The food sector has played a significant role in the growth of the enzyme industry, as carbohydrases and proteases are widely used in various industrial processes (Patel et al., 2023). Over the past decade, amylolytic enzymes such as CGTases have garnered increasing attention from both the scientific community and industrial sectors, particularly in the food and pharmaceutical industries, due to their ability to produce CDs in three primary configurations: α-CD, β-CD, and γ-CD. Currently, the two commercially available CGTases are Toruzyme® and Amano® (Ngo et al., 2021).

Structural and functional characteristics of cyclodextrins glucanotransferase

CGTase is a biological catalyst classified within the glycoside hydrolase family 13, or the α-amylase family, with the Enzyme Commission (EC) number 2.4.1.19 (Rabadiya et al., 2024). According to its EC classification, CGTase is categorized as a transferase (EC 2), belonging to the transglycosylase subclass (EC 2.4) and the hexosyltransferase subclass (EC 2.4.1) (De et al., 2024). Most CGTases are extracellular enzymes produced by various microorganisms, mainly bacteria, although certain archaea and fungi are also capable of secreting these enzymes (Lim et al., 2021).

CGTase is a highly versatile enzyme, capable of catalyzing both intramolecular transglycosylation reactions (cyclization, coupling, and disproportionation), as well as hydrolytic reactions (Cunha et al., 2024). While amylases efficiently catalyze hydrolysis reactions in starch, CGTases primarily catalyze transglycosylation reactions, with hydrolysis being a secondary and less frequent activity (Rabadiya et al., 2024). In this context, CGTase plays a key role in cleaving α-(1→4) glycosidic bonds and transferring glycosyl group found in starch and its derivatives to acceptor containing hydroxyl or other functional groups (De et al., 2024).

The product of this reaction is CDs, which belong to a family of cyclic oligosaccharides typically containing six (α-CD), seven (β-CD), or eight (γ-CD) glucose units (Cunha et al., 2024). These bioproducts perform several important functions, such as the extraction and stabilization of bioactive compounds and drugs, cholesterol removal, protection of volatile compounds, elimination of undesirable flavors, acting as emulsifying agents, solubilization of hydrophobic compounds, use in controlled drug release systems, and immobilization of toxic compounds (Del Valle, 2004; dos Santos et al., 2017) (see Fig. 3.1).

CGTases are classified as α-, β-, or γ-CGTases, depending on the type of CD produced in the highest concentration (Ibrahim et al., 2012; Pardhi et al., 2023).

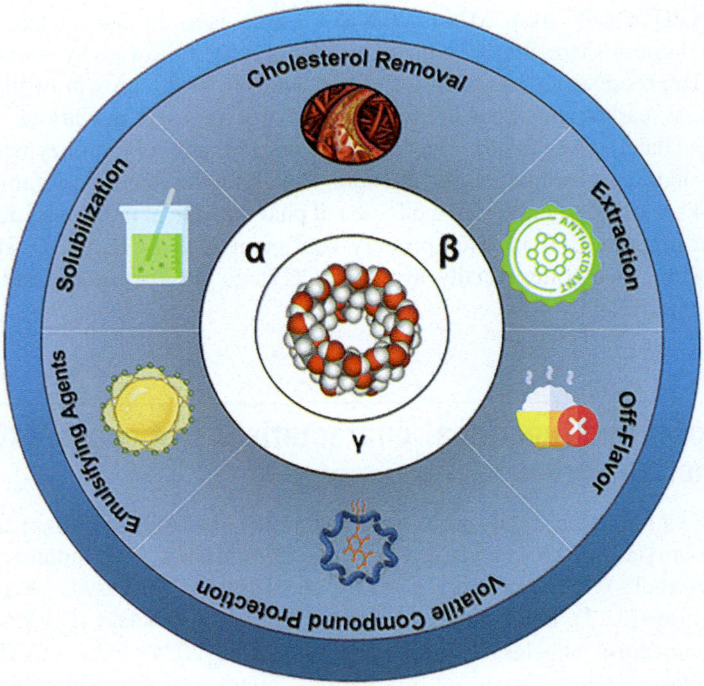

FIGURE 3.1

Applications of cyclodextrins (CDs). The figure illustrates the main applications of (CDs) (α-, β-, and γ-forms) in food and related industries, highlighting their ability to form inclusion complexes due to their hydrophobic cavity and hydrophilic outer surface. This unique structure allows CDs to encapsulate various guest molecules, improving their functionality. Key applications include: cholesterol removal—used to reduce cholesterol content in food products, particularly dairy; extraction—facilitates the extraction of bioactive compounds, such as antioxidants, improving yield and stability; off-flavor masking—reduces or eliminates undesirable odors or tastes from formulations; volatile compound protection—stabilizes aroma compounds and essential oils, minimizing losses during processing and storage; emulsifying agents—enhances emulsion stability by forming complexes at oil–water interfaces; solubilization—improves the solubility and bioavailability of poorly water-soluble substances. At the center, the three main types of CDs (α, β, and γ) are represented, each differing in the number of glucose units and cavity size, which influence their interactions with guest molecules.

Table 3.1 Physicochemical properties of α-, β-, and γ-cyclodextrins (*CDs*) (Varan et al., 2017; Connors, 1997).

Property/type of CD	α	β	γ
Number of glucose units	6	7	8
Molar mass (g/mol)	972	1135	1297
Cavity length (Å)	8	8	8
Central cavity diameter (Å)	4,7–5,3	6,0–6,5	7,5–8,3
Solubility in water at 25°C (mg/100 mL)	145,0	18,0	232,0

CDs have a hollow, truncated cone-shaped internal cavity with hydrophobic characteristics, primarily because of the presence of carbon chains (C–H groups) and oxygen atoms from glucose residues. Conversely, the hydroxyl groups present on the outside of the cone confer a hydrophilic nature to this region (Pereva et al., 2019).

Due to their internal hydrophobicity, CDs enhance the solubility of bioactive compounds, such as polyphenols, by complexing with them primarily through hydrophobic interactions and van der Waals forces (Li et al., 2024; Chen et al., 2024). This mechanism prevents the aggregation of hydrophobic molecules, making CDs more efficient than many conventional solubilizing agents (Li et al., 2024). CDs exhibit distinct physicochemical properties that determine their applications in food, medical, cosmetic, pharmaceutical, and agricultural fields (Pardhi et al., 2023). The physicochemical properties of different CDs are summarized in Table 3.1.

Compared with α- and β-CD, γ-CD exhibits superior properties with respect to water solubility and bioavailability, which enables its broader applicability. However, the versatility and efficiency of β-CD in the food and pharmaceutical industries reinforce its market dominance and solid growth prospects compared with other CDs (van der Veen et al., 2000). To illustrate this scenario, the CD market in the pharmaceutical sector is expected to grow by USD 211.8 million and exhibit a CAGR of 3.3% between 2024 and 2032. In the food and beverage sector, the market is expected to grow by USD 85.5 million over the same period, with a slightly higher CAGR of 4% (Global Market Insights, 2024). According to Global Market Insights (2024), North America is projected to maintain its leading position in the CD market, with an average annual growth rate of 3.5% from 2024 to 2032.

Relationship between microbial origin and structural properties of cyclodextrins glucanotransferase

As previously described, CGTases are enzymes produced extracellularly by a wide variety of microorganisms, including bacteria, archaea, and fungi. The first CGTase was isolated from *B. macerans* in 1974 (Okada, 1974), after which approximately 30 CGTase-coding genes were identified (Takano et al., 1986). The screening of CGTase-producing microbial strains is typically based on three main properties: (1) direct screening for starch hydrolysis capability; (2) visual detection methods employing chromogenic dye assays; and (3) hybrid methodologies combining dye-based detection with instrumental analytical techniques (Qi & Zimmermann, 2005).

In this context, extensive research efforts aimed at discovering novel CGTases have yielded promising results, with these enzymes being identified in a variety of sources, especially in bacterial genera such as *Bacillus, Paenibacillus, Thermoanaerobacterium*, and *Actinomyces* (Han et al., 2014). These findings demonstrate that the biochemical characteristics and catalytic properties of CGTases are closely associated with their microbial origin. For instance, most CGTases derived from *Bacillus* species exhibit optimal activities at temperatures ranging from 40°C to 65°C (Nakapong et al., 2022), whereas those from *Thermoanaerobacter* species demonstrated optimal functionality at a temperature of 80°C (Song et al., 2021). While many CGTases exhibit optimal activity under near-neutral pH, those produced by *E. clarkii* and *Brevibacterium* species show the highest activity in alkaline environments (Saini et al., 2022a). For example, Bautista et al. (2012) isolated and characterized the first CGTase from the halophilic archaeon *Haloferax mediterranei*. The enzyme had a molecular weight of 77 kDa and exhibited optimal activity at 55°C, pH 7.5, and 1.5 M NaCl. Under these conditions, the cyclization reaction was enhanced, resulting in the production of a mixture of CDs, with α-CD being the predominant product. In another study, Ibrahim et al. (2012) reported a new CGTase produced by an alkaliphilic bacterium. The purified enzyme was a monomeric protein with an estimated molecular weight of 92 kDa and demonstrated stability over a broad pH range (5.0 to 11.0) and temperature range (45°C to 70°C), with maximum activity observed at 50°C and pH 8.0. Recently, a γ-CGTase with high thermostability (up to 70°C) and salt tolerance (up to 10%) was identified from *Evansella caseinilytica* (Saini et al., 2022b).

According to literature data, CGTase monomers have molecular weights ranging from 27 to 110 kDa (Agnes Ariwaodo et al., 2022a). Additionally, the amino acid composition and sequence length of the CGTase polypeptide chain also vary depending on the microbial source (Lim et al., 2021), influencing the enzymes' catalytic activity, substrate specificity, and overall mechanisms of action (Pardhi et al., 2023). For example, CGTases from *Anaerobranca gottschalkii* and *H. mediterranei* comprise 679 and 713 amino acids, respectively (Bautista et al., 2012; Thiemann et al., 2004). Therefore both the amino acid composition and molecular weight of CGTase monomers are closely linked to the producing microorganism (see Table 3.2).

In summary, the differences observed in the molecular weight and amino acid composition of CGTases show the wide structural diversity among enzymes from different microbial sources. This variability is directly associated with the specific catalytic characteristics of each CGTase, impacting both the efficiency and the CD production profile. The CGTase-producing source is considered the determining factor for the type of CD generated (see Table 3.2). However, other parameters related to the reaction conditions, such as the type of starch to be modified, buffer composition, temperature, and pH, also exert a significant influence on the CD profile obtained (Leemhuis et al., 2010).

Based on the aforementioned findings, the scientific community has devoted considerable efforts to identifying CGTases with enhanced performance and reduced sensitivity to reaction parameters. These efforts include genetic engineering approaches such as cloning environmental DNA into expression vectors and screening recombinant clones, as well as bioinformatics and prospecting techniques such as genomic mining and the search for native microorganisms capable of expressing CGTase activity (Leemhuis et al., 2010). In a recent study, Song et al. (2021) isolated a recombinant CGTase from the metagenome of marine microorganisms and expressed it in *E. coli*. The recombinant enzyme exhibited superior physicochemical properties, indicating its potential for industrial applications. The reported optimum temperature was 80°C, a favorable condition for substrate solubilization (Song et al., 2021). Similarly, Lee et al. (2007) expressed the CGTase-encoding gene in *E. coli*, and the purified enzyme, with a molecular weight

Table 3.2 Molecular weight of cyclodextrins glucanotransferase (*CGTase*), and type of cyclodextrin (*CD*) produced according to the producing microorganism.

Producing microorganism	Molecular weight of CGTase (kDa)	CD produced	References
Bacillus pseudalcaliphilus 20RF	70	β/γ	Atanasova et al. (2011)
Microbacterium terrae KNR 9	27,72	β	Rajput et al. (2016)
Bacillus sp. T1	75	α/β	Liu et al. (2022)
Thermoactinomyces vulgaris TA1	66	-	Alsharawi et al. (2018)
Bacillus sp. NR5 UPM	30	β	Nik-Pa et al. (2020)
Evansella caseinilytica	68	γ	Saini et al. (2022b)
Paenibacillus macerans	74	α	Jiang et al. (2018)
B.agaradhaerens WN-I	85	-	Abdelnasser Salah Sheble Ibrahim, (2011)
P. pabuli	70	β	Jemli et al. (2007)
Pyrococcus furiosus	65	β	Lee et al. (2007)
Bacillus sp. Strain 7–12	69	β/γ	Cao et al. (2005)
Thermococcus kodakaraensis	77	α/β/γ	Rashid et al. (2002)

of 65 kDa, showed optimal activity at 95°C and pH 5.0 (conditions more extreme than those previously reported). In addition, the enzyme maintained 97% of its activity at 100°C, evidencing its excellent thermostability.

The use of thermostable enzymes enhances the efficiency of the CD production process by reducing processing time and energy consumption. Moreover, it eliminates the need for pretreatment with α-amylases, minimizing the formation of low-molecular-weight oligosaccharides (Starnes et al., 1991). Therefore the search for new CGTases from diverse microbial sources represents a highly promising strategy with the potential to drive significant advancements in biotechnology and industrial applications. The exploration of these biocatalysts can reveal more efficient catalytic properties, improve reaction specificity, and enable the targeted and customized production of CDs (Rabadiya et al., 2024).

Furthermore, the overall yield of CDs, as well as the selectivity toward specific CD types, can be significantly improved by the addition of small amounts of low-molecular-weight polar solvents (Blackwood & Bucke, 2000). Lee & Kim (1991) demonstrated that the incorporation of acetone, propanol, and ethanol into the reaction medium increased the CD-forming activity of CGTase, with the most pronounced effect observed upon the addition of 10% ethanol, resulting in a twofold increase in CD production. This increase was attributed to the enhanced hydrolytic activity of CGTase in the presence of ethanol, which in turn promoted the efficient cyclization of the generated oligosaccharides. Similarly, Tomita et al. (1990) reported a proportional increase in the yield of γ-CD with increasing ethanol concentrations up to 30% (v/v). This improvement was attributed to the inhibitory effect of ethanol on γ-CD degradation, likely by suppressing intermolecular transglycosylation reactions between γ-CD and small dextrins derived from potato starch during the reaction.

Primary structure of cyclodextrins glucanotransferase

The primary structure of CGTase is composed of a specific linear sequence of amino acids linked by peptide bonds, which determines the enzyme's structural and functional properties (Rathnakumar et al., 2023). This sequence varies among different CGTase-producing species but is highly conserved in critical functional regions such as the active site and substrate-binding domains (Liu et al., 2022). CGTase usually contains between 650 and 800 amino acid residues distributed among domains that participate in distinct activities, such as starch binding and cyclodextrinization catalysis (Sonnendecker et al., 2019). The order and type of amino acids present in the primary structure are fundamental for the correct folding of the protein and for the formation of catalytic centers responsible for the conversion of starch into CDs (Rathnakumar et al., 2023). Variations in the primary structure between CGTases of different origins can influence the specificity for the type of CD produced (α-, β-, or γ-CD), thermal stability, and the optimal pH for enzyme activity (Gautam & Arya, 2022), as demonstrated in the previous section.

In this context, computational, biophysical, and biochemical research have been successfully conducted to investigate and characterize these structures (Yagi et al., 2021; Vennelakanti et al., 2022). CGTases from different sources show clear similarity in their amino acid sequences, ranging from 47% to 99% (Qi & Zimmermann, 2005). A study conducted by Martins et al. (2003) sequenced the CGTase gene from *B. agaradhaerens*, revealing a sequence composed of 679 amino acids. A comparative analysis of the primary chain of the enzyme with other sequences from *Bacillus* strains showed an identity of 99% (Martins et al., 2003).

Highly conserved regions between CGTase sequences, particularly within the α-amylase family, have been extensively characterized. A comparison of 31 CGTase sequences in different studies revealed that at least 51 amino acids are highly conserved (Qi & Zimmermann, 2005). Previous studies by Nakajima et al. (1986) and Janecek et al. (1995) identified five specific regions shared among all members of the α-amylase family. One of these conserved sequences corresponds to a short region similar to the fifth conserved region (leucine–proline–aspartic acid–leucine–aspartic acid) of the α-amylase from *Aspergillus oryzae*, containing the calcium ligand Asp-175 (aspartic acid). Although Asp-175 is not invariably conserved, the sequence is easily recognizable and is located 26 or 28 amino acid residues before Asp-206 (Janecek et al., 1995). The presence of Asp is crucial for catalysis, being important both in binding with the substrate and in the interaction with calcium (Qi & Zimmermann, 2005).

The α- and β-CGTase groups do not show a clear difference in relation to the sequence of the primary structure (Pardhi et al., 2023). van der Veen et al. (2000) reported that most CGTases have approximately nine subsites, identified as −7 to +2. Several conserved amino acid residues have been identified at subsites −1 and −2 and are important in cooperative engagement with the substrate through hydrogen bonds and stacking. The −3 subsite shows variation in amino acid composition depending on the type of CGTase (Kelly et al., 2009). For example, in α-CGTases produced by *Thermococcus* sp. B1001 and *K. pneumoniae*, a lysine residue is found at position 47. In contrast, a threonine is present at the same site in γ-CGTases (Kelly et al., 2009). Another specific region in subsite −3 is loop 87–93 and the residue Asn-94 (asparagine). These residues interact hydrophobically with glucose bound to this subsite and are notably different in the three types of CGTase (Qi & Zimmermann, 2005). The Asn residue at subsite −3 is present only in certain α- and β-CGTases but is absent in all γ-CGTases (Kelly et al., 2009). In subsite +2, Ala (alanine) is present in γ-CGTases at position 232, while Lys (lysine) is found in the same location in α-CGTases and β-CGTases (Saini et al., 2022a).

The analysis of several CGTase crystal structures revealed that the active site, located at the bottom of a barrel-shaped structure $(\beta/\alpha)_8$, has a conserved architecture, being composed mainly of the catalytic triad formed by Glu257, Asp229, and Asp328 (Saini et al., 2022a). The evolutionary conservation of these residues can be attributed to the fact that starch contains the α-glycosidic bond. This bonding occurs in different configurations, such as α(1→4), α(1→6), or α(1→1), according to the position of the C1 carbon of a glucose unit in relation to the oxygen atoms O4, O6, or O1 of the adjacent glucose, respectively (Uitdehaag et al., 2000). In addition, Asp135, His140, Arg227, and His327 residues are almost completely conserved in CGTases (Uitdehaag et al., 2000). Structural studies have shown that residues conserved around the active site, such as Arg-47, can assist in the precise orientation of the substrate and in the stabilization of reaction intermediates, reinforcing the importance of these residues for enzymatic efficiency (Uitdehaag et al., 1999).

Three-dimensional structure of cyclodextrins glucanotransferase

In 1991 Klein & Schulz (1991) published the first report providing substantial information on the 3D structure of CGTase and revealed that the enzyme has five domains, labeled A, B, C, D, and E (De et al., 2024). Domains A, B, and C are structurally homologous in all α-amylases, while domains D and E belong exclusively to CGTases (Saini et al., 2022a). Furthermore, domains A and B are considered catalytic, and domains C and E are implicated in substrate binding (Saini et al., 2022a; Lim et al., 2021). The function of domain D remains unclear, but research has been conducted to fill this gap. For instance, Zhou et al. (2021), in a study on a novel antimicrobial CGTase from *B. cereus* YUPP-10, suggested that domain D is a key domain for hydrolytic activity.

Domain A is characterized by a barrel-shaped arrangement comprising a highly symmetric fold formed by eight parallel β-strands surrounded by eight α-helices (Lim et al., 2021). This catalytic domain, commonly referred to as the TIM barrel, consists of 300–400 amino acid residues and is conserved across all members of the α-amylase family (van der Veen et al., 2000). The enzyme's active site is located at the bottom of domain A (Saini et al., 2022a) and comprises nine subsites, designated from −7 to +2, which organize the starch substrate into a ring-like conformation, with each subsite accommodating a single glucose unit (Rabadiya et al., 2024). Negative subsite numbers correspond to the nonreducing end of the substrate, whereas positive numbers indicate the reducing end (Lim et al., 2021). The central catalytic cleavage occurs between subsites −1 and +1, mediated by the highly conserved catalytic acid triad, as mentioned previously. Residues present in subsites +2, −2, and −3 modulate the catalytic activity of CGTase, while subsites −4 to −7 determine the size specificity of the generated CDs. Subsites −6 and −7 are essential for efficient substrate anchoring (Rabadiya et al., 2024).

Domain B consists of a highly visible 60-residue loop between the β3 strand and the α3 helix of domain A. This domain provides several amino acid side chains that contribute to substrate recognition and binding on the enzyme surface (Saini et al., 2022a; De et al., 2024). Thus substrate binding is facilitated by a groove formed by domains A and B on the surface of CGTase, which accommodates at least 10 glucose residues across subsites −7 to +2 (Leemhuis et al., 2010). The stability of CGTase is influenced by a salt bridge in domain B that establishes a calcium-binding site at the interface between domains A and B (Agnes Ariwaodo et al., 2022b; Goh et al., 2012). Leemhuis et al. (2004) improved the stability of CGTase 251 from *B. circulans* by engineering a salt bridge within domain B. Therefore domain B is crucial not only for the enzyme's overall stability and thermostability but also for its cyclization activity and product specificity (De et al., 2024).

Comprising approximately 100 amino acids, the C domain contains a maltose-binding site and plays a crucial role in substrate interaction (van der Veen et al., 2000; Penninga et al., 1996). Together with the D and E domains, it forms the C-terminal region of CGTases, which exhibits a β-leaf structure (Qi & Zimmermann, 2005).

The E domain, classified within carbohydrate-binding module family 20, specializes in binding native starch and contains two specific maltose-binding sites (Lim et al., 2021; Leemhuis et al., 2010; Li et al., 2021). In recent years, researchers have placed considerable emphasis on the structural characterization of CGTases, resulting in the publication of more than 50 distinct crystallographic models (Han et al., 2014). These 3D structures serve as an essential foundation for upcoming studies focused on understanding how CGTases function and facilitating targeted molecular engineering (see Table 3.3).

The structural organization of CGTase can vary among different microbial strains, reflecting differences in the number and distribution of residues across its five functional domains. For instance, in the CGTase of the alkalophilic strain *Bacillus* sp. 1011.14, the residues are distributed among the domains as follows: domain A (1–406), domain B (139–203), domain C (407–496), domain D (497–584), and domain E (585–686), as described by Harata et al. (1996). In comparison, the enzyme produced by *B. circulans* strain eight exhibits a different configuration, with residues 80–108 in domain A, 185–192 in domain B, 407–494 in domain C, 495–580 in domain D, and 581–684 in domain E, according to Klein & Schulz (1991).

Reactions catalyzed by cyclodextrin glucanotransferase

CGTase uses a dual α-retention displacement mechanism to catalyze four different types of reactions: three transglycosylation reactions (disproportionation, cyclization, and coupling) and one hydrolysis reaction (Cunha et al., 2024; Leemhuis et al., 2002). The hydrolytic activity is considerably lower than the transglycosylation reactions because of hydrophobicity and the presence of phenylalanine (Phe) residues at positions 183 and 259 (Cunha et al., 2024; Mathew & Adlercreutz, 2012). Transglycosylation reactions are characterized by the transfer of sugar fractions or glycosyl group from donor to acceptor, resulting in glycosidic bonds between the two compounds involved in the reaction (Lim et al., 2021). There are

Table 3.3 Identification in the database and structural description of proteins based on their microbial source.

Producing source	Identification in theprotein database	Structural description	References
Bacillus circulans strain 251	1KCL	Mutant G179L	Leemhuis et al. (2002)
	1TCM	W616A mutant	Penninga et al. (1996)
	1OT1	D135A mutant	Leemhuis et al. (2003)
	1PJ9	Loop mutant 183–195	Leemhuis et al. (2004)
	1DTU	Complexed with maltononaose inhibitor	van der Veen et al. (2000)
	1CXE	Complex with alpha-cyclodextrin (CD)	Knegtel et al. (1995)
B. circulans strain 8	6CGT	Complex with hoxa	Parsiegla et al. (1998)
	1CGT	Wild type	Klein & Schulz, (1991)
	1CXE	Wild type	Knegtel et al. (1995)
Bacillus sp.	1V3J	Mutant F283L	Kanai et al. (2004)
Thermoanaerobacterium thermosulfurigenes	3BMV	Mutant S77P	Kelly et al. (2008)
	1A47	Complex with maltohexaose	Wind et al. (1998)
Geobacillus stearothermophillus	1QHP	Maltose complex	Kanai et al. (2004); Dauter et al. (1999)
Bacillus sp. 1011	1V3J	F283Y mutant	Kanai et al. (2004)
Niallia circulans	2CXG	Arcabous complex	Strokopytov et al. (1995)
	3CGT	β-CD complex	Schmidt et al. (1998)

two types of glucanotransferases: Leloir glucanotransferase and non-Leloir glucanotransferase (Lim et al., 2021). However, CGTases are classified as non-Leloir glucanotransferases, as they use nonactivated sugars, such as starch and maltooligosaccharides, as donors (Rather et al., 2015). Generally, non-Leloir enzymes are more accessible and cost-effective, use readily available commercial donors (in contrast to high-cost activated sugar nucleotides), and exhibit compatibility with a broader range of acceptor molecules (Singh et al., 2016; Ali et al., 2023).

Among transglycosylation reactions, disproportionation is the predominant process. This reaction involves the cleavage of a linear maltooligosaccharide, where one of the resulting fragments is transferred to the linear acceptor through a double-displacement mechanism (Rodríguez Gastón et al., 2012). According to Goh et al. (2012), starch and maltodextrin serve as more efficient donors to CGTase compared to lower-molecular-weight carbohydrates, such as CD, maltotriose, maltose, and dextrin. For effective donation, the substrate must contain at least two glucose units. In addition, both the substrate chain length and its concentration directly influence the size of the glycosides produced (Mathew & Adlercreutz, 2013). Smaller sugars, such as glucose and maltose, act primarily as acceptors and promote disproportionation reactions (Rodríguez Gastón et al., 2012).

Cyclization is an intramolecular transglycosylation reaction in which the nonreducing end of a linear maltooligosaccharide is transferred to its own reducing end, resulting in the formation of a cyclic structure known as CD (Leemhuis et al., 2002). The coupling reaction happens inversely to the cyclization reaction; that is, CDs are opened and converted into linear oligosaccharides through a random ternary-complex mechanism (Mathew & Adlercreutz, 2012). A schematic overview of the reactions catalyzed by CGTase is summarized in Fig. 3.2.

All reactions catalyzed by CGTase follow a fundamental mechanism shared by members of the α-amylase family, differing primarily in the nature of the acceptor molecule (Saini et al., 2022a; Leemhuis et al., 2002).

Catalytic mechanisms

In general, the functioning of CGTase can be understood from fundamental principles of enzymatic catalysis. Enzymes catalyze reactions by binding to free substrate to form enzyme–substrate complex. During this process, the enzyme stabilizes the transition state, which reduces the activation energy of the reaction and allows the substrate to more easily convert into the product. At the end, the complex dissociates, releasing the product and regenerating the free enzyme, ready to start a new catalytic cycle (Copeland, 2000). Although the catalytic mechanism of CGTase is not yet fully elucidated, the most widely accepted model is known as the double-displacement mechanism with retention of the α-anomeric configuration, originally proposed by Koshland (1953).

CGTase is an enzyme that acts on starch degradation, using simple sugars, such as dextrin, maltodextrin, maltose, and CD, as glucose donors. These sugars interact with the enzyme through noncovalent bonds, including hydrogen bonds, hydrophobic interactions, and electrostatic forces (De et al., 2024). For catalysis to occur in a targeted and efficient manner, the Asp229, Asp328, and Glu257 residues present in the active site are strategically organized to recognize the substrate

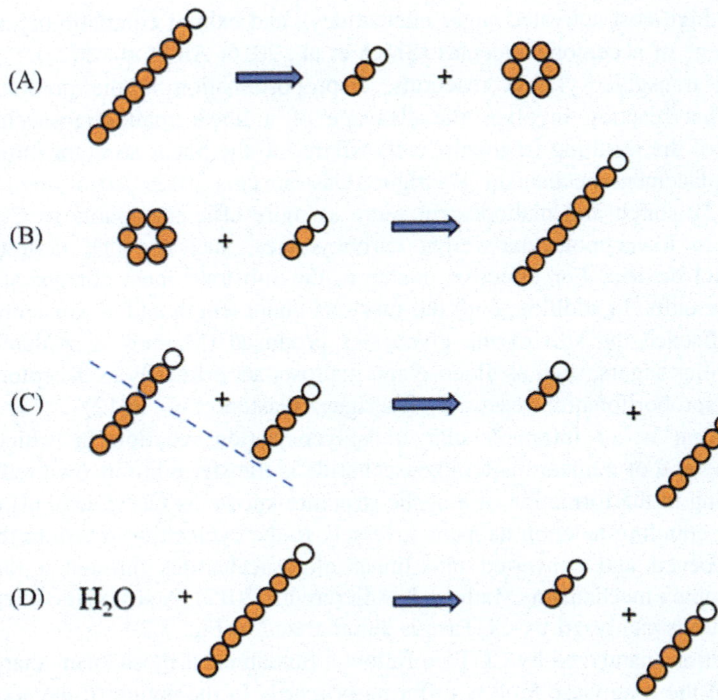

FIGURE 3.2

Enzymatic reactions catalyzed by cyclodextrin glycosyltransferase (*CGTase*): (A) cyclization, (B) coupling, (C) disproportionation, and (D) hydrolysis. This figure illustrates the main types of reactions catalyzed by CGTase, an enzyme responsible for the transformation of starch and linear oligosaccharides into cyclodextrins (CDs) and other derivatives. The schematic representations below describe each mechanism: (A) cyclization—CGTase cleaves a linear starch chain and forms a CD (circular structure) by joining the terminal ends of a segment, releasing a linear by-product. (B) Coupling—a CD molecule reacts with a linear acceptor, opening the ring and forming a longer linear chain. (C) Disproportionation—two linear oligosaccharide chains interact, transferring glucosyl units between them, resulting in chains of different lengths. (D) Hydrolysis—CGTase catalyzes the cleavage of a glycosidic bond in the presence of water (H_2O), yielding shorter linear oligosaccharides. Each reaction highlights the enzyme's versatility in modifying starch derivatives for applications in food, pharmaceuticals, and biotechnology through the production or modification of CDs and related compounds.

(Lim et al., 2021). The functions of these three carboxylic amino acids were clarified through the interaction of the enzyme with acarbose (Strokopytov et al., 1995; Qian et al., 1994). Glu257 acts as an acid/base catalyst, Asp229 acts as a nucleophile for intermediate stabilization, and Asp328 functions in substrate binding (Nakamura et al., 1994).

In summary, when the substrate binds, the Glu257 residue acts as an acid and donates a proton (H^+) to the oxygen of the glycosidic bond, promoting its cleavage. In this process, the sugar at the −1 position goes through a transition state with characteristics similar to an oxocarbenium ion, marked by a partial double bond between the O5 and C1 atoms and a delocalized positive charge. Thereafter, Asp229 acts as a nucleophile, forming a covalent intermediate between the enzyme and the substrate. This glycosylated complex interacts with subsites ranging from −7 to −1 (Lim et al., 2021). Following protonation, the glucose molecule at subsite +1 is released from the catalytic pocket, either as part of a linear oligosaccharide or as a by-product of the cyclization reaction. Subsequently, an acceptor glucose attacks the covalent intermediate between Asp229 and the C1 of glucose in the −1 subsite, again generating a new oxocarbenium ion-like transition state. At this stage, Glu257 acts as a catalytic base by capturing a proton from the glucose positioned at the +1 subsite. This deprotonation enables the oxygen atom to displace the oxocarbenium intermediate, resulting in the formation of a new hydroxyl group at the C1 position and ultimately leading to the synthesis of CD (De et al., 2024; Leemhuis et al., 2010).

Subsites +2, −2, and −3 contain residues that modulate the catalytic activity of CGTase, whereas subsites −4 to −7 primarily influence the enzyme's product-size specificity. Although absent in amylases, subsites −6 and −7 are essential for substrate anchoring to CGTase. Tyrosine (Y195) is a central amino acid residue whose function is closely related to the enzyme's substrate specificity (Saini et al., 2022a; Leemhuis et al., 2010). The reaction mechanism is similar among α-CGTases, β-CGTases, and γ-CGTases, differing only in the distance between the acceptor glucose and the glucose positioned at the −1 subsite, which is 8, 9, and 10 glucose units for α-, β-, and γ-CGTases, respectively (Qi & Zimmermann, 2005).

Applications of cyclodextrins glucanotransferase

CGTase has demonstrated remarkable versatility by catalyzing four distinct types of reactions, which are associated with its complex 3D structure and highly specialized catalytic mechanism. Although its main use is linked to the synthesis of CDs, the multifunctionality of this enzyme expands its potential for application in various bioprocesses.

In the bakery sector, several studies have shown the positive impact of CGTase on the technological quality of baked products (G. M & Mutsaers, 1999; Lee et al., 2002; Gujral et al., 2003; Shim et al., 2004; Yang et al., 2022). For instance, Jemli et al. (2007) demonstrated that the addition of β-CGTase isolated from *P. pabuli* US132 led to an 11% increase in bread volume compared with the control without enzyme. This improvement was attributed to the formation of β-CD, which enhances starch granule solubilization. Moreover, the enzyme contributed to delaying bread aging during storage, an effect attributed to its hydrolytic and cyclization activities, which reduce amylose retrogradation and prevent amylopectin crystallization. Basso et al. (2015) used commercial CGTase and CGTase produced by *B. firmus* cepa 37 in gluten-free bread formulations. Both enzymes increased bread volume (up to 44%) and reduced hardness (up to 51%). Additionally, they improved the rheological and sensory properties of the products, highlighting the potential of CGTase for producing breads with superior technological quality. Zhao et al. (2022) applied CGTase to gluten-free bread and observed a volume increase of 134%. The addition of CGTase also enhanced bread crumb quality, resulting in less stickiness and greater softness and elasticity.

Another promising application of CGTase is the production of glycosylated compounds with relevant biological and therapeutic properties (Kometani et al., 1996; Wang et al., 2010; Lee et al., 2017). For instance, Chen et al. (2023) used CGTase to catalyze hesperidin glycosylation. This flavonoid, found in citrus peels, possesses important antioxidant and antiinflammatory activities. The authors reported that this process significantly increased the compound's solubility in water and consequently enhanced its biological properties. Similarly, Zhang et al. (2023) explored the potential of β-CGTase-mediated glycosylation to improve the properties of steviol glycosides. The glycosylated products showed improved sensory quality attributes.

Furthermore, other notable applications include the glycosylation of alkyl glycosides for the production of biosurfactants (Svensson et al., 2009), as well as enhancing the solubility of poorly water-soluble bioactive compounds (Gao et al., 2022), thereby expanding the enzyme's applicability across diverse industrial processes. Based on the above, this chapter has emphasized the extensive application potential of CGTases, by detailing their structures, catalytic reactions, and mechanisms of action. The versatility of these enzyme enables their utilization in different industrial and scientific fields, positioning them as promising candidates for the development of value-added products with enhanced functionalities in alignment with the principles of green chemistry.

Conclusion

Enzymes have emerged as key catalysts driving the advancement of sustainable industrial processes in the food, pharmaceutical, biofuel, and textile sectors. Among them, CGTase is distinguished by its multifunctional catalytic capabilities, particularly its role in the cyclization of oligosaccharides to produce CDs. The enzyme's microbial origin significantly influences its structural characteristics, especially the composition of amino acids, which in turn impacts its catalytic activity and substrate selectivity during CDs synthesis. Comprehensive structural analyzes have identified conserved and variable regions within the protein, including distinct functional domains and the catalytic site, thereby providing a solid basis for understanding its mechanism of action. These insights are crucial for guiding targeted enzyme-engineering strategies aimed at improving the efficiency and specificity of CD production.

Acknowledgments

The authors are grateful to the Coordination of Improvement of Higher Education Personnel - Brazil (CAPES) - Financing Code 001, and the Minas Gerais Research Foundation (FAPEMIG, Brazil) for funding the project APQ-00388-21/ APQ-00785-23/ RED-00157-23 and FAPEMIG-CNPq for the productivity grant awarded to B. R. C. Leite Júnior (APQ-06600-24).

References

Abdelnasser Salah Sheble Ibrahim, A. A. A.-S. H. B. (2011). An alkaliphilic cyclodextrin glycosyltransferase from a new *Bacillus agaradhaerens* WN-I strain isolated from an Egyptian soda lake: Purification and properties. *African Journal of Biotechnology, 10*(32), 6107–6119.

Agnes, A. C., Felix, E. C., & Ugochukwu, N. T. (2022a). α- and β-cyclodextrin production by cyclodextrin glucosyl transferase from *Bacillus* species using different carbon substrate [Online]. www.aujst.com.

Agnes, A. C., Ezeama, C., Agnes, A. C., Felix, E. C., & Ugochukwu, N. T. (2022b). α- and β-cyclodextrin production by cyclodextrin glucosyl transferase from *Bacillus* species using different carbon substrate [Online]. www.aujst.com

Ali, M. Y., Liaqat, F., Khazi, M. I., Sethupathy, S., & Zhu, D. (2023). Utilization of glycosyltransferases as a seamless tool for synthesis and modification of the oligosaccharides-A review. *International Journal of Biological Macromolecules, 249*, 125916. https://doi.org/10.1016/J.IJBIOMAC.2023.125916.

Alsharawi, S., ElShatoury, E., Mohamed, S., & Abdalla, N. (2018). Optimization and characterization of cyclodextrin glycosyl- transferase produced by *Thermoactinomyces vulgaris* TA1. *The Egyptian Journal Of Experimental Biology (Botany), 14*(1), 1. https://doi.org/10.5455/egyjebb.20180321064543.

Atanasova, N., Kitayska, T., Bojadjieva, I., Yankov, D., & Tonkova, A. (2011). A novel cyclodextrin glucanotransferase from alkaliphilic *Bacillus pseudalcaliphilus* 20RF: Purification and properties. *Process Biochemistry, 46*(1), 116–122. https://doi.org/10.1016/j.procbio.2010.07.027.

Basso, F. M., Camila Sampaio, M., Maria Fernanda Alves, A., Antonio Roberto Giriboni, M., Rosane Marina, P., & Mandioli, G. (2015). Potential use of cyclodextrin-glycosyltransferase enzyme in bread-making and the development of gluten-free breads with pinion and corn flours. *International Journal of Food Sciences and Nutrition, 66*(3), 275–281. https://doi.org/10.3109/09637486.2015.1007450.

Bautista, V., Esclapez, J., Pérez-Pomares, F., Martínez-Espinosa, R. M., Camacho, M., & Bonete, M. J. (2012). Cyclodextrin glycosyltransferase: A key enzyme in the assimilation of starch by the halophilic archaeon *Haloferax mediterranei*. *Extremophiles: Life Under Extreme Conditions, 16*(1), 147–159. https://doi.org/10.1007/s00792-011-0414-z.

Blackwood, A. D., & Bucke, C. (2000). Addition of polar organic solvents can improve the product selectivity of cyclodextrin glycosyltransferase: Solvent effects on cgtase. *Enzyme and Microbial Technology, 27*(9), 704–708. https://doi.org/10.1016/S0141-0229(00)00270-2.

Cao, X., Jin, Z., Wang, X., & Chen, F. (2005). A novel cyclodextrin glycosyltransferase from an alkalophilic Bacillus species: Purification and characterization. *Food Research International, 38*(3), 309–314. https://doi.org/10.1016/j.foodres.2004.06.015.

Chen, H., et al. (2023). Rational design of cyclodextrin glycosyltransferase with improved hesperidin glycosylation activity. *Catalysts, 13*(5), 885. https://doi.org/10.3390/catal13050885.

Chen, S., et al. (2024). Improved stability and antibacterial activity of cinnamaldehyde emulsion co-stabilized by cyclodextrin glycosyltransferase-catalyzed starch products and Tween 80. *Food Hydrocoll, 154*, 110142. https://doi.org/10.1016/j.foodhyd.2024.110142.

Connors, K. A. (1997). The stability of cyclodextrin complexes in solution. *Chemical Reviews, 97*(5), 1325–1358. https://doi.org/10.1021/cr960371r.

Copeland, R. A. (2000). Chemical mechanisms in enzymatic catalysis. *Enzymes: A Practical Introduction to the Structure, Mechanism, and Analysis of Data*.

Cunha, J. S., Pacheco, F. C., Martins, C. C. N., Pacheco, A. F. C., Tribst, A. A. L., & Leite Júnior, B. R.de C. (2024). Use of ultrasound to improve the activity of cyclodextrin glycosyltransferase in the producing of β-cyclodextrins: Impact on enzyme activity, stability and insights into changes on enzyme macrostructure. *Food Research International, 191*, 114662. https://doi.org/10.1016/j.foodres.2024.114662.

Dauter, Z., et al. (1999). X-ray structure of novamyl, the five-domain 'maltogenic' α-Amylase from *Bacillus stearothermophilus*: Maltose and acarbose complexes at 1.7 Å resolution. *Biochemistry, 38*(26), 8385–8392. https://doi.org/10.1021/bi990256l.

De, I., Muniz, C. B., et al. (2024). Current advances in obtaining novel cyclodextrin glycosyltransferases for optimizing the synthesis of cyclodextrins. *Process Biochemistry, 145*, 195–209. https://doi.org/10.1016/j.procbio.2024.07.008.

Del Valle, E. M. M. (2004). Cyclodextrins and their uses: A review. *Process Biochemistry, 39*(9), 1033–1046. https://doi.org/10.1016/S0032-9592(03)00258-9.

dos Santos, C., Buera, P., & Mazzobre, F. (2017). Novel trends in cyclodextrins encapsulation. Applications in food science. *Current Opinion in Food Science, 16*, 106–113. https://doi.org/10.1016/j.cofs.2017.09.002.

dos Santos, F. R., et al. (2024). Improvement of the production of pequi almond (*Caryocar brasiliense* Camb.) protein hydrolysates through ultrasound-assisted enzymolysis: Impact on hydrolysis kinetics, structure and functional properties of hydrolysates. *Process Biochemistry, 147*, 381–390. https://doi.org/10.1016/j.procbio.2024.10.001.

Elisashvili, V., Kachlishvili, E., Asatiani, M., Darlington, R., & Kucharzyk, K. (2017). Physiological Peculiarities of lignin-modifying enzyme production by the white-rot basidiomycete *Coriolopsis gallica* strain BCC 142. *Microorganisms, 5*(4), 73. https://doi.org/10.3390/microorganisms5040073.

Gao, H.-Y., Liu, Y., Tan, F.-F., Zhu, L.-W., Jia, K.-Z., & Tang, Y.-J. (2022). The advances and challenges in enzymatic *C* -glycosylation of flavonoids in plants. *Current Pharmaceutical Design, 28*(18), 1466–1479. https://doi.org/10.2174/1381612828666220422085128.

Gautam, R., & Arya, S. K. (2022). Production of α, β, and γ-cyclodextrin gluconotransferase (CGTase) and their applications in food industry. In *Novel Food Grade Enzymes*. Singapore: Springer Nature Singapore, 107–138. http://doi.org/10.1007/978-981-19-1288-7_4.

Global Market Insights. Cyclodextrin Market Size - By Type (Alpha, Beta, Gamma), By Application (Pharmaceutical, Food and Beverage, Chemicals, Cosmetics and Personal Care) & Forecast, 2024-2032. 32. gminsights.com/industry-analysis/global-cyclodextrin-market?utm_source=globenews%20wire.com&utm_medium=referral&utm_campaign=Paid_globenewswire.

G. M, J. H., & Mutsaers, V. E. J. H. (1999). Patent and trademark office. *U.S. Patent, 5*(916), 607.

Goh, P. H., Illias, R. Md, & Goh, K. M. (2012). Rational mutagenesis of cyclodextrin glucanotransferase at the calcium binding regions for enhancement of thermostability. *International Journal of Molecular Sciences, 13*(5), 5307–5323. https://doi.org/10.3390/ijms13055307.

Gujral, H. S., Guardiola, I., Carbonell, J. V., & Rosell, C. M. (2003). Effect of cyclodextrinase on dough rheology and bread quality from rice flour. *Journal of Agricultural and Food Chemistry, 51*(13), 3814–3818. https://doi.org/10.1021/jf034112w.

Han, R., et al. (2014). Recent advances in discovery, heterologous expression, and molecular engineering of cyclodextrin glycosyltransferase for versatile applications. *Biotechnology Advances, 32*(2), 415–428. https://doi.org/10.1016/j.biotechadv.2013.12.004.

Han, R., Chai, B., Jiang, Y., Ni, J., & Ni, Y. (2022). Engineering of cyclodextrin glycosyltransferase from *Paenibacillus macerans* for enhanced product specificity of long-chain glycosylated sophoricosides. *Molecular Catalysis, 519*, 112147. https://doi.org/10.1016/j.mcat.2022.112147.

Han, R., Ni, J., Zhou, J., Dong, J., Xu, G., & Ni, Y. (2020). Engineering of cyclodextrin glycosyltransferase reveals pH-regulated mechanism of enhanced long-chain glycosylated sophoricoside specificity. *Applied and Environmental Microbiology, 86*(7). https://doi.org/10.1128/AEM.00004-20.

Harata, K., Haga, K., Nakamura, A., Aoyagi, M., & Yamane, K. (1996). X-ray structure of cyclodextrin glucano-transferase from alkalophilic *Bacillus* Sp. 1011. Comparison of two independent molecules at 1.8 Å resolution. *Acta Crystallographica. Section D, Biological Crystallography, 52*(6), 1136–1145. https://doi.org/10.1107/S0907444996008438.

Hatem, S., et al. (2024). Nano-vesicular systems for melanocytes targeting and melasma treatment: In-vitro characterization, ex-vivo skin retention, and preliminary clinical appraisal. *International Journal of Pharmaceutics, 665*, 124731. https://doi.org/10.1016/j.ijpharm.2024.124731.

Huang, W., et al. (2019). An efficient one-pot enzymatic synthesis of cardiac glycosides with varied sugar chain lengths. *Advanced Synthesis & Catalysis, 361*(13), 3114–3119. https://doi.org/10.1002/adsc.201900227.

Ibrahim, A. S. S., Al-Salamah, A. A., El-Tayeb, M. A., El-Badawi, Y. B., & Antranikian, G. (2012). A novel cyclodextrin glycosyltransferase from alkaliphilic *Amphibacillus* sp. NPST-10: Purification and properties. *International Journal of Molecular Sciences, 13*(8), 10505–10522. https://doi.org/10.3390/ijms130810505.

Janecek, Š., MacGregor, E. A., & Svensson, B. (1995). Characteristic differences in the primary structure allow discrimination of cyclodextrin glucanotransferases from α -amylases. *Biochemical Journal, 305*(2), 685–686. https://doi.org/10.1042/bj3050685.

Jemli, S., Messaoud, E. Ben, Ayadi-Zouari, D., Naili, B., Khemakhem, B., & Bejar, S. (2007). A β-cyclodextrin glycosyltransferase from a newly isolated Paenibacillus pabuli US132 strain: Purification, properties and potential use in bread-making. *Biochemical Engineering Journal, 34*(1), 44–50. https://doi.org/10.1016/j.bej.2006.11.016.

Ji, H., et al. (2022). Deciphering external chain length and cyclodextrin production with starch catalyzed by cyclodextrin glycosyltransferase. *Carbohydrate Polymers, 284*, 119156. https://doi.org/10.1016/j.carbpol.2022.119156.

Jiang, Y., et al. (2018). Heterologous expression of cyclodextrin glycosyltransferase from *Paenibacillus macerans* in *Escherichia coli* and its application in 2-O-α-D-glucopyranosyl-L-ascorbic acid production. *BMC Biotechnology, 18*(1), 53. https://doi.org/10.1186/s12896-018-0463-9.

Jung, S.-W., Kim, T.-K., Lee, K.-W., & Lee, Y.-H. (2007). Catalytic properties of β-cyclodextrin glucanotransferase from alkalophilic*Bacillus* sp. BL-12 and intermolecular transglycosylation of stevioside. *Biotechnology and Bioprocess Engineering, 12*(3), 207–212. https://doi.org/10.1007/BF02931094.

Kanai, R., Haga, K., Akiba, T., Yamane, K., & Harata, K. (2004). Role of Phe283 in enzymatic reaction of cyclodextrin glycosyltransferase from alkalophilic *Bacillus* sp.1011: Substrate binding and arrangement of the catalytic site. *Protein Science, 13*(2), 457–465. https://doi.org/10.1110/ps.03408504.

Kelly, R. M., Dijkhuizen, L., & Leemhuis, H. (2009). The evolution of cyclodextrin glucanotransferase product specificity. *Applied Microbiology and Biotechnology, 84*(1), 119–133. https://doi.org/10.1007/s00253-009-1988-6.

Kelly, R. M., Leemhuis, H., Rozeboom, H. J., van Oosterwijk, N., Dijkstra, B. W., & Dijkhuizen, L. (2008). Elimination of competing hydrolysis and coupling side reactions of a cyclodextrin glucanotransferase by directed evolution. *Biochemical Journal, 413*(3), 517–525. https://doi.org/10.1042/BJ20080353.

Khummanee, N., Rudeekulthamrong, P., & Kaulpiboon, J. (2019). Cyclodextrin glycosyltransferase-catalyzed synthesis of pinoresinol-α-D-glucoside having antioxidant and anti-inflammatory activities. *Appl Biochem Microbiol, 55*(4), 360–370. https://doi.org/10.1134/S0003683819040070.

Klein, C., & Schulz, G. E. (1991). Structure of cyclodextrin glycosyltransferase refined at 2.0 Å resolution. *Journal of Molecular Biology, 217*(4), 737–750. https://doi.org/10.1016/0022-2836(91)90530-J.

Knegtel, R. M. A., et al. (1995). Crystallographic studies of the interaction of cyclodextrin glycosyltransferase from *Bacillus circulans* strain 251 with natural substrates and products (*). *Journal of Biological Chemistry, 270*(49), 29256–29264. https://doi.org/10.1074/jbc.270.49.29256.

Kometani, T., Nishimura, T., Nakae, T., Takii, H., & Okada, S. (1996). Synthesis of neohesperidin glycosides and naringin glycosides by cyclodextrin glucano-transferase from an alkalophilic *Bacillus* Species. *Bioscience, Biotechnology, and Biochemistry, 60*(4), 645–649. https://doi.org/10.1271/bbb.60.645.

Koshland, D. E. (1953). Stereochemistry and the mechanism of enzymatic reactions. *Biological Reviews, 28*(4), 416–436. https://doi.org/10.1111/j.1469-185X.1953.tb01386.x.

Large, R. V. (n.d.). Enzyme market size, share, and trends analysis report by product (carbohydrase, proteases, polymerases, and nucleases), by type (industrial, special), by source (plants, animals), by region, and segment forecasts. https://www.grandviewresearch.com/industry-analysis/enzymes-industry.

Lawson, C. L., et al. (1994). Nucleotide sequence and x-ray structure of cyclodextrin glycosyltransferase from *Bacillus circulans* strain 251 in a maltose-dependent crystal form. *Journal of Molecular Biology, 236*(2), 590–600. https://doi.org/10.1006/jmbi.1994.1168.

Lee, M.-H., Yang, S.-J., Kim, J.-W., Lee, H.-S., Kim, J.-W., & Park, K.-H. (2007). Characterization of a thermostable cyclodextrin glucanotransferase from *Pyrococcus furiosus* DSM3638. *Extremophiles: Life Under Extreme Conditions, 11*(3), 537–541. https://doi.org/10.1007/s00792-007-0061-6.

Lee, S.-H., et al. (2002). Modulation of cyclizing activity and thermostability of cyclodextrin glucanotransferase and its application as an antistaling enzyme. *Journal of Agricultural and Food Chemistry, 50*(6), 1411–1415. https://doi.org/10.1021/jf010928q.

Lee, Y. Bin, Kyun, M.-L., Lee, Y. J., Shim, H.-E., Huh, K. M., & Kang, S.-W. (2025). Cyclodextrins as multifunctional tools for advanced biomaterials in tissue repair and regeneration. *Bioactive Material, 49*, 627–651. https://doi.org/10.1016/j.bioactmat.2025.03.018.

Lee, Y. D., & Kim, H. S. (1991). Enhancement of enzymatic production of cyclodextrins by organic solvents. *Enzyme and Microbial Technology, 13*(6), 499–503. https://doi.org/10.1016/0141-0229(91)90009-Y.

Lee, Y.-S., Woo, J.-B., Ryu, S.-I., Moon, S.-K., Han, N. S., & Lee, S.-B. (2017). Glucosylation of flavonol and flavanones by *Bacillus* cyclodextrin glucosyltransferase to enhance their solubility and stability. *Food Chemistry, 229*, 75–83. https://doi.org/10.1016/j.foodchem.2017.02.057.

Leemhuis, H., Kelly, R. M., & Dijkhuizen, L. (2010). Engineering of cyclodextrin glucanotransferases and the impact for biotechnological applications. *Applied Microbiology and Biotechnology, 85*(4), 823–835. https://doi.org/10.1007/s00253-009-2221-3.

Leemhuis, H., Rozeboom, H. J., Dijkstra, B. W., & Dijkhuizen, L. (2003). The fully conserved Asp residue in conserved sequence region I of the α-amylase family is crucial for the catalytic site architecture and activity. *FEBS Letters, 541*(1), 47–51. https://doi.org/10.1016/S0014-5793(03)00286-2.

Leemhuis, H., Rozeboom, H. J., Dijkstra, B. W., & Dijkhuizen, L. (2004). Improved thermostability of *bacillus circulans* cyclodextrin glycosyltransferase by the introduction of a salt bridge. *Proteins: Structure, Function, and Bioinformatics, 54*(1), 128–134. http://doi.org/10.1002/prot.10516.

Leemhuis, H., Uitdehaag, J. C. M., Rozeboom, H. J., Dijkstra, B. W., & Dijkhuizen, L. (2002). The remote substrate binding subsite 6 in cyclodextrin-glycosyltransferase controls the transferase activity of the enzyme via an induced-fit mechanism. *Journal of Biological Chemistry, 277*(2), 1113–1119. http://doi.org/10.1074/JBC.M106667200.

Li, C., et al. (2021). Maltose binding site 2 mutations affect product inhibition of *Bacillus circulans* STB01 cyclodextrin glycosyltransferase. *International Journal of Biological Macromolecules, 175*, 254–261. https://doi.org/10.1016/j.ijbiomac.2021.02.033.

Li, X., Jin, Z., Bai, Y., & Svensson, B. (2024). Progress in cyclodextrins as important molecules regulating catalytic processes of glycoside hydrolases. *Biotechnology Advances, 72*, 108326. https://doi.org/10.1016/j.biotechadv.2024.108326.

Lim, C. H., Rasti, B., Sulistyo, J., & Hamid, M. A. (2021). Comprehensive study on transglycosylation of CGTase from various sources. *Heliyon, 7*(2). https://doi.org/10.1016/j.heliyon.2021.e06305.

Liu, Z., Wu, G., & Wu, H. (2022). Molecular cloning, and optimized production and characterization of recombinant cyclodextrin glucanotransferase from *Bacillus* sp. T1. *3 Biotech, 12*(3), 58. https://doi.org/10.1007/s13205-022-03111-8.

Magalhães, I. S., Guimarães, A. D. B., Tribst, A. A. L., Oliveira, E. B.de, & Leite Júnior, B. R.de C. (2022). Ultrasound-assisted enzymatic hydrolysis of goat milk casein: Effects on hydrolysis kinetics and on the solubility and antioxidant activity of hydrolysates. *Food Research International, 157*, 111310. https://doi.org/10.1016/j.foodres.2022.111310.

Maghraby, Y. R., El-Shabasy, R. M., Ibrahim, A. H., & Azzazy, H. M. E.-S. (2023). Enzyme immobilization technologies and industrial applications. *ACS Omega, 8*(6), 5184–5196. https://doi.org/10.1021/acsomega.2c07560.

Martins, R. F., Delgado, O., & Hatti-Kaul, R. (2003). Sequence analysis of cyclodextrin glycosyltransferase from the alkalophilic *Bacillus agaradhaerens*. *Biotechnology Letters, 25*(18), 1555–1562. https://doi.org/10.1023/A:1025430532333.

Mathew, S., & Adlercreutz, P. (2012). Cyclodextrin glucanotransferase (CGTase) catalyzed synthesis of dodecyl glucooligosides by transglycosylation using α-cyclodextrin or starch. *Carbohydrate Polymers, 87*(1), 574–580. https://doi.org/10.1016/j.carbpol.2011.08.023.

Mathew, S., & Adlercreutz, P. (2013). Regioselective glycosylation of hydroquinone to α-arbutin by cyclodextrin glucanotransferase from *Thermoanaerobacter* sp. *Biochemical Engineering Journal, 79*, 187–193. https://doi.org/10.1016/j.bej.2013.08.001.

Muñoz-Labrador, A., et al. (2020). Transglycosylation of steviol glycosides and rebaudioside a: Synthesis optimization, structural analysis and sensory profiles. *Foods, 9*(12), 1753. https://doi.org/10.3390/foods9121753.

Nakajima, R., Imanaka, T., & Aiba, S. (1986). Comparison of amino acid sequences of eleven different ?-amylases. *Applied Microbiology and Biotechnology, 23*(5). https://doi.org/10.1007/BF00257032.

Nakamura, A., Haga, K., & Yamane, K. (1994). The transglycosylation reaction of cyclodextrin glucanotransferase is operated by a ping-pong mechanism. *FEBS Letters, 337*(1), 66–70. https://doi.org/10.1016/0014-5793(94)80631-4.

Nakapong, S., Tumhom, S., Kaulpiboon, J., & Pongsawasdi, P. (2022). Heterologous expression of 4α-glucanotransferase: Overproduction and properties for industrial applications. *World J Microbiol Biotechnol, 38*(2), 36. https://doi.org/10.1007/s11274-021-03220-1.

Ngo, N. T. N., Linares-Pastén, J. A., Grey, C., & Adlercreutz, P. (2021). Synthesis of novel oligomeric anionic alkyl glycosides using laccase/TEMPO oxidation and cyclodextrin glucanotransferase (CGTase)-catalyzed transglycosylation. *Biotechnology and Bioengineering, 118*(7), 2548–2558. https://doi.org/10.1002/BIT.27770;WGROUP:STRING:PUBLICATION.

Nik-Pa, N. I. M., et al. (2020). Combined optimization of codon usage and glycine supplementation enhances the extracellular production of a β-cyclodextrin glycosyltransferase from *Bacillus* sp. NR5 UPM in *Escherichia coli*. *International Journal of Molecular Sciences, 21*(11), 3919. https://doi.org/10.3390/ijms21113919.

Ogunbadejo, B. A., & Al-Zuhair, S. (2023). Bioconversion of starch to cyclodextrin using cyclodextrin glycosyltransferase immobilized on metal organic framework. *Biocatalysis and Agricultural Biotechnology, 53*, 102878. https://doi.org/10.1016/j.bcab.2023.102878.

Okada, S. (1974). Method of converting starch to /J-cyclodextrin. *Patents, 3812011*.

Pacheco, A. F. C., et al. (2025). Impact of ultrasonic pretreatment on pumpkin seed protein: Effect on protease activities, protein structure, hydrolysis kinetics and functional properties. *Food Research International, 201*, 115538. https://doi.org/10.1016/j.foodres.2024.115538.

Pardhi, D. S., Rabadiya, K. J., Panchal, R. R., Raval, V. H., Joshi, R. G., & Rajput, K. N. (2023). Cyclodextrin glucanotransferase: Fundamentals and biotechnological implications. *Applied Microbiology and Biotechnology, 107*(19), 5899–5907. https://doi.org/10.1007/s00253-023-12708-9.

Parsiegla, G., Schmidt, A. K., & Schulz, G. E. (1998). Substrate binding to a cyclodextrin glycosyltransferase and mutations increasing the γ-cyclodextrin production. *European Journal of Biochemistry / FEBS, 255*(3), 710–717. https://doi.org/10.1046/j.1432-1327.1998.2550710.x.

Patel, A. K., Dong, C.-D., Chen, C.-W., Pandey, A., & Singhania, R. R. (2023). Chapter 3 - Production, purification, and application of microbial enzymes. In G. Brahmachari (Ed.). *Biotechnology of microbial enzymes (Second Edition)* (pp. 25–57). Academic Press. https://doi.org/10.1016/B978-0-443-19059-9.00019-0.

Penninga, D., et al. (1996). The raw starch binding domain of cyclodextrin glycosyltransferase fromem *Bacillus circulansem* strain 251 * *Journal of Biological Chemistry, 271*(51), 32777–32784. http://doi.org/10.1074/jbc.271.51.32777.

Pereva, S., Nikolova, V., Angelova, S., Spassov, T., & Dudev, T. (2019). Water inside β-cyclodextrin cavity: Amount, stability and mechanism of binding. *Beilstein Journal of Organic Chemistry, 15*, 1592–1600. https://doi.org/10.3762/bjoc.15.163.

Priyanka, K., Umesh, M., & Preethi, K. (2022). Distillery effluent valorization through cost effective production of polyhydroxyalkanoate: Optimization and characterization. *Biomass Conversion and Biorefinery* https://doi.org/10.1007/s13399-022-03333-z.

Qi, Q., & Zimmermann, W. (2005). Cyclodextrin glucanotransferase: From gene to applications. *Applied Microbiology and Biotechnology, 66*(5), 475–485. https://doi.org/10.1007/s00253-004-1781-5.

Qian, M., Haser, R., Buisson, G., Duee, E., & Payan, F. (1994). The active center of a mammalian .alpha.-Amylase. Structure of the complex of a pancreatic .alpha.-Amylase with a carbohydrate inhibitor refined to 2.2-.ANG. resolution. *Biochemistry, 33*(20), 6284–6294. https://doi.org/10.1021/bi00186a031.

Rabadiya, K., Pardhi, D., Thaker, K., Patoliya, J., Rajput, K., & Joshi, R. (2024). A review on recent upgradation and strategies to enhance cyclodextrin glucanotransferase properties for its applications. *International Journal of Biological Macromolecules, 259*, 129315. https://doi.org/10.1016/j.ijbiomac.2024.129315.

Rajput, K. N., Patel, K. C., & Trivedi, U. B. (2016). A novel cyclodextrin glucanotransferase from an alkaliphile Microbacterium terrae KNR 9: Purification and properties. *3 Biotechnology, 6*(2), 168. https://doi.org/10.1007/s13205-016-0495-6.

Rashid, N., Cornista, J., Ezaki, S., Fukui, T., Atomi, H., & Imanaka, T. (2002). Characterization of an archaeal cyclodextrin glucanotransferase with a novel C-terminal domain. *Journal of Bacteriology, 184*(3), 777–784. https://doi.org/10.1128/JB.184.3.777-784.2002.

Rather, M. Y., Ara, K. Z. G., Nordberg Karlsson, E., & Adlercreutz, P. (2015). Characterization of cyclodextrin glycosyltransferases (CGTases) and their application for synthesis of alkyl glycosides with oligomeric head group. *Process Biochemistry, 50*(5), 722–728. https://doi.org/10.1016/j.procbio.2015.02.016.

Rathnakumar, K., et al. (2023). Applications of ultrasonication on food enzyme inactivation-recent review report (2017–2022). *Ultrasonics Sonochemistry, 96*, 106407. https://doi.org/10.1016/j.ultsonch.2023.106407.

Rodríguez Gastón, J. A., Costa, H., Rossi, A. L., Krymkiewicz, N., & Ferrarotti, S. A. (2012). Maltooligosaccharides production catalysed by cyclodextrin glycosyltransferase from Bacillus circulans DF 9R in batch and continuous operation. *Process Biochemistry, 47*(12), 2562–2565. https://doi.org/10.1016/j.procbio.2012.08.008.

Saini, K., Pathak, V. M., Tyagi, A., & Gupta, R. (2022a). Microbial cyclodextrin glycosyltransferases: Sources, production, and application in cyclodextrin synthesis. *Catalysis Research, 02*(03), 1–69. https://doi.org/10.21926/cr.2203029.

Saini, K., Kashyap, A., Saini, M., & Gupta, R. (2022b). Gamma cyclodextrin glycosyltransferase from evansella caseinilytica: Production, characterization and product specificity. *3 Biotechnology, 12*(1), 16. https://doi.org/10.1007/s13205-021-03077-z.

Schmidt, A. K., Cottaz, S., Driguez, H., & Schulz, G. E. (1998). Structure of cyclodextrin glycosyltransferase complexed with a derivative of its main product β-cyclodextrin. *Biochemistry, 37*(17), 5909–5915. https://doi.org/10.1021/bi9729918.

Schöffer, J.da N., et al. (2017). Effects of immobilization, pH and reaction time in the modulation of α-, β- or γ-cyclodextrins production by cyclodextrin glycosyltransferase: Batch and continuous process. *Carbohydrate Polymers, 169*, 41–49. https://doi.org/10.1016/j.carbpol.2017.04.005.

Shim, J., et al. (2004). Improvement of cyclodextrin glucanotransferase as an antistaling enzyme by error-prone PCR. *Protein Engineering, Design and Selection, 17*(3), 205–211. https://doi.org/10.1093/protein/gzh035.

Singh, R., Kumar, M., Mittal, A., & Mehta, P. K. (2016). Microbial enzymes: Industrial progress in 21st century. *3 Biotech, 6*(2), 174. https://doi.org/10.1007/s13205-016-0485-8.

Song, K., Sun, J., Wang, W., & Hao, J. (2021). Heterologous expression of cyclodextrin glycosyltransferase my20 in *Escherichia coli* and its application in 2-O-α-D-glucopyranosyl-L-ascorbic acid production. *Frontiers Microbiology, 12*. https://doi.org/10.3389/fmicb.2021.664339.

Sonnendecker, C., Melzer, S., & Zimmermann, W. (2019). Engineered cyclodextrin glucanotransferases from *Bacillus* sp. G-825-6 produce large-ring cyclodextrins with high specificity. *Microbiologyopen, 8*(6). https://doi.org/10.1002/mbo3.757.

Starnes, R. L., Hoffman, C. L., Flint, V. M., Trackman, P. C., Duhart, D. J., & Katkocin, D. M. (1991). Starch liquefaction with a highly thermostable cyclodextrin glycosyl transferase from *Thermoanaerobacter* species. In *Enzymes in biomass conversion* (ACS Symposium Series No. 460, pp. 384–393). American Chemical Society. https://doi.org/10.1021/bk-1991-0460.ch029.

Strokopytov, B., Penninga, D., Rozeboom, H. J., Kalk, K. H., Dijkhuizen, L., & Dijkstra, B. W. (1995). X-ray Structure of cyclodextrin glycosyltransferase complexed with acarbose. Implications for the catalytic mechanism of glycosidases. *Biochemistry, 34*(7), 2234–2240. https://doi.org/10.1021/bi00007a018.

Svensson, D., & Adlercreutz, P. (2011). Characterisation of a glycosylated alkyl polyglycoside produced by a cyclodextrin glycosyltransferase by HPLC–ELSD and –MS. *Journal of Chromatography B, 879*(21), 1857–1860. https://doi.org/10.1016/j.jchromb.2011.04.037.

Svensson, D., Ulvenlund, S., & Adlercreutz, P. (2009). Enzymatic route to alkyl glycosides having oligomeric head groups. *Green Chemistry, 11*(8), 1222–1226. https://doi.org/10.1039/B904849A.

Takano, T., Fukuda, M., Monma, M., Kobayashi, S., Kainuma, K., & Yamane, K. (1986). Molecular cloning, DNA nucleotide sequencing, and expression in *Bacillus subtilis* cells of the Bacillus macerans cyclodextrin glucanotransferase gene. *Journal of Bacteriology, 166*(3), 1118–1122. https://doi.org/10.1128/jb.166.3.1118-1122.1986.

Thiemann, V., Donges, C., Prowe, S., Sterner, R., & Antranikian, G. (2004). Characterisation of a thermoalkali-stable cyclodextrin glycosyltransferase from the anaerobic thermoalkaliphilic bacterium *Anaerobranca gottschalkii*. *Archives of Microbiology, 182*(2–3). https://doi.org/10.1007/s00203-004-0717-x.

Tomita, K., Tanaka, T., Fujita, Y., & Nakanishi, K. (1990). Some factors affecting the formation of γ-cyclodextrin using cyclodextrin glycosyltransferase from *Bacillus* sp. AL-6. *Journal of Fermentation and Bioengineering, 70*(3), 190–192. https://doi.org/10.1016/0922-338X(90)90185-Y.

Uitdehaag, J. C. M., Van Alebeek, G. J. W. M., Van Der Veen, B. A., Dijkhuizen, L., & Dijkstra, B. W. (2000). Structures of maltohexaose and maltoheptaose bound at the donor sites of cyclodextrin glycosyltransferase give insight into the mechanisms of transglycosylation activity and cyclodextrin size specificity. *Biochemistry, 39*(26), 7772–7780. https://doi.org/10.1021/BI000340X;PAGE:STRING:ARTICLE/CHAPTER.

Uitdehaag, J. C. M., Kalk, K. H., van der Veen, B. A., Dijkhuizen, L., & Dijkstra, B. W. (1999). The cyclization mechanism of cyclodextrin glycosyltransferase (CGTase) as Revealed by a γ-cyclodextrin-CGTase complex at 1.8-Å resolution * *Journal of Biological Chemistry, 274*(49), 34868–34876. https://doi.org/10.1074/jbc.274.49.34868.

Varan, G., Varan, C., Erdoğar, N., Hıncal, A. A., & Bilensoy, E. (2017). Amphiphilic cyclodextrin nanoparticles. *International Journal of Pharmaceutics, 531*(2), 457–469. https://doi.org/10.1016/j.ijpharm.2017.06.010.

van der Veen, B. A., Uitdehaag, J. C. M., Dijkstra, B. W., & Dijkhuizen, L. (2000). Engineering of cyclodextrin glycosyltransferase reaction and product specificity1centre for carbohydrate bioengineering (CCB), TNO/University of Groningen, Kerklaan 30, 9751 NN Haren, The Netherlands.1. *Biochimica et Biophysica Acta (BBA) - Protein Structure and Molecular Enzymology, 1543*(2), 336–360. https://doi.org/10.1016/S0167-4838(00)00233-8.

Vennelakanti, V., Nazemi, A., Mehmood, R., Steeves, A. H., & Kulik, H. J. (2022). Harder, better, faster, stronger: Large-scale QM and QM/MM for predictive modeling in enzymes and proteins. *Current Opinion in Structural Biology, 72*, 9–17. https://doi.org/10.1016/j.sbi.2021.07.004.

Wang, J., Wang, B., Zhou, J., Dong, J., Ni, Y., & Han, R. (2025). Engineering of cyclodextrin glucosyltransferase from *Paenibacillus macerans* for improved regioselectivity and product specificity toward hydroxyflavone glycosylation. *Catalysts, 15*(2), 120. https://doi.org/10.3390/catal15020120.

Wang, Y., et al. (2010). Glucosylation of steroidal saponins by cyclodextrin glucanotransferase. *Planta Medica, 76*(15), 1724–1731. https://doi.org/10.1055/s-0030-1249938.

Wind, R. D., Uitdehaag, J. C. M., Buitelaar, R. M., Dijkstra, B. W., & Dijkhuizen, L. (1998). Engineering of cyclodextrin product specificity and pH optima of the thermostable cyclodextrin glycosyltransferase from *Thermoanaerobacterium thermosulfurigenes* EM1 * *Journal of Biological Chemistry, 273*(10), 5771–5779. https://doi.org/10.1074/jbc.273.10.5771 Mar.

Xiao, Y., et al. (2022). Molecular mechanism of acceptor selection in cyclodextrin glycosyltransferases catalyzed ginsenoside transglycosylation. *Bioorganic Chemistry, 128*, 106094. https://doi.org/10.1016/j.bioorg.2022.106094.

Yagi, K., Ito, S., & Sugita, Y. (2021). Exploring the minimum-energy pathways and free-energy profiles of enzymatic reactions with QM/MM calculations. *The Journal of Physical Chemistry. B, 125*(18), 4701–4713. https://doi.org/10.1021/acs.jpcb.1c01862.

Yang, L., et al. (2022). Effect of cyclodextrin glucosyltransferase extracted from *Bacillus xiaoxiensis* on wheat dough and bread properties. *9*, 1026678. 10.3389/FNUT.2022.1026678/BIBTEX.

Zhang, R., et al. (2023). "Engineering of cyclodextrin glycosyltransferase improves the conversion efficiency of rebaudioside A to glucosylated steviol glycosides and increases the content of short-chain glycosylated steviol glycoside. *Microbial Cell Factories, 22*(1), 1–12. https://doi.org/10.1186/S12934-023-02121-2/FIGURES/1.

Zhao, F., Li, Y., Li, C., Ban, X., Gu, Z., & Li, Z. (2022). Glycosyltransferases improve breadmaking quality by altering multiscale structure in gluten-free bread. *Food Hydrocoll, 133*, 107951. https://doi.org/10.1016/j.foodhyd.2022.107951.

Zhou, J., et al. (2021). CGTase, a novel antimicrobial protein from *Bacillus cereus* YUPP-10, suppresses *Verticillium dahliae* and mediates plant defence responses. *Molecular Plant Pathology, 22*(1), 130–144. https://doi.org/10.1111/mpp.13014.

SECTION

II

Mechanism and cyclodextrin glucanotransferase production

CHAPTER

Catalytic mechanisms of cyclodextrin glucanotransferases

4

Preetibala Solanki[1], Isha Jain[2], Ajay Malviya[1] and Tushar Banerjee[2]
[1]Department of Microbiology, Barkatullah University, Bhopal, Madhya Pradesh, India,
[2]School of Life Sciences, Devi Ahilya Vishwavidyalaya, Indore, Madhya Pradesh, India

Introduction

Enzymes catalyze biochemical reactions with remarkable efficiency and specificity, enabling complex metabolic transformations to occur under mild physiological conditions. At the heart of enzymatic catalysis lies the ability to lower activation energy and stabilize high-energy transition states, allowing reactants to be converted into products at rates that far exceed uncatalyzed reactions. Most enzymatic mechanisms follow well-defined principles, including proximity and orientation effects, acid–base catalysis, covalent catalysis, metal-ion assistance, and transition-state stabilization. Together, these strategies enable enzymes to achieve rate enhancements of several orders of magnitude (Fersht, 1999).

A central feature of many catalytic processes is the formation of transient intermediates, which may be covalent or noncovalent depending on the class of enzyme involved. Retaining and inverting glycosidases, proteases, and other hydrolases often employ a two-step double-displacement (Ping–Pong) or single-displacement pathway to cleave and reform bonds with high stereochemical precision. In such mechanisms, key amino acid residues—commonly nucleophiles, general acids/bases, or stabilizing side chains—play coordinated roles in forming transition states and intermediates that guide the reaction forward (Sinnott, 1990).

Additionally, enzyme structure profoundly influences catalytic outcome. Active-site architecture, substrate-binding clefts, and the spatial arrangement of catalytic residues determine reaction specificity, regioselectivity, and stereoselectivity. Modern structural biology has revealed that subtle changes in active-site geometry, substrate binding loops, or residue positioning can shift catalytic pathways or alter product profiles, a principle that underscores the evolutionary adaptability of enzyme families (Henrissat, 1991).

Understanding these fundamental concepts provides a strong foundation for exploring more specialized catalytic systems, such as the α-retaining, double-displacement mechanism employed by cyclodextrin glycosyltransferase (CGTase).

Cyclodextrin Glucanotransferase. DOI: https://doi.org/10.1016/B978-0-443-41694-1.00015-0

Catalytic mechanism of cyclodextrin glycosyltransferase

The catalytic mechanism of CGTase is central to its ability to perform various transglycosylation reactions, including cyclisation, coupling, disproportionation, and limited hydrolysis. First proposed by Koshland, (1953), this mechanism is conserved among members of the α-amylase enzyme family and is classified as an α-retaining, double-displacement reaction. This type of mechanism is a hallmark of many glycosyltransferases and glycosidases, as it preserves the original stereochemistry of the anomeric carbon during both glycosidic bond cleavage and reformation.

Catalysis begins when the linear carbohydrate substrate binds to the enzyme's active site (Van Der Maarel et al., 2002). This region is highly conserved across CGTase enzymes and includes three essential carboxylate residues: Asp229, Glu257, and Asp328, particularly well-studied in CGTase from *Bacillus circulans*. Asp229, often termed the C2 carboxylate, serves as the nucleophile. It initiates the reaction by attacking the anomeric carbon (C1) of the glycosidic bond within the substrate. Simultaneously, Glu257 functions as the general acid catalyst (C1) and donates a proton to the glycosidic oxygen, thereby facilitating the departure of the leaving group, which is typically the nonreducing end of the maltooligosaccharide. This leads to the cleavage of the glycosidic bond, during which reaction the -1 sugar goes through an oxocarbonium ion-like transition state. This intermediate is characterised by a delocalised positive charge, partial double bond character between O5 and C1, and a nearly planar arrangement involving atoms C2, C1, O5, and C5 of the sugar ring. The oxocarbonium ion-like transition state is highly unstable and is stabilised by interactions within the enzyme's active site, including hydrogen bonding and van der Waals forces. The transition state quickly collapses to form a stable covalent glycosyl-enzyme intermediate where the sugar is β-glycosidically linked to the nucleophilic Asp229. Meanwhile, the protonated glucose molecule at subsite +1 leaves the active site (McCarter & Withers 1994,1996a).

In the second phase of the reaction, an appropriate acceptor molecule binds to the active site and targets the covalent linkage between the enzyme and the glucose residue located at subsite -1. Glu257, which previously served as a proton donor, now acts as a general base. It deprotonates the acceptor molecule (often another carbohydrate), thereby activating it for nucleophilic attack. The acceptor then displaces the sugar from Asp229 via a second oxocarbonium ion-like transition state (Uitdehaag et al., 1999). In this final step, the oxygen of the activated acceptor replaces the bond between the sugar and the aspartate residue, resulting in the formation of a new α-glycosidic linkage. This two-step process preserves the α-configuration at the anomeric carbon, hence the term "α-retaining" mechanism (Fig. 4.1) (Heightman & Vasella, 1999; Kirby, 1987; Ly & Withers 1999; McCarter et al., 1992; Sinnott, 1990; Vocadlo et al., 2001; White & Rose, 1997).

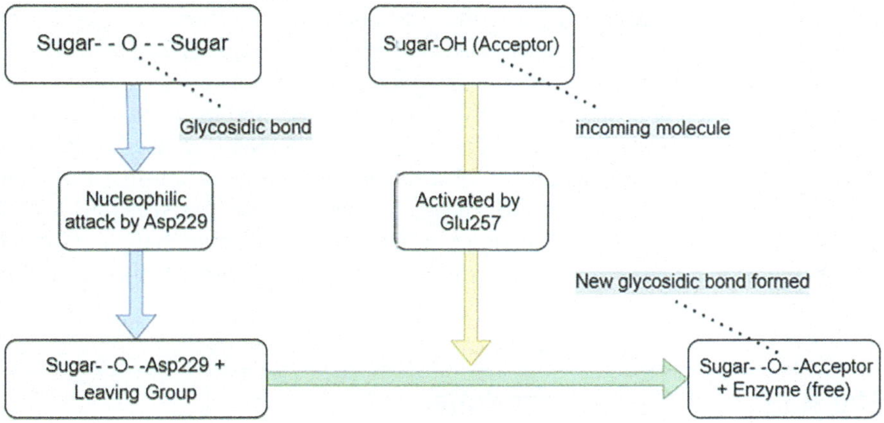

FIGURE 4.1

α-retaining, double displacement reaction mechanism of cyclodextrin glycosyltransferase.

The roles of these critical amino acid residues have been elucidated by X-ray crystallographic studies on alpha-amylase (Henrissat, 1991) and through structural studies involving inhibitors like acarbose, a pseudotetraose that mimics the transition state and binds tightly within the enzyme's active site (Klein et al. 1992; Nakamura et al., 1994a). Such studies confirm the roles of Asp229, Glu257, and Asp328 in positioning and catalysis. While Asp229 directly participates in the nucleophilic attack, Glu257 modulates the acid–base chemistry essential for bond cleavage and formation, and Asp328 ensures proper alignment and stabilisation of the substrate.

In addition to the classical double displacement (Ping-Pong) mechanism, which involves two sequential nucleophilic substitutions (SN2-type reactions), CGTase may follow a random ternary complex mechanism under certain conditions. In this alternate pathway, both the glycosyl donor and acceptor can bind simultaneously at the active site, forming a transient ternary complex. Once both substrates are properly positioned, the glycosyl group is transferred directly from the donor to the acceptor. Although this pathway is less commonly observed, it may offer greater flexibility in substrate use, particularly when working with noncarbohydrate acceptors or engineered CGTase variants tailored for broader substrate specificity (Lim et al., 2021).

The nature of the transition state

In enzymes belonging to the α-amylase family, the transition state has been studied extensively using kinetic isotope effect experiments (Tanaka et al., 1994). These investigations have revealed that the transition state closely resembles an oxocarbonium ion, with almost complete double-bond character between the O5 and C1 atoms. The positive charge is predominantly concentrated on the O5 oxygen, while the O4 position is nearly fully protonated, and the bond between C1 and O4 is nearly entirely cleaved (Sinnott, 1990; Tanaka et al., 1994). This structural configuration supports the idea of an oxocarbonium ion-like transition state. Additional evidence for this model is provided by the structure of the transition state analog inhibitor acarbose, which exhibits a planar configuration that mirrors the geometry expected of an oxocarbonium intermediate (Mosi et al., 1998).

The nature of the intermediate

The exact chemical nature of the reaction intermediate in glycosidic bond cleavage has long been debated in enzymology (Kirby, 1996; White & Rose, 1997). For retaining enzymes, the intermediate could either be an oxo-carbonium ion, which is electrostatically stabilised by a carboxylate, or involve the formation of a covalent bond, in which one of the catalytic aspartates is presumed to act as a nucleophile (Uitdehaag et al., 2002). Koshland's original model for both α- and β-retaining enzymes proposed the formation of a covalent glycosyl-enzyme intermediate (Koshland, 1953). However, when the crystal structure of the β-retaining enzyme hen egg white lysozyme was resolved, it was speculated that an oxocarbonium ion-like species might persist long enough in the active site to function as the true intermediate (Blake et al., 1967). This interpretation gained traction and was widely adopted in academic literature and textbooks, often being extended to enzymes of the α-amylase family as well (Kuriki & Imanaka, 1999; Lawson et al., 1994; Matsuura et al., 1984; Seddon, 1988; Svensson, 1994).

Definitive experimental support for the presence of a covalent glycosyl-enzyme intermediate in family 13 emerged from low-temperature ^{13}C NMR experiments (Tao et al., 1989). Further evidence came from successful attempts to trap these covalent intermediates with 4-deoxymaltotriosyl alpha-fluoride as a substrate in the virtually inactive Glu257Gln mutant of *B. circulans* 251 CGTase and elucidation of the X-ray crystallographic structure of a trisaccharide covalently bound to the catalytic nucleophile. (Braun et al., 1995; Braun et al., 1996; McCarter & Withers, 1996a, 1996b; Mooser et al., 1991; Mosi et al., 1997). This structural insight provided direct confirmation of covalent catalysis, supporting Koshland's original theory (Kirby, 1987; McCarter et al., 1992; Sinnott, 1990).

As a result of this accumulating body of evidence, the covalent nature of the intermediate is now widely accepted for enzymes in the α-amylase family. More recently, similar findings have reaffirmed that even hen egg white lysozyme, previously considered an exception, also employs covalent catalysis during its enzymatic mechanism. (Heightman & Vasella, 1999; Ly & Withers 1999; Vocadlo et al., 2001; White & Rose,1997).

Cyclodextrin glycosyltransferase-catalyzed reactions

The formation of cyclodextrins (CDs) by CGTase proceeds through cyclisation, an intramolecular transglycosylation reaction in which the nonreducing end of a linear malto-oligosaccharide is transferred to its own reducing end. However, two other transglycosylation reactions are also catalyzed by the enzyme: coupling which is the reverse of cyclization, opens the CD ring and transfers glycosyl groups to external acceptor molecules, and disproportionation, the major transferase reaction, involves the transfer of glycosyl groups between linear oligosaccharides, producing oligosaccharides of varying lengths while improving the solubility, bioavailability, absorption, and bioactivity of acceptors (Van der Veen et al., 2000). In addition, the enzyme has a weak hydrolytic activity, which involves the transfer of part of a linear oligosaccharide to water, leading to sugar chain cleavage. Despite their differences, all four reactions operate through the same fundamental catalytic mechanism, which is highly conserved within the amylase family, with the primary variation being the nature of the acceptor molecule. These diverse reactions make CGTase valuable for the enzymatic synthesis of compounds used in food additives, dietary supplements, detergents, pharmaceuticals, and cosmetics.

Intramolecular transglycosylation

Cyclisation activity of cyclodextrin glycosyltransferase

One of the most distinct enzymatic activities exhibited by CGTases is cyclisation. It enables the conversion of linear starch fragments into cyclic oligosaccharides known as CDs. In this reaction, the enzyme uses the nonreducing end of the same sugar chain as the acceptor rather than an external molecule, thereby forming a closed ring structure. The primary products of this reaction are α-, β-, and γ-CDs, composed of six, seven, and eight glucose units, respectively.

In *B. circulans* strain 251, a CGTase that predominantly synthesises β-CD, detailed structural studies revealed at least seven sugar-binding subsites that interact with the substrate during the reaction (Strokopytov et al., 1996; Uitdehaag et al., 1999b).

These subsites (ranging from -7 to +1) help position the linear starch molecule correctly within the active site. Initially, crystallographic analysis revealed that the nonreducing end of the donor chain bound at subsite -7 is approximately 23 Å from the catalytic centre (Bender, 1990). This unexpectedly large distance initially raised questions about its significance, but follow-up site-directed mutagenesis experiments confirmed that this remote positioning was genuine and functionally relevant (Parsiegla et al., 1998; Van der Veen et al., 2000).

Computational pathway modelling (stochastic path calculations) was employed to further understand how CGTases manage to bring the two ends of the sugar chain together to form a ring. They found that specific amino acids help guide and bend the sugar chain into a loop as it moves along the enzyme. The sugars bind step-by-step from subsite -7 toward subsite +1 (Uitdehaag et al., 2001).

Along the way, different amino acids stabilise and guide the sugar. Gly179, Gly180, Tyr167, and Asn193 help bind residues at subsite 6. Tyr195, Phe183, and Phe259 help move the chain forward. Leu194, Leu197, and Ala230 create a pocket that traps the sugar in the right shape. As the sugar chain folds, Asp196 and Tyr89 release their grip, and new interactions form with Arg47 and Asp371. This whole process bends the sugar chain into a ring, allowing the nonreducing end to react with the part near the active site, forming a CD (Uitdehaag et al., 2002).

The type of CD formed is determined by the length of the sugar chain required to bridge subsite -1 (donor) to subsite +1 (acceptor). Typically, this spacing corresponds to 8 glucose residues for α-CDs, 9 residues for β-CDs, and 10 residues for γ-CDs. However, in some cases, longer starch chains can be used as donors, leading to the formation of large-ring CDs. These oversized rings are often secondary substrates for CGTase's other transglycosylation reactions, such as coupling and hydrolysis, which convert them into standard-sized CDs over time (Leemhuis et al., 2010).

Extensive protein engineering efforts have targeted CGTases to optimise product specificity. To alter amino acids around critical subsites, approaches such as site-directed mutagenesis, random mutagenesis, molecular imprinting, and domain shuffling have been used. For instance, modifications at subsites -3 and -7 have been shown to significantly alter the enzyme's product profile, allowing certain engineered CGTases to selectively synthesise γ-CD (Wang et al., 2020).

Intermolecular transglycosylation

Coupling activity of cyclodextrin glycosyltransferase

The coupling activity of CGTases is the reverse of their cyclisation function. CGTases can form cyclic products, but they can also hydrolyse the ring structure of CDs and attach the resulting linear oligosaccharide to another sugar chain. This reaction plays a key role in balancing the product distribution during CGTase-mediated processes and contributes significantly to the dynamic conversion between different types of CDs (Uitdehaag et al., 2001).

The coupling reaction begins when CGTase cleaves a CD molecule at one of its glycosidic bonds, forming a linear glycosyl intermediate. The nonreducing end of this intermediate then acts as a donor and transfers the sugar unit to an external acceptor molecule, which may be a monosaccharide, disaccharide, or linear oligosaccharide. This process converts the ring-shaped CD into a linear molecule by elongating the acceptor chain (Lim et al., 2021).

In CGTase, the same catalytic site facilitates both coupling and cyclisation. The substrate-binding subsites, particularly subsites -1, +1, and beyond, play a major role in determining whether the reaction proceeds toward the formation of a linear chain or a cyclic product. Mutations at these sites, especially those that change acceptor specificity, can enable control over product profiles by significantly shifting the balance between coupling and cyclisation (Penninga et al., 1996).

The coupling activity of CGTase is not just a supporting feature; rather, it represents a critical step in the interconversion of CDs. For example, α-CDs can be rearranged into β-CDs, and β-CDs can subsequently be cleaved and recombined into γ-CDs. However, factors such as temperature, pH, the presence of additives or solvents, the type and concentration of available acceptor molecules, and the specific CGTase isoform or mutant used all influence the reaction outcome (Tonkova, 1998; Qi & Zimmermann, 2005). By carefully controlling the reaction, the coupling activity allows the precise modification of oligosaccharides and the production of desired sugar chains with specific lengths or functional properties.

Disproportionation activity of cyclodextrin glycosyltransferase

Another transglycosylation reaction of CGTases is disproportionation, in which the enzyme transfers a portion of a linear maltooligosaccharide (donor) onto another linear maltooligosaccharide (acceptor), without forming or breaking a ring. Disproportionation strictly occurs between linear carbohydrates, unlike the other two transglycosylation reactions, which involve cyclic substrates or products (Qi & Zimmermann 2005).

In a typical disproportionation reaction, the donor oligosaccharide is cleaved by CGTase to form a glycosyl-enzyme intermediate. This glycosyl unit is then transferred to the nonreducing end of an acceptor oligosaccharide, thereby elongating the chain. The efficiency and outcome of the reaction are influenced by both the length and structure of donor and acceptor molecules. The disproportionation reaction forms novel linear oligosaccharides by enabling CGTase to rearrange sugar chains of varying lengths (Leemhuis & Dijkhuizen, 2003).

Disproportionation is an important reaction from a biochemical standpoint because it modifies the size distribution of oligosaccharides in solution. This activity can also be used in industrial processes to tailor starch hydrolysates into products with specific degrees of polymerisation.

The substrate-binding region, particularly the donor and acceptor subsites (typically spanning from subsite -1 to +n) of CGTase, plays a critical role in disproportionation. These subsites guide the positioning of both donor and acceptor chains and enable efficient glycosyl transfer without the need to reenter a cyclic intermediate state.

CGTases have high catalytic efficiency and broad substrate tolerance, which makes them particularly attractive for the disproportionation reaction. They can act on a wide range of maltooligosaccharides, from maltotriose to much longer chains, allowing precise modulation of carbohydrate structure.

Disproportionation can be used for the synthesis of functional oligosaccharides, for debranching and chain restructuring in starch processing, and for the preparation of low-molecular-weight sugars for use in food or pharma formulations (Lim et al., 2021).

Hydrolytic activity of cyclodextrin glycosyltransferase

Although CGTases are best known for their transglycosylation activities, they also exhibit limited hydrolytic activity. CGTase cleaves the glycosidic bonds in starch without forming new glycosidic linkages, resulting in shorter linear sugars (Qi & Zimmermann, 2005).

CGTases are not primarily hydrolytic enzymes, in contrast to specialized amylases or glucosidases. Hydrolysis occurs under certain conditions, such as when the acceptor is a water molecule instead of a carbohydrate molecule (Tonkova, 1998). Here, the glycosyl-enzyme intermediate formed during catalysis is attacked by a water molecule, leading to the release of a linear sugar chain rather than a cyclic or transferred product.

The same catalytic residues and active site are used in both hydrolysis and transglycosylation reactions. The only difference is the acceptor molecule. For transglycosylation to occur, a sugar molecule occupies the acceptor subsite, while for hydrolysis to occur, water fills this position. The balance between transglycosylation and hydrolysis is strongly influenced by substrate concentration, enzyme specificity, and solvent composition.

Generally, hydrolytic activity is minor compared to the enzyme's transglycosylation functions, and its extent depends on the absence or scarcity of suitable carbohydrate acceptors, the availability of water in the reaction medium, and the specific structure and sequence of the CGTase used (Tonkova, 1998).

During CD production, hydrolytic activity competes with cyclisation and is considered undesirable. Despite its lower significance, hydrolytic activity plays a supporting role in determining the final products of CGTase reactions.

For example, during prolonged reactions, hydrolysis can break down CDs into smaller sugars. This can be beneficial in processes such as functional sugar preparation, bioethanol production, and the conversion of starch into useful products (Qi & Zimmermann 2005). Efforts have also been made in enzyme engineering to enhance hydrolytic activity. For example, boosting hydrolysis can be advantageous for starch liquefaction or saccharification (Leemhuis et al., 2009).

The role of conserved amino acid residue in cyclodextrin glycosyltransferase

The term "conserved" describes parts of a sequence that are remarkably identical or unaltered among various species or individuals. The amino acids in the active core of an enzyme that speed up the reaction it catalyzes are known as the catalytic residues. Conserved residues play a variety of roles in catalysis, affecting substrate interaction, stability, and enzyme activity. These residues, which are frequently found at or close to the active site, can be essential for preserving the enzyme's structural integrity and promoting its catalytic activity. These residues play a number of catalytic roles that reduce the activation energy of processes (Ribeiro et al., 2020). In general, catalytic residues are extremely conserved.

The many functions of these conserved residues in enzyme catalysis are described in the following sections

The CGTase enzyme is to be evolved from the α-amylase family. The five domains of CGTase are A through E (see Fig. 4.2). Domain A is the catalytic $(\alpha/\beta)_8$ domain, which CGTase has in common with the other amylase family members. This catalytic $(\alpha/\beta)_8$ domain is important, where the most conserved amino acids are present.

At the bottom of the substrate binding groove at the $(\alpha/\beta)_8$ barrel's surface, the seven most conserved amino acid residues in the amylase family group together. The catalytic site is situated in the middle of the conserved areas. The names of these conserved amino acids are aspartic acid (Asp), glutamic acid (Glu), arginine (Arg), and histidine (His). The initials of these amino acids, Asp135, His140, Arg227, and His327, are nearly entirely conserved, whereas Asp229, Glu257, and Asp328 are fully conserved.

The glycosidic link, which is the sole fundamental chemical bond found in their common substrate, starch, explains the evolutionary conservation of these residues. There are many forms of this bond, such as (134), (136), and (131), depending on whether the C1 atom of a glucose is joined to the O4, O6, or O1 atom of a nearby glucose molecule, respectively (Saini et al., 2022).

The functions of conserved residues in catalysis

Studies on various crystal structures suggested that CGTase has three conserved active site residues: Glu257, Asp229, and Asp328 (Saini et al., 2022). In the amylase family, Asp229 is the first conserved residue and is suggested as the catalytic nucleophile. Glu257 is the second conserved residue and is suggested as the catalytic acid/base. A mutation in Glu257 causes a 5000-fold decrease in enzyme activity (Nakamura et al., 1994b).

Prior to CD production, Asp229 forms a covalent intermediate with the cleaved substrate, Asp328 stabilizes the reaction intermediates, and Glu257 serves as both a

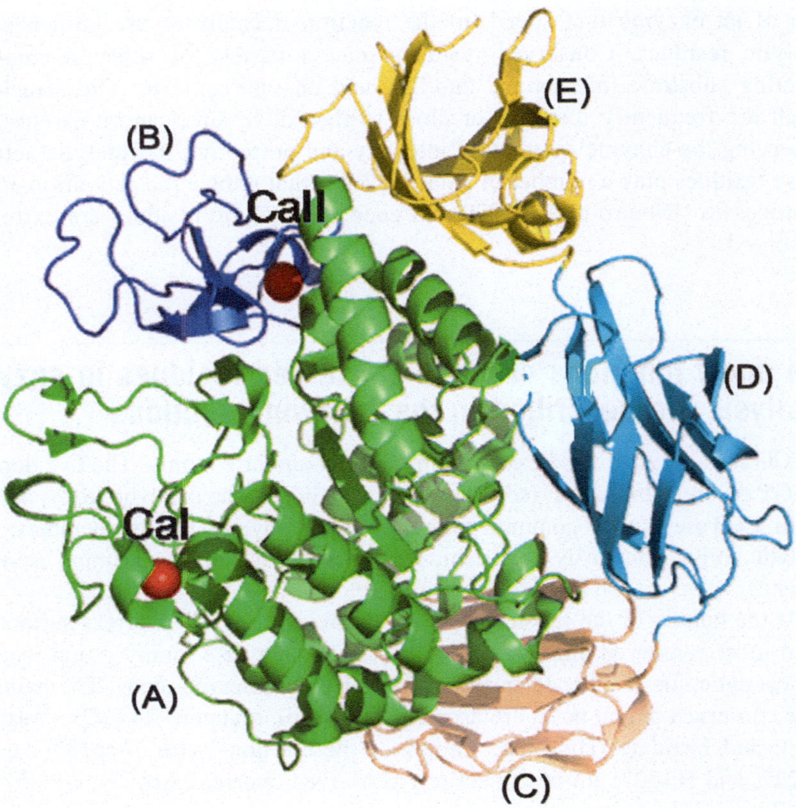

FIGURE 4.2

The 3D structure modelling of β-CGTase from *B. circulans* STB01 (mutants) has five domains. The red ball represents Ca^{2+} binding sites (Chen et al., 2019, and Kelly et al., 2009).

Chen, S., Li, Z., Gu, Z., Hong, Y., Cheng, L., & Li, C. (2019). Variants at position 603 of the CGTase from Bacillus circulans STB01 for reducing product inhibition. International Journal of Biological Macromolecules, 136, *460–468; Kelly, R.M., Dijkhuizen, L., & Leemhuis, H. (2009). The evolution of cyclodextrin glucanotransferase product specificity.* Applied Microbiology and Biotechnology, 84*(1), 119–133.*

proton donor and an acceptor. Mutations in Asp229 cause a 25,000-fold decrease in activity. Tyrosine (Y195) is the main core residue of CGTases active sites and is in charge of substrate selectivity. Seven conserved amino acid residues, or conserved sequence areas, were found in all α-amylase family enzymes after a sequence alignment of a few chosen CGTases (one α-CGTase, two β-CGTases, and one γ-CGTase) was carried out by Saini et al. (2022). This study suggested that differences in amino acid sequence may alter product specificity, underscoring their importance in establishing CD specificity.

The role of substrate in cyclodextrin glycosyltransferase-catalyzed reactions

The substrate plays an important role in reaction outcome, catalytic efficiency, and product distribution. This subheading provides a comprehensive analysis of how the four fundamental reactions that CGTase catalyzes—cyclization, coupling, disproportionation, and hydrolysis—are affected by substrate characteristics such as chain length, branching pattern, concentration, and structural features. In order to optimize CD formation while establishing CGTases with customized specificities, it is essential to figure out these substrate-dependent mechanisms. The catalytic machinery is provided by the enzyme's structure, whereas the substrate serves as both the reactant and, frequently, the allosteric regulator of reaction specificity.

The CGTase enzyme performs two essential functions: the first is catalyzing a reaction, and the second is binding to the substrate in the right orientation (Saini et al., 2022). The substrate serves as both the reactant and, frequently, the allosteric regulator of reaction specificity, whereas the catalytic machinery is provided by the enzyme's structure.

Enzyme structure and substrate characteristics both affect the distribution of α-, β-, and γ-CDs in the resultant mixture (Uitdehaag et al., 1999a; 2002) (see Table 4.1).

The characteristics of the substrate control CGTase catalysis: the type of reaction pathway that occurs, such as cyclization, coupling, disproportionation, and hydrolysis. Furthermore, the type of substrate used, structural characteristics, chain length of substrate, and substrate modification affect the type of CD produced. CGTase reactivity is strongly influenced by the physical type of starch. Native granular starch is less reactive because of its crystalline form, whereas soluble, gelatinized starch offers accessible glucan chains for enzyme action. The cyclization process is a crucial step for the formation of CDs; the enzyme from *K. pneumoniae* M5 al catalyzes both cyclization and disproportionation reaction (Bender, 1982). They observed that the helical shape of the substrate is essential for the reaction, as maximum cyclization rates are achieved with glucopyranosyl chain lengths ranging from glc16 to glc80. Knowing the structural changes that starches undergo during the CGTase reaction, with the concurrent generation of CDs, would enable the best choice of substrates to increase reaction yields (Benavent-Gil et al., 2021). The conversion rate and the first binding of the high molecular mass substrate in the cyclization reaction may influence the enzyme's product selectivity (Van der Veen et al., 2000).

Donor and acceptor specificity for cyclodextrin glycosyltransferase

CGTase catalyzes a variety of transglycosylation processes, such as coupling, hydrolysis, cyclization, and disproportionation. The process of moving glycosyl

Table 4.1 Comparative overview of α-, β-, and γ-cyclodextrin formation, highlighting differences in ring size, production behavior, and substrate preferences as influenced by cyclodextrin glycosyltransferase (*CGTase*) catalytic properties.

Product	Ring size	Common observations	Substrate choice
α-cyclodextrin (CD)	6 glucoses	Produced easily by a variety of CGTases; typically generated early in the initial phases of the reaction.	α-CD can be produced by short-to-moderate external chains; smaller ring closure can be less limited geometrically (Bender, 1982 and Uitdehaag et al., 1999a).
β-CD	7 glucoses	For most naturally occurring CGTases, it is frequently the main product that builds up over time.	Although formation might begin slower than α-CD, thermodynamic stability and reverse reactions cause it to become dominant (Bender, 1982, and Uitdehaag et al., 1999a).
γ-CD	8 glucoses	Less often generated; it needs bioengineering or certain enzyme variations.	Production is enhanced by active site mutations and is preferred when substrate binding allows for longer closure loops (Uitdehaag et al., 1999a and Kelly et al., 2009).

groups from a donor, such as carbohydrates, to an acceptor that has hydroxyl or other functional groups to create glycoconjugates is known as the transglycosylation reaction (Lim et al., 2021).

A complex subsite design controls the enzyme's donor and acceptor selectivity by identifying substrates through certain molecular interactions. On one side of the active site, donor substrates (mostly linear malto-oligosaccharides, starch polymers, and CDs) bind across negative subsites (−7 to −1), whereas acceptor substrates occupy positive subsites (+1, +2) and peripheral binding pockets on the other side. Key amino acid residues at these subsites govern steric complementarity, hydrophobic interactions, and precise hydrogen-bonding networks that dictate specificity for both donors and acceptors (Han et al., 2020; Uitdehaag et al., 2002). Tyrosine 195 and tyrosine 260, these residues' hydroxyl groups and aromatic rings affect donor and acceptor binding. Acceptor accessibility was impacted by mutations at these locations (Y195S, Y260R).

Different transglycosylation processes, such as cyclization, disproportionation, coupling, and hydrolysis, are catalyzed by the amazing enzyme CGTase. A complex subsite design that detects substrates via specific molecular interactions controls the enzyme's donor and acceptor selectivity. Acceptor substrates occupy positive subsites (+1, +2) and peripheral binding pockets on one side of the active site, whereas donor substrates—mainly linear malto-oligosaccharides, starch polymers, and CDs—bind across negative subsites (−7 to −1) on the other. At these subsites, specialized hydrogen bonding networks, hydrophobic interactions, and steric complementarity mediated by important amino acid residues control the selectivity for both donors and acceptors.

The type and availability of acceptor molecules determine the fate of the glycosyl-enzyme intermediate and, consequently, the type of reaction.

Factors affecting cyclodextrin glycosyltransferase-catalysed reactions

Effects of pH on transglycosylation reaction

A variety of environmental and biochemical factors play a crucial role in the effective catalytic functioning of CGTase. One of the most critical parameters is the pH range of the buffer, since extreme acidity or alkalinity can denature the enzyme. Most CGTases exhibit optimal activity at slightly alkaline conditions (pH 6.0–8.5), although this may vary by microbial source. For instance, CGTase isolated from alkaliphilic, halophilic, mesophilic, and thermophilic *Bacilli* exhibits optimal activity at pH 6.0–7.5 (Abelyan et al., 2004). Likewise, the optimum pH for transglycosylation of the piceid catalyzed by CGTase from *Bacillus macerans* is in the range of 5.5–7.0, but when the reaction was carried out at a lower pH (4.5–5) (Makino et al., 2009), the conversion yield was reduced. Variation in the optimal pH can reduce enzyme-substrate binding efficiency and transition-state stabilisation, ultimately lowering catalytic turnover.

Effects of temperature on transglycosylation reaction

Temperature is another crucial factor in the enzymatic activity. The temperature used in most studies on CGTase has been in the range of 40°C–60°C (Choung et al., 2017; Rather et al., 2015), but enzymes derived from different microorganisms may exhibit their optimal activity at varying temperatures. For instance, CGTase extracted from mesophilic and halophilic strains shows the optimum temperature in the 45°C–50°C range, while the CGTase isolated from alkalophilic and thermophilic strains is 50°C–55°C (Abelyan et al., 2004).

Effects of metal ions on transglycosylation reaction

The production yield of the transglycosylation reaction by CGTase is also affected by the addition of metal ions. This is because some metal ions, such as calcium ions, can help stabilize enzymes. For example, calcium ions can influence the enzyme's activity and stability. CGTase enzymes have two Ca^{2+} binding sites at their N-terminal end and the active site region, which acts as a protective agent for CGTase against heat (Li, Ban et al., 2013). Studies have reported an increase of up to 123% in β-CGTase transglycosylation yield in the presence of Ca^{2+} (Sugimoto et al., 2007). Other metal ions, like Mn^{2+}, have also exhibited a positive effect on CGTase transglycosylation—alkalophilic *Bacillus sp.* BL-31 showed an increased yield of glycosylated naringin from 80.2% to 92.1% in the presence of manganese ions (Go et al., 2007). However, not all metal enzymes positively affect production; metal ions such as Zn^{2+}, Cu^{2+}, and Fe^{2+} are inhibitors of most CGTase (Tonkova, 1998).

Effects of cyclodextrin glycosyltransferase concentration on transglycosylation reaction

The amount of CGTase in the reaction mixture and the enzyme source also affect the transglycosylation activity of the enzyme. Generally, the initial reaction rate of transglycosylation is directly proportional to the amount of CGTase added into the reaction mixture, but the catalytic reaction would perform unwanted side reactions, such as the formation of extra CDs, if an excessive amount of CGTase is used (Go et al., 2007; Li et al., 2013).

Sources of cyclodextrin glycosyltransferase

The microorganism from which the CGTase has been isolated can also affect the conversion yield. Abelyan et al. (2004) in their study used CGTase extracted from alkaliophilic (*B. alcalophilus*), halophilic (*B. halophilus*), mesophilic (*B. macerans* and *B. circulans*), and thermophilic (*B. stearothermophilus*) *bacilli* to synthesise stevioside glycoside. The productivity of the enzyme in stevioside glycoside formation was in ascending order with CGTase from *B. stearothermophilus* and *B. macerans* producing 8–9 U/g and 11–11.5 U/g, respectively, *B. circulans* producing 15–16 U/g, *B. alcalophilus* and *B. halophilus* producing 15–16 U/g (Jaitak et al., 2009). Other studies have shown that CGTase from *Thermoanaerobacter sp.* produces a higher yield of resveratrol glucoside when compared with *B. macerans* (Torres et al., 2011). This might be due to the higher thermostability of CGTase isolated from *Thermoanaerobacter sp.*, which exhibits high disproportionation activity (Martin et al., 2004). Moreover, studies on the elongation of carbohydrate groups in alkyl glycosides found that CGTase from *Paenibacillus macerans* yields the highest β-dodecyl maltooctaoside concentration compared with the CGTase from *Paenibacillus thermoanaerobacter sp., Carboxydocella sp.,* and *B. stearothermophilus* (Rather et al., 2015).

Approaches for increasing cyclodextrin glycosyltransferase efficacy

Apart from temperature, pH, the presence of metal ions, and the concentration of substrate and enzyme, the effect of organic solvents and several protein modification techniques, including immobilisation and chemical modification, have also been used to improve the transglycosylation activity of CGTase.

Immobilization

Immobilisation of CGTase can improve its transglycosylation activity by enhancing the enzyme's stability and shortening the purification process, as the enzyme is not mixed with the final product. Other advantages of this method include the reusability of enzymes and the simple procedure for enzyme immobilisation in insoluble matrices (Datta et al., 2013). Adsorption, covalent binding, and entrapment are some of the different techniques applied for CGTase immobilisation. Supports used for immobilisation of CGTase through the absorption method are purified sand (Iyer et al., 2003) and synthetic adsorption resin (DIAION® HP-20) (Kato & Horikoshi, 1984). CGTase has been covalently immobilised on porous glass (Steighardt & Kleine, 1993), polyvinyl chloride (Abdel-Naby, 1999), polyethylene film (Choi et al., 2002), alumina (Prousoontorn & Pantatan, 2007), and chitosan (Rashid et al., 2002). Meanwhile, alginate, polyacrylamide (Lee et al., 2007), and vegetable sponge natural supports (Fenelon et al., 2018) have been used in the entrapment immobilisation technique.

Immobilised CGTase from *Paenibacillus* sp. A11 used for transglycosylation of L-ascorbic acid showed 20% higher thermal stability than free CGTase and can be stored for a longer period of up to two months (Prousoontorn & Pantatan, 2007). Furthermore, xylitol produced by CGTase from *Thermoanaerobacter* sp. immobilised on Amberlite® IRA-900 showed longer half-life and higher optimal temperature (by 20°C) (Kim et al., 2000). In addition, 2-O-α-glucopyranosyl L-ascorbic acid produced by α-CGTase immobilised on sodium alginate has shown a higher optimal temperature (45°C) and twofold yield compared with free α-CGTase (Zhang et al., 2011). Some other products synthesised by CGTase immobilisation are xylitol (Kim et al., 2000), maltooligosyl fructofuranosides (Martin et al., 2004), 2-O-α-glucopyranosyl L-ascorbic acid (Prousoontorn & Pantatan, 2007; Zhang et al., 2011), and alkyl glycosides (Svensson & Adlercreutz, 2011).

Site-directed mutagenesis

An alternative method for improving enzyme efficacy is protein engineering, which has been studied using approaches such as directed evolution, site-directed mutagenesis, and combinations of X-ray crystallography with site-directed mutagenesis (Leemhuis et al., 2003; Wind et al., 1998). The site-directed mutagenesis has been used for the replacement of both aromatic amino acids, like Tyr 195 in CGTase from *B. circulans* Strain 251 (Penninga et al., 1995) and Tyr 188 in CGTase from *Bacillus ohbensis* (Sin et al., 1994), as well as of nonaromatic amino acids like lysine 47 in CGTase from *P. macerans* strain JFB05-01 (Li et al., 2009) and Ala 223 and Gly 255 in CGTase from *Bacillus clarkia* 7364 (Nakagawa et al., 2006). Protein engineering of CGTase is mainly used to improve specificity in CD production, such as enhanced production of α-CD.

Donor Specificity

The conversion yield of transglycosylated products is the most important factor in the industry, and donor specificity also plays a role. Mathew et al. (Mathew & Adlercreutz, 2013) had used α-CD as a glycosyl donor and the CGTase extracted from *B. macerans* to report 93% conversion yield. Similarly, Li et al. (2013) reported the use of CGTae from *B. macerans* to transfer the glycosyl group from starch, thereby synthesizing diglucosylated stevioside with a 76% conversion yield. Svensson et al. (2009) also reported an 80% conversion yield of dodecyl-β-D-maltoside to dodecyl-β-D-maltooctaoside using CGTase and α-CD as a donor.

Common donors used by CGTase are starch, maltose, dextrin, maltodextrin, and CDs. Since CGTase degrades starch, its donors are generally carbohydrate sources. The effectiveness of donor substrates used for glycoside synthesis in decreasing order is- α-CD > maltodextrin > maltotriose > maltose. Using CDs as donors has the advantage of enhancing acceptor solubility (Yadav et al., 2013), but the high cost makes them less attractive than cheaper alternatives such as glucose, sucrose, and maltose. It has been reported that the conversion yield of trans-glycosylated products can be improved by increasing the donor concentration. Maltodextrin, when used in high concentration, can be an efficient alternative to α-CD (Mathew & Adlercreutz 2013).

Donor substrates for CGTase require at least two glucose units; therefore, glucose cannot act as a donor (Mathew & Adlercreutz 2013). Furthermore, the donor's chain length and concentration also influence the glycosides' chain length. For instance, long-chain donors like starch can perform better than low molecular weight carbohydrates such as maltose, because high concentrations of starch would result in glycosides with longer chain length, while donors such as maltose would form shorter glycosides (Mathew & Adlercreutz, 2012; Mathew & Adlercreutz, 2013; Lim et al., 2021).

Effects of organic reagents

The addition of small amounts of organic solvents increases the production yield of the transglycosylation reaction carried out by CGTase by dissolving the hydrophobic acceptor, like

polyphenols, in water-miscible organic solvents or co-solvents, such as dimethyl sulfoxide (DMSO), before being added to the buffer solution. (Blackwood & Bucke, 2000). To ensure a good level of solubility of the hydrophobic acceptor, the buffer-to-organic solvent ratio must be adjusted so that the organic solvent does not exert denaturing or inhibitory effects on the enzyme. DMSO, ethanol, isopropanol, n-propanol, sulfolane, dioxane, glycerol, dimethylformamide, and ethyl lactate are a few examples of organic solvents or cosolvents used to achieve better solubility for hydrophobic acceptors. DMSO mixed with buffers exhibits the best performance in terms of solubility of the hydrophobic acceptors and product's yield as compared with other cosolvents (Torres et al., 2011).

Conclusion

CGTases represent one of the most versatile enzyme systems within the glycoside hydrolase family 13, owing to their unique ability to catalyze multiple reactions: cyclization, coupling, disproportionation, and limited hydrolysis through a unified α-retaining, double-displacement mechanism. Their catalytic efficiency is rooted in the highly conserved residues Asp229, Glu257, and Asp328, which orchestrate nucleophilic attack, acid–base chemistry, and transition-state stabilization. Structural studies have revealed an intricate arrangement of substrate-binding subsites that precisely position donor and acceptor molecules, enabling CGTases to transform linear starch chains into valuable cyclic or modified oligosaccharides with remarkable control over product specificity.

The interplay among enzyme architecture, substrate characteristics, and reaction conditions determines the distribution of α-, β-, and γ-CDs and the efficiency of transglycosylation. Factors such as pH, temperature, metal ions, enzyme concentration, and the molecular structure of donors and acceptors significantly influence catalytic outcomes. Advances in protein engineering—including site-directed mutagenesis, domain modifications, and directed evolution have further expanded the functional diversity of CGTases, enabling tailored product profiles, enhanced stability, and improved industrial applicability.

Immobilization strategies have emerged as powerful tools to increase enzyme reusability, operational stability, and overall catalytic yield, making CGTases more economically viable for large-scale applications. Collectively, these developments underline the importance of CGTases as robust biocatalysts in food, pharmaceutical, cosmetic, and chemical industries. Continued integration of structural biology, computational modelling, and enzyme engineering is expected to drive the next generation of CGTase-based bioprocesses, widening their utility in specialized glycoside synthesis and sustainable biotechnological production.

References

Abdel-Naby, M. A. (1999). Immobilization of *Paenibacillus macerans* NRRL B-3186 cyclodextrin glucosyltransferase and properties of the immobilized enzyme. *Process Biochemistry, 34*(4), 399–405.

Abelyan, V. A., Balayan, A. M., Ghochikyan, V. T., & Markosyan, A. A. (2004). Transglycosylation of stevioside by cyclodextrin glucanotransferases of various groups of microorganisms. *Applied Biochemistry and Microbiology, 40*, 129–134.

Benavent-Gil, Y., Benavent-Gil, Y., Rosell, C. M., Gilbert, E. P., & Gilbert, E. P. (2021). Understanding CGTase action through the relationship between starch structure and cyclodextrin formation. *Food Hydrocolloids, 112*, 106316. https://doi.org/10.1016/j.foodhyd.2020.106316.

Bender, H. (1982). Enzymology of the cyclodextrins. In *Proceedings of the First International Symposium on Cyclodextrins: Budapest, Hungary, 30 September–2 October, 1981* (pp. 77–87). Dordrecht: Springer Netherlands.

Bender, H. (1990). Studies of the mechanism of the cyclisation reaction catalysed by the wildtype and a truncated α-cyclodextrin glycosyltransferase from *Klebsiella pneumoniae* strain M 5 al, and the β-cyclodextrin glycosyltransferase from *Bacillus circulans* strain 8. *Carbohydrate Research, 206*(2), 257–267.

Blackwood, A. D., & Bucke, C. (2000). Addition of polar organic solvents can improve the product selectivity of cyclodextrin glycosyltransferase: Solvent effects on cgtase. *Enzyme and Microbial Technology, 27*(9), 704–708.

Blake, C. C. F., Johnson, L. N., Mair, G. A., North, A. C. T., Phillips, D. C., & Sarma, V. R. (1967). Crystallographic studies of the activity of hen egg-white lysozyme. *Proceedings of the Royal Society of London. Series B. Biological Sciences, 167*(1009), 378–388.

Braun, C., Brayer, G. D., & Withers, S. G. (1995). Mechanism-based inhibition of yeast α-glucosidase and human pancreatic α-amylase by a new class of inhibitors: 2-deoxy-2, 2-difluoro-α-glycosides. *Journal of Biological Chemistry, 270*(45), 26778–26781.

Braun, C., Lindhorst, T., Madsen, N. B., & Withers, S. G. (1996). Identification of Asp 549 as the catalytic nucleophile of glycogen-debranching enzyme via trapping of the glycosyl– enzyme intermediate. *Biochemistry, 35*(17), 5458–5463.

Chen, S., Li, Z., Gu, Z., Hong, Y., Cheng, L., & Li, C. (2019). Variants at position 603 of the CGTase from *Bacillus circulans* STB01 for reducing product inhibition. *International Journal of Biological Macromolecules, 136*, 460–468.

Choi, S. H., Kim, M. S., Ryoo, J. J., Lee, K. P., Shin, H. D., Kim, S. H., & Lee, Y. H. (2002). Immobilization of a cyclodextrin glucanotransferase (CGTase) onto polyethylene film with a carboxylic acid group and production of cyclodextrins from corn starch using CGTase-immobilized PE film. *Journal of Applied Polymer Science, 85*(11), 2451–2457.

Choung, W. J., Hwang, S. H., Ko, D. S., Kim, S. B., Kim, S. H., Jeon, S. H., Choi, H. D., Lim, S. S., & Shim, J. H. (2017). Enzymatic synthesis of a novel kaempferol-3-O-β-D-glucopyranosyl-(1→ 4)-O-α-D-glucopyranoside using cyclodextrin glucanotransferase and its inhibitory effects on aldose reductase, inflammation, and oxidative stress. *Journal of Agricultural and Food Chemistry, 65*(13), 2760–2767.

Datta, S., Christena, L. R., & Rajaram, Y. R. S. (2013). Enzyme immobilization: An overview on techniques and support materials. *3 Biotechnology, 3*(1), 1–9.

Fenelon, V. C., Miyoshi, J. H., Mangolim, C. S., Noce, A. S., Koga, L. N., & Matioli, G. (2018). Different strategies for cyclodextrin production: Ultrafiltration systems, CGTase immobilization and use of a complexing agent. *Carbohydrate polymers, 192*, 19–27.

Fersht, A. (1999). *Structure and mechanism in protein science: A guide to enzyme catalysis and protein folding*. Macmillan.

Go, Y. H., Kim, T. K., Lee, K. W., & Lee, Y. H. (2007). Functional characteristics of cyclodextrin glucanotransferase from alkalophilic *Bacillus* sp. BL-31 highly specific for intermolecular transglycosylation of bioflavonoids. *Journal of Microbiology and Biotechnology, 17*(9), 1550–1553.

Han, R., Ni, J., Zhou, J., Liu, L., Huang, Y., & Chen, G. (2020). Engineering of cyclodextrin glycosyltransferase reveals pH-regulated mechanism of enhanced long-chain glycosylated sophoricoside specificity. *Applied and Environmental Microbiology, 86*(7), e00004–e00020.

Heightman, T. D., & Vasella, A. T. (1999). Recent insights into inhibition, structure, and mechanism of configuration-retaining glycosidases. *Angewandte Chemie International Edition, 38*(6), 750–770.

Henrissat, B. (1991). A classification of glycosyl hydrolases based on amino acid sequence similarities. *Biochemical journal, 280*(2), 309–316.

Iyer, J. L., Shetty, P., & Pai, J. S. (2003) Immobilisation of cyclodextrin glucanotransferase from *Bacillus circulans* ATCC 21783 on purified seasand. *Journal of Industrial Microbiology and Biotechnology, 30*(1), 47–51.

Jaitak, V., Kaul, V. K., Bandna, Kumar, N., Singh, B., Savergave, L. S., Jogdand, V. V., & Nene, S. (2009). Simple and efficient enzymatic transglycosylation of stevioside by β-cyclodextrin glucanotransferase from *Bacillus firmus*. *Biotechnology Letters, 31*, 1415–1420.

Kato, T., & Horikoshi, K. (1984). Immobilized cyclomaltodextrin glucanotransferase of an alkalophilic *Bacillus* sp. No. 38-2. *Biotechnology and Bioengineering, 26*(6), 595–598.

Kelly, R. M., Dijkhuizen, L., & Leemhuis, H. (2009). The evolution of cyclodextrin glucanotransferase product specificity. *Applied Microbiology and Biotechnology, 84*(1), 119–133.

Kim, P. S., Shin, H. D., Park, J. K., & Lee, Y. H. (2000). Immobilization of cyclodextrin glucanotransferase on amberlite IRA-900 for biosynthesis of transglycosylated xylitol. *Biotechnology and Bioprocess Engineering, 5*(3), 174–180.

Kirby, A. J. (1987). Mechanism and stereoelectronic effects in the lysozyme reaction. *Critical Reviews in Biochemistry, 22*(4), 283–315.

Kirby, A. J. (1996). Illuminating the ancient retainer. *Nature Structural Biology, 3*(2), 107–108.

Klein, C., Hollender, J., Bender, H., & Schulz, G. E. (1992). Catalytic center of cyclodextrin glycosyltransferase derived from X-ray structure analysis combined with site-directed mutagenesis. *Biochemistry, 31*(37), 8740–8746.

Koshland, D. E., Jr (1953). Stereochemistry and the mechanism of enzymatic reactions. *Biological rEviews, 28*(4), 416–436.

Kuriki, T., & Imanaka, T. (1999). The concept of the α-amylase family: Structural similarity and common catalytic mechanism. *Journal of Bioscience and Bioengineering, 87*(5), 557–565.

Lawson, C. L., van Montfort, R., Strokopytov, B., Rozeboom, H. J., Kalk, K. H., de Vries, G. E., Penninga, D., Dijkhuizen, L., & Dijkstra, B. W. (1994). Nucleotide sequence and X-ray structure of cyclodextrin glycosyltransferase from *Bacillus circulans* strain 251 in a maltose-dependent crystal form. *Journal of Molecular Biology, 236*(2), 590–600.

Lee, M. H., Yang, S. J., Kim, J. W., Lee, H. S., Kim, J. W., & Park, K. H. (2007). Characterization of a thermostable cyclodextrin glucanotransferase from *Pyrococcus furiosus* DSM3638. *Extremophiles: Life Under Extreme Conditions, 11*(3), 537–541.

Leemhuis, H., & Dijkhuizen, L. (2003). Hydrolysis and transglycosylation reaction specificity of cyclodextrin glycosyltransferases. *Journal of Applied Glycoscience, 50*(2), 263–271.

Leemhuis, H., Kelly, R. M., & Dijkhuizen, L. (2009). Directed evolution of enzymes: Library screening strategies. *IUBMB Life, 61*(3), 222–228.

Leemhuis, H., Kelly, R. M., & Dijkhuizen, L. (2010). Engineering of cyclodextrin glucanotransferases and the impact for biotechnological applications. *Applied Microbiology and Biotechnology, 85*, 823–835.

Leemhuis, H., Rozeboom, H. J., Wilbrink, M., Euverink, G. J. W., Dijkstra, B. W., & Dijkhuizen, L. (2003). Conversion of cyclodextrin glycosyltransferase into a starch hydrolase by directed evolution: The role of alanine 230 in acceptor subsite+ 1. *Biochemistry, 42*(24), 7518–7526.

Li, C., Ban, X., Gu, Z., & Li, Z. (2013). Calcium ion contribution to thermostability of cyclodextrin glycosyltransferase is closely related to calcium-binding site CaIII. *Journal of Agricultural and Food Chemistry, 61*(37), 8836–8841.

Li, S., Li, W., Xiao, Q. Y., & Xia, Y. (2013). Transglycosylation of stevioside to improve the edulcorant quality by lower substitution using cornstarch hydrolyzate and CGTase. *Food Chemistry, 138*(2-3), 2064–2069.

Li, Z. F., Zhang, J. Y., Sun, Q., Wang, M., Gu, Z. B., Du, G. C., & Chen, J. (2009). Mutations of lysine 47 in cyclodextrin glycosyltransferase from *Paenibacillus macerans* enhance β-cyclodextrin specificity. *Journal of Agricultural and Food Chemistry, 57*(18), 8386–8391.

Lim, C. H., Rasti, B., Sulistyo, J., & Hamid, M. A. (2021). Comprehensive study on transglycosylation of CGTase from various sources. *Heliyon, 7*(2).

Ly, H. D., & Withers, S. G. (1999). Mutagenesis of glycosidases. *Annual Review of Biochemistry, 68*(1), 487–522.

Makino, T., Shimizu, R., Kanemaru, M., Suzuki, Y., Moriwaki, M., & Mizukami, H. (2009). Enzymatically modified isoquercitrin, α-oligoglucosyl quercetin 3-O-glucoside, is absorbed more easily than other quercetin glycosides or aglycone after oral administration in rats. *Biological and Pharmaceutical Bulletin, 32*(12), 2034–2040.

Martin, M. T., Cruces, M. A., Alcalde, M., Plou, F. J., Bernabé, M., & Ballesteros, A. (2004). Synthesis of maltooligosyl fructofuranosides catalyzed by immobilized cyclodextrin glucosyltransferase using starch as donor. *Tetrahedron, 60*(3), 529–534.

Mathew, S., & Adlercreutz, P. (2012). Cyclodextrin glucanotransferase (CGTase) catalyzed synthesis of dodecyl glucooligosides by transglycosylation using α-cyclodextrin or starch. *Carbohydrate polymers, 87*(1), 574–580.

Mathew, S., & Adlercreutz, P. (2013). Regioselective glycosylation of hydroquinone to α-arbutin by cyclodextrin glucanotransferase from *Thermoanaerobacter* sp. *Biochemical Engineering Journal, 79*, 187–193.

Matsuura, Y., Kusunoki, M., Harada, W., & Kakudo, M. (1984). Structure and possible catalytic residues of taka-amylase A. *The Journal of Biochemistry, 95*(3), 697–702.

McCarter, J. D., & Withers, G. S. (1994). Mechanisms of enzymatic glycoside hydrolysis. *Current Opinion in Structural Biology, 4*(6), 885–892.

McCarter, J. D., & Withers, S. G. (1996a). Unequivocal identification of Asp-214 as the catalytic nucleophile of *Saccharomyces cerevisiae* α-glucosidase using 5-fluoro glycosyl fluorides. *Journal of Biological Chemistry, 271*(12), 6889–6894.

McCarter, J. D., & Withers, S. G. (1996b). 5-Fluoro glycosides: A new class of mechanism-based inhibitors of both α-and β-glucosidases. *Journal of the American Chemical Society, 118*(1), 241–242.

McCarter, J. D., Adam, M. J., & Withers, S. G. (1992). Binding energy and catalysis. Fluorinated and deoxygenated glycosides as mechanistic probes of *Escherichia coli* (lacZ) β-galactosidase. *Biochemical Journal, 286*(3), 721–727.

Mooser, G., Hefta, S. A., Paxton, R. J., Shively, J. E., & Lee, T. D. (1991). Isolation and sequence of an active-site peptide containing a catalytic aspartic acid from two *Streptococcus sobrinus* alpha-glucosyltransferases. *Journal of Biological Chemistry, 266*(14), 8916–8922.

Mosi, R., He, S., Uitdehaag, J., Dijkstra, B. W., & Withers, S. G. (1997). Trapping and characterization of the reaction intermediate in cyclodextrin glycosyltransferase by use of activated substrates and a mutant enzyme. *Biochemistry, 36*(32), 9927–9934.

Mosi, R., Sham, H., Uitdehaag, J. C., Ruiterkamp, R., Dijkstra, B. W., & Withers, S. G. (1998). Reassessment of acarbose as a transition state analogue inhibitor of cyclodextrin glycosyltransferase. *Biochemistry, 37*(49), 17192–17198.

Nakagawa, Y., Takada, M., Ogawa, K., Hatada, Y., & Horikoshi, K. (2006). Site-directed mutations in alanine 223 and glycine 255 in the acceptor site of γ-cyclodextrin glucanotransferase from alkalophilic *Bacillus clarkii* 7364 affect cyclodextrin production. *Journal of Biochemistry, 140*(3), 329–336.

Nakamura, A., Haga, K., & Yamane, K. (1994a). Four aromatic residues in the active center of cyclodextrin glucanotransferase from alkalophilic *Bacillus* sp. 1011: effects of replacements on substrate binding and cyclization characteristics. *Biochemistry, 33*(33), 9929–9936.

Nakamura, A., Haga, K., & Yamane, K. (1994b). The transglycosylation reaction of cyclodextrin glucanotransferase is operated by a ping-pong mechanism. *FEBS Letters, 337*(1), 66–70.

Parsiegla, G., Schmidt, A. K., & Schulz, G. E. (1998). Substrate binding to a cyclodextrin glycosyltransferase and mutations increasing the γ-cyclodextrin production. *European Journal of Biochemistry, 255*(3), 710–717.

Penninga, D., Strokopytov, B., Rozeboom, H. J., Lawson, C. L., Dijkstra, B. W., Bergsma, J., & Dijkhuizen, L. (1995). Site-directed mutations in tyrosine 195 of cyclodextrin glycosyltransferase from *Bacillus circulans* strain 251 affect activity and product specificity. *Biochemistry, 34*(10), 3368–3376.

Penninga, D., van der Veen, B. A., Knegtel, R. M., van Hijum, S. A., Rozeboom, H. J., Kalk, K. H., Dijkstra, B. W., & Dijkhuizen, L. (1996). The raw starch binding domain of cyclodextrin glycosyltransferase from *Bacillus circulans* strain 251. *Journal of Biological Chemistry, 271*(51), 32777–32784.

Prousoontorn, M. H., & Pantatan, S. (2007). Production of 2-O-α-glucopyranosyl L-ascorbic acid from ascorbic acid and β-cyclodextrin using immobilized cyclodextrin glycosyltransferase. *Journal of Inclusion Phenomena and Macrocyclic Chemistry, 57*(1), 39–46.

Qi, Q., & Zimmermann, W. (2005). Cyclodextrin glucanotransferase: From gene to applications. *Applied Microbiology and Biotechnology, 66*, 475–485.

Rashid, N., Cornista, J., Ezaki, S., Fukui, T., Atomi, H., & Imanaka, T. (2002). Characterization of an archaeal cyclodextrin glucanotransferase with a novel C-terminal domain. *Journal of Bacteriology, 184*(3), 777–784.

Rather, M. Y., Ara, K. Z. G., Karlsson, E. N., & Adlercreutz, P. (2015). Characterization of cyclodextrin glycosyltransferases (CGTases) and their application for synthesis of alkyl glycosides with oligomeric head group. *Process Biochemistry, 50*(5), 722–728.

Ribeiro, A. J., Tyzack, J. D., Borkakoti, N., Holliday, G. L., & Thornton, J. M. (2020). A global analysis of function and conservation of catalytic residues in enzymes. *Journal of Biological Chemistry, 295*(2), 314–324.

Saini, K., Pathak, V. M., Tyagi, A., & Gupta, R. (2022). Microbial cyclodextrin glycosyltransferases: Sources, production, and application in cyclodextrin synthesis. *Catalysis Research, 2*(3), 1.

Seddon, N. R. (1988). Biochemistry by Lubert Stryer. pp 1136. WH Freeman, New York. 1988.£ 39.95 or£ 24.95 (pbk) ISBN 0-7167-1843-X or-1920-7.

Sin, K. A., Nakamura, A., Masaki, H., Matsuura, Y., & Uozumi, T. (1994). Replacement of an amino acid residue of cyclodextrin glucanotransferase of *Bacillus ohbensis* doubles the production of γ-cyclodextrin. *Journal of Biotechnology, 32*(3), 283–288.

Sinnott, M. L. (1990). Catalytic mechanism of enzymic glycosyl transfer. *Chemical Reviews, 90*(7), 1171–1202.

Steighardt, J., & Kleine, R. (1993). Production and immobilization of a proteinase-reduced cyclodextrin glycosyltransferase preparation. *Applied Microbiology and Biotechnology, 39*(1), 63–68.

Strokopytov, B., Knegtel, R. M., Penninga, D., Rozeboom, H. J., Kalk, K. H., Dijkhuizen, L., & Dijkstra, B. W. (1996). Structure of cyclodextrin glycosyltransferase complexed with a maltononaose inhibitor at 2.6 Å resolution. Implications for product specificity. *Biochemistry, 35*(13), 4241–4249.

Sugimoto, K., Nishimura, T., & Kuriki, T. (2007). Development of alpha-arbutin: Production at industrial scale and application for a skin-lightening cosmetic ingredient. *Trends in Glycoscience and Glycotechnology, 19*(110), 235.

Svensson, B. (1994). Protein engineering in the α-amylase family: Catalytic mechanism, substrate specificity, and stability. *Plant Molecular Biology, 25*, 141–157.

Svensson, D., & Adlercreutz, P. (2011). Immobilisation of CGTase for continuous production of long-carbohydrate-chain alkyl glycosides: Control of product distribution by flow rate adjustment. *Journal of Molecular Catalysis B: Enzymatic, 69*(3-4), 147–153.

Svensson, D., Ulvenlund, S., & Adlercreutz, P. (2009). Efficient synthesis of a long carbohydrate chain alkyl glycoside catalyzed by cyclodextrin glycosyltransferase (CGTase). *Biotechnology and Bioengineering, 104*(5), 854–861.

Tanaka, Y., Tao, W., Blanchard, J. S., & Hehre, E. J. (1994). Transition state structures for the hydrolysis of alpha-D-glucopyranosyl fluoride by retaining and inverting reactions of glycosylases. *Journal of Biological Chemistry, 269*(51), 32306–32312.

Tao, B. Y., Reilly, P. J., & Robyt, J. F. (1989). Detection of a covalent intermediate in the mechanism of action of porcine pancreatic α-amylase by using 13C nuclear magnetic resonance. *Biochimica et Biophysica Acta (BBA)-Protein Structure and Molecular Enzymology, 995*(3), 214–220.

Tonkova, A. (1998). Bacterial cyclodextrin glucanotransferase. *Enzyme and Microbial Technology, 22*(8), 678–686.

Torres, P., Poveda, A., Jimenez-Barbero, J., Parra, J. L., Comelles, F., Ballesteros, A. O., & Plou, F. J. (2011). Enzymatic synthesis of α-glucosides of resveratrol with surfactant activity. *Advanced Synthesis & Catalysis, 353*(7), 1077–1086.

Uitdehaag, J. C. M., Mosi, R., Kalk, K. H., van der Veen, B. A., Dijkhuizen, L., Withers, S. G., & Dijkstra, B. W. (1999a). Catalytic mechanism and product specificity of cyclodextrin glycosyltransferase, a prototypical transglycosylase from the α-amylase family. *Nature Structural Biology, 6*(5), 432–436.

Uitdehaag, J. C. M., van der Veen, B. A., Dijkhuizen, L., & Dijkstra, B. W. (2002). Catalytic mechanism and product specificity of cyclodextrin glycosyltransferase, a prototypical transglycosylase from the α-amylase family. *Enzyme and Microbial Technology, 30*(3), 295–304.

Uitdehaag, J., Mosi, R., Kalk, K. H., van der Veen, B. A., Dijkhuizen, L., Withers, S. G., & Dijkstra, B. W. (1999b). X-ray structures along the reaction pathway of cyclodextrin glycosyltransferase elucidate catalysis in the α-amylase family. *Nature Structural Biology, 6*(5), 432–436.

Uitdehaag, J. C., Kalk, K. H., van der Veen, B. A., Dijkhuizen, L., & Dijkstra, B. W. (1999). The cyclization mechanism of cyclodextrin glycosyltransferase (CGTase) as revealed by a γ-cyclodextrin-CGTase complex at 1.8-Å resolution. *Journal of Biological Chemistry, 274*(49), 34868–34876.

Uitdehaag, J. C., van der Veen, B. A., Dijkhuizen, L., Elber, R., & Dijkstra, B. W. (2001). Enzymatic circularization of a malto-octaose linear chain studied by stochastic reaction path calculations on cyclodextrin glycosyltransferase. *Proteins: Structure, Function, and Bioinformatics, 43*(3), 327–335.

Van Der Maarel, M. J., Van der Veen, B., Uitdehaag, J. C., Leemhuis, H., & Dijkhuizen, L. (2002). Properties and applications of starch-converting enzymes of the α-amylase family. *Journal of Biotechnology, 94*(2), 137–155.

Van der Veen, B. A., Uitdehaag, J. C., Penninga, D., van Alebeek, G. J. W., Smith, L. M., Dijkstra, B. W., & Dijkhuizen, L. (2000). Rational design of cyclodextrin glycosyltransferase from *Bacillus circulans* strain 251 to increase α-cyclodextrin production. *Journal of Molecular Biology, 296*(4), 1027–1038.

Van der Veen, B. A., van Alebeek, G. J. W., Uitdehaag, J. C., Dijkstra, B. W., & Dijkhuizen, L. (2000). The three transglycosylation reactions catalyzed by cyclodextrin glycosyltransferase from *Bacillus circulans* (strain 251) proceed via different kinetic mechanisms. *European Journal of Biochemistry, 267*(3), 658–665.

Vocadlo, D. J., Davies, G. J., Laine, R., & Withers, S. G. (2001). Catalysis by hen egg-white lysozyme proceeds via a covalent intermediate. *Nature, 412*(6849), 835–838.

Wang, L., Xia, Y., Su, L., & Wu, J. (2020). Modification of *Bacillus clarkii* γ-cyclodextrin glycosyltransferase and addition of complexing agents to increase γ-cyclodextrin production. *Journal of Agricultural and Food Chemistry, 68*(43), 12079–12085.

White, A., & Rose, D. R. (1997). Mechanism of catalysis by retaining β-glycosyl hydrolases. *Current Opinion in Structural Biology, 7*(5), 645–651.

Wind, R. D., Uitdehaag, J. C., Buitelaar, R. M., Dijkstra, B. W., & Dijkhuizen, L. (1998). Engineering of cyclodextrin product specificity and pH optima of the thermostable cyclodextrin glycosyltransferase from *Thermoanaerobacterium thermosulfurigenes* EM1. *Journal of Biological Chemistry, 273*(10), 5771–5779.

Yadav, S., Yadava, S., & Yadav, K. D. S. (2013). Purification and characterization of α-L-rhamnosidase from *Penicillium corylopholum* MTCC-2011. *Process Biochemistry, 48*(9), 1348–1354.

Zhang, Z., Li, J., Liu, L., Sun, J., Hua, Z., Du, G., & Chen, J. (2011). Enzymatic transformation of 2-O-α-d-glucopyranosyl-l-ascorbic acid (AA-2G) by immobilized α-cyclodextrin glucanotransferase from recombinant *Escherichia coli. Journal of Molecular Catalysis B: Enzymatic, 68*(3-4), 223–229.

CHAPTER

Cyclodextrin glucanotransferase from *Bacillus* spp.

5

Anjali D. Solanki, Tejalba R. Rathod, Paras C. Mevada, Vikram H. Raval and Kiransinh N. Rajput

Department of Microbiology & Biotechnology, University School of Sciences, Gujarat University, Ahmedabad, Gujarat, India

Introduction

The genus *Bacillus* stands as a cornerstone of industrial biotechnology, particularly in the enzymatic transformation of starch. Renowned for its robust secretory capacity and its ability to thrive in diverse environmental conditions, *Bacillus* species, most notably *B. licheniformis*, *B. subtilis*, and *B. macerans*, are primary sources for high-yield extracellular enzymes (Mehta & Satyanarayana, 2016). While its production of α-amylase is fundamental for the initial liquefaction of starch into shorter dextrins (Park et al., 2017), the genus is equally significant for its production of cyclodextrin (CD) glycosyltransferase (CGTase). Unlike standard amylases that merely hydrolyze glycosidic bonds, CGTase possesses the unique ability to catalyze an intramolecular transglycosylation reaction, converting linear starch chains into cyclic oligosaccharides known as CDs (Saini et al., 2022). These cyclic molecules are invaluable in the pharmaceutical and food industries for their "molecular encapsulation" capabilities—forming inclusion complexes that enhance the solubility and stability of hydrophobic guest molecules—marking *Bacillus* as an indispensable tool for advanced starch modification.

The *Bacillus* genus is widely considered an excellent source of CGTase due to its extraordinary physiological diversity and industrial robustness. These gram-positive, spore-forming microbes are a major source of extracellular CD CGTase belonging to the GH13 family. Many *Bacillus* species, such as *B. circulans, B. firmus, B. sphaericus, B. halophilus*, and *B. megaterium,* are reported to produce all three types of CGTase. This enzyme catalyzes the transglycosylation of starch and related malto-oligosaccharides to produce cyclic CDs via cyclization reaction. CGTase from *Bacillus* accounts for approximately 90% of the industrially relevant producers identified to date. This dominance is primarily attributed to the genus's robust biological characteristics, including its high capacity for extracellular secretion, which simplifies the downstream

Cyclodextrin Glucanotransferase. DOI: https://doi.org/10.1016/B978-0-443-41694-1.00002-2

recovery of enzymes, and its ability to form resilient endospores that facilitate long-term stability in industrial settings (Horikoshi, 1971).

Of particular importance are alkalophilic *Bacillus* species, such as *B. circulans* and *B. macerans*, which thrive in high-pH environments. The CGTases secreted by these organisms often exhibit superior stability and activity across broad pH and temperature ranges compared to those from neutralophilic bacteria, making them ideal for the harsh conditions of industrial starch processing (Jiang et al., 2025). Furthermore, the generally nonpathogenic (generally regarded as safe [GRAS]) status of many *Bacillus* species allows for their widespread application in the production of CDs for the food and pharmaceutical sectors.

CGTase (EC 2.4.1.19) is a multifunctional extracellular enzyme of the α-amylase superfamily (glycoside hydrolase family 13). It catalyzes the cleavage and transfer of glycosidic bonds between carbohydrate molecules, or between carbohydrate and noncarbohydrate moieties. CGTase is also known as CGTase, cyclomaltodextrin glucanotransferase, cyclomaltodextrin glycosyltransferase, or *konchizaimu*, and its systematic name is 1,4-α-D-glucan 4-α-D-(1,4-α-D-glucano)-transferase (Schomburg et al., 2006). A distinctive characteristic of CGTases is their ability to catalyze transglycosylation reactions. These enzymes convert starch and other α-1,4-linked glucans into nonreducing malt oligosaccharides known as CDs through an intramolecular transglycosylation (cyclization) process. CGTases can also open CDs rings and transfer the glucosyl units to linear acceptor substrates, as well as interconvert linear oligosaccharides into products of different chain lengths via intermolecular transglycosylation reactions such as coupling and disproportionation (Pardhi et al., 2023).

CGTase from different sources mainly uses starch as the substrate to catalyze the formation of mixtures in three different ratios, which are classified as α-CD, β-CD, and γ-CD based on the number of sugar units they contain (Liu & Yu, 2025; Saini et al., 2022).

History of cyclodextrin glucanotransferase-producing *Bacillus*

Historically, *B. macerans* is the first and most frequently reported source of CGTase. It is the primary choice for producing α-CD. It typically operates at a neutral pH (around 7.0) and moderate temperatures (37°C–40°C). Unlike many alkalophilic *Bacillus* species, *B. macerans* produces an enzyme with high specificity for the 6 glucose ring (α-CD) (Tonkova, 1998).

First reported crystallized structure of CGTase from *B. circulans* 8 with a resolution of 3.4 Å. *B. circulans* is a cornerstone of the β-CD industry. Most wild-type *Bacillus* strains favor β-CD production, but *B. circulans* (specifically strain ATCC21783) is notable for its high yields and wide pH stability. The enzyme is highly efficient at converting starch into β-CD, often reaching conversion rates of up to without expensive additives. It exhibits a dual pH optimum (often near 6.0 and 8.6), 40%–41%, and is relatively thermostable up to 65°C (Hofmann et al., 1989).

Table 5.1 outlines the historical background of CGTase research. Villiers provided the first recorded observation of these molecules in 1891, which were later identified as products of *B. macerans* by Schardinger in 1911. Tilden and Hudson were the first to successfully isolate the enzyme from *B. macerans*, officially

Table 5.1 Historical milestones in ***Bacillus*** cyclodextrin glucanotransferase (*CGTase*) research.

Era	Focus	Key milestone	Reference(s)
Discovery (1891–1911)	Identification of "Cellulosine"	Villiers first observed cyclic dextrins from *B. amylobacter*. Schardinger later identified *B. macerans* as the primary producer.	Villiers (1891), Schardinger (1911)
Enzymology (1939–70)	Enzyme naming and isolation	Tilden and Hudson successfully isolated the enzyme from *B. macerans* and officially named it CGTase.	Tilden and Hudson (1939)
Industrial expansion (1971–85)	Alkaliphilic strains	Horikoshi discovered alkaliphilic *Bacillus* strains (e.g., *B. alkaliphilus*), which allowed for large-scale, high-pH industrial production.	Horikoshi (1971), Szejtli (1998)
Structural era (1986–99)	Domain characterization	The gene was first cloned in 1986. The first 3D crystal structure (*B. circulans*) was solved in 1991, identifying the five-domain structure.	Binder et al. (1986), Klein and Schulz (1991)
Engineering era (2000–15)	Product specificity	Researchers used site-directed mutagenesis to "shift" *Bacillus* enzymes from producing mixtures to specific α-, β-, or γ-types.	van der Veen, Uitdehaag, et al. (2000), van der Veen, van Alebeek, et al. (2000), Li et al. (2007)
Modern era (2016–26)	Green synthesis and nano-tech	Shift toward using agricultural waste for production and recombinant *B. subtilis* as high-yield "cell factories."	Liu and Yu (2025)

designating it as "CGTase" and establishing it as a unique class of amylolytic enzyme. Work by Horikoshi identified species such as *B. alkaliphilus* that could thrive in high-pH environments, a discovery that directly enabled large-scale industrial production by allowing enzymes to function under the alkaline conditions typical of starch processing. The gene encoding CGTase was first cloned in 1986, and by 1991, the first 3D crystal structure derived from *B. circulans* was solved. This resolution revealed a complex five-domain structure (domains A through E), providing the structural basis necessary to understand how the enzyme binds starch and catalyzes cyclization.

Wild-type *Bacillus* strains typically produce mixtures of CDs—α-, β-, and γ-CDs. Researchers employed site-directed mutagenesis to "shift" these ratios. By altering specific amino acid residues within the binding pocket, it became possible to engineer enzymes that preferentially produce a single ring size, significantly reducing downstream purification costs. Industrial efforts are now focused on utilizing agricultural waste as a sustainable carbon source for fermentation. Furthermore, recombinant *B. subtilis* has emerged as a preferred "cell factory," optimized for the high-yield secretion of tailor-made CGTases. These advancements align with the integration of CGTase products into nanotechnology and green synthesis, where precise molecular encapsulation is essential for stabilizing high-value compounds.

Isolation and screening techniques of cyclodextrin glycosyltransferase producing *Bacillus* strains

CGTase–producing *Bacillus* strains are frequently isolated from soil, especially from alkaline and starch-rich environments. The isolation process typically includes an enrichment step in which soil suspensions are subjected to heat treatment (80°C for 10–15 min) to selectively enrich spore-forming *Bacillus* species, followed by plating on alkaline starch agar media with pH values ranging from 9.0 to 10.5. Bacteria remain the primary producers of CGTase, as *B. macerans* was the first organism identified to synthesize this enzyme. CGTase production has since been reported in a broad spectrum of bacterial groups, including anaerobic thermophiles, aerobic thermophiles, aerobic mesophiles, and aerobic alkalophiles (Lim et al., 2021). CGTase-producing microorganisms are predominantly isolated from diverse environmental sources such as soil, lake water, hot springs, flour-industry wastewater, and decaying potatoes (Freitas et al., 2004; Ibrahim et al., 2012).

Table 5.2 provides a comprehensive overview of the environmental origins of CGTase-producing *Bacillus* species, underscoring the remarkable ecological adaptability of this genus. A majority of the reported strains have been isolated from soil, which remains the most common habitat for CGTase producers due to its rich organic content and the continuous availability of starch and other polysaccharides that support microbial communities capable of CD formation (Goel & Nene, 1995; Kitahata & Okada, 1974; Sivakumar & Shakilabanu, 2013).

Table 5.2 Environmental sources and isolation localities of cyclodextrin glucanotransferase (*CGTase*)-producing *Bacillus* strains.

Bacillus	Source of isolation	References
B. megaterium	Soil	Kitahata and Okada (1974)
B. circulans A110	Soil	Pongsawasdi and Yagisawa (1987)
B. circulans	Wheat bran	Hofmann et al. (1989)
B. lentus	Soil	Sabioni and Park (1992)
B. autolyticus 11149	Soil	Tomita et al. (1993)
B. firmus	Soil	Goel and Nene (1995)
B. stearothermophilus	Soil	Chung et al. (1998)
B. firmus strain *37*	Cassava plantation soil	Matioli et al. (1998)
B. stearothermophilus	Hyperthermal springs	Stefanova et al. (1999)
B. circulans DF 9 R	Rotten potatoes	Rosso et al. (2002)
B. agaradhaerens	Soda lake	Martins and Hatti-Kaul (2002)
B. clarkii (7364)	Alkaline soil (Japan)	Takada et al. (2003)
B. licheniformis	Cassava peels	Bonilha et al. (2006)
B. clausii E16	Crop soil	Alves-Prado et al. (2007)
B. cereus N1	Soil	Ismail et al. (2009)
B. sphaericus strain *41*	Soybean soil culture	Moriwaki, Ferreira, et al. (2009), Moriwaki, Mazzer, et al. (2009)
B. agaradhaerens KSU-A11	Soil and water of the soda lake	Ibrahim et al. (2010)
B. pseudalcaliphilus 8SB	Mineral water and soil	Kitayska et al. (2011)
B. pseudalcaliphilus 20RF	Rice soil	Atanasova et al. (2011)
B. pseudalcaliphilus	Highly alkaline Bulgarian soil	Atanasova et al. (2011)
B. halodurans	Sugarcane field soil	More et al. (2012)
B. lehensis SV1	Soil	Vidya et al. (2012)
B. megaterium	Soil	Sivakumar and Shakilabanu (2013)
B. thuringiensis GU-2	Soil	Goo et al. (2014)
B. sphaericus strain *41*	Soybean soil	Moriwaki, Ferreira, et al. (2009)
B. lehensis G1	Soil	Noor et al. (2014)
B. lehensis	Soil of a biodiversity-rich vegetation	Elbaz et al. (2015)
B. firmus strain 37	Cassava plantation soil	Fenelon et al. (2015)
B. sp. *SM-02*	Soil of the cassava industries wastewater lake	de Araújo Coelho et al. (2016)

(*Continued*)

Table 5.2 Environmental sources and isolation localities of cyclodextrin glucanotransferase (*CGTase*)-producing *Bacillus* strains. (*Continued*)

Bacillus	Source of isolation	References
B. circulans strain *B-65*	Soil	Stanković et al. (2016)
B. amyloliquefaciens	Soil	Sahoo (2017)
B. flexus SV1	Cornfield soil	Reddy et al. (2017)
B. cereus	Beach soil	Kulshreshtha et al. (2020)
B. flexus	Lonar lake	Vinod and More (2016)
B. sp. *PBS1*	Potato rhizosphere	Solanki et al. (2022)
B. paramycoides MZ005634	Soil	Saleh et al. (2023)

Beyond general soils, agro-based and starch-rich environments, including wheat bran, cassava plantation soils, cassava peels, and soybean-cultivated soils, have frequently been identified as important reservoirs of CGTase-producing *Bacillus*, where prolonged exposure to complex carbohydrate substrates likely exerts selective pressure favoring efficient CGTase expression (Bonilha et al., 2006; Hofmann et al., 1989; Matioli et al., 1998; Moriwaki, Ferreira, et al., 2009; Moriwaki, Mazzer, et al., 2009). Notably, several CGTase-producing strains have been recovered from extreme and alkaline environments such as soda lakes, highly alkaline soils, hyperthermal springs, and alkaline lakes like Lonar Lake, indicating the potential of these microorganisms to produce enzymes with inherent tolerance to high pH and elevated temperatures, traits that are highly desirable for industrial CD production processes (Atanasova et al., 2011; Martins & Hatti-Kaul, 2002; Stefanova et al., 1999; Vinod & More, 2016). In addition, the isolation of CGTase producers from less conventional habitats, including beach soils and industrial or wastewater-affected environments, further highlights the metabolic versatility and resilience of *Bacillus* species under diverse ecological conditions (de Araújo Coelho et al., 2016; Kulshreshtha et al., 2020). Collectively, the diversity of isolation sources summarized in Table 5.2 emphasizes the continued importance of extensive environmental screening as a strategy for discovering novel CGTase-producing *Bacillus* strains with improved catalytic efficiency, stability, and suitability for industrial applications.

Screening methods for cyclodextrin glucanotransferase

The primary challenge in screening is distinguishing CGTase activity from general α-amylase activity. While both degrade starch, only CGTase produces CDs.

Direct screening based on starch hydrolysis is a primary method used to identify potential CD-producing bacteria by cultivating isolates on starch-containing agar plates. During incubation, bacteria capable of secreting extracellular starch-degrading enzymes hydrolyze the starch present in the medium, which is visualized after flooding the plates with iodine solution as clear halos around the colonies against a dark background. Colonies showing prominent zones of clearance are selected as efficient starch degraders and subsequently inoculated into liquid media containing starch as the carbon source for further cultivation under controlled conditions. The resulting culture broths are then analyzed for CD production using specific colorimetric assays, allowing confirmation of CGTase activity and differentiation from general amylase producers.

Dye-based screening methods are commonly used because they offer a quick and simple way to detect CD production by microorganisms. These methods are based on the principle that certain indicator dyes change their color or absorbance when they form inclusion complexes with specific types of CDs. For example, under alkaline conditions, phenolphthalein preferentially interacts with γ-cyclodextrin (CD_7), leading to a reduction in absorbance at around 550 nm, which can be easily observed. In a similar way, bromocresol dye forms complexes with δ-cyclodextrin (CD_8) under acidic conditions, while methyl orange shows a distinct color change at neutral pH when it complexes with α-cyclodextrin (CD_6). When these dyes are incorporated into starch-containing agar plates, CD production by CGTase-producing colonies becomes visible as clear or discolored zones around the growth, making the method suitable for rapid visual screening (Qi & Zimmermann, 2005). However, despite their convenience, these assays are not highly specific, as the dyes can also interact with other metabolites or starch degradation products released by the microorganisms. This often results in false-positive reactions, meaning that positive colonies identified by dye-based screening must be further confirmed using more precise and quantitative analytical methods.

To overcome the limitations of dye-based screening, combining these preliminary assays with instrumental analytical techniques provides a much more reliable and accurate approach for detecting CD production. Dye-based methods are useful for rapid initial selection, but instrumental analyses allow clear confirmation and precise quantification of the CDs formed. Capillary electrophoresis has been successfully applied for the separation, detection, and quantification of CDs produced by CGTase-producing microorganisms, offering high resolution, sensitivity, and relatively fast analysis with minimal sample preparation (Larsen et al., 1998). In addition, high-performance anion-exchange chromatography coupled with pulsed amperometric detection is a powerful technique that enables direct analysis of culture supernatants without extensive purification steps. This method allows accurate identification and quantification of the major CDs, namely α-, β-, and γ-CDs, while also being capable of detecting larger-ring CDs with degrees of polymerization ranging from 9 up to 60 (Terada et al., 1997). By integrating dye-based screening with such advanced analytical tools, researchers can efficiently screen large numbers of isolates while ensuring reliable confirmation of CGTase activity and detailed characterization of the CD profiles produced.

Production of cyclodextrin glucanotransferase from *Bacillus* species

CGTase production by *Bacillus* species has been extensively investigated, as these microorganisms are recognized as efficient enzyme producers, particularly under alkaline conditions. Optimization of culture media composition and environmental parameters is essential for improving CGTase production and for understanding the biochemical characteristics of the enzyme.

CGTase is an extracellular enzyme produced by many *Bacillus* spp. and is typically produced either by submerged fermentation (SmF) or solid-state fermentation (SSF). Process choice depends on desired scale, substrate availability (raw starch/agro-waste), downstream needs, and cost (Saini et al., 2022). SmF is the most widely adopted method for laboratory and industrial applications. In SmF, *Bacillus* strains are cultivated in liquid media containing starch or starch-rich substrates under controlled conditions of pH, temperature, aeration, and agitation. The enzyme is secreted extracellularly, facilitating its recovery from the fermentation broth, and maximum CGTase production is generally observed within 18–48 h, depending on the strain and culture conditions. Although SSF using agro-industrial residues has been reported as a cost-effective alternative, SmF remains the preferred approach due to better process control, reproducibility, and scalability (Solanki & Banerjee, 2022).

In general, media used for CGTase production by *Bacillus* species contain a common set of components, including a starch-based carbon source, organic nitrogen sources such as yeast extract, polypeptone, or corn steep liquor, buffering salts like K_2HPO_4, and divalent cations supplied through $MgSO_4{\cdot}7\ H_2O$, with Na_2CO_3 frequently added to maintain alkaline conditions (Moriwaki, Ferreira, et al., 2009; Moriwaki, Mazzer, et al., 2009). While these basic components remain largely consistent across studies, the primary variation among different production media lies in the type of starch substrate employed, such as soluble starch or corn-derived starch, which significantly influences CGTase yield and CD formation. Similarly, Horikoshi Medium II, one of the most widely adopted media for CGTase production, incorporates soluble starch as the principal carbon source along with yeast extract and polypeptone, and is extensively used for alkalophilic *Bacillus* strains. Variations in starch source not only affect enzyme productivity but also influence the pattern of CD formation. Studies using CGTase from *B. macerans* demonstrated that the enzyme initially favors α-CD production, with overall CD yields typically ranging between 35% and 50%, depending on substrate availability and reaction conditions (Rendleman, 1999).

Table 5.3 shows that a wide variety of starch substrates have been employed for CGTase production by *Bacillus* species, including corn starch, sago starch, cassava starch, soluble starch, and potato starch. Corn starch has been frequently used to

Table 5.3 Comparative optimal growth parameters and cyclodextrin glucanotransferase (*CGTase*) production across different starch substrates.

Substrate type	*Bacillus*	Optimal pH	Optimal temp. (°C)	Major type of CGTase	Time (h)	CGTase (U/mL)	References
Corn starch	*B. subtilis* WB800	8	37	β	24	10.40	Crozatti et al. (2023)
	B. megaterium	7.4	37	β	72	2.54	Pishtiyski et al. (2008)
	B. macerans 314	-	37	α	72	2.5	Ismail et al. (1996)
Sago starch	*B. stearothermophilus* HR1	7.54	50	-	24	14.20	Abd Rahman et al. (2004)
	B. lehensis S8	10.7	37	-	-	18.9	Yap PuiWoon et al. (2010)
	Bacillus sp. *TS1-1*	-	37	-	24	84.32	Mahat et al. (2004)
Cassava starch	*B. circulans* DF 9 R	8.3	37	-	48	5.8	Rosso et al. (2002)
Soluble starch	*Bacillus* sp. DSM 2523	10	37	-	12	-	Sanjari and Vahabzadeh (2014)
	Bacillus sp. ND1	8.9	30.55	β	-	8.28	Upadhyay et al. (2019)
	B. autolyticus 11149	6	30	β	24	9.52	Tomita et al. (1993)
	Bacillus. sp. G1	-	37	α	24	-	Sian et al. (2005)
	B. thuringiensis GU-2	8.5	37	γ	72	2.54	Goo et al. (2014)
Potato starch	*Bacillus* sp. SS2	8.8	-	-	50	38.31	Kulshreshtha et al. (2020)
	B. licheniformis	6–9	37	-	12	-	Bonilha et al. (2006)

support CGTase production by *B. subtilis* WB800, *B. megaterium*, and *B. macerans*, although enzyme yield and production time varied significantly among strains (Crozatti et al., 2023; Ismail et al., 1996; Pishtiyski et al., 2008). Sago starch was found to be particularly effective for CGTase production by several *Bacillus* strains, with some reports indicating markedly higher enzyme activities, suggesting efficient substrate utilization (Abd Rahman et al., 2004; Mahat et al., 2004; Yap PuiWoon et al., 2010). Cassava starch and potato starch were also suitable substrates, supporting moderate to high CGTase yields depending on the strain, while soluble starch served as a commonly used reference substrate for controlled fermentation studies (Kulshreshtha et al., 2020; Rosso et al., 2002; Sanjari & Vahabzadeh, 2014).

Production temperature plays a critical role in CGTase synthesis by *Bacillus* species. Most studies reported optimal CGTase production at temperatures around 37°C, which is consistent with the mesophilic nature of many *Bacillus* strains. However, thermophilic strains such as *B. stearothermophilus* HR1 exhibited higher CGTase activity at elevated temperatures up to 50°C, indicating the advantage of thermophilic fermentation for enhanced enzyme stability and productivity (Abd Rahman et al., 2004). These findings demonstrate that optimal production temperature is strain-specific and must be carefully optimized to achieve maximum CGTase yield.

The pH of the fermentation medium significantly influences CGTase production and activity. The table indicates that many *Bacillus* strains produce CGTase optimally under neutral to alkaline conditions, with reported optimal pH values ranging from 6.0 to 10.7. Alkaliphilic strains such as *B. lehensis* S8 showed maximum CGTase production at strongly alkaline pH, highlighting their potential for industrial applications requiring alkaline processing conditions (Yap PuiWoon et al., 2010). In contrast, some strains, including *B. autolyticus* 11149, exhibited optimal CGTase production at near-neutral or slightly acidic pH values (Tomita et al., 1993). Overall, these observations confirm that pH optimization is a crucial parameter for enhancing CGTase production and CD specificity.

Incubation time is an important factor influencing CGTase production by *Bacillus* species, as it determines the growth phase during which maximum enzyme secretion occurs. As shown in Table 5.3, the time required to achieve maximum CGTase production varies considerably among different *Bacillus* strains and substrates, ranging from as short as 12 h to as long as 72 h. Several strains, including *B. subtilis* WB800 and *Bacillus* sp. DSM 2523 reached peak CGTase production within 12–24 h, indicating rapid enzyme synthesis during the early to mid-exponential growth phase (Crozatti et al., 2023; Sanjari & Vahabzadeh, 2014). In contrast, other strains such as *B. megaterium*, *B. macerans*, and *B. thuringiensis* GU-2 required extended incubation periods of up to 72 h to achieve maximum CGTase production, suggesting slower growth kinetics or delayed enzyme expression (Ismail et al., 1996; Pishtiyski et al., 2008). These observations highlight that incubation time is highly strain-dependent and must be optimized in conjunction with substrate type, pH, and temperature to maximize CGTase production efficiency.

The production of CGTase is governed by several nutritional and physicochemical factors that rarely act independently. While the one-factor-at-a-time (OFAT) approach is useful for gaining an initial understanding of individual variables, it does not account for interactions between parameters and often leads to suboptimal outcomes. To address these limitations, statistically based optimization techniques such as Plackett–Burman design and response surface methodology (RSM) have been increasingly adopted in enzyme production studies. These approaches enable simultaneous evaluation of multiple factors, reduce experimental workload, and provide reliable mathematical models to predict optimal production conditions. Consequently, statistical optimization has become an indispensable tool for improving CGTase yield and process efficiency in microbial fermentation systems. Several studies have successfully applied statistical optimization strategies to enhance CGTase production by *Bacillus* species. In a *B. cereus* group strain (SS2), central composite design-based RSM significantly improved CGTase activity by optimizing key process variables, resulting in more than a two-fold increase compared to the unoptimized condition (Kulshreshtha et al., 2020). Likewise, CGTase production by *B. macerans* NCIM 2131 was enhanced through RSM-guided optimization of nutritional components, and subsequent scale-up further increased enzyme yield, demonstrating the industrial relevance of this approach (Dalmotra et al., 2016). Similar improvements were reported when RSM was employed using sago starch and mineral supplements, where close agreement between predicted and experimental values validated the robustness of the statistical model (Abd Rahman et al., 2004). In addition, Plackett–Burman screening followed by the steepest ascent method successfully identified critical variables for CGTase production in *B. megaterium*, leading to substantial enhancement in enzyme activity (Ahmed & El-Refai, 2010). Collectively, these studies establish that the integration of OFAT screening with statistical optimization is a robust and efficient approach for maximizing CGTase production by *Bacillus* species.

Purification and characterization of *Bacillus* cyclodextrin glucanotransferase

CGTase from *Bacillus* species is typically produced under SmF using starch-based alkaline media such as Horikoshi II broth, followed by removal of cells through centrifugation to obtain the crude extracellular enzyme. The cell-free supernatant serves as the primary source of CGTase for downstream purification.

Purification strategies for CGTase generally exploit its strong affinity for starch and CDs, combined with chromatographic techniques. Initial concentration of the crude enzyme is commonly achieved by rotary evaporation or ultrafiltration. Selective adsorption of CGTase onto corn starch in the presence of ammonium sulfate is frequently employed as an efficient primary purification step. The bound enzyme is subsequently eluted using β-CD, yielding a partially purified preparation with significantly enhanced specific activity (Martins & Hatti-Kaul, 2002).

Further purification is accomplished using chromatographic techniques such as gel filtration, hydrophobic interaction chromatography, and ion exchange chromatography. Gel filtration on matrices like Fractogel EMD BioSec enables separation based on molecular size, while hydrophobic interaction chromatography using phenyl-sepharose exploits surface hydrophobicity under high salt conditions. Ion exchange chromatography, particularly with diethylaminoethyl (DEAE)-Sepharose, is widely applied to achieve higher purity by separating CGTase based on charge properties through salt-gradient elution (Savergave et al., 2008).

Protein concentration during purification is routinely estimated using the Bradford method, and enzyme activity is assessed by standard CD-forming assays. The efficiency of purification is evaluated in terms of specific activity, purification fold, and recovery yield. Characterization of purified CGTase typically includes determination of molecular weight by SDS-PAGE, assessment of optimum pH and temperature, evaluation of thermal and pH stability, and analysis of substrate specificity and CD production profiles. These studies collectively provide insights into the catalytic efficiency and industrial applicability of CGTase derived from *Bacillus* species.

Biochemical and kinetic characterization of *Bacillus* cyclodextrin glucanotransferase

At present, it is generally accepted that CGTase has four activities: cyclization, coupling, disproportionation, and hydrolysis. As for the main reaction (cyclization), the enzymes are classified into 3 types: α-CD, β-CD, and γ-CD. CGTase, which produces mainly β-CD, seems to exist widely. The number of γ-CD-producing enzymes is increasing, partly because of γ-CD's high solubility, large cavity, and potential usefulness.

The genus *Bacillus* serves as a prolific biological factory for the production of CGTase, exhibiting a remarkable diversity in functional characteristics and kinetic properties across different species. As shown in Table 5.4, while the majority of these enzymes, such as those from *B. flexus SV1*, *B. lehensis*, and *B. sphaericus*, are primarily beta-CGTases, others like *B. megaterium* and alkalophilic *Bacillus* sp. *7–12* demonstrate the versatile ability to produce varying ratios of α, β, and γ CDs. This diversity in product specificity is economically vital, as it allows for the targeted production of specific ring sizes, such as the highly soluble γ-CDs produced by *B. thuringiensis GU-2* for pharmaceutical use. Structurally, these enzymes range in molecular weight from approximately 59 to 97.4 kDa, typically showing optimal catalytic activity at temperatures around 60°C and within a neutral to alkaline pH range of 6.0 to 9.0. These robust physicochemical profiles,

Table 5.4 Kinetic properties and physicochemical characterization of purified ***Bacillus*** cyclodextrin glucanotransferase (*CGTases*).

Bacteria	Type	K_m	V_{max}	M.W (KD)	Temperature (°C)	pH	References
B. flexus SV1	β-CGTase	0.08976 μmol/mL	585.1 μmol/mL/min	96.68	60	8	Reddy et al. (2017)
B. cereus	β-CGTase	2.613 ± 0.5 mg/mL	0.309 ± 0.05 μg/min	97.4	50	9	Upadhyay et al. (2019)
B. licheniformis	α-CGTase	-	-	70	40	6	Jamil and Man, 2017
B. lehensis	β-CGTase	2.2 ± 0.17 mg/mL	7.8 ± 0.38 mg β-CD/mL/min	74.1	60	7	Elbaz et al. (2015)
B. thuringiensis GU-2	γ-CGTase	-	-	-	37	8.5	Goo et al. (2014)
B. megaterium	α-, β-, and γ-CDs	0.152 mM	13.4 μM/min	66	50	8	Arce-Vázquez et al (2016)
B. megaterium	α/β/γ:27:68:5	3.4 g/L	0.117 μg/min/mL	68.2	-	-	Pishtiyski et al. (2008)
B. aminoliqueficiantes L4–6	β-CGTase	25 μg/mL	144 U/mL	81	60	8	
Bacillus sp.	β-CGTase	4.1 g/L	5.2 μM β-CD/min mL	-	60	7.5	Ramirez Pérez (2014)
Bacillus sp.	β-CGTase	0.15 mg/mL	60.39 mg/mL min	75	60	6	Sian et al. (2005)
B. sphaericus	β-CGTase	0.0008 mol/L	0.0631 mol of β-CD/(L h)	59	65	6	Moriwaki, Ferreira, et al. (2009), Moriwaki, Mazzer, et al. (2009)
B. licheniformis	-	1.77 mg/mL	-	85.2	55	-	Bonilha et al. (2006)
B. autolyticus	β-CGTase	-	-	70	60	5–6	Tomita et al. (1993)
Alkalophilic Bacillus sp. *7–12*	α-, β- and γ-cyclodextrins	-	-	69	60	8.5	Cao et al. (2005)

particularly the thermostability seen in *B. sphaericus*, are significant for industrial "green chemistry" as they enhance starch solubility and prevent microbial contamination during processing. The wide variation in substrate affinity (K_m) and reaction velocity (V_{max}) evidenced by the high-affinity *B. flexus* and the robust *B. lehensis* underscores the significant industrial potential of *Bacillus*-derived CGTases; specifically, low K_m values allow these enzymes to remain highly efficient even at low substrate concentrations, making them ideal candidates for cost-effective, large-scale carbohydrate modification and CD production.

Cyclodextrin glucanotransferase classification by product specificity

α-CGTases are characterized by their ability to preferentially synthesize α-CD, which consists of 6 glucose units. These enzymes are relatively less common in nature compared to β-CGTases, with the most well-known producer being *B. macerans* (Endo et al., 2007). On a molecular level, specific amino acid residues in the active site (Domain A) create a smaller "binding pocket" that favors the cyclization of six glucose units (Leemhuis et al., 2010). In industrial settings, the production of α-CD often requires the use of "complexing agents" like decanol to shift the equilibrium toward the six-membered ring and prevent further conversion into larger rings (Del Valle, 2004).

β-CGTases are the most abundant and extensively studied type of CGTase, predominantly producing β-CD, which consists of 7 glucose units. These enzymes are commonly found in alkaliphilic *Bacillus* species such as *B. circulans*, *B. megaterium*, and *B. ohbensis* (Tonkova, 2006). β-CD is of particularly high interest due to the specific size of its nonpolar cavity, which is suitable for encapsulating several guest molecules, and its low solubility in water, which facilitates its separation from the reaction mixture (Li et al., 2007). Furthermore, β-CD inclusion complexes are easily prepared and more stable than those of other CDs. These enzymes typically exhibit high thermal stability and high conversion rates; for instance, the CGTase from *B. circulans* 251 serves as a model for studying the catalytic mechanism where the distance between the catalytic nucleophile and the subsite +1 is optimized for a 7-unit chain (van der Veen, Uitdehaag, et al., 2000; van der Veen, van Alebeek, et al., 2000).

γ-CGTases specifically produce γ-CD, which contains 8 glucose units, and are considered the rarest and most industrially sought-after due to the high market value of γ-CD. These enzymes are found in specialized strains like *B. clarkii*, *B. subtilis*, and certain alkaliphilic *Bacillus* sp. G1 (Wu et al., 2012). Structurally, γ-CGTases possess a wider active-site cleft, allowing for the accommodation of a longer malto-oligosaccharide chain before cyclization occurs (Szejtli, 1998). However, γ-CD presents unique product challenges; because it is the largest of the three, it is often quickly degraded by the enzyme back into linear sugars through a coupling reaction, making the timing of the reaction critical for achieving a high yield (Moriwaki, Ferreira, et al., 2009; Moriwaki, Mazzer, et al., 2009).

Table 5.5 illustrates the diversity of product specificity among various bacterial CGTases, as detailed in the provided data. These enzymes are capable of synthesizing multiple ring sizes; the ratio of α-, β-, and γ-CD is highly strain-specific. This variability is a critical factor in industrial selection, as the "default" production profile of a strain determines the complexity of downstream purification.

A primary observation from the data is the overwhelming natural preference for β-CD production. Many *Bacillus* strains, such as *B. lentus* (Sabioni & Park, 1992) and *Alkaliphilic B.* sp. *ATCC21783* (Nakamura & Horikoshi, 1976), exhibit a strong bias toward the seven-unit ring. In the case of *B. lentus*, the ratio of α:β:γ-CDs 1:67:1.6 highlights an extreme level of specificity. This dominance suggests that the 7-unit ring is often the most thermodynamically stable and structurally favored product for the majority of wild-type CGTase active sites.

While β-CD is the common standard, certain strains have evolved to prioritize the smaller 6-unit ring. *B. amyloquefaciens AL35* stands out as a highly efficient producer, achieving up to 95% α-CD (Yu et al., 1988). Similarly, *B. macerans IFO*

Table 5.5 Specificity ratios of α, β, and γ-cyclodextrin (*CD*) production by ***Bacillus*** strains.

Origin	CD formation ratio α-CD:β-CD:γ-CD	References
B. megaterium	1:2.1:1	Kitahata and Okada (1974)
B. stearothermophilus	4.2:5.9:1	Chung et al. (1998)
B. circulans E 192	1:7:3	Bovetto et al. (1992)
B. circulans DF9R	1:0.9:0.4	Marechal et al. (1996)
B.amyloliquefaciens AL35	up to 95% α-CD	Yu et al. (1988)
Alkaliphlic B. sp. *ATCC21783*	2:22:3	Nakamura and Horikoshi (1976)
B. coagulans A-147	2:2.6:1	Kaneko et al. (1990)
B. coagulans	8.1:8.9:1	Akimaru et al. (1991)
B. lentus	1:67:1.6	Sabioni and Park (1992)
B. firmus	1:11:5	Yim et al. (1997)
B. autolyticus	10:30:5	Tomita et al. (1993)
B. obhensis	0:4:1	Sato et al. (1985)
B. licheniformis	5:1:0	Aoki et al. (1988)
B. subtilis	Mainly-γ-CD	Starnes (1990)
B. agaradhaerens LS-3C	0.12:1:0	Martins and Hatti-Kaul (2002)
B. macerans IFO 3490	71% α-CD	Fujiwara et al. (1992), Kaneko et al. (1990)

3490 maintains a clear preference for the smaller ring at 71% (Fujiwara et al., 1992; Kaneko et al., 1990). From a structural biology perspective, these enzymes typically feature a more constricted binding pocket that physically limits the glucose chain to six units before the cyclization reaction is triggered.

The table further confirms that γ-CD (the 8-unit ring) remains the most elusive product. In most profiles, it represents the smallest fraction of the total yield. However, specific exceptions like *B. subtilis* (Starnes, 1990) and *B. obhensis* (Sato et al., 1985) demonstrate a shift toward these larger cavities. Interestingly, *B. obhensis* produces no α-CD at all (0:4:1), focusing exclusively on the larger rings. Achieving high yields of γ-CD is a significant engineering challenge because the larger ring is often more susceptible to reopening and degradation by the enzyme itself through coupling reactions.

The data highlights that even within the same species, enzymatic behavior can vary significantly. Comparing *B. circulans E 192* (Bovetto et al., 1992), which favors β-CD at a 1:7:3 ratio, with *B. circulans DF9R* (Marechal et al., 1996), which produces a nearly 1:1 ratio of α to β, underscores the impact of minor genetic variations.

Conclusion

The genus *Bacillus* remains the unrivaled cornerstone of CGTase production, as explored throughout this chapter. The preeminence of *Bacillus* species—ranging from the classic *B. macerans* to specialized alkaliphilic and thermophilic strains like *B. lehensis* and *B. stearothermophilus*—is rooted in their extraordinary secretory capacity, environmental resilience, and the diverse catalytic profiles of their GH13 family enzymes. The industrial significance of these microorganisms is underscored by their ability to transform starch into α-, β-, and γ-CDs with varying degrees of specificity. While the natural tendency of most wild-type strains, such as *B. lentus* and *B. circulans*, is to produce β-CD, the discovery and engineering of strains like *B. amyloquefaciens* AL35 have enabled high-purity α-CD synthesis. Furthermore, the increasing demand for the larger, more soluble γ-CD has shifted focus toward specialized producers like *B. subtilis* and *B. thuringiensis*, despite the inherent catalytic challenges of large-ring synthesis and degradation. From an operational perspective, the transition from traditional SmF to "green" bioprocessing—utilizing agricultural waste and recombinant "cell factories"—marks a new era in CGTase production. The robust kinetic properties of *Bacillus* CGTases, characterized by high thermal stability (up to 65°C–75°C) and broad pH tolerance, ensure their viability in the demanding conditions of industrial starch liquefaction. As we move forward, the synergy between metagenomic screening of extreme environments and site-directed mutagenesis will likely continue to refine these enzymes, moving away from product mixtures toward tailor-made biocatalysts with 100% specificity. The *Bacillus* genus does not merely offer a source of enzymes; it provides a versatile toolkit for advanced carbohydrate modification. Its continued development is essential to meet the growing needs of the pharmaceutical, food, and nanotechnology sectors, where molecular encapsulation remains a critical hurdle to the stability and delivery of high-value molecules.

References

Abd Rahman, R., Illias, R. M., Nawawi, M. G. M., Ismail, A. F., Hassan, O., & Kamaruddin, K. (2004). Optimisation of growth medium for the production of cyclodextrin glucanotransferase from *Bacillus stearothermophilus* HR1 using response surface methodology. *Process Biochemistry, 39*(12), 2053–2060.

Ahmed, E. M., & El-Refai, H. A. (2010). Cyclodextrin glucosyltransferase production by *Bacillus megaterium* NCR: Evaluation and optimization of culture conditions using factorial design. *Indian Journal of Microbiology, 50*(3), 303–308.

Akimaru, K., Yagi, T., & Yamamoto, S. (1991). Purification and properties of *Bacillus coagulans* cyclomaltodextrin glucanotransferase. *Journal of Fermentation and Bioengineering, 71*(5), 322–328.

Alves-Prado, H. F., Bocchini, D. A., Gomes, E., Baida, L. C., Contiero, J., Roberto, I. C., & Da Silva, R. (2007). Optimization of cyclodextrin glucanotransferase production from *Bacillus clausii E16* in submerged fermentation using response surface methodology. *Applied Biochemistry and Biotechnology, 137–140*(1–12), 27–40. https://doi.org/10.1007/s12010-007-9037-3.

Aoki, H., Yu, E. K. C., & Misawa, M. (1988). Novel cyclodextrin glycosyltransferase and its use in manufacturing cyclodextrin. *US Patent,* 87–35952.

Arce-Vázquez, M. B., Ponce-Alquicira, E., Delgado-Fornué, E., Pedroza-Islas, R., Díaz-Godínez, G., & Soriano-Santos, J. C. R. G. E. (2016). Integral use of Amaranth starch to obtain cyclodextrin glycosyltransferase, by *Bacillus megaterium,* to produce β-cyclodextrin. *Frontiers in Microbiology, 7,* 1513.

Atanasova, N., Kitayska, T., Bojadjieva, I., Yankov, D., & Tonkova, A. (2011). A novel cyclodextrin glucanotransferase from alkaliphilic *Bacillus pseudalcaliphilus* 20RF: Purification and properties. *Process Biochemistry, 46*(1), 116–122.

Binder, F., Huber, O., & Böck, A. (1985). Cyclodextrin-glycosyltransferase from *Klebsiella pneumoniae M5a1*: Cloning, nucleotide sequence, and expression. *Gene, 47*(2–3), 269–277. https://doi.org/10.1016/0378-1119(86)90070-3.

Bonilha, P. R. M., Menocci, V., Goulart, A. J., Polizeli, M. D. L. T. D. M., & Monti, R. (2006). Cyclodextrin glycosyltransferase from *Bacillus licheniformis*: Optimization of production and its properties. *Brazilian Journal of Microbiology, 37*(3), 317–323. https://doi.org/10.1590/S1517-83822006000300022.

Bovetto, L., Backer, D., Villette, J., Sicard, P., & Bouquelet, S. (1992). Cyclomaltodextrin glucanotransferase from *Bacillus circulans E 192*. I. Purification and characterization of the enzyme. *Biotechnology and Applied Biochemistry, 15*(1), 48–58. https://doi.org/10.1111/j.1470-8744.1992.tb00196.x.

Cao, X., Jin, Z., Wang, X., & Chen, F. (2005). A novel cyclodextrin glycosyltransferase from an alkalophilic *Bacillus species*: Purification and characterization. *Food Research International, 38*(3), 309–314. https://doi.org/10.1016/j.foodres.2004.06.015.

Chung, H.-J., Yoon, S.-H., Lee, M.-J., Kim, M.-J., Kweon, K.-S., Lee, I.-W., Kim, J.-W., Oh, B.-H., Lee, H.-S., Spiridonova, V. A., & Park, K.-H. (1998). Characterization of a thermostable cyclodextrin glucanotransferase isolated from *Bacillusstearothermophilus* ET1. *Journal of Agricultural and Food Chemistry, 46*(3), 952–959. https://doi.org/10.1021/jf970707d.

Crozatti, T. T. D. S., Larentis, P. V., Fenelon, V. C., Miyoshi, J. H., Brito, J. R. D., Salinas, G. D. S., Mazzotti, B. D. O., Teles, G. C., Lima Neto, Q. A. D., & Matioli, G. (2023). Challenges and alternatives for the production of cyclodextrins from the CGTase enzyme from recombinant *Bacillus subtilis WB800. Food Science and Technology, 43*. https://doi.org/10.1590/fst.104122.

Dalmotra, N., Tripathi, A. D., Srivastava, S. K., Arya, S. K., & Naik, B. (2016). Statistical optimization of cyclodextrin glycosyltransferase (CGTase) production from *Bacillus macerans* in batch cultivation and its purification. *International Journal of Food and Fermentation Technology, 6*(2), 261.

de Araújo Coelho, S. L., Magalhães, V. C., Marbach, P. A. S., & Cazetta, M. L. (2016). A new alkalophilic isolate of *Bacillus* as a producer of cyclodextrin glycosyltransferase using cassava flour. *Brazilian Journal of Microbiology, 47*(1), 120–128.

Del Valle, E. M. (2004). Cyclodextrins and their uses: A review. *Process Biochemistry, 39*(9), 1033–1046.

Elbaz, A. F., Sobhi, A., & ElMekawy, A. (2015). Purification and characterization of cyclodextrin β-glucanotransferase from novel alkalophilic *Bacilli. Bioprocess and Biosystems Engineering, 38*(4), 767–776.

Endo, T., Ogawa, N., Nagase, H., Sambe, H., Takaha, T., Terada, Y., & Ueda, H. (2007). Production of large-ring cyclodextrins composed of 9~ 21 α-D-glucopyranose units by cyclodextrin glucano-transferase, effects of incubation temperature and molecular weight of amylose. *Heterocycles, 74*, 991–997.

Fenelon, V. C., Aguiar, M. F. A., Miyoshi, J. H., Martinez, C. O., & Matioli, G. (2015). Ultrafiltration system for cyclodextrin production in repetitive batches by CGTase from *Bacillus firmus* strain *37. Bioprocess and Biosystems Engineering, 38*(7), 1291–1301.

Freitas, T. L. D., Monti, R., & Contiero, J. (2004). Production of CGTase by a *Bacillus alkalophilic CGII* strain isolated from the wastewater of a manioc flour industry. *Brazilian Journal of Microbiology, 35*(3), 255–260.

Fujiwara, S., Kakihara, H., Woo, K. B., Lejeune, A., Kanemoto, M., Sakaguchi, K., & Imanaka, T. (1992). Cyclization characteristics of cyclodextrin glucanotransferase are conferred by the NH2-terminal region of the enzyme. *Applied and Environmental Microbiology, 58*(12), 4016–4025. https://doi.org/10.1128/aem.58.12.4016-4025.1992.

Goel, A., & Nene, S. (1995). A novel cyclomaltodextrin glucanotransferase from *Bacillus firmus* that degrades raw starch. *Biotechnology Letters, 17*(4), 411–416. https://doi.org/10.1007/BF00130799.

Goo, B. G., Hwang, Y. J., & Park, J. K. (2014). *Bacillus thuringiensis*: A specific gamma-cyclodextrin producer strain. *Carbohydrate Research, 386*, 12–17. https://doi.org/10.1016/j.carres.2013.12.005.

Hofmann, B. E., Bender, H., & Schulz, G. E. (1989). Three-dimensional structure of cyclodextrin glycosyltransferase from *Bacillus circulans* at 3·4 Å resolution. *Journal of Molecular Biology, 209*(4), 793–800. https://doi.org/10.1016/0022-2836(89)90607-4.

Horikoshi, K. (1971). Production of alkaline enzymes by alkalophilic microorganisms. *Agricultural and Biological Chemistry, 35*(9), 1407–1414. https://doi.org/10.1080/00021369.1971.10860094.

Ibrahim, A. S., Al-Salamah, A. A., El-Tayeb, M. A., El-Badawi, Y. B., & Antranikian, G. (2012). A novel cyclodextrin glycosyltransferase from alkaliphilic *Amphibacillus sp. NPST-10*: purification and properties. *International Journal of Molecular Sciences, 13*(8), 10505–10522.

Ibrahim, A. S., El-Tayeb, M. A., & Al-Salamah, A. A. (2010). Characterization of immobilized alkaline cyclodextrin glycosyltransferase from a newly isolated *Bacillus agaradhaerens KSU-A11. African Journal of Biotechnology, 9*(44), 7550–7559.

Ismail, A. M. S., Sobieh, U. I., & Abdel-Fattah, A. F. (1996). Biosynthesis of cyclodextrin glucosyltransferase and β-cyclodextrin by *Bacillus macerans 314* and properties of the crude enzyme. *The Chemical Engineering Journal and The Biochemical Engineering Journal, 61*(3), 247–253.

Ismail, S. A., El-Sayed, O. H., Helmy, S. A., Mohammad, Y., & Mohammad, N. R. (2009). Production, purification, and biochemical characterization of cyclodextrin glucanotransferase from *Bacillus cereus N1. Egypt. Journal of Microbiology (Seoul, Korea), 6*, 370–377.

Jamil, N., Man, R. C., Shaarani, S. M., Sulaiman, S. Z., Mudalip, S. A., & Arshad, Z. M. (2017). Characterization of α-cyclodextrin glucanotransferase from *Bacillus licheniformis. Indian Journal of Science and Technology, 10* 10.17485.

Jiang, Z., Zhu, T., Gu, Z., Li, Z., Ban, X., Hong, Y., & Li, C. (2025). Cyclodextrin glycosyltransferases for food innovation: From molecular mechanism to industrial application. *Critical Reviews in Food Science and Nutrition,* 1–17.

Kaneko, T., Yoshida, M., Yamamoto, M., & Nakamura, N. (1990). Production of cyclodextrins by simultaneous actions of two CGTases from three strains of *Bacillus. Starch - Stärke, 42*(7), 277–281. https://doi.org/10.1002/star.19900420709.

Kitahata, S., & Okada, S. (1974). Action of cyclodextrin glycosyltransferase from *Bacillus megaterium* strain no. 5 on starch. *Agricultural and Biological Chemistry, 38*(12), 2413–2417. https://doi.org/10.1080/00021369.1974.10861521.

Kitayska, T., Petrova, P., Ivanova, V., & Tonkova, A. I. (2011). Purification and properties of a new thermostable cyclodextrin glucanotransferase from *Bacillus pseudalcaliphilus 8SB. Applied Biochemistry and Biotechnology, 165*(5), 1285–1295.

Klein, C., & Schulz, G. E. (1991). Structure of cyclodextrin glycosyltransferase refined at 2.0 Å resolution. *Journal of Molecular Biology, 217*(4), 737–750. https://doi.org/10.1016/0022-2836(91)90530-J.

Kulshreshtha, N. M., Upadhyay, D., Shrivastava, D., Chauhan, A., & Singh, N. K. (2020). Production and optimization of cyclodextrin glucanotranferase from *bacillus SP. SS2. Journal of Microbiology, Biotechnology and Food Sciences, 10*(2), 159–165. https://doi.org/10.15414/jmbfs.2020.10.2.159-165.

Larsen, K. L., Christensen, H. J. S., Mathiesen, F., Pedersen, L. H., & Zimmermann, W. (1998). Production of cyclomaltononaose (δ-cyclodextrin) by cyclodextrin glycosyltransferases from *Bacillus spp.* and bacterial isolates. *Applied Microbiology and Biotechnology, 50*(3), 314–317. https://doi.org/10.1007/s002530051297.

Leemhuis, H., Kelly, R. M., & Dijkhuizen, L. (2010). Engineering of cyclodextrin glucanotransferases and the impact for biotechnological applications. *Applied Microbiology and Biotechnology, 85*(4), 823–835. https://doi.org/10.1007/s00253-009-2221-3.

Li, Z., Wang, M., Wang, F., Gu, Z., Du, G., Wu, J., & Chen, J. (2007). γ-Cyclodextrin: A review on enzymatic production and applications. *Applied Microbiology and Biotechnology, 77*(2), 245–255. https://doi.org/10.1007/s00253-007-1166-7.

Lim, C. H., Rasti, B., Sulistyo, J., & Hamid, M. A. (2021). Comprehensive study on transglycosylation of CGTase from various sources. *Heliyon, 7*(2).

Liu, Z.-Y., & Yu, X.-Z. (2025). Engineering *Bacillus subtilis* for high-value bioproduction: Recent advances and applications. *Microbial Cell Factories, 24*(1), 182. https://doi.org/10.1186/s12934-025-02818-6.

Mahat, M. K., Illias, R. M., Rahman, R. A., Rashid, N. A. A., Mahmood, N. A. N., Hassan, O., Aziz, S. A., & Kamaruddin, K. (2004). Production of cyclodextrin glucanotransferase (CGTase) from alkalophilic *Bacillus sp. TS1-1*: Media optimization using experimental design. *Enzyme and Microbial Technology, 35*(5), 467–473. https://doi.org/10.1016/j.enzmictec.2004.07.008.

Marechal, L. R., Rosso, A. M., Marechal, M. A., Krymkiewicz, N., & Ferrarotti, S. A. (1996). Some properties of a cyclomaltodextrin-glucanotransferase from *Bacillus circulans DF 9 R* type. *Cellular and Molecular Biology (Noisy-le-Grand, France), 42*(5), 659–664.

Martins, R. F., & Hatti-Kaul, R. (2002). A new cyclodextrin glycosyltransferase from an alkaliphilic *Bacillus agaradhaerens* isolate: Purification and characterisation. *Enzyme and Microbial Technology, 30*(1), 116–124. https://doi.org/10.1016/S0141-0229(01)00461-6.

Matioli, G., Zanin, G. M., Gljimarães, M. F., & De moraes, F. F. (1998). Production and purification of CGTase of alkalophilic *Bacillus* isolated from Brazilian soil. *Applied Biochemistry and Biotechnology, 70–72*(1), 267–275. https://doi.org/10.1007/BF02920143.

Mehta, D., & Satyanarayana, T. (2016). Bacterial and archaeal α-amylases: Diversity and amelioration of the desirable characteristics for industrial applications. *Frontiers in Microbiology, 7*. https://doi.org/10.3389/fmicb.2016.01129.

More, S. S., Niraja, R., Evelyn, C., Byadgi, A. M., Shwetha, V., & Das Mangaraj, S. (2012). Isolation, purification and biochemical characterization of CGTase from *Bacillus halodurans*. *Hrvatski časopis za prehrambenu tehnologiju, biotehnologiju i nutricionizam, 7*(1-2), 90–97.

Moriwaki, C., Ferreira, L. R., Rodella, J. R. T., & Matioli, G. (2009). A novel cyclodextrin glycosyltransferase from *Bacillus sphaericus strain 41*: Production, characterization, and catalytic properties. *Biochemical Engineering Journal, 48*(1), 124–131.

Moriwaki, C., Mazzer, C., Pazzetto, R., & Matioli, G. (2009). Produção, purificação e aumento da performance de ciclodextrina glicosiltransferases para produção de ciclodextrinas. *Química Nova, 32*(9), 2360–2366. https://doi.org/10.1590/S0100-40422009000900024.

Nakamura, N., & Horikoshi, K. (1976). Characterization of acid-cyclodextrin glycosyltransferase of an alkalophilic *Bacillus sp. Agricultural and Biological Chemistry, 40*(8), 1647–1648. https://doi.org/10.1271/bbb1961.40.1647.

Noor, Y. M., Samsulrizal, N. H., Jema'on, N. A., Low, K. O., Ramli, A. N. M., Alias, N. I., Damis, S. I. R., Fuzi, S. F. Z. M., Isa, M. N. M., Murad, A. M. A., Raih, M. F. M., Bakar, F. D. A., Najimudin, N., Mahadi, N. M., & Illias, R. M. (2014). A comparative genomic analysis of the alkalitolerant soil bacterium *Bacillus lehensis G1*. *Gene, 545*(2), 253–261. https://doi.org/10.1016/j.gene.2014.05.012.

Pardhi, D. S., Rabadiya, K. J., Panchal, R. R., Raval, V. H., Joshi, R. G., & Rajput, K. N. (2023). Cyclodextrin glucanotransferase: Fundamentals and biotechnological implications. *Applied Microbiology and Biotechnology, 107*(19), 5899–5907.

Park, S. H., Na, Y., Kim, J., Kang, S. D., & Park, K.-H. (2017). Properties and applications of starch-modifying enzymes for use in the baking industry. *Food Science and Biotechnology*. https://doi.org/10.1007/s10068-017-0261-5.

Pishtiyski, I., Popova, V., & Zhekova, B. (2008). Characterization of cyclodextrin glucanotransferase produced by *Bacillus megaterium. Applied Biochemistry and Biotechnology, 144*(3), 263–272. https://doi.org/10.1007/s12010-007-8009-y.

Pongsawasdi, P., & Yagisawa, M. (1987). Screening and identification of a cyclomaltodextrin glucanotransferase-producing bacteria [*Bacillus circulans*]. *Journal of Fermentation and Bioengineering, 65*.

Qi, Q., & Zimmermann, W. (2005). Cyclodextrin glucanotransferase: From gene to applications. *Applied Microbiology and Biotechnology, 66*(5), 475–485. https://doi.org/10.1007/s00253-004-1781-5.

Ramirez Pérez, H. L. (2014). Production and characterization of cyclodextrin glycosyltransferase from *Bacillus. sp* isolated from Cuban Soil. *ScienceOpen Research.*

Reddy, S. V., More, S. S., & Annappa, G. S. (2017). Purification and properties of beta-cyclomaltodextrin glucanotransferase from *Bacillus flexus SV 1. Journal of Basic Microbiology, 57*(11), 974–981.

Rendleman, J. A., Jr (1999). The production of cyclodextrins using CGTase from *Bacillus macerans*. In *Carbohydrate Biotechnology Protocols*. Totowa, NJ: Humana Press, 89–101.

Rosso, A. M., Ferrarotti, S. A., Krymkiewicz, N., & Nudel, B. (2002). Optimisation of batch culture conditions for cyclodextrin glucanotransferase production from *Bacillus circulans DF 9R. Microbial Cell Factories, 1*(1), 3. https://doi.org/10.1186/1475-2859-1-3.

Sabioni, J.G., & Park, Y. K. (1992). Cyclodextrin glycosyltransferase production by alkalophilic *Bacillus lentus*.

Sahoo, S. (2017). Screening, partial purification and stability studies of cyclodextrin glycosyl transferases from *Bacillus amyloliquefaciens Strain L4-6. Deleted Journal.*

Saini, K., Pathak, V. M., Tyagi, A., & Gupta, R. (2022). Microbial cyclodextrin glycosyltransferases: Sources, production, and application in cyclodextrin synthesis. *Catalysis Research, 02*(03), 1–69. https://doi.org/10.21926/cr.2203029.

Saleh, A. M., Zaki, R. A., Gomaa, S. K., Elrefai, H. A., Shafei, M. S., Abd-Elsalam, I. S., & Elrefaey, A. H. (2023). Optimization and scaling-up of cyclodextrin glucosyltransferase production by the local isolate *Bacillus paramycoides MZ005634. Egyptian Journal of Chemistry, 66*(9), 343–351.

Sanjari, S., & Vahabzadeh, F. (2014). Experimental studies and kinetic modeling of cyclodextrin glucanotransferase production from starch with the use of *Bacillus sp. DSM 2523. Starch-Stärke, 66*(7-8), 606–614.

Sato, M., Yagi, Y., Nagano, H., & Ishikura, T. (1985). Determination of CGTase from *Bacillus ohbensis* and its optimum pH using HPLC. *Agricultural and Biological Chemistry, 49*(4), 1189–1191. https://doi.org/10.1080/00021369.1985.10866871.

Savergave, L. S., Dhule, S. S., Jogdand, V. V., Nene, S. N., & Gadre, R. V. (2008). Production and single step purification of cyclodextrin glycosyltransferase from alkalophilic *Bacillus firmus* by ion exchange chromatography. *Biochemical Engineering Journal, 39*(3), 510–515. https://doi.org/10.1016/j.bej.2007.09.020.

Schardinger, F. (1911). Bildung kristallisierter Polysaccharide (dextrine) aus Stärkekleister durch microben. *Zentralbl. Bakteriol. Parasitenk. Abt. II, 29*, 188–197.

Class 2: Cyclomaltodextrin glucanotransferase. In D. Schomburg, I. Schomburg, & A. Chang (Vol. Eds.), *Springer Handbook of Enzymes: 31*Berlin, Heidelberg: Springerhttps://doi.org/10.1007/3-540-37718-2_19 Transferases IV.

Sian, H. K., Said, M., Hassan, O., Kamaruddin, K., Ismail, A. F., Rahman, R. A., & Illias, R. M. (2005). Purification and characterization of cyclodextrin glucanotransferase from alkalophilic *Bacillus sp. G1. Process Biochemistry, 40*(3-4), 1101–1111.

Sivakumar, N., & Shakilabanu, S. (2013). Production of cyclodextrin glycosyl transferase by *Bacillus megaterium. International Journal of Current Microbiology and Applied Sciences, 2*, 44–55.

Solanki, P., & Banerjee, T. (2022). Medium optimization for submerged fermentative production of β-cyclodextrin glucosyltransferase by isolated novel alkalihalophilic *Bacillus sp. NCIM 5799* using statistical approach. *Letters in Applied Microbiology, 75*(2), 431–441. https://doi.org/10.1111/lam.13746.

Solanki, P., Awadhiya, P., & Banerjee, T. (2022). Conventional production optimization of cyclodextrin glucosyl transferase by a novel isolate of *Bacillus sp. PBS1* from potato rhizosphere. *Journal of Microbiology, Biotechnology and Food Sciences, 12*(1), e5130.

Stanković, S., Berić, T., Simić, D., & Pešić, D. (2016). Determination of cyclodextrin production by cyclodextrin glycosyltransferase from alkalophilic *Bacillus circulans strain B-65. Botanica Serbica, 40*(1).

Starnes, R.L. (1990). Industrial potential of cyclodextrin glycosyl transferases.

Stefanova, M. E., Tonkova, A. I., Miteva, V. I., & Dobreva, E. P. (1999). Characterization and cultural conditions of a novel cyclodextrin glucanotransferase-producing *Bacillus stearothermophilus* strain. *Journal of Basic Microbiology: An International Journal on Biochemistry, Physiology, Genetics, Morphology, and Ecology of Microorganisms, 39*(4), 257–263.

Szejtli, J. (1998). Introduction and general overview of cyclodextrin chemistry. *Chemical Reviews, 98*(5), 1743–1754.

Takada, M., Nakagawa, Y., & Yamamoto, M. (2003). Biochemical and genetic analyses of a novel γ-cyclodextrin glucanotransferase from an alkalophilic *Bacillus clarkii 7364. Journal of Biochemistry, 133*(3), 317–324.

Terada, Y., Yanase, M., Takata, H., Takaha, T., & Okada, S. (1997). Cyclodextrins are not the major cyclic α-1,4-glucans produced by the initial action of cyclodextrin glucanotransferase on amylose. *Journal of Biological Chemistry, 272*(25), 15729–15733. https://doi.org/10.1074/jbc.272.25.15729.

Tilden, E. B., & Hudson, C. S. (1939). The conversion of starch to crystalline dextrins by the action of a new type of amylase separated from cultures of *Aerobacillus macerans. Journal of the American Chemical Society, 61*(10), 2900–2902.

Tomita, K., Kaneda, M., Kawamura, K., & Nakanishi, K. (1993). Purification and properties of a cyclodextrin glucanotransferase from *Bacillus autolyticus 11149* and selective formation of β-cyclodextrin. *Journal of Fermentation and Bioengineering, 75*(2), 89–92. https://doi.org/10.1016/0922-338X(93)90215-T.

Tonkova, A. (1998). Bacterial cyclodextrin glucanotransferase. *Enzyme and Microbial Technology, 22*(8), 678–686. https://doi.org/10.1016/S0141-0229(97)00263-9.

Tonkova, A. (2006). Microbial starch converting enzymes of the α-amylase family. In *Microbial Biotechnology in Horticulture, Vol. 1*, CRC Press, 421–472.

Upadhyay, D., Sharma, S., Shrivastava, D., & Kulshreshtha, N. M. (2019). Production and characterization of β-cyclodextrin glucanotransferase from *Bacillus sp. ND1. Journal of Basic Microbiology, 59*(2), 192–205.

van der Veen, B. A., Uitdehaag, J. C., Dijkstra, B. W., & Dijkhuizen, L. (2000). Engineering of cyclodextrin glycosyltransferase reaction and product specificity. *Biochimica et Biophysica Acta (BBA)-Protein Structure and Molecular Enzymology, 1543*(2), 336–360.

van der Veen, B. A., van Alebeek, G. J. W., Uitdehaag, J. C., Dijkstra, B. W., & Dijkhuizen, L. (2000). The three transglycosylation reactions catalyzed by cyclodextrin glycosyltransferase from *Bacillus circulans* (strain 251) proceed via different kinetic mechanisms. *European Journal of Biochemistry, 267*(3), 658–665.

Vidya, A. S., Veena, S. M., & More, S. S. (2012). Isolation, purification, and characterization of a novel CGTase from alkalophilic *Bacillus lehensis SV1*. *BTAIJ, 6*(7), 226–234.

Villiers, A. (1891). Sur la fermentation de la fécule par l'action du ferment butyrique. *Compt. Rend. Acad. Sci, 112*, 536–538.

Vinod, A. S., & More, S. M. (2016). Purification and characterization of a novel CGTase from alkaliphilic *Bacillus flexus* Isolated from Lonar lake, India. *Int. J. Curr. Microbiol. App. Sci, 5*(7), 164–170.

Wu, D., Chen, S., Wang, N., Chen, J., & Wu, J. (2012). Gamma-cyclodextrin production using cyclodextrin glycosyltransferase from *Bacillus clarkii 7364*. *Applied Biochemistry and Biotechnology, 167*(7), 1954–1962. https://doi.org/10.1007/s12010-012-9741-5.

Yap PuiWoon, Y. P., Ariff, A. B., Woo KwanKit, W. K., & Hii SiewLing, H. S. (2010). Production of cyclodextrin glycosyltransferase (CGTase) by *Bacillus lehensis S8* using sago starch as carbon source.

Yim, D. G., Sato, H. H., Park, Y. H., & Park, Y. K. (1997). Production of cyclodextrin from starch by cyclodextrin glycosyltransferase from *Bacillus firmus* and characterization of purified enzyme. *Journal of Industrial Microbiology and Biotechnology, 18*(6), 402–405. https://doi.org/10.1038/sj.jim.2900400.

Yu, E. K. C., Aoki, H., & Misawa, M. (1988). Specific alpha-cyclodextrin production by a novel thermostable cyclodextrin glycosyltransferase. *Applied Microbiology and Biotechnology, 28*(4–5), 377–379. https://doi.org/10.1007/BF00268199.

CHAPTER

Cyclodextrin glucanotransferase from other than *Bacillus* genera

6

Nurul Nabila Huda Baharudin, Rohaida Che Man, Siti Kholijah Abdul Mudalip, Siti Zubaidah Sulaiman and Nurul Aini Mohd Azman

Faculty of Chemical and Process Engineering Technology, Universiti Malaysia Pahang Al-Sultan Abdullah Lebuhraya Tun Khalil Yaakob, Gambang, Pahang, Malaysia

Introduction

Cyclodextrin glucanotransferase (CGTase; EC 2.4.1.19) is a multifunctional enzyme that catalyzes the conversion of starch to form a product named cyclodextrin (CD) (Ogunbadejo & Al-zuhair, 2024). These CDs are cyclic oligosaccharides consisting of six, seven, or eight glucose units and are referred to as α-, β-, and γ-CDs, respectively (Bautista et al., 2012; Jailani et al., 2022). CGTase belongs to the glycoside hydrolase family 13 (GH13) and works via a double displacement mechanism (Graebin et al., 2016; Leija et al., 2022). It can perform cyclization, coupling, disproportionation, and limited hydrolysis reactions due to its catalytic versatility, all of which are of considerable industrial interest (Lee et al., 2007; Li, Sun, et al., 2023).

Structurally, CGTase consists of five distinct domains, labeled A to E. The central domain A forms a (β/α)_8-barrel structure (TIM barrel) in which the active center is located (Uitdehaag et al., 2002). Domain B is inserted into domain A and contributes to substrate-binding, while domains C, D, and E play an important role in the stability of the enzyme and specificity for different CD products (Kelly et al., 2009). Structural variations between CGTases from different microbial sources affect both the substrate affinity and the ratio of α-, β-, and γ-CDs produced during catalysis (Rahman et al., 2006).

CGTase is predominantly produced by *Bacillus* spp. However, other bacterial genera such as *Paenibacillus*, *Klebsiella*, *Thermoanaerobacter*, alkalophilic *Microbacterium,* and *Geobacillus* have also been identified as CGTase producers. These alternative sources provide CGTase yield with different biochemical properties, including thermal stability, pH optimum, and product selectivity (Li, Sun, et al., 2023; Saini et al., 2022). The choice of microbial

Cyclodextrin Glucanotransferase. DOI: https://doi.org/10.1016/B978-0-443-41694-1.00007-1

strain, substrate, and process parameters has a great influence on the distribution and yield of CD, so strain selection and enzyme engineering are crucial for targeted production.

The primary industrial importance of CGTase lies in its ability to produce CD under mild reaction conditions. CDs are widely used in pharmaceutical formulations to improve the solubility and stability of active ingredients, in the food industry to encapsulate flavors and remove cholesterol, in cosmetics as carriers for active ingredients, and in environmental applications to remove pollutants and for controlled release systems (Kamaraj et al., 2024; Nicolaescu et al., 2025; Pereira et al., 2021). Future research directions include the exploration of agro-industrial residues as renewable substrates, the development of high-yield recombinant expression systems, and the integration of CGTase into continuous or semicontinuous bioreactor platforms. These innovations are expected to support the transition to sustainable and scalable CD production.

Scope of the chapter

The main objective of this chapter is to systematically discover the CGTase enzymes derived from bacterial genera other than *Bacillus*, emphasizing their increasing importance for industrial biotechnology. While CGTases from *Bacillus* spp. have been widely studied and commercialized due to their ease of cultivation and high productivity, there are numerous studies showing that enzymes from alternative microbial sources possess distinctive, and in some cases superior, biochemical properties. These include improved thermostability, broader pH tolerance, substrate specificity, and CD yield profiles, which are particularly valuable for specific industrial applications.

This chapter is structured to provide both comprehensive detail and conceptual insight. It begins with an introduction to major non-*Bacillus* genera reported to produce CGTase, including *Paenibacillus*, *Amphibacillus*, *Brevibacillus*, *Klebsiella*, *Thermoanaerobacterium*, and others. Each genus is discussed in terms of its ecological niche, fermentation behavior, enzyme kinetics, and product profile. In a comparative analysis, the differences in catalytic efficiency, optimal production conditions, and the predominant type of CD produced (α-, β-, or γ-CDs) are evaluated.

By the end of this chapter, readers will gain a comprehensive understanding of the importance, potential, and current limitations of CGTases from non-*Bacillus* spp. This scope not only highlights untapped microbial resources but also encourages innovation in CD production for next-generation bioprocesses.

CGTase was first identified in the early 20th century as the key enzyme responsible for the conversion of starch to CD, a discovery that dates back to the work of Villiers in 1891 and was further characterized in the 1950s (Aiassa et al., 2021; Fouda-mbanga et al., 2024; Martina et al., 2025). Over time, *Bacillus* spp.

emerged as the dominant CGTase producers due to their ease of cultivation, high yields, and well-characterized fermentation parameters. However, limitations such as product inhibition, thermal instability, and limited specificity prompted researchers to investigate CGTase-producing organisms from alternative environments (Rabadiya et al., 2024). Non-*Bacillus* sources, particularly those from extremophilic or anaerobic ecosystems, have shown novel catalytic mechanisms, broader stability profiles, and unique product specificities (Ayse et al., 2014).

The demand for high-performance enzymes in all industrial sectors has sparked interest in CGTases from unconventional microbial sources. While *Bacillus* spp. have been extensively studied for CGTase production, several other bacterial genera have demonstrated the ability to synthesize CGTase with desirable enzymatic properties. These non-*Bacillus* producers are often isolated from unique ecological niches such as thermophilic environments, alkaline lakes, or anaerobic fermenters that confer adaptive enzymatic properties such as high thermostability, pH tolerance, resistance to inhibitors, or anaerobic functionality (Leija et al., 2022; Rashid et al., 2002). The discovery of CGTases from such genera expands the spectrum of available biocatalysts and could provide superior alternatives for certain industrial conditions.

Cyclodextrin glucanotransferase-producing genera

Although *Bacillus* remains the most extensively studied CGTase-producing genus, increasing attention has been directed toward non-*Bacillus* microorganisms due to their unique enzymatic properties, environmental adaptations, and distinct CD product profiles. These organisms frequently exhibit specialized physiological traits such as thermophilicity, alkaliphilicity, halophilicity, or anaerobiosis, that translate into valuable industrial advantages. This section introduces and discusses each non-*Bacillus* CGTase-producing genus, focusing on their characteristics and roles in CGTase production.

Paenibacillus

Paenibacillus sp. are facultative anaerobic, gram-positive rods found in water, soil, and plant-associated environments (Aizawa, 2014; Han et al., 2025). Some *Paenibacillus* sp., such as Paenibacillus macerans and *Paenibacillus graminis,* are known to produce industrially relevant enzymes, including CGTase. *P. macerans* and *P. graminis* are good CGTase producers, active at neutral to slightly acidic pH (Vollu et al., 2008). Unlike CGTases from *Bacillus* spp., some *Paenibacillus*-derived enzymes are more stable and retain activity at lower temperatures, making them suitable for mild processing conditions (Jiang et al., 2018). It is also suitable for temperature-sensitive bioprocesses.

The CGTase from *P. macerans* is very efficient in converting starch into CD, thus it has potential to be applied in the food and pharmaceutical industries (Chen et al., 2017). Vollu et al. (2008) showed that the CGTase obtained from *P. graminis* was similar to *P. macerans* at the gene level. Thus, *P. graminis* presented the same potential for α- and β-CDs production as *P. macerans*. Some *Paenibacillus* CGTases also produce more γ-CDs than their *Bacillus* counterparts, which may be beneficial for specific drug delivery applications (Rimphanitchayakit et al., 2005). More research is needed to scale up *Paenibacillus* CGTases.

A specific example is *Paenibacillus pabuli* US132, which produced CGTase optimally at pH 8.0 and 25°C using potato starch (Jemli et al., 2007). The CD distribution was reported as α-CD (22%), β-CD (63%), and γ-CD (15%), indicating β-CD as the dominant product. The relatively low optimal temperature, compared with that of many thermophilic producers, highlights its suitability for mild industrial processes with reduced energy consumption.

Amphibacillus

The genus *Amphibacillus* comprises gram-positive, rod-shaped, spore-forming bacteria that are facultatively anaerobic and belong to the *Bacillaceae* family. Members of this genus have been isolated from extreme and alkaline environments such as hot springs, volcanic soils, and soda lakes (Allode et al., 2024). These environmental conditions appear to be conducive to the production of industrially important enzymes, including CGTase.

Two strains in this genus are known for producing CGTase, such as *Amphibacillus* sp. NPST-10 and *Amphibacillus* sp. NRC-WN, both isolated from soda lakes in Egypt (Ibrahim et al., 2012). The CGTase from *Amphibacillus* sp. NPST-10 showed optimal production at pH 10.5 and 30°C using corn starch, demonstrating strong alkaliphilic behavior. The enzyme exhibited marked β-CD selectivity with a product ratio of α-CD (10.2%), β-CD (86.4%), and γ-CD (3.4%) (Ibrahim et al., 2012).

Chief among their distinguishing characteristics are alkaline and thermal stability, desirable for industrial processes operating under elevated pH (Al-Sharawi et al., 2013). Their consistent production of β-CD further increases industrial relevance. The resilience of *Amphibacillus* CGTases under harsh conditions and their high β-CD specificity emphasize their potential as robust industrial biocatalysts.

Brevibacillus

The genus *Brevibacillus* consists of gram-positive, rod-shaped, endospore-forming bacteria that typically exhibit aerobic metabolism (Rasool et al., 2022). Members of this genus are widely distributed in natural environments, including soil, compost, and decomposing organic matter. They are generally characterized by thermophilic and alkaliphilic tendencies, traits that confer robustness under industrial processing conditions (Kumari and Satpal, 2014). Among them, *Brevibacillus borstelensis* has been extensively studied for its ability to degrade complex polymers and persist in high-temperature environments, indicating strong potential for industrial enzyme production (Hadad et al., 2005).

Although CGTase production in *Brevibacillus* has not been as comprehensively documented as in genera such as *Bacillus* or *Amphibacillus*, emerging evidence indicates that certain strains are capable of synthesizing CGTase with favorable catalytic properties. For example, a *Brevibacillus* sp. reported by Fan et al. (2018) produced CGTase optimally at pH 6.0 and 50°C using soluble starch, predominantly yielding β-CD. This moderate thermophilic behavior suggests compatibility with industrial processes that require elevated but not extreme temperatures.

Another study by Kim et al. (1998) was reported for *Brevibacillus brevis* CD162, in which the mature enzyme (~75 kDa) exhibited optimal activity at 55°C and pH 8.0, with stability over the pH range 5.5–9.0. Product analysis revealed predominant β-CD formation (33.9%), accompanied by lower γ-CD (9.9%) and minimal α-CD (1.3%) production. Recombinant expression in *Escherichia coli* retained comparable enzymatic properties and product distribution, indicating structural conservation and secretion efficiency. These findings support the classification of *Brevibacillus*-derived CGTase as a moderate thermotolerant, β-CD–dominant enzyme with appreciable γ-CD forming capability.

In related studies of GH13 family enzymes, calcium ions have been shown to enhance structural stability and catalytic performance, particularly within pH 6–8 and temperatures up to 50°C (Mori et al., 1994). Such calcium-mediated stabilization is commonly observed in starch-degrading enzymes, where metal ions contribute to maintaining tertiary structure and active-site conformation. It is therefore reasonable to infer that similar stabilization mechanisms may operate in *Brevibacillus*-derived CGTases.

Depending on enzyme structure and reaction parameters, these CGTases may produce α-, β-, or γ-CDs. However, current evidence suggests that β-CD predominates under the reported fermentation conditions. Comprehensive kinetic characterization, including K_m, V_{max}, and product specificity ratios, remains limited and warrants further investigation. In addition, the environmental resilience of enzymes derived from *Brevibacillus* suggests promising applicability in bioremediation and waste treatment processes (Rasool et al., 2022). Continued biochemical and structural characterization of CGTase-producing strains within this genus is expected to strengthen their industrial relevance, particularly in applications demanding moderate thermotolerance and stable β-CD production.

Klebsiella

Klebsiella is a genus of gram-negative, facultatively anaerobic, rod-shaped bacteria within the *Enterobacteriaceae* family. These bacteria are commonly found in various environments, including soil, water, and the gastrointestinal tract of humans and animals (Allode et al., 2024). While *Klebsiella* sp. are often known for their clinical importance, certain strains have shown industrial potential, particularly in the production of CGTase, an enzyme that catalyzes the conversion of starch to CD.

One notable strain, *Klebsiella pneumoniae* AS-22, has been identified for its ability to produce CGTase, which predominantly synthesizes α-CD from starch substrates (Lee and Choi, 1992). This enzyme exhibits unique properties, including the ability to degrade raw starch, which is beneficial for industrial processes aimed at reducing energy consumption associated with starch gelatinization. The CGTase from *K. pneumoniae* AS-22 is characterized by its preference for the production of α-CD, which is particularly valuable for pharmaceutical applications due to α-CD's ability to form inclusion complexes with various drug molecules, thereby improving their solubility and stability. It had been proven from a study by Gawande & Patkar (2001b), where *K. pneumoniae* AS-22 produced CGTase active at pH 6.0–9.0 and 45°C using soluble starch and the enzyme predominantly synthesized α-CD (81%), β-CD (12%), and γ-CD (7%), indicating strong α-CD selectivity.

Microbacterium

Microbacterium is a genus of gram-positive, non-spore-forming, rod-shaped bacteria in the family *Microbacteriaceae*. These organisms are predominantly aerobic and have high GC content in their genomic DNA. They are widely distributed across diverse environments, including soil, water, and plant-associated niches, reflecting their metabolic versatility and adaptability. Members of this genus are known for producing various enzymes of industrial significance, including CGTase.

Among the species, *Microbacterium terrae* KNR 9 has been identified as a notable CGTase producer. This alkaliphilic strain produces CGTase optimally at pH 6.0 and 55.6°C using potato starch as substrate (Rajput et al., 2016a). The enzyme demonstrates moderate thermotolerance and exhibits the ability to degrade both gelatinized and raw starch substrates, significantly enhancing its industrial applicability by reducing the need for extensive pretreatment steps. Product specificity analysis revealed that the enzyme predominantly synthesizes β-CD, consistent with its suitability for pharmaceutical and starch-processing industries.

In industrial applications, β-CD produced by *Microbacterium*-derived CGTase finds extensive use in the pharmaceutical, food, and cosmetic sectors (Rajput et al., 2016b). The enzyme's stability under alkaline conditions, combined with its efficiency in processing raw starch and its preference for β-CD, makes it a promising candidate for sustainable and cost-effective CD production. These characteristics highlight the potential of *M. terrae* KNR 9 for further optimization and scale-up in industrial CD manufacturing processes.

Thermoanaerobacter

Thermoanaerobacter is a genus of gram-positive, rod-shaped, endospore-forming bacteria within the family *Thermoanaerobacteraceae*. These thermophilic anaerobes are predominantly isolated from geothermal habitats such as hot springs, geothermal soils, and compost heaps (Allode et al., 2024). Their metabolism is strictly anaerobic, relying on fermentative pathways for energy generation. A particularly well-studied strain, *Thermoanaerobacter* sp., and *Thermococcus* sp., produce a CGTase with remarkable thermostability (Ayse & Sedat, 2012; Tachibana et al., 1999).

CGTase from *Thermoanaerobacterium* sp. exhibits optimal catalytic activity at pH 6.0 and 80°C using soluble potato starch (Ayse & Sedat, 2012). Although cellular growth typically occurs at 60°C–65°C, maximum enzyme activity is observed at higher temperatures, indicating strong intrinsic thermostability. Product distribution analysis revealed predominant α-CD production (62.6%), followed by β-CD (37.4%), with no detectable γ-CD formation. This α-CD dominance differentiates it from many mesophilic CGTases that preferentially synthesize β-CD.

The high thermostability of *Thermoanaerobacterium*-derived CGTase offers significant industrial advantages. Elevated operating temperatures enhance reaction kinetics, improve substrate solubility, and reduce contamination risks, making these enzymes particularly suitable for high-temperature continuous systems. Some strains demonstrate enzyme activity peaks at 80°C–95°C, while growth and enzyme production occur at lower temperatures (60°C–65C) under anaerobic conditions (Abelyan and Balayan, 2002; Thiemann and Dönges, 2004).

Optimizing fermentation parameters remains essential for maximizing CGTase productivity. Key variables include carbon sources (e.g., soluble starch and sago), incubation period, temperature, pH, substrate concentration, nitrogen sources (yeast extract, peptone, urea), agitation speed, and inoculum size. Proper optimization enhances catalytic efficiency and overall CD yield. In summary, *Thermoanaerobacterium* sp. represent significant producers of highly thermostable CGTase, characterized by α-CD dominance and suitability for high-temperature industrial applications in pharmaceuticals, food processing, and biotechnology.

Carboxydocella

Carboxydocella is a genus of strictly anaerobic, thermophilic, gram-positive bacteria commonly isolated from geothermal environments, including volcanic hot springs and hydrothermal systems (Omae et al., 2019; Yoneda et al., 2015). Members of this genus are known for their ability to oxidize carbon monoxide under anaerobic conditions, reflecting a highly specialized metabolism adapted to extreme habitats (Kochetkova et al., 2011; Omae et al., 2019).

Although reports of CGTase production from *Carboxydocella* remain limited, one study reported CGTase production at 37°C using wheat starch, yielding α-CD

(40%), β-CD (50%), and γ-CD (10%) (Ara et al., 2015). This indicates a relatively balanced CD profile with moderate β-CD dominance. Genomic analyzes of thermophilic anaerobes closely related to this genus have also revealed genes encoding GH13 enzymes, the same family to which CGTase belongs, suggesting potential CD-forming activity in certain strains. The thermophilic origin of *Carboxydocella* implies that any CGTase produced would likely exhibit high thermal stability, possibly retaining activity above 70–80°C.

From an industrial perspective, a thermophilic anaerobic CGTase offers two major advantages: compatibility with high-temperature starch liquefaction processes and reduced contamination risk due to operation under elevated temperatures and oxygen-limited conditions (Harnvoravongchai et al., 2020; Szijártó and Horan, 2011). Additionally, enzymes from strictly anaerobic thermophiles often demonstrate enhanced structural rigidity, contributing to prolonged half-life during continuous processing (Li, Yuan, et al., 2023). However, experimental characterization of CGTase activity, product specificity (α-, β-, γ-CD ratio), and kinetic parameters in *Carboxydocella* remains scarce and warrants systematic investigation.

Geobacillus

Geobacillus thermoglucosidans is a thermophilic, gram-positive, endospore-forming bacterium belonging to the genus *Geobacillus* (Torres-Guzman et al., 2021). It is widely distributed in compost, oil reservoirs, and geothermal environments, with optimal growth temperatures typically ranging from 55°C to 65°C (Javed et al., 2017). The species is well recognized for its efficient starch-degrading enzymatic system, including α-amylases and other GH13 enzymes (Khaswal et al., 2022). Genome mining studies have identified CGTase-like genes in certain *Geobacillus* strains, suggesting the potential to form CDs. Enzymes from this species generally exhibit strong thermostability, often maintaining catalytic function at temperatures exceeding 70°C, with moderate to high tolerance to alkaline pH (Hammadi et al., 2025; Mitra et al., 2015).

Experimental evidence further supports its CD-forming potential. CGTase production at pH 8.0 and 60°C using soluble starch as substrate yielded α-CD (18.75%), β-CD (34.66%), and γ-CD (6.89%) (Jia et al., 2018). Although β-CD was the predominant product, the overall conversion level suggests moderate total CD production under the tested conditions. Product distribution appears condition-dependent, as temperature and pH significantly influence cyclization versus hydrolysis reactions.

Thermostable CGTases derived from *Geobacillus* sp. are particularly attractive for industrial CD production integrated with starch liquefaction systems. High operational temperatures enhance starch solubility, reduce viscosity, and improve mass transfer, thereby increasing reaction efficiency (Jia et al., 2018). Additionally, alkaline tolerance supports compatibility with liquefaction processes typically performed under slightly alkaline conditions. Thermophilic enzymes also tend to exhibit improved structural rigidity and resistance to denaturation or proteolytic degradation, facilitating prolonged operation in batch and continuous bioreactors (Bosch and Sanchez-freire, 2021; Khazaal et al., 2025; Synowiecki, 2010).

Overall, the inherent thermostability, alkaline adaptability, and moderate yet balanced CD production profile of *G. thermoglucosidans* suggest strong potential for integration into high-temperature, energy-efficient starch liquefaction–cyclization systems. Nevertheless, further kinetic characterization, structural analysis, and process optimization studies are necessary to fully assess its industrial feasibility and maximize CD yield.

Haloferax

Haloferax mediterranei is an extremely halophilic archaeon belonging to the genus *Haloferax*. It thrives in hypersaline environments such as salt lakes and salterns, requiring high NaCl concentrations (typically 2–5 M) for optimal growth (Matarredona & Guío, 2024). Unlike the bacterial genera discussed previously, *H. mediterranei* represents an archaeal source of carbohydrate-active enzymes. Enzymes from *halophilic archaea* are structurally adapted to high-salt environments, often displaying unique surface charge distributions and an excess of acidic residues that maintain protein solubility and catalytic activity under extreme ionic strength (Dassarma & Dassarma, 2015). Reports of amylolytic enzymes in *H. mediterranei* suggest the presence of GH13 family proteins, including CGTase-like enzymes capable of cyclization reactions (Perez-Pomares et al., 2009).

Experimental studies have demonstrated CD production under fermentation conditions at pH 7.2 and 37°C using potato starch as substrate, where α-CD was the predominant product (Bautista et al., 2012). This α-CD dominance indicates a strain-specific product preference distinct from many β-CD–biased *Bacillus* CGTases. The high-salt tolerance and structural adaptation to ionic stress further suggest potential applicability in saline or solvent-containing systems where conventional bacterial CGTases may lose stability.

A potential CGTase from *H. mediterranei* would therefore be expected to demonstrate:

- High-salt tolerance, retaining activity in hypersaline media (Bautista et al., 2012).
- Moderate thermostability (consistent with optimal growth at 37°C–50°C) (Montero-Lobato et al., 2018).
- Structural stability under low water activity and high ionic strength conditions (Griffiths et al., 2025).

These properties are industrially relevant for processes conducted in high-salinity streams, including specialized food fermentations, saline waste bioremediation, and pharmaceutical formulations requiring controlled water activity (Delgado-García et al., 2012; Karan et al., 2012; Sharma and Goel, 2025). Additionally, halophilic enzymes often exhibit notable solvent tolerance, which may enhance CD synthesis in biphasic or organic–aqueous reaction systems (Chang et al., 2013).

Anaerobranca gottschalkii

Anaerobranca gottschalkii is an anaerobic, thermoalkaliphilic bacterium isolated from the hot alkaline soda lake Lake Bogoria in Kenya, thriving optimally at 55°C and pH 9–10. This gram-positive organism efficiently utilizes starch and is notable for secreting highly thermostable and alkali-stable amylolytic enzymes, particularly CGTase. Its ability to convert starch into α-, β-, and γ-CDs and larger cyclic glucans under extreme conditions makes it an important extremophile with strong potential for industrial applications in starch bioconversion, CD production, and other high-temperature, high-pH bioprocesses.

A. gottschalkii plays a crucial role as a robust producer of extracellular CGTase. When grown on starch at 55°C and pH 9.0, it secretes a highly active, thermo- and alkali-stable CGTase capable of efficiently converting amylose, amylopectin, and native starch into α-, β-, γ-CDs and even larger CDs (> DP 8). Its enzyme displays broad pH (5–10) and temperature (55°C–70°C) activity ranges, making it the first reported CGTase from an anaerobic thermoalkaliphile and an excellent candidate for industrial CD production under harsh processing conditions (Thiemann and Dönges, 2004).

Nevertheless, detailed biochemical characterization remains limited. Quantitative kinetic constants (K_m, K_{cat}, and catalytic efficiency), comprehensive product specificity ratios (α:β:γ distribution), and structural analyzes have not been extensively reported. Systematic enzymatic studies and heterologous expression strategies are therefore necessary to evaluate its catalytic efficiency and determine its feasibility as a non-*Bacillus* CGTase source for saline or unconventional processing environments.

Comparative enzyme activity and production yields

CGTase from non-*Bacillus* spp. displays distinct biochemical characteristics, catalytic behaviors, and production capacities that influence their industrial relevance. Detailed comparative analyzes of representative non-*Bacillus* CGTase-producing microorganisms reveal notable differences in optimal operating conditions, thermostability, catalytic efficiency, and CD product profiles.

Enzyme activity and catalytic efficiency

A classic example of non-*Bacillus* CGTase is from *P. macerans* (formerly *Bacillus macerans*). The enzyme exhibits broad pH tolerance (pH 5.5–9.5) with optimum activity typically around neutral to slightly alkaline conditions. Reported catalytic efficiency is relatively high, with significant coupling and disproportionation activities, rendering it suitable for mild industrial processing environments such as food and pharmaceutical applications (Gawande & Patkar, 2001b). The enzyme also demonstrates good stability under moderate temperatures (45°C–60°C), balancing activity and operational feasibility.

Thermophilic representatives such as *Thermoanaerobacterium thermosulfurigenes* produce CGTase with remarkable thermostability. Ding et al. (2010) reported that the enzyme retained more than 80% of its activity at temperatures exceeding 80°C. Such thermal resilience significantly enhances industrial applicability, particularly in starch processing, where high temperatures reduce viscosity and contamination risks. The enzyme also demonstrates strong β-cyclization activity at elevated temperatures, indicating structural robustness in its catalytic domain.

In contrast, CGTase from *P. pabuli* shows relatively high activity at lower temperatures (30°C–45°C), which may reduce energy consumption in bioconversion systems (Gawande & Patkar, 2001b). Although less thermostable than its thermophilic counterparts, its catalytic performance under mesophilic conditions makes it advantageous for cost-effective operations. Additionally, CGTase isolated from *Carboxydocella thermautotrophica* has been reported to possess high coupling activity, particularly toward γ-CD formation (Rather et al., 2015). This unique catalytic behavior suggests structural adaptations favoring larger CD ring formation, broadening its industrial significance for specialty CD production.

Production yields across genera

Production yield varies considerably depending on microbial source, strain genetics, and fermentation strategy. Under optimized submerged fermentation conditions, *P. macerans* has been reported to produce CGTase levels exceeding 150 U/mL (Rambu and Dumitru, 2024), demonstrating its suitability for large-scale enzyme production. Optimization parameters typically include soluble starch concentration, nitrogen source supplementation, and controlled pH during fermentation.

Moderate CGTase production has been observed in *Klebsiella oxytoca*, although this organism is more notable for its product selectivity rather than volumetric enzyme yield (Lee and Choi, 1992). Production titers are generally lower than those of *Paenibacillus* sp. However, the enzyme's high α-CD specificity compensates for its moderate activity in targeted applications. For *M. terrae*, reported CGTase production is comparatively low (typically <50 U/mL) (Rajput et al., 2016c). Despite limited volumetric yield, its CGTase exhibits distinctive substrate-binding features and potential for selective cyclization reactions. Such characteristics suggest opportunities to improve strain through genetic or process engineering.

Thermophilic strains such as *T. thermosulfurigenes* generally produce moderate enzyme yields. However, their industrial value lies in their pronounced thermal robustness. Supporting this, a novel thermophilic anaerobic bacterium from the genus *Thermoanaerobacter*, phylogenetically closely related to *T. thermosulfurigenes*, has been reported to produce a CGTase with optimal catalytic activity at 80°C and pH 7.0–8.0 (Avci & Do, 2009). Such an elevated temperature optimum directly substantiates the operational stability characteristic of thermophilic CGTases and highlights their suitability for high-temperature starch conversion processes, where substrate solubility and reaction kinetics are enhanced.

From a process engineering perspective, the thermostability of these enzymes confers tangible economic advantages. Operation at elevated temperatures reduces the need for cooling systems, lowers contamination risks by suppressing mesophilic microbial growth, and permits more consistent long-duration reactions without rapid enzyme denaturation. Furthermore, enhanced thermal stability often translates to prolonged catalytic half-life, which can decrease enzyme replenishment frequency and potentially simplify downstream purification steps due to improved structural resilience (Wu et al., 2021; Wulandari et al., 2025). Collectively, although volumetric enzyme yields may be moderate, the superior operational stability of thermophilic CGTases substantially improves overall process efficiency and economic feasibility in industrial CD production systems.

Cyclodextrin yield and specificity

Product specificity remains a critical determinant of CGTase industrial value. Distinct non-*Bacillus* genera demonstrate characteristic α-, β-, and γ-CDs production profiles. CGTase from *P. pabuli* predominantly produces α-CD, which is particularly valuable in pharmaceutical applications due to its superior solubility and inclusion complex formation with hydrophobic drug molecules (Vollu et al., 2008). Similarly, *K. oxytoca* exhibits exceptionally high α-CD selectivity. Lee and Choi, 1992 reported a production ratio of α:β:γ-CD of 96.5:3.5:0, representing one of the highest α-selective profiles documented among non-*Bacillus* CGTases. This specificity significantly reduces downstream separation requirements. In contrast, *Amphibacillus xylanus* favors β-CD production (Ibrahim et al., 2012). β-CD is widely applied in food stabilization and emulsification processes due to its moderate cavity size and inclusion capability.

Thermophilic organisms such as *T. thermosulfurigenes* are recognized for producing a relatively balanced distribution of α-, β-, and γ-CDs, thereby enhancing their versatility across multiple industrial sectors. Reported product profiles indicate that CGTase from *T. thermosulfurigenes* converts starch into CDs at an approximate ratio of 34:62:4 (α:β:γ), demonstrating a predominance of β-CD while still generating appreciable amounts of α- and γ-forms (Leija et al., 2022). Such a distribution is industrially advantageous because it reduces reliance on highly selective strains when a broader CD spectrum is desirable. The simultaneous formation of all three CD types expands applicability in pharmaceutical, food, and chemical formulations, where different cavity sizes are required for encapsulating diverse guest molecules. Combined with the inherent thermostability of thermophilic CGTases, this balanced product specificity strengthens the suitability of *T. thermosulfurigenes* for integrated and high-temperature CD production processes.

Industrial implications

The biochemical and catalytic diversity observed among non-*Bacillus* CGTases strongly influences their industrial applicability and process integration strategies. High α-CD–producing strains such as *K. oxytoca* and certain isolates of *P. pabuli* are particularly attractive for pharmaceutical and drug delivery systems, where α-CD is preferred due to its superior aqueous solubility and favorable inclusion complex formation with hydrophobic compounds (Vollu et al., 2008). The exceptionally high α-selectivity reported for *K. oxytoca* 19-1 significantly reduces downstream purification requirements, thereby improving economic feasibility in large-scale production.

Thermostable CGTases derived from thermophilic organisms such as *T. thermosulfurigenes* enable operation at elevated temperatures, which enhances starch solubility, accelerates reaction kinetics, and minimizes microbial contamination risks during industrial processing (Bonch-osmolovskaya et al., 2017; Peng et al., 2016; Synowiecki, 2010). Such thermal robustness is particularly advantageous in continuous bioreactor systems and high-temperature starch conversion processes. Conversely, CGTases favoring β-CD production, such as those from *A. xylanus*, are more suitable for applications in food, cosmetic, and flavor stabilization industries, where β-CD is predominantly utilized due to its balanced cavity size and complexation capability (Pardhi et al., 2023).

Emerging γ-CD–producing strains, such as *C. thermautotrophica*, present considerable potential for the development of advanced pharmaceutical formulations. The larger internal cavity of γ-CD enables effective encapsulation of bulky or structurally complex bioactive compounds that may not readily fit within α- or β-CD cavities (Jansook & Loftsson, 2022; Saokham & Loftsson, 2017). This structural advantage contributes to improved aqueous solubility, enhanced molecular stability against degradation, and increased bioavailability of therapeutic agents. Consequently, γ-CD–biased CGTases from such non-*Bacillus* thermophilic organisms represent a promising platform for next-generation drug delivery systems and high-value pharmaceutical applications (Ara et al., 2015).

Collectively, these examples demonstrate that the selection of non-*Bacillus* CGTase sources should be guided by a comprehensive evaluation of catalytic efficiency, stability profile, production yield, and CD specificity to ensure alignment with targeted industrial objectives and process economics. Comparing CGTase enzymes from non-*Bacillus* spp. highlights their potential in diverse industrial applications. Optimizing fermentation strategies and enzyme engineering approaches could further enhance their commercial viability. Table 6.1 shows the summary of CGTase-producing genera and enzymatic properties.

Bioprocessing and purification of cyclodextrin glucanotransferase

To enhance the production of CGTase, optimization of fermentation techniques must be incorporated into bioreactor systems. Both *K. pneumoniae* and *Thermoanaerobacterium* sp. have been shown to benefit from controlled fed-

Table 6.1 Summary of cyclodextrin glucanotransferase-producing genera and enzymatic properties.

Genus	Source	Optimum pH	Optimum temperature	Enzyme production	Major cyclodextrin (CD) produced	Notable enzyme traits
Paenibacillus	Soil	Neutral to alkalic	Moderate	High	β-CD	Moderate temperature stable
Amphibacillus	Alkaline soda lake	Alkalic	Moderate	Moderate	β-CD	Alkaliphilic, β-CD producer
Brevibacillus	Compost	Slightly acidic	Higher	Moderate	β-CD	Thermotolerant, stable
Klebsiella	Wastewater	Slightly acidic	Moderate	Low to moderate	α-CD	Moderate yield, α-CD specific
Microbacterium	Diverse environment	Acidic	High	Low	β-CD	Unusual substrate-binding sites
Thermoanaerobacterium	Hot spring	Neutral	High	High	Balanced CDs	Highly thermostable, anaerobic
Carboxydoxella	Geothermal environments	-	Moderate	Moderate	β-CD	Highly thermostable
Geobacillus thermoglucosidans	Oilfields, hay compost, hydrothermal vents, and soils	Alkalic	High	Moderate	β-CD	Highly thermostable, pH stability, metal ions resistance
Haloferax mediterranei	Hypersaline environments	Neutral	Moderate	-	α-CD	Adaptability to extreme environments

batch fermentation, with favorable nutrient concentrations and lowered substrate inhibition (Kubiak & Wawrzyniak, 2023; Oh et al., 2018). Continuous culture systems offer stable enzyme production, although contamination control and system complexity can pose challenges (Xie, 2022).

Some researchers have also shown that utilizing inexpensive carbon sources such as agricultural residues, potato peel, molasses, and even corn steep liquor can significantly lower production cost while maintaining the quality of the enzyme (Islam and Mamun, 2020; Spigno & Marco, 2015).

After fermentation, CGTase is typically secreted into the extracellular medium and then recovered from the culture supernatant using a multistep purification protocol:

- Cell removal: achieved through centrifugation or microfiltration to separate biomass from the supernatant containing CGTase (Sun et al., 2011; Wang et al., 2006).
- Primary concentration: via ammonium sulfate precipitation or ultrafiltration (Ibrahim et al., 2012).
- Desalting/dialysis: to remove small impurities (Sun et al., 2011).
- Chromatographic purification: ion-exchange chromatography (e.g., DEAE-cellulose) to separate based on charge (Li & Vanderah, 2025); gel filtration chromatography (e.g., Sephadex G-100) for molecular weight-based separation (Ibrahim et al., 2012); affinity chromatography (e.g., starch-based columns) for high-purity recovery.

Following purification, detailed characterization is essential for defining the functional limits of the CGTase enzyme. Characterization involves biochemical and biophysical analysis to determine enzyme properties and confirm activity. Standard SDS-PAGE is used to confirm molecular weight, usually in the 65–85 kDa range, depending on the genus, and for purity assessment (Vassileva and Atanasova, 2023). However, biochemical characterization and advanced techniques provide deeper insights.

Cyclodextrin production

CGTase is a starch-degrading enzyme that catalyzes the conversion of linear maltooligosaccharides into cyclic oligosaccharides known as CD. The enzyme typically utilizes carbohydrate-based donors, such as dextrin, maltose, maltodextrin, maltoheptaose, starch, CD, and maltotriose (Lim et al., 2021). These substrates interact with CGTase through non-covalent forces, including hydrophobic, electrostatic, and hydrogen-bonding interactions. Three conserved carboxylate residues, Glu257, Asp229, and Asp328 within the active site, are important for substrate selectivity and catalytic activity. CGTase catalyzes four key reactions: cyclization, coupling, disproportionation, and hydrolysis (Ara et al., 2021; Qi & Zimmermann, 2005; van der Veen et al., 2000).

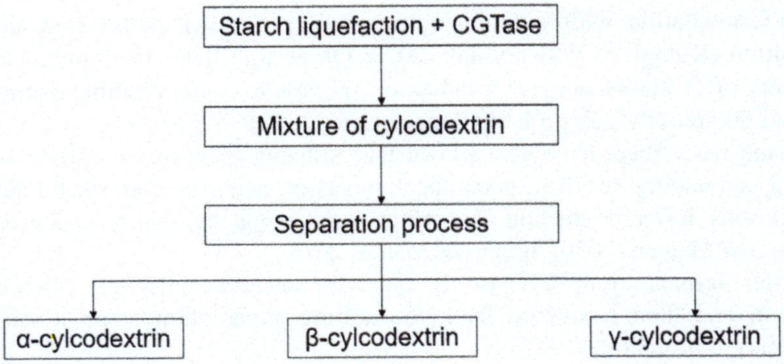

FIGURE 6.1

Cyclodextrin (*CD*) production process using starch liquefaction and cyclodextrin glucanotransferase (*CGTase*), followed by separation into α-, β-, and γ-CDs.

The efficiency of CGTase-mediated transglycosylation reactions is influenced by enzyme concentration and source. Increasing the amount of enzyme loading can improve the initial reaction rate, but too much CGTase can lead to overproduction of CD and additional by-products (Saini et al., 2022). The CGTase from *Thermoanaerobacter* sp. showed more disproportionation activity and thermostability, hence resulting in a higher amount of transglycosylated compound (like resveratrol glucoside) than that from *B. macerans* (Avci & Dönmez, 2012). Besides, an investigation on alkyl glycoside elongation has exposed that CGTase from *P. macerans* produced the highest β-dodecyl maltooctaoside concentration compared to CGTase from *Bacillus stearothermophilus*, *Carboxydocella* sp., and *Thermoanaerobacter* sp. (Lim et al., 2021).

Fig. 6.1 shows the formation of α-, β-, and γ- CDs using CGTase and starch as the substrate. The examples of numerous CD types formed by diverse organisms under different conditions are shown in Table 6.2 β-CD is of particular industrial interest due to several favorable properties: its hydrophobic cavity is optimally sized to form inclusion complexes with a broad range of molecules, such as aromatic compounds and pharmaceuticals. Additionally, its relatively low solubility in water facilitates its separation from the reaction mixture. β-CD complexes are also known for their stability and ease of preparation (Doukyu et al., 2003; Szejtli, 1990).

It should be noted that the CD types and the yield of the CD depend on the microbial source of the CGTases and the bioconversion conditions used (Martins & Hatti-Kaul, 2002; Penninga et al., 1995; Sian et al., 2005). In order to obtain higher yield and better selectivity of CD, various organic solvents were added into the reaction system (Payamifar & Marjani, 2025). However, organic solvents suffer some restrictions, especially when used with food and pharmaceutical products due to regulatory concerns and safety (Yue et al., 2023). This has led to increased concern in attempts to develop selectively efficient CGTases that do not require the

Table 6.2 Different types of cyclodextrin (*CD*) are produced from different organisms and fermentation conditions.

Organism	Fermentation conditions			CD production	References
	pH	Temperature (°C)	Substrate		
Paenibacillus pabuli US132	8.0	25	Potato starch	α-CD (22%): β-CD (63%): γ-CD (15%)	(Jemli et al., 2007)
Amphibacillus sp. NPST-10	10.5	30	Corn starch	α-CD (10.2%): β-CD (86.4%): γ-CD (3.4%)	(Ibrahim et al., 2012)
Brevibacillus	6.0	50	Soluble starch	Mostly β-CD	(Fan et al., 2018)
Klebsiella pneumoniae AS- 22	6.0–9.0	45	Soluble starch	α-CD (81%): β-CD (12%): γ-CD (7%)	(Gawande and Patkar, 2001)
Microbacterium terrae KNR 9	6.0	55.6	Potato starch	Mostly β-CD	(Rajput et al., 2016a)
Thermoanaebacterium	6.0	80	Soluble potato starch	α-CD (62.6%): β-CD (37.4%): γ-CD (0%)	(Ayse & Sedat, 2012)
Carboxydocella sp.	-	37	Wheat starch	α-CD (40%):β-CD (50%): γ-CD (10%)	(Ara et al., 2015)
Geobacillus thermoglucosidans	8.0	60	Soluble starch	α-CD (18.75%): β-CD (34.66%): γ-CD (6.89%)	(Jia et al., 2018)
Haloferax mediterranei	7.2	37	Potato starch	Mostly α-CD	(Bautista, et al., 2012)

Table 6.3 Effect of various parameters on the cyclodextrin (*CD*) production.

Parameter	Examples	Significance
Types of substrate	Pretreatment and types of starch	Conversion and accessibility
CGTase	Concentration, source of CGTase, and immobilization process	Affect the type and quantity of cyclodextrin produced
Operational conditions	Agitation rate, time, and design of the reactor	Influences the reaction rate and overall yield
Reaction parameters	Temperature, ions, and pH	Modifies enzyme activity and stability
Additives	Complexing agents and organic solvent	Alter the distribution of CD produced

addition of such organic solvents, especially for the production of high concentrations of β-CD.

Achieving high starch conversion rates into CD endures a significant limitation. An example of a limitation is the amylopectin's complicated structure, especially the α-1,6-branch, which can prevent the CGTase activity. Using debranching enzymes like isoamylase or pullulanase increases substrate accessibility, hence leading to the production of CD (Rajput et al., 2016c; Rashid et al., 2002). Furthermore, CD accumulation in the reaction mixture can inhibit CGTase activity. This matter can be prevented by choosing suitable complexing agents according to the target type of CD. For instance, cyclohexane, acetonitrile, and *n*-decanol were used as a complexing agent for α-CD type, toluene for β-CD type, and glycyrrhizic acid or bromobenzene was used for γ-CD type (Leemhuis et al., 2010; Saini et al., 2022).

CD production using an enzymatic approach is complex and affected by numerous parameters. Table 6.3 shows the efficiency and specificity of products formed that are significantly influenced by numerous parameters. These parameters include the chemical and physical characteristics of the substrate, the enzyme source and activity, and operational conditions such as the agitation rate, time, and reactor design. Besides, reaction parameters like temperature, ions, and pH can significantly affect the yield and quality of the CD. A comprehensive understanding and optimization of these parameters significantly increases the CD production, enhances selectivity to the intended form of CD, and cost-effectiveness for industrial applications.

The type of starch substrate plays a critical role in the yield and composition of CD. Starches originating from many plant sources, such as rice, potato, corn, and cassava, differ in particle size, branching structure, and amylose and amylopectin concentrations (Wang & Guo, 2020). These features affected the accessibility of the enzyme and its efficiency throughout the cyclization and hydrolysis process. Amylopectin-rich starches are frequently preferable for the CGTase activity because of their branched and high solubility features. The raw starch also requires pretreatment, such as gelatinization, to break up its structure and increase substrate accessibility.

The concentration of substrate used during the reaction process is an important parameter that affects the production of CD. An appropriate amount of starch permits an adequate supply of reactants for the cyclization process while preventing mixing and diffusion restrictions. Excessive starch concentration resulted in high viscosity, reduced enzyme mobility, and substrate inhibition (Karim et al., 2024; Ona et al., 2025). To increase substrate accessibility, the pretreatment processes such as enzymatic liquefaction, thermal gelatinization, and acid hydrolysis must be implemented. However, a severe pretreatment process would lead to degradation and alteration in product distribution, resulting in smaller oligosaccharides or unwanted by-products.

The CGTase source influences its properties such as stability, kinetics, and specificity, thus affecting the synthesis of CD. Microbial strains yield different forms of CGTases, including α-, β-, and γ-CD. The source of CGTase also affected the catalytic efficiency of the enzyme, tolerance to environmental pressure, and susceptibility to inhibition. Recombinant enzymes synthesized mutations provide high selectivity and improved thermal stability, but usually have disadvantages, such as reduced substrate strength and altered substrate affinity (Teufl et al., 2022).

The concentration of CGTase has a substantial impact on the synthesis rate and the CD yield. The rapid transformation of starch into CD occurred at an optimal CGTase concentration. However, a high concentration of CGTase leads to hydrolysis of CD into linear oligosaccharides (Jicsinszky et al., 2025). Enzyme specificity also affected the ratio of α-, β-, and γ-CD. Certain applications and processes may require a predominance type of CD, necessitate process modifications or enzyme engineering to alter the ratio of produced CD.

Reactor design significantly affects mass and heat transfer, especially in viscous or high-solid media (Rebrov et al., 2011). Inadequate mixing may lead to uneven enzyme distribution, substrate stratification, and local pH or temperature gradients. Batch, fed-batch, or continuous stirred-tank reactors each offer advantages and challenges in scaling CD production. In particular, immobilized enzyme systems require specially designed reactors to ensure efficient substrate access and minimize diffusional resistance.

Reaction parameters such as pH and temperature are decisive for the stability and activity of CGTase. Most CGTases work optimally at temperatures between 45°C–60°C and a pH of 5.5–8.0 (Saini et al., 2022). Differences from these optimal values can lead to reduced enzyme denaturation and efficiency. For industrial processes, a constant and optimal pH and temperature are needed to avoid loss of activity and allow reproducibility. Additionally, fluctuations can also influence CD size distribution, so the smaller CD is preferable at the higher temperatures or conformation of the binding site changes in acidic or alkaline conditions.

Complexing agents and certain organic solvents can shift the CD product distribution or stabilize specific CD forms during the reaction (Mura et al., 2025). For example, toluene, ethanol, or certain ketones have been shown to increase β-CD selectivity by reducing its solubility and promoting crystallization. Similarly, the addition of guest molecules (e.g., cholesterol, aromatic compounds) may enhance the formation of larger CD-like γ-CD. However, the use of such additives must be carefully controlled to avoid enzyme denaturation or environmental and safety concerns.

Industrial applications of non-*Bacillus* cyclodextrin glycosyltransferase

Beyond the well-studied *Bacillus* spp., several **non-***Bacillus* microorganisms have been reported as valuable producers of CGTase, demonstrating diverse biochemical properties and industrial relevance. Among these, *Paenibacillus* sp. represent an important group of CGTase-producing bacteria with promising applications.

The CDs produced using CGTase from *Paenibacillus campinasensis* are applied in multiple sectors, including the food, cosmetic, pharmaceutical, and chemical industries. Their ability to form inclusion complexes with hydrophobic compounds enhances solubility, stability, and controlled release properties, thereby expanding their commercial utility (Alves-Prado et al., 2006). Similarly, *Paenibacillus barengoltzii* produces a CGTase distinguished by its product specificity and catalytic versatility. The CGTase from this species is particularly relevant in the production of maltooligosaccharides, compounds recognized for their potential prebiotic and health-promoting effects (Castillo et al., 2018). Another significant example is *Paenibacillus* sp. L55, which was isolated from the alkaline soda lake of Lonar, India (Krishna Kumari et al., 2013). The resulting CDs are valuable in food, pharmaceutical, and cosmetic industries, particularly where stability under alkaline processing conditions is advantageous.

In the food sector, CGTase is used to produce CD, which improves the quality and shelf life of products. A study has shown that β-CGTase from *P. pabuli* US132 strain improves bread quality by increasing the volume of bread and reducing the firmness of the bread during storage, which is due to the formation of inclusion complexes with starch components, thereby changing the dough properties and slowing down retrogradation (Jemli et al., 2007). In addition, CGTase has been used to produce CD, which encapsulates flavor and aroma compounds and stabilizes them during processing and storage (Allode et al., 2024). This encapsulation technique has been shown to be effective in preserving the sensory qualities of various foods, including dairy products and baked goods.

Species of the genus *Klebsiella* have been reported to produce functional CGTase enzymes capable of synthesizing CDs from starch substrates. CGTase derived from *Klebsiella* spp. has attracted attention due to its catalytic efficiency and potential adaptability to different processing conditions (Pardhi et al., 2023). Although research on *Klebsiella*-derived CGTase is comparatively less extensive than that on *Bacillus* spp., it demonstrated enzymatic capability and broad applicability, highlighting its potential as an alternative microbial source for industrial CD production.

Another promising non-*Bacillus* source of CGTase is the thermophilic bacterial genus *Thermoanaerobacter*. The CGTases derived from *Thermoanaerobacter* sp. are of considerable industrial interest due to their intrinsic thermostability and operational robustness at elevated temperatures. Enzymes from thermophilic microorganisms generally exhibit enhanced structural stability, prolonged half-life, and resistance to thermal denaturation, which are highly advantageous for

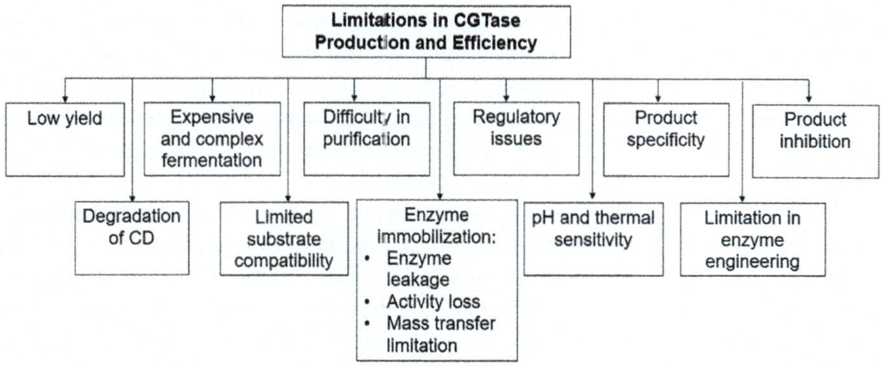

FIGURE 6.2

Limitations in cyclodextrin glycosyltransferase (*CGTase*) production and efficiency.

large-scale starch conversion processes (Adedeji et al., 2024). High-temperature processing reduces the risk of microbial contamination, improves substrate solubility, and can enhance reaction kinetics, thereby increasing overall production efficiency. In many cases, thermophilic CGTases also demonstrate activity under alkaline conditions, further broadening their suitability for industrial applications. Advances in molecular biology, including gene cloning, heterologous expression, and protein engineering, have enabled optimization of these enzymes to achieve higher yields, improved catalytic efficiency, and tailored CD product specificity. Consequently, *Thermoanaerobacter*-derived CGTase represents a valuable alternative platform for sustainable and high-performance CD production in food, pharmaceutical, and chemical industries.

Collectively, these examples highlight that non-*Bacillus* CGTase sources significantly contribute to the diversification of CD production platforms. Their distinct enzymatic characteristics, including altered product specificity, enhanced thermostability, and adaptability to extreme conditions, support expanding industrial applications across food, pharmaceutical, cosmetic, and chemical sectors.

Challenges and future perspectives

The CGTase production and usage face challenges that prevent its complete biotechnological potential from being achieved, even though it has become a significant role in industrial (Fig. 6.2). The major challenge in the production of CGTase is the inherent low enzyme yield from wild-type microbial strains such as *K. oxytoca* and *Thermoanaerobacter* (Muniz et al., 2024). These organisms commonly need to be grown under optimized fermentation conditions in order to produce CGTase at sufficient levels to be detected, further complicating the production process. The repressive nature of the enzyme expression and poor regulation in the native hosts limit scalability for industrial production. Instead, recombinant DNA technologies have created alternatives, but expression in other heterologous systems, such as *E. coli*, and *Pichia pastoris,* can pose folding problems or post-translational incompatibilities.

To grow CGTase-producing microbes, strict control of environmental parameters such as pH, temperature, aeration, and carbon/nitrogen source is required (Amiri et al., 2015). This leads to high energy consumption and high equipment requirements, especially in industry. Additionally, downstream purification is labor-intensive and costly due to the need to isolate CGTase from a mixture of extracellular proteins, pigments, and polysaccharides. Affinity and ion-exchange chromatography are common methods, but may not be economically viable for bulk enzyme production unless process intensification strategies are adopted.

CGTase typically produces a mixture of α-, β-, and γ-CDs from starch, with the distribution depending on the enzyme source and reaction conditions. However, industrial applications often require a high yield of a specific CD type, most commonly β-CD. Achieving selective production remains difficult, as the enzyme's active site is inherently flexible and accommodates a variety of intermediate ring sizes. While protein engineering has enabled some improvement in specificity, fully tailoring CGTase toward a single product type without compromising its activity remains a significant hurdle.

The other limitation is that the synthesized CD can block the action of CGTase by competing with the substrate for binding to the active site of the enzyme (Leemhuis et al., 2003). Among them, β-CD is particularly inhibitory owing to its strong affinity. Besides, some of the by-products like maltose, glucose, and linear dextrins can accumulate in the reaction and harm the enzyme activity and stability. This inhibition can decrease reaction kinetics and overall yield, and demands in situ product removal strategies or the use of a chelating agent for chelating the CD formed.

The enzymatic action profiles of CGTase generally have narrow pH and temperature action ranges with an optimal pH of 5.5–8.0 and thermostability at temperatures of 45°C–60°C. Deviation from these conditions can lead to partial unfolding or denaturation, resulting in a permanent loss of activity (Pardhi et al., 2023). This sensitivity reduces the enzyme's activity in variable industrial processes, particularly those that require extreme conditions or long reaction times. Some thermostable variants have been identified, but they are catalytically inferior to the mesophilic counterparts.

CGTase prefers gelatinized or soluble starch, as native starch granules are often inaccessible due to their semicrystallinity. Therefore, pretreatment steps such as thermal gelatinization or enzymatic liquefaction are necessary to improve the accessibility of the substrate (Singh et al., 2007). These steps are very energy-intensive and increase the cost and complexity of the entire process. The type and form of starch used can also influence the enzyme's performance and the CD distribution profile.

Although CGTase catalyzes cyclisation, it can also exhibit additional side activities, such as hydrolysis or coupling reactions. Such side effects can break or rebuild CD that has already been formed and reduce the overall yield of CD. The extent of this unwanted activity is dependent on enzyme concentration, reaction time, and substrate-to-product ratios, and can be problematic in prolonged batch or continuous operations without feedback control.

Immobilizing CGTase on solid supports can enhance its reusability and operational stability (Schöffer et al., 2017). However, immobilization often leads to reduced enzyme activity due to mass transfer limitations, improper orientation, or partial denaturation during the binding process. Furthermore, immobilization materials and procedures can be costly, and enzyme leaching remains a concern in repeated use or continuous-flow systems. To attain a balance between stability and catalytic performance, immobilization parameters and support materials need to be optimized.

Scaling up CGTase-catalyzed operation from laboratory to industrial application comprises several limitations, such as consistent control of reaction parameters, preserving enzyme stability over a lengthy period, and preventing microbial contamination. However, continuous operation leads to a poor performance because of enzyme inactivation, fouling, or accumulation of undesirable by-products (Rabadiya et al., 2024). Therefore, reactor design must account for mass transfer requirements and enzyme limitations to gain optimal performance during the production stage.

Conclusion

CGTase from non-*Bacillus* sources has become a promising biocatalyst in the field of industrial biotechnology due to its unique enzymatic properties and its ability to respond to a variety of production systems. Although commercially available CGTases exist with Toruzyme and Amano CGTase, problems such as product inhibition, low yield, thermolability, and complexity of CD separation are still critical barriers to commercialization on an industrial scale. To resolve these problems, great advancements have been made in heterologous gene expression, molecular modification, and protein engineering to improve the enzyme specificity, stability, and catalytic performance. These actions assist to form high thermostable CGTase that are more resistant to product inhibition and denaturation, which is essential for high production of CD.

Research into new microbial sources and the use of synthetic biology and gene editing technologies are constantly adding to the selection of CGTase with the desired industrial properties. With the global market for CD expected to reach approximately $390 million by 2027, there is an even greater need for progress towards more efficient, economical, and environmentally friendly CGTase-based processes. Continuous innovation in enzyme engineering, bioprocess engineering, and biocatalysts will be critical to realizing the full potential of CGTase to meet the increasing demand for CD in the market and expanding its applications across various industries.

AI use declaration

This chapter contains text parts that were created or refined using generative artificial intelligence (AI) tools, including ChatGPT (OpenAI), InstaText, and Copilot. These tools were used exclusively to assist with language reformulation, summarization, and clarity improvement. All scientific content, data interpretations, and conclusions are the result of the authors' own analysis, who take full responsibility for the integrity and accuracy of the final manuscript. No AI tool was used to generate the original research results or to perform a critical analysis.

AI disclosure

During the preparation of this work, the author(s) used ChatGPT (OpenAI), InstaText, and Copilot in order to assist with language reformulation, summarization, and clarity improvement. All scientific content, data interpretations, and conclusions are the result of the authors' own analysis, who take full responsibility for the integrity and accuracy of the final manuscript. No AI tool was used to generate the original research results or to perform a critical analysis. After using this tool/service, the author(s) reviewed and edited the content as needed and take(s) full responsibility for the content of the publication.

References

Abelyan, V. A., Balayan, A. M., Manukyan, L. S., Afyan, K. B., Meliksetyan, V. S., Andreasyan, N. A., & Markosyan, A. A. (2002). Characteristics of cyclodextrin production using cyclodextrin glucanotransferases from various groups of microorganisms. *38*(6), 616–624.

Adedeji, O. S., Egwim, E. C., Kabiru, A. Y., Adebayo Oyewole, O., & Contiero, J. (2024). Production and purification of cyclodextrin glucanotransferase from *Alkalihalobacillus lehensis* JO-26; [Producción y purificación de la glucanotransferasa de ciclodextrina de *Alkalihalobacillus lehensis* JO-26]. *Biotecnologia Aplicada, 41*(1), 1201–1207.

Aiassa, V., Garnero, C., Longhi, M. R., & Zoppi, A. (2021). *Cyclodextrin multicomponent complexes: Pharmaceutical applications.*

Aizawa, S. I. (2014). Paenibacillus alvei—Flagella-dependent SOcial motility. *The Flagellar World,* 3–4. https://doi.org/10.1016/B978-0-12-417234-0.00017-7.

Allode, T. D., Vodounon, A. C., & Dossou, A. G. (2024). Applications of cyclodextrin-gluconotransferase in the biosynthesis of cyclodextrins: Characteristics, sources and production. *American Journal of Microbiological Research, 12*(3), 63–78. https://doi.org/10.12691/ajmr-12-3-4.

Al-Sharawi, S. Z. R., Ibrahim, A. S. S., El-Shatoury, E. H., Gebreel, H. M., & Eldiwany, A. (2013). A new low molecular mass alkaline cyclodextrin glucanotransferase from *Amphibacillus* sp. NRC-WN isolated from an Egyptian soda lake. *Electronic Journal of Biotechnology, 16*, 1–13. https://doi.org/10.2225/vol16-issue6-fulltext-18.

Alves-Prado, H. F., Gomes, E., & Da Silva, R. (2006). Evaluation of solid and submerged fermentations for the production of cyclodextrin glycosyltransferase by *Paenibacillus campinasensis* H69-3 and characterization of crude enzyme. *Applied Biochemistry and Biotechnology, 129*(1–3), 234–246. https://doi.org/10.1007/978-1-59745-268-7_18.

Amiri, A., Rosfarizan, M., Raha Abdul, R., Rosli Md, I., Farideh, N., Joo Shun, T., & Abbasiliasi, S. (2015). Cyclodextrin glycosyltransferase biosynthesis improvement by recombinant *Lactococcus lactis* NZ:NSP:CGT: Medium formulation and culture condition optimization. *Biotechnology & Biotechnological Equipment, 29*(3), 555–563. https://doi.org/10.1080/13102818.2015.1009713.

Ara, K. Z. G., Lundemo, P., Fridjonsson, O. H., Hreggvidsson, G. O., Adlercreutz, P., & Karlsson, E. N. (2015). A CGTase with high coupling activity using γ-cyclodextrin isolated from a novel strain clustering under the genus *Carboxydocella*. *Glycobiology, 25*(5), 514–523. https://doi.org/10.1093/glycob/cwu182.

Ara, K. Z. G., Linares-Pastén, J. A., Jönsson, J., Viloria-Cols, M., Ulvenlund, S., Adlercreutz, P., & Karlsson, E. N. (2021). Engineering CGTase to improve synthesis of alkyl glycosides. *Glycobiology, 31*(5), 603–612. https://doi.org/10.1093/glycob/cwaa109.

Avci, A., & Do, S. (2009). A novel thermophilic anaerobic bacteria producing cyclodextrin glycosyltransferase. *Process Biochemistry, 44*, 36–42. https://doi.org/10.1016/j.procbio.2008.09.006.

Avci, A., & Dönmez, S. (2012). Purification and characterization of a thermostable cyclodextrin glycosyltransferase from *Thermoanaerobacter* sp. P4. *African Journal of Biotechnology, 11*(45), 10407–10415.

Ayse, A., Nichols, N. N., Saha, B. C., Frazer, S. E., Cotta, M. A., & Donmez, S. (2014). A thermostable cyclodextrin glycosyltransferase from *Thermoanaerobacter*. *Current Biotechnology, 3*(October 2023), 305–312. https://doi.org/10.2174/2211550103666141027220452.

Ayse, A., & Sedat, D. (2012). Purification and characterization of a thermostable cyclodextrin glycosyltransferase from *Thermoanaerobacter* sp. P4. *African Journal of Biotechnology, 11*(45), 10407–10415. https://doi.org/10.5897/ajb11.3513.

Bautista, V., Esclapez, J., PErez-Pomares, F., MartInez-Espinosa, R. M., Camacho, M., & Bonete, M. J. (2012). Cyclodextrin glycosyltransferase: A key enzyme in the assimilation of starch by the halophilic archaeon *Haloferax mediterranei*. *Extremophiles, 16*, 147–159. https://doi.org/10.1007/s00792-011-0414-z.

Bonch-osmolovskaya, E., Elcheninov, A., Zayulina, K., & Kublanov, I. (2017). New thermophilic prokaryotes with hydrolytic activities. *Microbiology Australia, 2013*, 18–21. https://doi.org/10.2495/WP080271.

Bosch, S., Sanchez-freire, E., Morana, C., Quesada, J., Berenguer, J., & Hidalgo, A. (2021). Thermostability engineering of a class II pyruvate aldolase from *Escherichia coli* by in vivo folding interference. *ACS Sustainable Chemistry & Engineering*. https://doi.org/10.1021/acssuschemeng.1c00699.

Castillo, J. D. L. M., Caminata Landriel, S., Sànchez Costa, M., Taboga, O. A., Berenguer, J., Hidalgo, A., Ferrarotti, S. A., & Costa, H. (2018). A single mutation in cyclodextrin glycosyltransferase from *Paenibacillus barengoltzii* changes cyclodextrin and maltooligosaccharides production. *Protein Engineering, Design and Selection, 31*(10), 399–407. https://doi.org/10.1093/protein/gzy034.

Chang, J., Lee, Y., & Fang, S. (2013). Recombinant expression and characterization of an organic-solvent-tolerant ! -Amylase from *Exiguobacterium* sp. DAU5. *Applied Biochemistry and Biotechnology, 169*, 1870–1883. https://doi.org/10.1007/s12010-013-0101-x.

Chen, S., Xiong, Y., Su, L., Wang, L., & Wu, J. (2017). Position 228 in *Paenibacillus macerans* cyclodextrin glycosyltransferase is critical for 2-O- d -glucopyranosyl- l -ascorbic acid synthesis. *Journal of Biotechnology, 247*, 18–24. https://doi.org/10.1016/j.jbiotec.2017.02.011.

Dassarma, S., & Dassarma, P. (2015). Halophiles and their enzymes: Negativity put to good use. *Current Opinion in Microbiology, 25*, 120–126. https://doi.org/10.1016/j.mib.2015.05.009 (Table 6.1),.

Delgado-García, M., Valdivia-Urdiales, B., Aguilar-González, C. N., Contreras-Esquivel, J. C., & Rodríguez-Herrera, R. (2012). Halophilic hydrolases as a new tool for the biotechnological industries. *Journal of the Science of Food and Agriculture, 92*(13), 2575–2580. https://doi.org/10.1002/jsfa.5860.

Ding, R., Li, Z., Chen, S., Wu, D., Wu, J., & Chen, J. (2010). Enhanced secretion of recombinant -cyclodextrin glucosyltransferase from *E. coli* by medium additives. *Process Biochemistry, 45*(6), 880–886. https://doi.org/10.1016/j.procbio.2010.02.009.

Doukyu, N., Kuwahara, H., & Aono, R. (2003). Isolation of *Paenibacillus illinoisensis* that produces cyclodextrin glucanotransferase resistant to organic solvents. *Bioscience, Biotechnology, and Biochemistry, 67*(2), 334–340.

Fan, B., Duan, X., & Wu, J. (2018). Studies on preparation of β-cyclodextrin production using *Bacillus stearothermophilus* β-cyclodextrin glycosyltransferase and reaction conditions optimization. *Journal of Food Science and Biotechnology, 37*(9), 897–902. https://doi.org/10.3969/j.issn.1673-1689.2018.09.001.

Fouda-mbanga, B. G., Tywabi-ngeva, Z., Badawy, W. M., Ebite, C., Onotu, O. P., Abogidi, C., & Uzordinma, A. P. (October 2024). Green cyclodextrins-derivatives for sustainable remediation of pesticides and heavy metals. *A review,* 1328.

Gawande, B. N., & Patkar, A. Y. (2001). Purification and properties of a novel raw starch degrading-cyclodextrin glycosyltransferase from *Klebsiella pneumoniae* AS- 22. *28*, 735–743.

Gawande, B. N., & Patkar, A. Y. (2001b). Studies on enzyme cyclodextrin glycosyKransferase from *Klebsiella pneumoniae* pneumoniae AS-22. *Enzyme and Microbial Technology, 28*, 735–743.

Graebin, N. G., Schöffer, J. D. N., De Andrades, D., Hertz, P. F., Ayub, M. A. Z., & Rodrigues, R. C. (2016). Immobilization of glycoside hydrolase families GH1, GH13, and GH70: State of the art and perspectives. *Molecules, 21*(8). https://doi.org/10.3390/molecules21081074.

Griffiths, D. B., Tiwari, R. P., Murphy, D. V., & Scott, C. (2025). *Haloferax* and the Haloferacaceae: Potential role in bioindustry. *Biotechnology Advances, 84*(July), 108666. https://doi.org/10.1016/j.biotechadv.2025.108666.

Hadad, D., Geresh, S., & Sivan, A. (2005). Biodegradation of polyethylene by the thermophilic bacterium *Brevibacillus borstelensis*. *Journal of Applied Microbiology, 98*, 1093–1100. https://doi.org/10.1111/j.1365-2672.2005.02553.x.

Hammadi, A. I., Merzoug, M., Aireche, M., & Zater, Z. Y. (2025). Integrated proteomic and molecular identification of thermophilic geobacillus strains from algerian desert sands and their enzymatic potential. *Life, 1*, 19.

Han, F., Tong, D., Guo, Z., Li, Z., Zhang, Z., Du, R., & Xu, Q. (2025). Food bioscience isolation and characterization of a novel *Paenibacillus* species from high-salt environment with broad-spectrum antimicrobial activity against food spoilage microorganisms. *Food Bioscience, 68*(April), 106660. https://doi.org/10.1016/j.fbio.2025.106660.

Harnvoravongchai, P., Singwisut, R., Ounjai, P., Aroonnual, A., Kosiyachinda, P., Janvilisri, T., & Id, S. C. (2020). *Isolation and characterization of thermophilic cellulose and hemicellulose degrading from tropical dry deciduous forest soil,* 1–16. https://doi.org/10.1371/journal.pone.0236518.

Ibrahim, A. S. S., Al-Salamah, A. A., El-Tayeb, M. A., El-Badawi, Y. B., & Antranikian, G. (2012). A novel cyclodextrin glycosyltransferase from alkaliphilic *Amphibacillus* sp. NPST-10: Purification and properties. *International Journal of Molecular Sciences, 13*(8), 10505–10522. https://doi.org/10.3390/ijms130810505.

Islam, F., Mamun, M. A. A., Azad, A. K., Yamashoji, S., & Bari, M. L. (2020). Use of potato peel waste-a cost effective alternative substrate for the production of industrially useful α -amylase enzyme use of potato peel waste- a cost effective alternative substrate for the production of industrially useful α -amylase enzyme. *Journal of Agriculture, Food and Environment.* https://doi.org/10.47440/JAFE.2020.1420 *October 2021.*

Jailani, N., Rizwana, N., Suhaimi, S., & Mohamed, M. (2022). International journal of biological macromolecules cross-linked cyclodextrin glucanotransferase aggregates from *Bacillus lehensis* G1 for cyclodextrin production: Molecular modeling, developmental, physicochemical, kinetic and thermodynamic properties. *International Journal of Biological Macromolecules, 213*(January), 516–533. https://doi.org/10.1016/j.ijbiomac.2022.05.170.

Jansook, P., & Loftsson, T. (2022). Self-assembled γ -cyclodextrin as nanocarriers for enhanced ocular drug bioavailability. *International Journal of Pharmaceutics, 618*(March), 121654. https://doi.org/10.1016/j.ijpharm.2022.121654.

Javed, M., Baghaei-yazdi, N., Qin, W., & Amartey, S. (2017). An improved agar medium for growth of *Geobacillus thermoglucosidarius* strains. *Journal of Microbiological Methods, 132*, 116–118. https://doi.org/10.1016/j.mimet.2016.11.014.

Jemli, S., Messaoud, E. B., Ayadi-Zouari, D., Naili, B., Khemakhem, B., & Bejar, S. (2007). A β-cyclodextrin glycosyltransferase from a newly isolated *Paenibacillus pabuli* US132 strain: Purification, properties and potential use in bread-making. *Biochemical Engineering Journal, 34*(1), 44–50. https://doi.org/10.1016/j.bej.2006.11.016.

Jia, X., Ye, X., Chen, J., Lin, X., Vasseur, L., & You, M. (2018). Purification and biochemical characterization of a cyclodextrin glycosyltransferase from *Geobacillus thermoglucosidans* CHB1. *Starch-Stärke, 70*(1–2), 1700016.

Jiang, Y., Zhou, J., Wu, R., Xin, F., Zhang, W., Fang, Y., & Ma, J. (2018). Heterologous expression of cyclodextrin glycosyltransferase from *Paenibacillus macerans* in *Escherichia coli* and its application in 2-O- α -D-glucopyranosyl-L- ascorbic acid production. *BMC Biotechnology, 18*(53), 1–10.

Jicsinszky, L., Martina, K., Boffa, L., & Cravotto, G. (2025). Cyclodextrin based artificial enzymes. *Sustainable Chemistry and Pharmacy, 44*, 101945.

Kamaraj, M., Suresh Babu, P., Shyamalagowri, S., Pavithra, M. K. S., Aravind, J., Kim, W., & Govarthanan, M. (2024). β-cyclodextrin polymer composites for the removal of pharmaceutical substances, endocrine disruptor chemicals, and dyes from aqueous solution- A review of recent trends. *Journal of Environmental Management, 351*(November 2023), 119830. https://doi.org/10.1016/j.jenvman.2023.119830.

Karan, R., Kumar, S., Sinha, R., & Khare, S. K. (2012). In T. Satyanarayana, & B. N. Johri (Eds.). *Halophilic microorganisms as sources of novel enzymes bt - microorganisms in sustainable agriculture and biotechnology* (pp. 555–579). Springer Netherlands. https://doi.org/10.1007/978-94-007-2214-9_25.

Karim, A., Mohammadi, L., Osse, E. F., Aider, M., Saqui-Salces, M., & Khalloufi, S. (2024). Effect of polysaccharide-induced viscosity on the digestion of proteins, fats, and carbohydrates in food: A comprehensive review and future perspectives. *Trends in Food Science & Technology,* 104757.

Kelly, R. M., Dijkhuizen, L., & Leemhuis, H. (2009). The evolution of cyclodextrin glucanotransferase product specificity. *Applied Microbiology and Biotechnology, 84*(1), 119–133. https://doi.org/10.1007/s00253-009-1988-6.

Khaswal, A., Chaturvedi, N., Kumar, S., Priya, M., Kumar, R., & Kumar, P. (2022). Current status and applications of genus *Geobacillus* in the production of industrially important products—A review. *Folia Microbiologica, 389*, 404. https://doi.org/10.1007/s12223-022-00961-w.

Khazaal, A., Al-mansoori, K., Yu, S., & Thangadurai, S. (2025). Exploring enzyme thermostability enhancement through molecular modification: What remains to be known ? *Journal of Molecular Structure, 1337*(February), 141852. https://doi.org/10.1016/j.molstruc.2025.141852.

Kim, M. H., Sohn, C. B., & Oh, T. K. (1998). Cloning and sequencing of a cyclodextrin glycosyltransferase gene from *Brevibacillus brevis* CD162 and its expression in *Escherichia coli*. *FEMS Microbiology Letters, 164*(313), 411–418.

Kochetkova, T. V., Rusanov, I. I., Pimenov, N. V., Kolganova, T. V., Lebedinsky, A. V., & Tatyana, E. A. B. (2011). Anaerobic transformation of carbon monoxide by microbial communities of Kamchatka hot springs. *Extremophiles, 15*(2), 319–325. https://doi.org/10.1007/s00792-011-0362-7.

Krishna Kumari, C., Shanmuga Reddy, C., Raja Ratna Reddy, Y., Narapa Reddy, D., & Damodar Reddy, C. (2013). In vitro antimicrobial activity of the leaf extracts of *Argemone mexicana* against selected pathogenic microorganisms. *International Journal of Pharma and Bio Sciences, 4*(1), B536–B541.

Kubiak, P., & Wawrzyniak, J. (2023). The impact of process parameters on 1, 3-propanediol production and 3-hydroxypropionaldehyde accumulation in fed-batch fermentation of glycerol with *Citrobacter freundii* AD119. *Applied Science, 13*, 1–15.

Kumari, A., Satpal, P., & Bisht, S. (2014). Brevibacillus as a biological tool : A short review. *Antonie Van Leeuwenhoek,* 623–639. https://doi.org/10.1007/s10482-013-0099-7.

Lee, J., Choi, K., Choi, J., & Lee, Y. (1992). Enzymatic production of a-cyclodextrin with the cyclomaltodextrin glucanotransferase of Klebsiella oxytoca 19-1. *Enzyme and Microbial Technology, 14*, 1–4.

Lee, M., Yang, S., Kim, J., Lee, H., Kim, J., & Park, K. (2007). Characterization of a thermostable cyclodextrin glucanotransferase from *Pyrococcus furiosus* DSM3638. *Extremophiles, 11*, 537–541. https://doi.org/10.1007/s00792-007-0061-6.

Leemhuis, H., Dijkstra, B. W., & Dijkhuizen, L. (2003). Thermoanaerobacterium thermosulfurigenes cyclodextrin glycosyltransferase: Mechanism and kinetics of inhibition by acarbose and cyclodextrins. *European Journal of Biochemistry, 270*(1), 155–162.

Leemhuis, H., Kelly, R. M., & Dijkhuizen, L. (2010). Engineering of cyclodextrin glucanotransferases and the impact for biotechnological applications. *Applied Microbiology and Biotechnology, 85*(4), 823–835. https://doi.org/10.1007/s00253-009-2221-3.

Leija, S. C., Barrera, L. E., Cruz, B. V., Conejo, Y. C., Ortíz, R. V., Cruz, G. V., Saenz, R. A., Tovar, Y. M., Manzo, S. G., & Ochoa, B. H. (2022). Mining for novel cyclomaltodextrin glucanotransferases unravels the carbohydrate metabolism pathway via cyclodextrins in *Thermoanaerobacterales*. *Scientific Reports,* 1–16. https://doi.org/10.1038/s41598-021-04569-x.

Li, B., Sun, Y., Zhu, X., Qian, S., Pu, J., Guo, Y., Wu, H., Zhang, L., & Xin, Y. (2023). Aggregation interface and rigid spots sustain the stable framework of a thermophilic N-demethylase. *Journal of Agricultural and Food Chemistry, 71*(14), 5614–5629. https://doi.org/10.1021/acs.jafc.3c00877.

Li, G., Yuan, Y., Jin, B., Zhang, Z., Murtaza, B., Zhao, H., Li, X., Wang, L., & Xu, Y. (2023). Feasibility insights into the application of *Paenibacillus pabuli* E1 in animal feed to eliminate non-starch polysaccharides. *Frontiers in Microbiology*. https://doi.org/10.3389/fmicb.2023.1205767 *August*.

Li, T., & Vanderah, D. (2025). Influence of the ion-exchange functional groups on protein anion-exchange chromatography. *Analytical Biochemistry, 702*(March), 115849. https://doi.org/10.1016/j.ab.2025.115849.

Lim, C. H., Rasti, B., Sulistyo, J., & Hamid, M. A. (2021). Comprehensive study on transglycosylation of CGTase from various sources. *Heliyon, 7*(2), e06305. https://doi.org/10.1016/j.heliyon.2021.e06305.

Martina, K., Boffa, L., & Cravotto, G. (2025). *Cyclodextrin based artificial enzymes ´ szl o, 44*(December 2024). https://doi.org/10.1016/j.scp.2025.101945.

Martins, R. F., & Hatti-Kaul, R. (2002). A new cyclodextrin glycosyltransferase from an *akaliphilic Bacillus agaradhaerens* isolate: purification and characterisation. *Enzyme and Microbial Technology, 30*(1), 116–124.

Matarredona, L., & Guío, J. (2024). Global Lrp regulator protein from *Haloferax mediterranei*: Transcriptional analysis and structural characterization. *International Journal of Biological Macromolecules, 260*(January). https://doi.org/10.1016/j.ijbiomac.2024.129541.

Mitra, S., Mukhopadhyay, B. C., Mandal, A. R., Arukha, A. P., Chakrabarty, K., Das, G. K., Chakrabartty, P. K., & Biswas, S. R. (2015). Cloning, overexpression, and characterization of a novel alkali-thermostable xylanase from *Geobacillus* sp. WBI. *Journal of Basic Microbiology, 55*(4), 527–537. https://doi.org/10.1002/jobm.201400495.

Montero-Lobato, Z., Ramos-Merchante, A., Fuentes, J. L., Sayago, A., Fernandez-Recamales, A., Martinez-Espinosa, R. M., Vega, J. M., Vilchez, C., & Garbayo, I. (2018). Optimization of growth and carotenoid production by haloferax mediterranei using response. *Marine drugs*. https://doi.org/10.3390/md16100372 *october*.

Mori, S., Hirose, S., Oya, T., & Kitahata, S. (1994). Purification and properties of cyclodextrin glucanotransferase from *Brevibacterium* sp. *Bioscience, Biotechnology, and Biochemistry, 58*(11), 1968–1972.

Muniz, I., de, C. B., Santos, J. B., de Oliveira, R. M., Santos, F. G., Souza Junior, E. C., de, Oyama, L., Fontan, R.da C. I., & Bonomo, R. C. F. (2024). Current advances in obtaining novel cyclodextrin glycosyltransferases for optimizing the synthesis of cyclodextrins. *Process Biochemistry, 145*(April), 195–209. https://doi.org/10.1016/j.procbio.2024.07.008.

Mura, P., Maestrelli, F., Gonçalves, L. M. D., Cirri, M., Mennini, N., & Almeida, A. J. (2025). Cyclodextrin complexation as a fruitful strategy for improving the performance of nebivolol delivery from solid lipid nanoparticles. *International Journal of Pharmaceutics, 668*, 124972.

Nicolaescu, O. E., Belu, I., Mocanu, A. G., Manda, V. C., Rău, G., Pîrvu, A. S., Ionescu, C., Ciulu-Costinescu, F., Popescu, M., & Ciocîlteu, M. V. (2025). Cyclodextrins: Enhancing drug delivery, solubility and bioavailability for modern therapeutics. *Pharmaceutics, 17*(3), 1–36. https://doi.org/10.3390/pharmaceutics17030288.

Ogunbadejo, B. A., & Al-zuhair, S. (2024). Enhanced performance of cyclodextrin glycosyltransferase by immobilization on amine-induced macroporous metal organic framework. *LWT, 200*(May), 116221. https://doi.org/10.1016/j.lwt.2024.116221.

Oh, B. R., Lee, S. M., Heo, S. Y., Seo, J. W., & Kim, C. H. (2018). Efficient production of 1, 3 - propanediol from crude glycerol by repeated fed - batch fermentation strategy of a lactate and 2, 3 - butanediol deficient mutant of *Klebsiella pneumoniae*. *Microbial Cell Factories*, 1–9. https://doi.org/10.1186/s12934-018-0921-z.

Omae, K., Fukuyama, Y., Yasuda, H., Mise, K., Yoshida, T., & Sako, Y. (2019). Diversity and distribution of thermophilic hydrogenogenic carboxydotrophs revealed by microbial community analysis in sediments from multiple hydrothermal environments in Japan. *Archives of Microbiology, 201*(7), 969–982. https://doi.org/10.1007/s00203-019-01661-9.

Ona, J. I., Halling, P. J., & Ballesteros, M. (2025). Effect of steam explosion on enzyme hydrolysis of cassava peels. *Journal of Chemical Society of Nigeria, 50*(2).

Pardhi, D. S., Rabadiya, K. J., Panchal, R. R., Raval, V. H., Joshi, R. G., & Rajput, K. N. (2023). Cyclodextrin glucanotransferase: Fundamentals and biotechnological implications. *Applied Microbiology and Biotechnology, 107*(19), 5899–5907.

Payamifar, S., & Marjani, A. P. (2025). A review of β-cyclodextrin-based catalysts system in the chemical fixation of carbon dioxide. *Journal of CO2 Utilization, 91*, 103011.

Peng, X., Su, H., Mi, S., & Han, Y. (2016). Biotechnology for biofuels a multifunctional thermophilic glycoside hydrolase from *Caldicellulosiruptor owensensis* with potential applications in production of biofuels and biochemicals. *Biotechnology for Biofuels*, 1–13. https://doi.org/10.1186/s13068-016-0509-y.

Penninga, D., Strokopytov, B., Rozeboom, H. J., Lawson, C. L., Dijkstra, B. W., Bergsma, J., & Dijkhuizen, L. (1995). Site-directed mutations in tyrosine 195 of cyclodextrin glycosyltransferase from *Bacillus circulans* strain 251 affect activity and product specificity. *Biochemistry, 34*(10), 3368–3376.

Pereira, A. G., Carpena, M., Oliveira, P. G., Mejuto, J. C., Prieto, M. A., & Gandara, J. S. (2021). Main applications of cyclodextrins in the food industry as the compounds of choice to form host–guest complexes. *International Journal of Molecular Sciences, 22*(3), 1–23. https://doi.org/10.3390/ijms22031339.

Perez-Pomares, F., Diaz, S., Bautista, V., Pire, C., Bravo, G., Esclapez, J., Zafrilla, B., & Bonete, M.-J. (2009). Identification of several intracellular carbohydrate-degrading activities from the halophilic archaeon *Haloferax mediterranei*. *Extremophiles*, 633–641. https://doi.org/10.1007/s00792-009-0246-2.

Qi, Q., & Zimmermann, W. (2005). Cyclodextrin glucanotransferase: From gene to applications. *Applied Microbiology and Biotechnology, 66*(5), 475–485.

Rabadiya, K., Pardhi, D., Thaker, K., Patoliya, J., Rajput, K., & Joshi, R. (2024). A review on recent upgradation and strategies to enhance cyclodextrin glucanotransferase properties for its applications. *International Journal of Biological Macromolecules, 259*(P2), 129315. https://doi.org/10.1016/j.ijbiomac.2024.129315.

Rahman, K., Illias, R. M., Hassan, O., Mahmood, N. A. N., & Rashid, N. A. A. (2006). Molecular cloning of a cyclodextrin glucanotransferase gene from alkalophilic *Bacillus* sp. TS1-1 and characterization of the recombinant enzyme. *Enzyme and Microbial Technology, 39*(1), 74–84. https://doi.org/10.1016/j.enzmictec.2005.09.014.

Rajput, K. N., Patel, K. C., & Trivedi, U. B. (2016a). β -Cyclodextrin production by cyclodextrin glucanotransferase from an alkaliphile *Microbacterium terrae* KNR 9 using different starch substrates. *Biotechnology Research International, 2016*, 1–7. https://doi.org/10.1155/2016/2034359.

Rajput, K. N., Patel, K. C., & Trivedi, U. B. (2016b). A novel cyclodextrin glucanotransferase from an alkaliphile *Microbacterium terrae* KNR 9: Purification and properties. *3 Biotechnology, 6*(2), 1–11. https://doi.org/10.1007/s13205-016-0495-6.

Rajput, K. N., Patel, K. C., & Trivedi, U. B. (2016c). Screening and selection of medium components for cyclodextrin glucanotransferase production by new alkaliphile *Microbacterium terrae* KNR 9 using plackett-burman design. *Biotechnology Research International, 2016*(1), 3584807.

Rambu, D., Dumitru, M., & Ciurescu, G. (2024). Solid-state fermentation using *Bacillus licheniformis* -Driven changes in composition, viability and in vitro protein digestibility of oilseed cakes. *1*, 17.

Rashid, N., Cornista, J., Ezaki, S., Fukui, T., Atomi, H., & Imanaka, T. (2002). Characterization of an archaeal cyclodextrin glucanotransferase with a novel C-terminal domain. *Journal of Bacteriology, 184*(3), 777–784. https://doi.org/10.1128/JB.184.3.777.

Rasool, M., Malik, A., Sabir, J. S. M., Ahmad, I., & Kim, C. (2022). Insights into the biodegradation and heavy metal resistance potential of the genus *Brevibacillus* through comparative genome analyses. *Gene, 846*(September), 146853. https://doi.org/10.1016/j.gene.2022.146853.

Rather, M. Y., Zubaida, K., Ara, G., Karlsson, E. N., & Adlercreutz, P. (2015). Characterization of cyclodextrin glycosyltransferases (CGTases) and their application for synthesis of alkyl glycosides with oligomeric head group. *Process Biochemistry, 50*(5), 722–728. https://doi.org/10.1016/j.procbio.2015.02.016.

Rebrov, E. V., Schouten, J. C., & De Croon, M. H. J. M. (2011). Single-phase fluid flow distribution and heat transfer in microstructured reactors. *Chemical Engineering Science, 66*(7), 1374–1393.

Rimphanitchayakit, V., Tonozuka, T., & Sakano, Y. (2005). Construction of chimeric cyclodextrin glucanotransferases from *Bacillus circulans* A11 and *Paenibacillus macerans* IAM1243 and analysis of their product specificity. *Carbohydrate Research, 340*, 2279–2289. https://doi.org/10.1016/j.carres.2005.07.013.

Saini, K., Pathak, V. M., Tyagi, A., & Gupta, R. (2022). Microbial cyclodextrin glycosyltransferases: Sources, production, and application in cyclodextrin synthesis. *Catalysis Research, 02*(03), 1–69. https://doi.org/10.21926/cr.2203029.

Saokham, P., & Loftsson, T. (2017). γ-Cyclodextrin. *International Journal of Pharmaceutics, 516*, 278–292. https://doi.org/10.1016/j.ijpharm.2016.10.062.

Schöffer, J.da N., Matte, C. R., Charqueiro, D. S., de Menezes, E. W., Costa, T. M. H., Benvenutti, E. V., Rodrigues, R. C., & Hertz, P. F. (2017). Directed immobilization of CGTase: The effect of the enzyme orientation on the enzyme activity and its use in packed-bed reactor for continuous production of cyclodextrins. *Process Biochemistry, 58*(April), 120–127. https://doi.org/10.1016/j.procbio.2017.04.041.

Sharma, K., Goel, Y., Warsa, K., Tewari, A., Khanduri, U., & Kumar, S. (2025). Harnessing enzymes from halophiles for sustainable biofuel production. *19*(3), 170–182.

Sian, H. K., Said, M., Hassan, O., Kamaruddin, K., Ismail, A. F., Rahman, R. A., Mahmood, N. A. N., & Illias, R. M. (2005). Purification and characterization of cyclodextrin glucanotransferase from alkalophilic *Bacillus* sp. G1. *Process Biochemistry, 40*(3–4), 1101–1111.

Singh, J., Kaur, L., & McCarthy, O. J. (2007). Factors influencing the physico-chemical, morphological, thermal and rheological properties of some chemically modified starches for food applications—A review. *Food Hydrocolloids, 21*(1), 1–22.

Spigno, G., & Marco, D. F. D. (2015). Glucose syrup and corn steep liquor as alternative to molasses substrates for production of baking-quality yeast glucose syrup and corn steep liquor as alternative to molasses substrates for production of baking-quality yeast. *Chem Eng Trans*. https://doi.org/10.3303/CET0917141 January.

Sun, T., Jiang, B., & Pan, B. L. (2011). Purification and characterization of novel cyclodextrin glucanotransferase from *Bacillus* sp. SK13. 002. *Asian Journal of Chemistry, 23*(11), 4977–4982.

Synowiecki, J. (2010). Some applications of thermophiles and their enzymes for protein processing. *African Journal of Biotechnology, 9*(42), 7020–7025. https://doi.org/10.5897/AJB10.966.

Szejtli, J. (1990). The cyclodextrins and their applications in biotechnology. *Carbohydrate Polymers, 12*(4), 375–392.

Szijártó, N., Horan, E., Zhang, J., Puranen, T., Siika-aho, M., & Viikari, L. (2011). Thermostable endoglucanases in the liquefaction of hydrothermally pretreated wheat straw. *Biotechnology for Biofuels,* 1–10.

Tachibana, Y., Kuramura, A., Shirasaka, N., Suzuki, Y., Yamamoto, T., Fujiwara, S., Takagi, M., & Imanaka, T. (1999). Purification and characterization of an extremely thermostable cyclomaltodextrin glucanotransferase from a newly isolated hyperthermophilic archaeon, a *Thermococcus* sp. *Applied and Environmental Microbiology, 65*(5), 1991–1997.

Teufl, M., Zajc, C. U., & Traxlmayr, M. W. (2022). Engineering strategies to overcome the stability–function trade-off in proteins. *ACS Synthetic Biology, 11*(3), 1030–1039.

Thiemann, V., Dönges, C., Prowe, S. G., Sterner, R., & Antranikian, G. (2004). Characterisation of a thermoalkali-stable cyclodextrin glycosyltransferase from the anaerobic thermoalkaliphilic bacterium *Anaerobranca gottschalkii*. *Archives of Microbiology,* 226–235. https://doi.org/10.1007/s00203-004-0717-x.

Torres-Guzman, J. C., Padilla-Guerrero, I. E., Cervantes-Quintero, K. Y., Martinez-Vazquez, A., Ibarra-Guzman, M., & Gonzalez-Hernandez, G. A. (2021). Peculiarities of nitronate monooxygenases and perspectives for in vivo and in vitro applications. *Applied Microbiology and Biotechnology, 105*(21), 8019–8032. https://doi.org/10.1007/s00253-021-11623-1.

Uitdehaag, J. C. M., Van Der Veen, B. A., Dijkhuizen, L., & Dijkstra, B. W. (2002). Catalytic mechanism and product specificity of cyclodextrin glycosyltransferase, a prototypical transglycosylase from the α-amylase family. *Enzyme and Microbial Technology, 30*(3), 295–304. https://doi.org/10.1016/S0141-0229(01)00498-7.

Vassileva, A., Atanasova, N., Ivanova, V., Dhulster, P., & Tonkova, A. (2023). Characterisation of cyclodextrin glucanotransferase from *Bacillus circulans* ATCC 21783 in terms of cyclodextrin production. *Annals of microbiology*. https://doi.org/10.1007/BF03175362 *December 2007*.

van der Veen, B. A., van Alebeek, G. W. M., Uitdehaag, J. C. M., Dijkstra, B. W., & Dijkhuizen, L. (2000). The three transglycosylation reactions catalyzed by cyclodextrin glycosyltransferase from *Bacillus circulans* (strain 251) proceed via different kinetic mechanisms. *European Journal of Biochemistry, 267*(3), 658–665.

Vollu, R. E., Mota, F. F., da, Gomes, E. A., & Seldin, L. (2008). Cyclodextrin production and genetic characterization of cyclodextrin glucanotranferase of *Paenibacillus graminis*. *Biotechnology Letters, 30*, 929–935. https://doi.org/10.1007/s10529-008-9635-3.

Wang, S., & Guo, P. (2020). Botanical sources of starch. In *Starch structure, functionality and application in foods*. Springer, 9–27.

Wang, Z., Qi, Q., & Wang, P. G. (2006). Engineering of cyclodextrin glucanotransferase on the cell surface of *Saccharomyces cerevisiae* for improved cyclodextrin production. *Applied and Environmental Microbiology, 72*(3), 1873–1877. https://doi.org/10.1128/AEM.72.3.1873.

Wu, M., Jiang, Y., Liu, Y., Mou, L., Zhang, W., Xin, F., & Jiang, M. (2021). Microbial application of thermophilic *Thermoanaerobacterium* species in lignocellulosic biorefinery. *Applied Microbiology and Biotechnology, 105*(14), 5739–5749. https://doi.org/10.1007/s00253-021-11450-4.

Wulandari, D., Budiharjo, A., Awalia, A., Putri, K., & Setiarto, R. H. B. (2025). Diversity of thermophilic bacteria isolated from extreme environments in Indonesia: A perspective in biotechnology applications. *Journal of Tropical Biodiversity and Biotechnology, 10*(03). https://doi.org/10.22146/jtbb.19548.

Xie, D. (2022). Continuous biomanufacturing with microbes—upstream progresses and challenges. *Current Opinion in Biotechnology, 78*, 102793. https://doi.org/10.1016/j.copbio.2022.102793.

Yoneda, Y., Kano, S. I., Yoshida, T., Ikeda, E., Fukuyama, Y., Omae, K., Kimura-sakai, S., Daifuku, T., Watanabe, T., & Sako, Y. (2015). Detection of anaerobic carbon monoxide-oxidizing thermophiles in hydrothermal environments. *FEMS Journals, 91*(July), 1–9. https://doi.org/10.1093/femsec/fiv093.

Yue, Y.-D., Gao, Y., Fu, S.-Y., Cui, Y.-Y., & Yang, C.-X. (2023). Synthesis of flower-shaped cyclodextrin microporous organic networks for selective recovery of acetophenone and 1-phenylethanol from petrochemical effluents. *Chemical Engineering Journal, 475*, 146344.

SECTION

III

Purification and engineering of cyclodextrin glucanotransferases

CHAPTER

7

Cloning and overexpression of cyclodextrin glucanotransferase

Krishnendu Mondal[1,2], Prayatna Sharma[3], Ishfaq Nabi Najar[4], Arijit Jana[1], Nagendra Thakur[3] and Keshab Chandra Mondal[1]

[1]*Department of Microbiology, Vidyasagar University, Midnapore, West Bengal, India,*
[2]*Department of Microbiology, Kalinga University, Raipur, Chhattisgarh, India,*
[3]*Department of Microbiology, Sikkim University, Gangtok, Sikkim, India,*
[4]*Fermentation and Microbial Biotechnology Division, CSIR-IIIM, Jammu and Kashmir, India*

Introduction

Cyclodextrin glucanotransferases (CGTases) are specialized enzymes belonging to the α-amylase family, known for their unique capability to catalyze the conversion of starch and related polysaccharides into cyclodextrins, that is, cyclic oligosaccharides composed of glucose units linked by α-1,4-glycosidic bonds (Leemhuis et al., 2010). Owing to their ability to form inclusion complexes with various guest molecules, cyclodextrins possess distinctive physicochemical properties that have made them invaluable in diverse industrial sectors, including food, pharmaceutical, cosmetic, and environmental industries (Saini et al., 2022). The growing global demand for cyclodextrins has spurred extensive research into improving production processes, particularly through molecular cloning and recombinant overexpression of CGTase enzymes (Muniz et al., 2024).

The study and manipulation of CGTases have become central to biotechnology due to the enzymes' broad utility and the demand for high-yield, stable, and catalytically efficient biocatalysts (Sonnendecker et al., 2017). Traditional methods for producing enzymes from wild-type microbial strains are often limited by low yields, complex downstream processing, and inconsistent enzyme properties (Low et al., 2012). Recombinant DNA technology, however, has revolutionized enzyme production by enabling the cloning of CGTase genes into heterologous expression systems, thus allowing controlled overproduction of the enzyme with tailor-made characteristics (Song et al., 2023). Cloning and overexpression not only enhance enzyme yields but also enable engineering of CGTases with improved thermostability, altered substrate specificity, and tolerance to industrial processing conditions such as extreme pH and high salinity (Zhang et al., 2017; Wang et al., 2018).

Cyclodextrin Glucanotransferase. DOI: https://doi.org/10.1016/B978-0-443-41694-1.00001-0

This chapter provides an in-depth overview of current methodologies and advances in the cloning and overexpression of CGTases. It begins by exploring the sources of CGTase genes, primarily from *Bacillus* species, which are known for robust industrial enzyme production, as well as from other bacterial genera that offer unique enzyme variants (Zhang et al., 2018). The cloning process involves isolating the gene of interest using molecular biology techniques, such as polymerase chain reaction (PCR), gene synthesis, and DNA fragment assembly (Goh et al., 2012). A highlight is placed on the design of cloning vectors—plasmids and expression systems—that are compatible with common microbial hosts such as *Escherichia coli* and *Bacillus subtilis* (Xu et al., 2024). The chapter also discusses codon optimization strategies, a crucial step to ensure efficient translation of CGTase genes in heterologous hosts, overcoming differences in codon usage bias and thus improving expression yields (Chung et al., 1998).

Following cloning, gene expression optimization is addressed extensively. *E. coli* remains the preferred host for CGTase expression due to its well-characterized genetics, fast growth, and ease of scaling bioprocesses (Liu et al., 2022). Induction systems, primarily using isopropyl β-D-1-thiogalactopyranoside (IPTG) under the T7 promoter control, are discussed with respect to tuning expression parameters such as inducer concentration, temperature, and induction time (Zhang et al., 2023). These factors significantly affect enzyme solubility and activity, with common challenges including inclusion body formation, where misfolded proteins aggregate into inactive forms. To mitigate these issues, methods such as lowering induction temperature, coexpression with molecular chaperones, and fusion tags are reviewed (Miao et al., 2024). Alternative hosts such as *B. subtilis* and yeast species (e.g., *Pichia pastoris*) are also considered for their ability to secrete active CGTases extracellularly, thereby simplifying downstream processing (Zhang et al., 2017).

Purification of recombinant CGTase enzymes represents a critical stage in obtaining active, high-purity protein suitable for industrial use and characterization. The chapter outlines common purification workflows, beginning with cell lysis techniques such as sonication and chemical disruption, followed by clarification steps to remove debris. It further details chromatographic purification strategies, including ion-exchange chromatography, affinity chromatography—especially his-tag-based nickel chelation—and size-exclusion chromatography, which collectively improve enzyme purity and specific activity (Xu et al., 2024). Purification performance indicators such as fold-purification, recovery yield, and enzyme activity units are highlighted alongside the use of SDS-PAGE for molecular weight determination and purity confirmation, typically ranging between 40–75 kDa for CGTases, depending on the source (Zhang et al., 2018).

Characterizing purified recombinant CGTases is essential to understanding their enzymatic properties and suitability for various applications. Parameters such as optimum temperature, pH range, thermal and pH stability, substrate specificity, and kinetic constants (K_m and V_{max}) are standardly assessed using relevant biochemical assays (Song et al., 2023). The chapter integrates examples from the literature of CGTases that exhibit remarkable thermostability, retaining activity at temperatures up to 80°C–95°C, which is advantageous for industrial starch processing and cyclodextrin production (Leemhuis et al., 2010). The influence of metal ions, inhibitors, and salt concentration on enzymatic performance is also discussed, given their relevance to processing conditions in food and pharmaceutical industries (Liu et al., 2022). Moreover, recombinant CGTases engineered through mutagenesis or directed evolution have shown enhanced stability and product specificity, emphasizing the dynamic potential of molecular engineering alongside cloning and expression (Zhang et al., 2023; Miao et al., 2024).

Lastly, the chapter underscores the industrial and biotechnological significance of cloning and overexpressing CGTases. By enabling large-scale production of enzymes with desired qualities and improving process consistency, these molecular approaches support sustainable and cost-effective manufacturing of cyclodextrins (Saini et al., 2022). Emerging applications in drug delivery systems, environmental remediation, and functional foods could greatly benefit from customized CGTase enzymes produced through advanced recombinant technologies (Muniz et al., 2024). This chapter thus serves as a comprehensive guide for researchers and practitioners seeking to use recombinant CGTases for diverse industrial and environmental applications, highlighting best practices, challenges, and frontier innovations in enzyme biotechnology.

Microbial sources of cyclodextrin glucanotransferase: exploration and strain selection

The quest for efficient CGTase production begins with the rigorous identification and exploration of diverse microbial sources capable of synthesizing this pivotal enzyme. Historically, members of the genus *Bacillus* have been the predominant and most extensively studied CGTase producers due to their ability to produce high enzyme yields and diverse cyclodextrin product profiles. Notable species such as *B. macerans*, *B. circulans*, *B. stearothermophilus*, and *B. firmus* exemplify the typical robust producers that have been optimized for industrial processes (Horikoshi, 1999). However, evolving industrial demands and an increased understanding of microbial diversity have expanded the spectrum of microorganisms investigated for CGTase production to include other genera such as *Paenibacillus* (Kitahata & Okada, 1974), *Alkaliphilus* (Qi et al., 2007), *Klebsiella* (Biwer et al., 2002), as well as thermophilic, alkaliphilic, and halophilic bacteria, archaea, and certain fungi (Larsbrink et al., 2011). These explorations continue to highlight novel enzymatic properties and broaden the industrial applicability of CGTases. Microbial sources of CGTase and their industrial significance are presented in Table 7.1.

Table 7.1 Microbial sources of cyclodextrin glucanotransferase (*CGTase*) with industrial relevance and key characteristics.

Organism	Description	Remarks	References
Bacillus sp. G1	Main source of CGTase, including species such as *B. circulans*, *B. stearothermophilus*, *B. macerans*, *B. firmus*, *B. ohbensis*	Most widely studied and robust industrial producers	Sian et al. (2005)
Paenibacillus spp.	Isolated from various soil samples; promising CGTase producers	Also used for recombinant expression	Lim et al. (2021)
Brevibacillus brevis CD162	Source of cloned CGTase genes	Potential for industrial expression	Kim et al. (1998)
Klebsiella pneumoniae AS-22	Gram-negative bacterium producing alpha-CGTase	Exhibits less sequence similarity to other CGTases	Gawande and Patkar (2001)
Thermoanaerobacter sp. 5K.	CGTases known for thermostable properties	High temperature activity and stability	Avci et al. (2014)
Haloferax mediterranei	Halophilic archaeon producing CGTase	Adapted to saline, high-salt environments	Bautista and Esclapez (2012)
Trichoderma viride	Fungal source of CGTase	Less common source compared to bacteria	Nazir et al. (2018)
Actinoplanes sp. strain SE50	Produces related enzymes with CGTase activity potential	Functionally similar glycosyltransferases	Hemker et al. (2001)

Exploration of natural and extreme environments

Microbial biodiversity serves as a rich reservoir for novel CGTase producers, often sourced from unique and extreme ecological niches. These environments provide selective pressures that drive the evolution of enzymes with enhanced thermostability, broad pH tolerance, salt tolerance, and unusual substrate specificity—attributes highly desirable in industrial biocatalysts (Vester et al., 2014). Common sources for isolating CGTase-producing microbes include soil, compost, decomposing plant material rich in starch, and rhizospheres where starch substrates are abundant. More specialized environments, such as hot springs, soda lakes, saline soils, thermal vents, and alkaline or acidic habitats, yield microbes adapted to harsh conditions (Ashcroft and Poma, 2025). These microbes possess enzymes that remain active and stable under extreme pH, temperature, or salinity, which are necessary attributes for robust industrial bioprocesses. Isolation methods range from classical culture-dependent plating on starch-containing agar media to identify zones of clearance or colorimetric changes indicative of CGTase activity, to advanced high-throughput microplate assays. For instance, phenolphthalein-based assays enable rapid screening by directly detecting cyclodextrin production (Usanov et al., 2007). The advent of modern metagenomic approaches bypasses the cultivation barrier by directly accessing environmental DNA from samples to identify novel CGTase genes. This culture-independent method has expanded the repertoire of known CGTases by uncovering genes from unculturable microbes inhabiting extreme or niche environments, thus vastly broadening enzyme diversity and prospecting new properties for industrial use.

Strain selection and characterization

Candidate strains exhibiting CGTase activity undergo a stringent evaluation based on multiple critical parameters. Enzyme activity and yield remain the primary criteria, wherein initial screening involves quantifying CGTase activity in crude extracts by assessing total cyclodextrin production from soluble starch and determining specific activity (units per mg protein); strains showing higher enzyme yields and activity are prioritized for subsequent analysis (Zheng et al., 2025).

Another key determinant is the product profile, since different CGTases preferentially synthesize α-, β-, or γ-cyclodextrins, each with distinct cavity sizes and complexation characteristics (Spiridon & Anghel, 2025). Strain choice, therefore aligns with the intended commercial application, as specific industries require the dominance of particular cyclodextrin types based on solubility, inclusion complex formation, or therapeutic applications (Topuz & Uyar, 2024). Additionally, stability attributes such as thermostability, broad pH tolerance, especially alkaline stability relevant to textile and detergent processes, resistance to proteolytic degradation, and tolerance to high-salt concentrations or organic solvents present in industrial effluents are critical indicators of enzyme robustness (Mesbah, 2022).

Genetic tractability significantly influences strain selection, as the ability to manipulate and clone cgt genes is essential for molecular engineering, overexpression, and process optimization, with microorganisms that are more amenable to recombinant techniques—such as *B. subtilis* and *E. coli*—being preferred host systems for production and detailed study (Zhang et al., 2020).

Characterization of selected strains extends to genomic sequencing, enzyme purification, substrate specificity profiling, and comparative assessment under simulated process conditions.

Comprehensive exploration and careful selection of microbial CGTase sources underpin the entire workflow for producing these valuable enzymes. By integrating classical microbiological techniques with modern molecular, metagenomic, and bioinformatic tools, researchers can continually identify, characterize, and improve strains for high-yield, robust production of CGTase tailored to current and future industrial needs.

Application of advanced genetic tools for cyclodextrin glucanotransferase gene isolation and manipulation

The field of CGTase gene isolation and engineering has been revolutionized by advanced genetic tools, including PCR innovations, CRISPR-based genome editing, and versatile molecular cloning strategies. These technologies collectively accelerate the discovery, modification, and optimization of CGTase enzymes for industrial and biotechnological applications. Fig. 7.1 shows a schematic workflow of CGTase gene isolation and engineering using advanced genetic tools.

Polymerase chain reaction

PCR remains the foundational technique for isolating and amplifying CGTase genes from diverse microbial genomes, and several modern PCR-based strategies significantly enhance gene recovery and downstream engineering applications. Degenerate primer design enables amplification of both known and

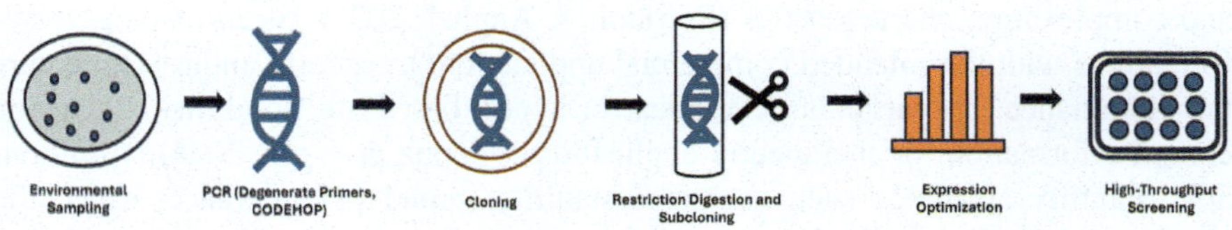

FIGURE 7.1

Workflow of cyclodextrin glucanotransferase gene isolation and engineering.

novel homologous CGTase genes by targeting conserved amino acid motifs across multiple enzyme sequences; primers directed toward conserved catalytic and starch-binding domains have proven effective in amplifying CGTase genes from *Bacillus* species as well as environmental isolates (Creer et al., 2016; Liu & Xiao, Li, et al., 2025).

In addition, consensus degenerate hybrid oligonucleotide primer strategies increase specificity for distantly related CGTase genes by accommodating sequence degeneracy while anchoring to conserved catalytic regions, a method that has facilitated gene recovery from taxa beyond *Bacillus*, including *Thermoanaerobacterium* and *Carboxydocella* (Rose et al., 1998; Rose et al., 2003; Makarani & Kaushal, 2025).

Quantitative real-time PCR further serves as a valuable tool for evaluating CGTase mRNA expression in engineered hosts, aiding in the selection of high-expression transformants and optimization of recombinant production systems (Nakapong et al., 2022). Moreover, overlap extension PCR provides a versatile recombination-based approach for constructing chimeric CGTases through seamless DNA fragment fusion, enabling modifications in substrate affinity or cyclodextrin product profiles to meet specific industrial requirements (Gupta et al., 2016).

CRISPR-based gene editing

The CRISPR-Cas9 system has transformed microbial genome editing in CGTase research by providing precise, efficient, and multiplexed manipulation capabilities:

- Gene knockout and knock-in: In industrial hosts like *B. subtilis*, CRISPR-Cas9 enables targeted knockout of extracellular protease genes or negative regulators, resulting in up to 30% higher extracellular CGTase yields. Additionally, knock-in strategies facilitate the stable integration of CGTase expression cassettes into high-expression chromosomal loci (e.g., *amyE*), ensuring robust enzyme production without plasmid maintenance burdens (Benites-Pariente et al., 2024).
- Iterative and multiplex genome editing: Advanced CRISPR variants such as Cas9 nickase paired with all-in-one plasmids permit multiple simultaneous genomic modifications with efficiencies up to 96% for insertions and near 100% for deletions. These capabilities expedite strain optimization cycles critical for metabolic engineering to maximize CGTase output and host robustness (Wang et al., 2021).
- Base editing and promoter engineering: CRISPR-mediated base editors enable targeted nucleotide modifications in promoter sequences, fine-tuning CGTase transcription with minimal fitness cost to the host. For example, promoter editing in *E. coli* achieved up to 2.5-fold higher expression compared to conventional plasmid-based systems (Taparia et al., 2022; Munjal et al., 2015).

Other genetic tools

In addition to PCR and CRISPR, modular cloning technologies and computational design platforms have further enhanced CGTase gene construction and expression tuning. Gibson assembly and Golden Gate cloning enable efficient multifragment assembly, with Gibson assembly joining DNA fragments through homologous overlaps in a single isothermal reaction to rapidly construct complex expression vectors containing CGTase genes and regulatory sequences (Bomfiglio et al., 2025), while Golden Gate cloning uses Type IIS restriction enzymes to achieve scarless, one-step, high-throughput combinatorial assembly (Engler et al., 2009).

Advances in modular expression systems and computational design further support precise transcriptional and translational tuning, as synthetic promoters and engineered ribosome-binding sites (RBSs) can be optimized using platforms such as PromoterHunter and RBS Calculator to maximize enzyme yield while minimizing host metabolic burden (Kent, 2020; Srivastava & Kumar, 2024). Emerging strategies in computational promoter and RBS design, including the use of machine learning models such as generative adversarial networks and variational autoencoders, enable the creation of synthetic promoters with enhanced regulatory performance, while in vivo RBS selection systems facilitate the identification of optimal translation initiation sequences across different host backgrounds (Shi & Barna, 2015).

Furthermore, automated high-throughput construction and screening—integrating automated DNA synthesis, liquid-handling robotics, and microfluidic platforms—have accelerated the generation and evaluation of large CGTase gene variant libraries, supporting rapid directed evolution and synthetic biology workflows (Ma et al., 2024).

Vector selection, promoter design, and host systems for efficient cyclodextrin glucanotransferase overexpression

Efficient overexpression of CGTases hinges on the synergistic optimization of vector systems, promoter elements, and host strains. Each component significantly influences yield, enzyme solubility, secretion, and posttranslational modifications, all of which are essential for meeting industrial demands (Schallmey et al., 2004). This section provides a comprehensive review of current state-of-the-art plasmid vectors, promoter engineering strategies, and host system choices optimized for high-level CGTase production. An overview of microbial expression systems and plasmid vectors for recombinant CGTase production is summarized in Table 7.2.

Table 7.2 Common microbial expression systems, vectors, and promoter types for recombinant protein production: advantages and limitations.

Host system	Plasmid vectors	Promoter type/ examples	Advantages	Limitations	Example reported yield	References
Escherichia coli	pET-28a, pET-22b (+), pET-24a, pUC19	Inducible: T7/lac, tac; constitutive: lac	Fast growth, low-cost media, high-copy vectors, easy genetic manipulation, and codon optimization possible	Inclusion body formation, limited secretion, no posttranslational modifications	Up to ~0.8 g/L (intracellular) in fermenters	Ojima-Kato (2025)
Pichia pastoris (*Komagataella phaffii*)	pPICZα, pPinkα-HC	Methanol-inducible AOX1; strong secretion signals (α-factor)	High-level secreted expression, proper folding, posttranslational modifications, easy purification	Requires methanol induction, glycosylation may alter enzyme properties in some cases	Up to ~1.2 g/L (secreted)	Pan et al. (2022)
Bacillus subtilis	pHT01, pHYCGT1 (shuttle vectors)	Constitutive: P43; inducible: IPTG; dual-promoter: HpaII–amyQ′	GRAS status, natural secretion, no endotoxins, easy downstream processing	Protease activity may degrade product unless engineered out, yields may be lower than yeast	Moderate–high (varies; dual-promoter ↑1.3×)	Liu and Yu (2025)
General Synthetic Systems	Modular/custom vectors with synthetic parts	Tunable/synthetic promoters; dual promoters; machine learning optimized	Fine control over expression, adaptable to any host, promoter strength customization	Requires advanced design and validation, higher initial setup cost	Host-dependent; up to 30%–50% yield improvement	Schütz et al. (2023)

Plasmid vector choices

E. coli remains the primary workhorse for CGTase gene expression due to its rapid growth, well-characterized genetics, simplicity, and low production cost. A variety of expression vectors have been optimized and widely used for CGTase expression.

pET series vectors [pET-28a, pET-22b(+), and pET-24a] are notable vectors carrying the strong T7 promoter coupled with lac operon elements. These offer high-copy plasmids that support robust, inducible expression regulated by IPTG. For instance, the expression of CGTase genes cloned into pET28b(+) under optimized conditions (e.g., induction at 20°C–28°C with 0.1–0.4 mM IPTG) has yielded high enzymatic activity with soluble and properly folded protein products (Liu et al., 2022). Codon optimization of CGTase genes to match *E. coli*'s codon usage biases often further enhances expression levels and solubility.

pUC19 is widely used for cloning and moderate expression. pUC19 plasmids carry the lac promoter, enabling inducible expression. Modifications and fusion with signal peptides facilitate secretion or periplasmic localization to augment soluble enzyme recovery (Klimova, 2024).

However, a significant bottleneck in *E. coli* expression is inclusion body formation—aggregation of misfolded proteins, leading to inactive enzymes that require solubilization and refolding. Strategies to overcome this include lowering induction temperatures (typically 16°C–25°C), optimizing IPTG concentrations, coexpression with molecular chaperones such as GroEL-GroES, DnaK-DnaJ-GrpE, and modifications in culture media composition. For example, coexpression of chaperones has been reported to improve soluble CGTase yields by up to 14-fold. Adjusting cultivation conditions, such as pH, salt concentration, and agitation, further enhances enzyme folding and secretion efficiency.

In *Pichia pastoris* (*Komagataella phaffii*)

The methylotrophic yeast *P. pastoris* is a favored eukaryotic host for secretory expression of CGTases, providing significant advantages over bacterial systems. Vectors such as pPICZα and pPinkα-HC combine the strong methanol-inducible AOX1 promoter with secretion signal sequences (e.g., α-factor prepro leader), facilitating high-level extracellular enzyme production (Baghban et al., 2018). These vectors also support easy integration into the yeast genome.

P. pastoris offers eukaryotic posttranslational modifications, proper protein folding, and robust protein secretion pathways, leading to higher quality, soluble active CGTase enzymes that are often more stable than bacterial-expressed counterparts.

Recent advancements, including fermentation optimization and gene copy number amplification, have enabled remarkable secretion titers exceeding 1.2 g/L of active CGTase in 2024 studies—a significantly higher yield compared to typical bacterial expression systems.

Codon optimization for yeast and fusion with native secretion signals are common strategies to maximize expression and secretion (O'Riordan et al., 2024).

B. subtilis serves as a generally recognized as safe (GRAS) host with natural protein secretion capabilities, advantageous for industrial-scale enzyme production without complex downstream processing. Shuttle vectors such as pHT01, pHYCGT1, and their derivatives allow propagation in *E. coli* for cloning and expression in *B. subtilis* for industrial production.

These plasmids often harbor strong, native *Bacillus* promoters such as P43 or engineered dual-promoter systems (e.g., HpaII–amyQ′), which have been shown to increase CGTase expression by up to 1.3-fold compared with single promoters (Zhang et al., 2017). The *B. subtilis* system benefits from direct extracellular enzyme secretion into the fermentation broth, eliminating the need for cell disruption, simplifying purification, and reducing costs. Strain engineering to delete proteases and stress-response genes has improved secretion stability and yields.

Promoter design: balancing expression and host fitness

Promoter engineering is a critical factor in regulating the expression of CGTase genes to maximize yield without compromising host viability or enzyme functionality. Strong constitutive promoters, including P43 from *B. subtilis* and HpaII from *Staphylococcus aureus*, can drive continuous high-level transcription, although excessive expression may impose a metabolic burden leading to growth inhibition, plasmid instability, or incorrect folding of the target protein (Kim et al., 2008).

In contrast, inducible promoters such as the T7/lac system regulated by IPTG in *E. coli* and the methanol-inducible AOX1 promoter in *P. pastoris* provide temporal control over expression, allowing induction at optimal cell densities to minimize metabolic stress and channel cellular resources toward protein production (Sarkar et al., 2010). Advances in synthetic and tunable promoter design have enabled the use of computational and machine learning approaches to generate promoter libraries with precisely adjustable strengths; spacer region optimization—particularly through specific dinucleotide combinations such as TG:CA or TA:TA—has been shown to significantly enhance CGTase transcription and enzyme output (Deng et al., 2018). For instance, hybrid promoters like P_tacI in *E. coli* exhibit four to sevenfold higher transcriptional activity relative to native promoters, while dual-promoter constructs combining HpaII and amyQ′ in *B. subtilis* have achieved approximately 30% increases in extracellular CGTase activity. Collectively, these tunable promoter architectures offer the flexibility to optimize CGTase expression for maximal enzyme yield while mitigating risks of host toxicity or inclusion body formation (Bhatwa et al., 2021).

Host system selection for industrial overexpression

The choice of host system affects not only the total enzyme yield but also enzyme solubility, secretion, posttranslational modifications, and cost of production:

Escherichia coli

Preferred for rapid intracellular expression due to fast growth rates, inexpensive media, and simple genetics (Gutiérrez-Granados et al., 2018). Yield optimization in fermenters can achieve up to ~0.8 g/L of CGTase. Limitations include cytoplasmic inclusion bodies, necessitating refolding protocols, and a lack of secretion pathways. Efforts to improve soluble expression focus on molecular chaperone coexpression and fermentation parameter optimization.

P. pastoris offers superior extracellular expression and secretion of active CGTase, with eukaryotic protein-processing capabilities, often achieving titers exceeding 1.2 g/L (Arsov et al., 2024). The secretion simplifies downstream processing and enhances enzyme stability by promoting correct folding and glycosylation. Its robustness and scalability make it highly suitable for industrial enzyme production.

B. subtilis, due to its GRAS status and inherent secretion system, is highly attractive for producing secreted enzymes without additional purification challenges from cytoplasmic proteins. The use of shuttle vectors allows convenient cloning in *E. coli* and expression in *Bacillus* (Yang et al., 2021). Optimized dual-promoter expression systems and strain protease deletions result in high enzyme titers and process reliability.

Exemplary system performance

P. pastoris's superior capacity for industrial CGTase production. For example, *P. pastoris* strains engineered to overexpress CGTase under the AOX1 promoter reached secretion titers of 1.2 g/L active enzyme in fermenter settings. Parallel *E. coli* expression systems under optimal IPTG induction and temperature control achieved up to 0.8 g/L intracellular enzyme, often requiring additional refolding and cell disruption steps. *B. subtilis* systems, while somewhat lower in yield compared to yeast, offer advantages in secretion and regulatory acceptance, supporting niche industrial applications.

With continuous improvements in synthetic promoter design, codon optimization, fusion tags, and host strain engineering—including protease gene knockouts, chaperone coexpression, and secretion pathway enhancements—the overexpression efficiency of CGTase from selected hosts is increasing steadily.

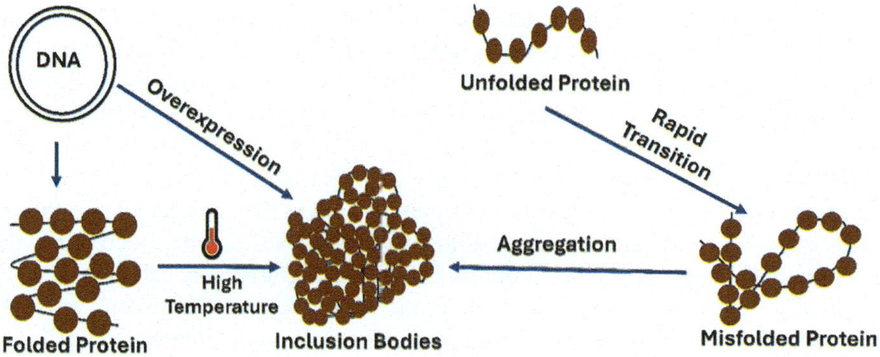

FIGURE 7.2

Schematic diagram of protein misfolding and inclusion body formation.

Issues of protein misfolding and inclusion body formation

Overexpression of CGTases in heterologous hosts is a cornerstone of recombinant enzyme production, but it is frequently hampered by protein misfolding and inclusion body formation. These issues pose significant challenges to achieving high yields of soluble, active CGTase complicate and increase the cost of downstream processing. This section explores the molecular basis and contributing factors of misfolding and aggregation, with an emphasis on host-specific issues, and reviews contemporary strategies to enhance soluble expression and functional enzyme recovery. Fig. 7.2 outlines a Schematic diagram of protein misfolding and inclusion body formation.

Protein misfolding and inclusion body formation: underlying causes

Inclusion bodies are dense aggregates of improperly folded or partially folded proteins that accumulate in the cytoplasm, predominantly in bacterial expression hosts like *E.coli*. The formation of inclusion bodies greatly reduces the amount of enzymatically active CGTase because the enzyme's complex tertiary and quaternary structures are disrupted, preventing critical domain interactions, substrate binding, and catalysis.

Several factors contribute to the tendency of CGTase—and other complex proteins—to misfold during overexpression. Rapid translation rates in bacterial hosts such as *E. coli* can overwhelm the cellular folding machinery, as strong promoters and high-copy plasmids accelerate protein synthesis to a level where nascent polypeptide chains emerge too quickly for molecular chaperones to assist properly, thereby increasing misfolding and aggregation (Niazi & Magoola, 2023).

Disulfide bond formation deficiencies also play a major role, as CGTases often require disulfide bonds for structural stability and catalytic performance; however, the reducing environment of the *E. coli* cytoplasm is unfavorable for proper disulfide bond formation compared to the oxidative conditions of the eukaryotic ER, resulting in unstable or misfolded proteins (Ma et al., 2020). Additionally, misfolding is exacerbated by the absence of native chaperones, since CGTases from certain bacterial or fungal species naturally fold in the presence of specialized chaperones that are absent in heterologous hosts like *E. coli*, increasing the likelihood of improper folding (Pouresmaeil & Azizi-Dargahlou, 2023).

Finally, high intracellular concentrations of overexpressed protein intensify crowding effects within the cytoplasm, promoting nonspecific interactions that lead to aggregation and inclusion body formation, particularly under heat stress or oxygen-limited bioprocessing conditions (Ahanger et al., 2024).

Strategies to enhance soluble expression

Overcoming protein misfolding and inclusion body formation is critical to optimize active CGTase yields. Multiple complementary strategies have demonstrated success based on molecular biology, genetic engineering, and fermentation process optimization:

Coexpression with molecular chaperones

Harnessing the native protein-folding network of the host by overexpressing chaperone proteins improves solubility by assisting in the correct folding of nascent CGTase polypeptides and preventing premature aggregation. In *E. coli*, coexpression of chaperone complexes such as GroEL/GroES and DnaK/DnaJ/GrpE is widely applied. GroEL/GroES form a chaperonin complex that provides an isolated environment for folding proteins and was shown to increase soluble CGTase yield by up to 40% in recent studies. They trap unfolded or partially folded protein intermediates, promoting proper conformational maturation before release.

DnaK/DnaJ/GrpE acts as a chaperone foldase system modulated by ATP hydrolysis, preventing improper intermolecular interactions. Coexpression with DnaK has improved folding efficiency for several industrial enzymes and can act synergistically with GroEL/GroES.

Optimization of culture conditions and expression parameters

Slowing down protein synthesis through modulation of growth and induction conditions gives the folding machinery more time to process emerging polypeptides, reducing aggregation. Reducing induction temperature from 37°C to 16°C–25°C is a common practice. This slows ribosomal translation and metabolism, greatly improving soluble protein fractions by balancing synthesis with folding capacity (Samatova et al., 2021).

Using lower IPTG concentrations for *E. coli* reduces the instantaneous production rate of CGTase, decreasing the propensity for inclusion body formation while maintaining sufficient expression (Man et al., 2016). Media formulations that gradually induce protein expression allow cells to reach sufficient biomass before production initiates, improving folding success (Castro et al., 2021).

Fusion tags for enhanced solubility

Fusion of CGTase genes to solubility-enhancing partners at the N- or C-terminus facilitates folding and increases protein solubility. Maltose-binding protein (MBP) is a highly effective solubility tag commonly employed in *E. coli* expression systems (De Marco, 2025). Beyond improving solubility, MBP can promote proper folding and stabilize the fusion protein. However, the tag usually needs to be cleaved postpurification to avoid interference with enzyme activity or industrial applications. Others include glutathione-S-transferase, thioredoxin, and small ubiquitin-related modifier tags, each with unique benefits depending on the enzyme and host context.

Refolding of inclusion bodies

When inclusion body formation is unavoidable, downstream strategies to recover active enzyme include denaturation followed by stepwise refolding under optimized buffer conditions, redox potential, and additives like osmolytes or detergents. This process, though labor-intensive and often costly, can salvage large quantities of misfolded protein and is used in industrial settings.

Host-specific solutions

Host choice greatly influences the frequency and severity of protein misfolding issues and determines the strategies required to overcome these challenges. In *E. coli*, although it remains a widely used expression host, its cytoplasmic environment is often unfavorable for proper folding of complex CGTases, necessitating interventions such as engineering oxidizing cytoplasm in strains like Origami™ and SHuffle®, which possess modified reducing pathways that facilitate disulfide bond formation and improve solubility (Chapman, 2021). Periplasmic expression using leader sequences such as pelB or OmpA can further enhance correct folding due to the more oxidative periplasmic environment, and combining these approaches with chaperone coexpression and optimized culture conditions significantly reduces inclusion body formation.

In contrast, *P. pastoris* provides a eukaryotic folding environment with an advanced ER machinery containing endogenous chaperones, including BiP, PDI, and calnexin, supporting proper protein maturation (Liu, Gong, et al., 2022). Its α-factor secretion signal efficiently directs recombinant CGTase through the secretory pathway, yielding an extracellular enzyme without intracellular aggregation, while the oxidative ER lumen promotes correct disulfide bond formation, collectively making misfolding events rare and simplifying downstream processing.

B. subtilis similarly offers advantages through natural extracellular secretion, where native signals such as those from the aprE alkaline protease gene enable efficient CGTase export and reduce cytoplasmic aggregation (Tjalsma et al., 2000); however, exposure to extracellular proteases requires the use of protease-deficient strains or fermentation-stage inhibitors. Despite having a simpler folding chaperone network than eukaryotic hosts, *B. subtilis* still manages proper folding through its secretory pathways and typically exhibits lower inclusion body formation than *E. coli*.

Emerging and complementary approaches

In addition to traditional methods, innovative technologies continue to improve folding efficiency and soluble expression of CGTase. Synthetic biology-based folding modulation employs engineered genetic circuits that fine-tune expression kinetics or regulate chaperone production under inducible promoters, enabling dynamic adjustment of the host's folding capacity in response to cellular stress (Müller et al., 2025).

Directed evolution and protein engineering further enhance CGTase performance by introducing mutations that accelerate intrinsic folding kinetics or eliminate aggregation-prone motifs, with strategies such as modifying surface residues to increase hydrophilicity, helping to reduce self-association and improve solubility (Daniel, 2022). Complementing these experimental approaches, computational protein folding prediction platforms such as AlphaFold and Rosetta assist in rational design by identifying flexible, unstable, or aggregation-prone regions suitable for stabilizing mutations, thereby streamlining the engineering of more robust CGTase variants (Vega et al., 2025).

Recent progress in directed evolution and synthetic biology

Directed evolution and synthetic biology have revolutionized the engineering of CGTases, enabling significant improvements in enzyme stability, activity, production efficiency, and tailored specificity to meet the growing industrial demand for these versatile biocatalysts. This section provides an in-depth exploration of recent advances in these fields, highlighting how cutting-edge techniques in molecular biology, computational design, and high-throughput screening are shaping the future of CGTase biotechnology.

Directed evolution

Directed evolution mimics natural selection in the laboratory by generating genetic diversity and screening for desirable phenotypes. For CGTases, this approach has become invaluable for enhancing enzyme properties such as thermal stability, catalytic efficiency, substrate specificity, and tolerance to industrial process conditions (Rabadiya et al., 2024). The foundation of directed evolution involves introducing mutations into CGTase genes using methods such as error-prone PCR, DNA shuffling, or saturation mutagenesis (Nakapong et al., 2022). These mutagenesis techniques create vast libraries of CGTase variants. Efficient and sensitive high-throughput screening methods are then employed to identify mutants exhibiting superior characteristics.

A notable breakthrough reported in 2024 demonstrated the use of error-prone PCR on B. *circulans* CGTase, yielding variants with up to 20% higher thermal stability at 60°C (Mao et al., 2024). Critical mutations were located in the calcium-binding site—a structural region whose integrity is essential for thermostability and activity. This enhanced stability broadens CGTase's operational temperature range, making it more suitable for industrial starch processing and cyclodextrin production at elevated temperatures. High-throughput screening platforms are crucial to the success of directed evolution. Fluorescence-based cyclodextrin detection assays allow rapid quantification of product formation directly in microtiter plates, significantly accelerating candidate variant identification. Such assays often use cyclodextrin-binding fluorescent dyes or enzyme-coupled reactions to detect cyclodextrin levels with high sensitivity and specificity. Technologies like fluorescence-activated cell sorting permit sorting of individual cells or droplets based on enzyme activity, facilitating the screening of enormous mutant libraries beyond the scale of traditional plate-based methods (Bowman & Alper, 2020).

Directed evolution combined with computational modeling increasingly directs mutagenesis efforts. Structural bioinformatics and docking simulations identify amino acid residues implicated in substrate binding, catalysis, and stability. For example, mutations at residues Asn33, Tyr119, and His255 in *Bacillus* CGTase variants demonstrated improved substrate affinity while maintaining catalytic function. Such rational integration with random mutagenesis expedites the identification of beneficial mutations (Miao, Chen, et al., 2024).

Moreover, iterative rounds of mutagenesis and screening enable cumulative improvements in enzyme traits, tailoring CGTase variants with enhanced disproportionation activity, altered product specificity favoring α-, β-, or γ-cyclodextrins, and improved resistance to inhibitors or denaturing conditions (Pardhi et al., 2023). Directed evolution thus provides a robust, generally applicable means to optimize CGTase enzymes without requiring detailed mechanistic knowledge.

Synthetic biology

Synthetic biology transcends traditional protein engineering by designing and constructing artificial genetic circuits, operons, and metabolic pathways to optimize CGTase expression and functionality holistically. One hallmark of synthetic biology is the construction of synthetic operons that integrate CGTase genes with molecular chaperones, secretion signal sequences, and regulatory elements into modular constructs (Abdi et al., 2024). These systems streamline enzymatic production by coordinating expression and folding machinery, thus maximizing yield and activity in recombinant hosts like *E. coli* or *B. subtilis*. Computational protein design tools, such as Rosetta, have been applied to predict stabilizing mutations within the CGTase tertiary structure (Goh, Hong, et al., 2012). By simulating protein folding and energetics, these platforms design substitutions that enhance overall enzyme stability or catalytic efficiency before experimental validation. Such in silico-guided design complements directed evolution by focusing mutation efforts on rational candidates.

Synthetic scaffolds represent another synthetic biology innovation, aiming to spatially colocalize CGTase enzymes with starch substrates or accessory proteins. By physically bringing enzyme-substrate complexes into proximity, these scaffolds increase local substrate concentration and catalytic efficiency. Modular protein—protein interaction domains or nucleic acid-based scaffolds have been engineered to accommodate multiple CGTase variants or synergistic enzymes, producing higher cyclodextrin yields than unattached systems (Childers & Daggett, 2017).

Dynamic regulatory circuits based on synthetic biology enable fine-tuned control over CGTase expression. Synthetic promoters, optimized RBSs, and inducible systems are integrated into operons, allowing expression modulation in response to environmental or metabolic signals. For instance, engineered strains can minimize metabolic burden during cell growth and induce high-level CGTase production at the optimal phase, enhancing overall productivity and maintaining cell viability. A landmark 2025 study demonstrated the power of synthetic biology by engineering an *E. coli* strain harboring an optimized CGTase operon designed through computational and combinatorial approaches. This strain produced 1.5 g/L of β-cyclodextrin—a 50% increase over native *E. coli* strains expressing wild-type CGTase (Jiang et al., 2025). The synthetic operon incorporated elements that promote efficient transcription, translation, and secretion, along with chaperone coexpression to improve protein folding. This achievement underscores how integrating synthetic biology design principles with classical enzyme engineering and heterologous expression can dramatically improve CGTase-based bioprocesses.

Finally, synthetic biology also facilitates metabolic pathway engineering, in which the entire host metabolism is rewired to funnel precursors toward cyclodextrin production. By coupling CGTase expression to pathways increasing starch or maltooligosaccharide availability, and optimizing energy and cofactor supply, synthetic biology expands the scope from isolated enzyme enhancement towards integrated biomanufacturing platforms.

Conclusion and future perspectives

Over the past two decades, the cloning and overexpression of CGTases have progressed from basic gene isolation to sophisticated molecular engineering. This remarkable advancement has been driven by the convergence of classical microbiology, modern genetic engineering, synthetic biology, and computational design. By tapping into diverse microbial sources—spanning *Bacillus*, *Paenibacillus*, *Klebsiella*, extremophilic archaea, and even certain fungi—researchers have constructed a broad genetic reservoir from which industrially relevant CGTases can be sourced and engineered. The deployment of advanced molecular tools such as PCR variants, degenerate primer design, metagenomic gene mining, and high-fidelity cloning methods has enabled the precise isolation and assembly of CGTase genes. Genome editing platforms, particularly CRISPR-Cas systems, have further revolutionized strain engineering, allowing targeted gene knockouts, promoter tuning, and integration of synthetic operons to optimize expression. Combined with codon optimization and signal peptide engineering, these tools have helped overcome long-standing bottlenecks in active enzyme production. Host system optimization has played an equally critical role. *E. coli*, long favored for its rapid growth and genetic tractability, has been complemented by eukaryotic hosts such as *P. pastoris* for secretory expression and *B. subtilis* for GRAS-certified extracellular production. Matching expression vectors and promoters to the host—whether strong constitutive, inducible, or synthetic tunable systems—has allowed fine control over transcriptional output, balancing productivity with host health. A recurring challenge in CGTase overexpression, particularly in *E. coli*, has been protein misfolding and inclusion body formation. This has been addressed through multiple strategies: coexpression with molecular chaperones, modulation of induction conditions, fusion to solubility-enhancing tags, and use of oxidizing cytoplasmic environments or periplasmic targeting to promote correct disulfide bond formation. In secretion-capable hosts, optimized signal sequences have minimized intracellular aggregation while streamlining purification. Recent breakthroughs in directed evolution and synthetic biology have taken CGTase engineering to a new level. Directed evolution, powered by error-prone PCR, DNA shuffling, and high-throughput screening, has yielded variants with improved thermostability, altered cyclodextrin product specificity, and enhanced resistance to process stressors. Synthetic biology complements these approaches by enabling the design of multi-gene operons, artificial scaffolds for substrate channeling, and computationally guided mutations that enhance stability and catalytic efficiency. The integration of both approaches has led to significant yield and performance improvements—with some engineered strains now capable of producing well over a gram per liter of active enzyme in optimized fermentation systems.

Looking forward, future research should integrate multiomics approaches—genomics, transcriptomics, proteomics, and metabolomics—to holistically optimize host cell metabolism for CGTase production. Systems biology models could guide dynamic regulation of metabolic fluxes, ensuring maximal precursor supply and energy efficiency. The exploration and characterization of extremophilic CGTases from thermophiles, halophiles, and alkaliphiles promise enzymes with superior stability under industrial conditions, reducing process costs and enabling novel applications. Furthermore, advances in continuous bioprocessing and immobilized enzyme technologies could transform CGTase production into steady, high-throughput manufacturing pipelines with lower operational costs and enhanced process sustainability. Such developments will expand cyclodextrin applications not only in drug delivery, flavor encapsulation, and food stabilization, but also in environmental biotechnology, such as wastewater purification and pollutant remediation.

References

Abdi, G., Patil, N., Tendulkar, R., Dhariwal, R., Mishra, P., Tariq, M., & Mudgal, G. (2024). Engineering genomic landscapes: Synthetic biology approaches in genomic rearrangement. In *Advances in genomics: Methods and applications*. Springer Nature Singapore, 227–264.

Ahanger, I. A., Ashraf, G. M., Sharma, A., & Islam, A. (2024). Comprehensive perspective towards the management of proteinopathies by elucidating protein misfolding and aggregation. *CNS & Neurological Disorders – Drug Targets, 23*(2), 153–180.

Arsov, A., Armenova, N., Gergov, E., Petrov, K., & Petrova, P. (2024). Cloning systems in *Bacillus*: Bioengineering of metabolic pathways for valuable recombinant products. *Fermentation, 10*(1), 50.

Ashcroft, E., Poma, M., Tischler, D., & Munoz-Munoz, J. (2025). Mining metagenomes from extremophiles as a resource for novel glycoside hydrolases for industrial applications. *Methods in Enzymology, 714*, 45–60.

Avci, A., Nichols, N. N., Saha, B. C., Frazer, S. E., Cotta, M. A., & Donmez, S. (2014). A thermostable cyclodextrin glycosyltransferase from *Thermoanaerobacter* sp. 5K. *Current Biotechnology, 3*(4), 305–312. https://doi.org/10.2174/2211550103666141027220452.

Baghban, R., Farajnia, S., Ghasemi, Y., Mortazavi, M., Zarghami, N., & Samadi, N. (2018). New developments in *Pichia pastoris* expression system: Review and update. *Current Pharmaceutical Biotechnology, 19*(6), 451–467.

Bautista, V., Esclapez, J., Pérez-Pomares, F., Martínez-Espinosa, R. M., Camacho, M., & Bonete, M. J. (2012). Cyclodextrin glycosyltransferase: A key enzyme in the assimilation of starch by the halophilic archaeon *Haloferax mediterranei*. *Extremophiles: Life Under Extreme Conditions, 16*(1), 147–159. https://doi.org/10.1007/s00792-011-0414-z.

Benites-Pariente, J. S., Samolski, I., Ludeña, Y., & Villena, G. K. (2024). CRISPR/Cas9-mediated targeted knock-in of eglA gene to improve endoglucanase activity of *Aspergillus fumigatus* LMB-35Aa. *Scientific Reports, 14*(1), 19661.

Bhatwa, A., Wang, W., Hassan, Y. I., Abraham, N., Li, X. Z., & Zhou, T. (2021). Challenges associated with the formation of recombinant protein inclusion bodies in *Escherichia coli* and strategies to address them for industrial applications. *Frontiers in Bioengineering and Biotechnology, 9*, 630551.

Biwer, A., Antranikian, G., & Heinzle, E. (2002). Enzymatic production of cyclodextrins. *Applied Microbiology and Biotechnology, 59*(6), 609–617. https://doi.org/10.1007/s00253-002-1063-y.

Bomfiglio, I. F., Mendes, I. S. D. M., & Bonatto, D. (2025). A review of DNA restriction-free overlapping sequence cloning techniques for synthetic biology. *Biotechnology Journal, 20*(7), e70084.

Bowman, E. K., & Alper, H. S. (2020). Microdroplet-assisted screening of biomolecule production for metabolic engineering applications. *Trends in Biotechnology, 38*(7), 701–714.

Castro, L. S., Lobo, G. S., Pereira, P., Freire, M. G., Neves, M. C., & Pedro, A. Q. (2021). Interferon-based biopharmaceuticals: Overview on the production, purification, and formulation. *Vaccines, 9*(4), 328.

Chapman, L. (2021). Manipulating sugar utilization in *Escherichia coli* through synthetic fusions to membrane transporters (Doctoral dissertation, University of York).

Childers, M. C., & Daggett, V. (2017). Insights from molecular dynamics simulations for computational protein design. *Molecular Systems Design & Engineering, 2*(1), 9–33.

Chung, H. J., Yoon, S. H., & Lee, S. Y. (1998). Engineering of host microorganisms for efficient production of enzymes. *Applied Microbiology and Biotechnology, 50*(5), 473–483. https://doi.org/10.1007/s002530051318.

Creer, S., Deiner, K., Frey, S., Porazinska, D., Taberlet, P., Thomas, W. K., & Bik, H. M. (2016). The ecologist's field guide to sequence-based identification of biodiversity. *Methods in Ecology and Evolution, 7*(9), 1008–1018.

Daniel, R. (2022). Topologies of synthetic gene circuit for optimal fold change activation in bacterial biosensors. *Biophysical Journal, 121*(3), 16a.

De Marco, A. (2025). Recent advances in recombinant production of soluble proteins in *E. coli*. *Microbial Cell Factories, 24*(1), 21.

Deng, C., Li, J., Shin, H. D., Du, G., Chen, J., & Liu, L. (2018). Efficient expression of cyclodextrin glycosyltransferase from *Geobacillus stearothermophilus* in *Escherichia coli* by promoter engineering and downstream box evolution. *Journal of Biotechnology, 266*, 77–83.

Engler, C., Gruetzner, R., Kandzia, R., & Marillonnet, S. (2009). Golden Gate shuffling: A one-pot DNA shuffling method based on type IIs restriction enzymes. *PLoS One, 4*(5), e5553.

Gawande, B. N., & Patkar, A. Y. (2001). Purification and properties of a novel raw starch degrading cyclodextrin glycosyltransferase from *Klebsiella pneumoniae* AS-22. *Enzyme and Microbial Technology, 28*(9–10), 735–743. https://doi.org/10.1016/S0141-0229(01)00347-7.

Goh, K. M., Hong, G. P., Chyi, N. H., Ng, P., Piaw, C. K., & Rahman, R. N. Z. R. A. (2012). Trends and tips in protein engineering: A review. *Jurnal Teknologi (Sciences & Engineering), 59*(1).

Goh, K. M., Mahadi, N. M., Hassan, O., Rahman, R. N. Z. R. A., & Illias, R. M. (2012). Cloning, expression, and characterization of cyclodextrin glucanotransferases: Current status and applications. *Applied Microbiology and Biotechnology, 93*(3), 967–979. https://doi.org/10.1007/s00253-011-3763-7.

Gupta, V., Sengupta, M., Prakash, J., & Tripathy, B. C. (2016). Fundamentals of recombinant DNA technology. In *Basic and Applied Aspects of Biotechnology*. Singapore: Springer, 23–58.

Gutiérrez-Granados, S., Cervera, L., Kamen, A. A., & Gòdia, F. (2018). Advancements in mammalian cell transient gene expression technology for accelerated production of biologics. *Critical Reviews in Biotechnology, 38*(6), 918–940.

Hemker, M., Stratmann, A., Goeke, K., Schröder, W., Lenz, J., Piepersberg, W., & Pape, H. (2001). Identification, cloning, expression, and characterization of the extracellular acarbose-modifying glycosyltransferase, AcbD, from *Actinoplanes* sp. strain SE50. *Journal of Bacteriology, 183*(15), 4484–4492. https://doi.org/10.1128/jb.183.15.4484-4492.2001.

Horikoshi, K. (1999). Alkaliphiles: Some applications of their products for biotechnology. *Microbiology and Molecular Biology Reviews, 63*(4), 735–750. https://doi.org/10.1128/MMBR.63.4.735-750.1999.

Jiang, Z., Liu, X., Ding, Y., Gu, Z., Li, Z., Ban, X., & Li, C. (2025). Enhance thermostability of β-cyclodextrin glycosyltransferase by hydrophobic interactions in the β-sheet structure. *Journal of Agricultural and Food Chemistry, 73*(32), 20317–20327.

Kent, R.W. (2020). Systematic design and optimisation of synthetic biology tools for regulating protein production (Doctoral dissertation, University of Manchester).

Kim, J. H., Hwang, B. Y., Roh, J., Lee, J. K., Kim, K., Wong, S. L., & Kim, B. G. (2008). Comparison of P_aprE, P_amyE, and P_P43 promoter strength for β-galactosidase and staphylokinase expression in *Bacillus subtilis*. *Biotechnology and Bioprocess Engineering, 13*(3), 313–318.

Kim, M. H., Sohn, C. B., & Oh, T. K. (1998). Cloning and sequencing of a cyclodextrin glycosyltransferase gene from *Brevibacillus brevis* CD162 and its expression in *Escherichia coli. FEMS Microbiology Letters, 164*(2), 411–418. https://doi.org/10.1111/j.1574-6968.1998.tb13117.x.

Kitahata, S., & Okada, S. (1974). Production of cyclodextrin glycosyltransferase by *Paenibacillus macerans* and its action on starch. *Agricultural and Biological Chemistry, 38*(2), 387–393. https://doi.org/10.1080/00021369.1974.10861116.

Klimova, K. (2024). Regulatory roles of MarA in flagellar biosynthesis in *Escherichia coli*: A novel mechanism for transcription initiation (Doctoral dissertation, University of Birmingham).

Larsbrink, J., Izumi, A., Igarashi, K., Samejima, M., & Stålbrand, H. (2011). Structural and functional studies of a glycoside hydrolase family 31 α-glucosidase from *Aspergillus niger*. *FEBS Journal, 278*(14), 2337–2349. https://doi.org/10.1111/j.1742-4658.2011.08161.x.

Leemhuis, H., Kelly, R. M., & Dijkhuizen, L. (2010). Engineering of cyclodextrin glucanotransferases and the impact for biotechnological applications. *Applied Microbiology and Biotechnology, 85*(4), 823–835. https://doi.org/10.1007/s00253-009-2328-4.

Lim, C. H., Rasti, B., Sulistyo, J., & Hamid, M. A. (2021). Comprehensive study on transglycosylation of CGTase from various sources. *Heliyon, 7*(2), e06305. https://doi.org/10.1016/j.heliyon.2021.e06305.

Liu, C., Gong, J. S., Su, C., Li, H., Rao, Z. M., & Shi, J. S. (2022). Pathway engineering facilitates efficient protein expression in *Pichia pastoris. Applied Microbiology and Biotechnology, 106*(18), 5893–5912.

Liu, J., Zhang, Y., & Xu, H. (2022). Advances in recombinant expression strategies for industrial enzymes. *Frontiers in Bioengineering and Biotechnology, 10*, 838912. https://doi.org/10.3389/fbioe.2022.838912.

Liu, M., Xiao, R., Li, X., Zhao, Y., & Huang, J. (2025). A comprehensive review of recombinant technology in the food industry: Exploring expression systems, application, and future challenges. *Comprehensive Reviews in Food Science and Food Safety, 24*(2), e70078.

Liu, Z., & Yu, X. (2025). Engineering *Bacillus subtilis* for high-value bioproduction: Recent advances and applications. *Microbial Cell Factories, 24*(1), 182.

Liu, Z., Wu, G., & Wu, H. (2022). Molecular cloning and optimized production and characterization of recombinant cyclodextrin glucanotransferase from *Bacillus* sp. *T1. 3 Biotech, 12*(3), 58.

Low, D., O'Connell, B., & Pesa, B. (2012). The evolving landscape of recombinant protein production in *Escherichia coli. Protein Expression and Purification, 83*(1), 1–9. https://doi.org/10.1016/j.pep.2012.02.009.

Ma, Y., Lee, C. J., & Park, J. S. (2020). Strategies for optimizing the production of proteins and peptides with multiple disulfide bonds. *Antibiotics, 9*(9), 541.

Ma, Y., Zhang, Z., Jia, B., & Yuan, Y. (2024). Automated high-throughput DNA synthesis and assembly. *Heliyon, 10*(6).

Makarani, N., & Kaushal, R. S. (2025). Advances in actinobacteria-based bioremediation: Mechanistic insights, genetic regulation, and emerging technologies. *Biodegradation, 36*(2), 24.

Man, R. C., Ismail, A. F., Fuzi, S. F. Z. M., Ghazali, N. F., & Illias, R. M. (2016). Effects of culture conditions of immobilized recombinant *Escherichia coli* on cyclodextrin glucanotransferase excretion and cell stability. *Process Biochemistry, 51*(4), 474–483.

Mao, S., Jiang, J., Xiong, K., Chen, Y., Yao, Y., Liu, L., & Li, X. (2024). Enzyme engineering: Performance optimization, novel sources, and applications in the food industry. *Foods, 13*(23), 3846.

Mesbah, N. M. (2022). Industrial biotechnology based on enzymes from extreme environments. *Frontiers in Bioengineering and Biotechnology, 10*, 870083.

Miao, B., Huang, D., Wang, T., Liu, H., Hao, Z., Yuan, H., & Jiang, Y. (2024). Enhancing trehalose production via *Bacillus* species G1 cyclodextrin glucanotransferase mutants: Modifying disproportionation characteristics and thermal stability. *Frontiers in Microbiology, 15*, 1500232.

Miao, S., Chen, L., & Zhao, H. (2024). Directed evolution of cyclodextrin glucanotransferase for improved thermostability and product specificity. *Biotechnology Advances, 68*, 108217. https://doi.org/10.1016/j.biotechadv.2023.108217.

Müller, M. M., Arndt, K. M., & Hoffmann, S. A. (2025). Genetic circuits in synthetic biology: Broadening the toolbox of regulatory devices. *Frontiers in Synthetic Biology, 3*, 1548572.

Muniz, H., Silva, C., & Santos, J. (2024). Industrial biotechnology of cyclodextrins: From microbial enzymes to pharmaceutical applications. *Critical Reviews in Biotechnology, 44*(2), 145–163.

Munjal, N., Jawed, K., Wajid, S., & Yazdani, S. S. (2015). A constitutive expression system for cellulase secretion in *Escherichia coli* and its use in bioethanol production. *PLoS One, 10*(3), e0119917.

Nakapong, S., Tumhom, S., Kaulpiboon, J., & Pongsawasdi, P. (2022). Heterologous expression of 4α-glucanotransferase: Overproduction and properties for industrial applications. *World Journal of Microbiology and Biotechnology, 38*(2), 36.

Nazir, S., Sulistyo, J., Hashmi, M. I., Ho, A. L., & Khan, M. S. (2018). Enzymatic synthesis of polyphenol glycosides catalyzed by transglycosylation reaction of cyclodextrin glucanotransferase derived from *Trichoderma viride*. *Journal of Food Science and Technology, 55*(8), 3026–3034.

Niazi, S. K., & Magoola, M. (2023). Advances in *Escherichia coli*-based therapeutic protein expression: Mammalian conversion, continuous manufacturing, and cell-free production. *Biologics, 3*(4), 380–401.

O'Riordan, N. M., Jurić, V., O'Neill, S. K., Roche, A. P., & Young, P. W. (2024). A yeast modular cloning toolkit expansion for optimization of heterologous protein secretion and surface display in *Saccharomyces cerevisiae*. *ACS Synthetic Biology, 13*(4), 1246–1258.

Ojima-Kato, T. (2025). Advances in recombinant protein production in microorganisms and functional peptide tags. *Bioscience, Biotechnology, and Biochemistry, 89*(1), 1–10.

Pan, Y., Yang, J., Wu, J., Yang, L., & Fang, H. (2022). Current advances of *Pichia pastoris* as cell factories for production of recombinant proteins. *Frontiers in Microbiology, 13*, 1059777.

Pardhi, D. S., Rabadiya, K. J., Panchal, R. R., Raval, V. H., Joshi, R. G., & Rajput, K. N. (2023). Cyclodextrin glucanotransferase: Fundamentals and biotechnological implications. *Applied Microbiology and Biotechnology, 107*(19), 5899–5907.

Pouresmaeil, M., & Azizi-Dargahlou, S. (2023). Factors involved in heterologous expression of proteins in *E. coli* host. *Archives of Microbiology, 205*(5), 212.

Qi, Q., Zimmermann, W., & Zhong, H. (2007). Production of cyclodextrins by a novel alkaliphilic *Alkaliphilus* sp. CGTase. *Journal of Applied Microbiology, 102*(2), 389–397.

Rabadiya, K., Pardhi, D., Thaker, K., Patoliya, J., Rajput, K., & Joshi, R. (2024). A review on recent upgradation and strategies to enhance cyclodextrin glucanotransferase properties for its applications. *International Journal of Biological Macromolecules, 259*, 129315.

Rose, T. M., Henikoff, J. G., & Henikoff, S. (2003). CODEHOP PCR primer design. *Nucleic Acids Research, 31*(13), 3763–3766.

Rose, T. M., Schultz, E. R., Henikoff, J. G., Pietrokovski, S., McCallum, C. M., & Henikoff, S. (1998). Consensus-degenerate hybrid oligonucleotide primers for amplification of distantly related sequences. *Nucleic Acids Research, 26*(7), 1628–1635.

Saini, R., Saini, A., & Dahiya, A. (2022). Applications of cyclodextrins in food, pharmaceutical, and environmental industries. *International Journal of Biological Macromolecules, 213*, 104–118.

Samatova, E., Daberger, J., Liutkute, M., & Rodnina, M. V. (2021). Translational control by ribosome pausing in bacteria. *Frontiers in Microbiology, 11*, 619430.

Sarkar, S., Pramanik, A., Mitra, A., & Mukherjee, J. (2010). Bioprocessing data for the production of marine enzymes. *Marine Drugs, 8*(4), 1323–1372.

Schallmey, M., Singh, A., & Ward, O. P. (2004). Developments in the use of *Bacillus* species for industrial production. *Canadian Journal of Microbiology, 50*(1), 1–17.

Schütz, A., Bernhard, F., Berrow, N., Buyel, J. F., Ferreira-da-Silva, F., Haustraete, J., & Remans, K. (2023). A concise guide to choosing suitable gene expression systems for recombinant protein production. *STAR Protocols, 4*(4), 102572.

Shi, Z., & Barna, M. (2015). Translating the genome in time and space. *Annual Review of Cell and Developmental Biology, 31*(1), 31–54.

Sian, H. K., Said, M., Hassan, O., Kamaruddin, K., Ismail, A. F., Rahman, R. A., & Illias, R. M. (2005). Purification and characterization of cyclodextrin glucanotransferase from alkalophilic *Bacillus* sp. G1. *Process Biochemistry, 40*(3–4), 1101–1111.

Song, L., Zhao, W., & Li, Y. (2023). Recombinant DNA technology for enzyme engineering. *Journal of Molecular Catalysis B: Enzymatic, 203*, 112167.

Sonnendecker, C., Zimmermann, W., & Schwarze, R. (2017). Enzyme engineering for improved performance of industrially relevant biocatalysts. *Journal of Biotechnology, 258*, 14–28.

Spiridon, I., & Anghel, N. (2025). Cyclodextrins as multifunctional platforms in drug delivery and beyond. *Molecules (Basel, Switzerland), 30*(14), 3044.

Srivastava, S., & Kumar, H. (2024). Recent advancements in ribosome binding site prediction, ribosome profiling, and structural analysis in prokaryotes. In *Microbial omics in environment and health*. Springer Nature Singapore, 367–382.

Taparia, Y., Dolui, A. K., Boussiba, S., & Khozin-Goldberg, I. (2022). Multiplexed genome editing via an RNA polymerase II promoter-driven sgRNA array in the diatom *Phaeodactylum tricornutum. Frontiers in Plant Science, 12*, 784780.

Tjalsma, H., Bolhuis, A., Jongbloed, J. D., Bron, S., & van Dijl, J. M. (2000). Signal peptide-dependent protein transport in *Bacillus subtilis. Microbiology and Molecular Biology Reviews, 64*(3), 515–547.

Topuz, F., & Uyar, T. (2024). Recent advances in cyclodextrin-based nanoscale drug delivery systems. *Wiley Interdisciplinary Reviews: Nanomedicine and Nanobiotechnology, 16*(6), e1995.

Usanov, N. G., Gil'vanova, E. A., Elizar'ev, P. A., Prutsakova, E. A., & Melent'ev, A. I. (2007). An improved method of photometric determination of cyclodextrin glucanotransferase activity. *Applied Biochemistry and Microbiology, 43*(1), 105–110.

Vega, A., Planas, A., & Biarnés, X. (2025). A practical guide to computational tools for engineering biocatalytic properties. *International Journal of Molecular Sciences, 26*(3), 980.

Vester, J. K., Glaring, M. A., & Stougaard, P. (2014). Discovery of novel enzymes with industrial potential from a cold and alkaline environment. *Microbial Cell Factories, 13*, 72.

Wang, Y., Liu, Y., Zheng, P., Sun, J., & Wang, M. (2021). Microbial base editing. *Trends in Biotechnology, 39*(2), 165–180.

Wang, Y., Xu, Z., & Chen, J. (2018). Improving enzyme thermostability through protein engineering. *Applied Microbiology and Biotechnology, 102*(6), 2481–2491.

Xu, H., Zhang, Y., & Liu, J. (2024). Recombinant protein purification strategies. *Frontiers in Bioengineering and Biotechnology, 12*, 1123456.

Yang, H., Qu, J., Zou, W., Shen, W., & Chen, X. (2021). An overview and future prospects of recombinant protein production in *Bacillus subtilis. Applied Microbiology and Biotechnology, 105*(18), 6607–6626.

Zhang, D., Li, H., & Wu, J. (2017). Heterologous expression of cyclodextrin glucanotransferase in *Bacillus subtilis*. *Enzyme and Microbial Technology, 100*, 55–62.

Zhang, K., Su, L., & Wu, J. (2020). Recent advances in recombinant protein production by *Bacillus subtilis*. *Annual Review of Food Science and Technology, 11*(1), 295–318.

Zhang, K., Su, L., Duan, X., Liu, L., & Wu, J. (2017). High-level extracellular protein production in *Bacillus subtilis* using an optimized dual-promoter expression system. *Microbial Cell Factories, 16*(1), 32.

Zhang, L., Wang, Q., & Zhao, Y. (2018). Advances in codon optimization strategies for heterologous gene expression. *Synthetic and Systems Biotechnology, 3*(2), 82–89.

Zhang, Y., Liu, J., & Xu, H. (2023). Overcoming challenges in recombinant protein expression. *Biotechnology Journal, 18*(4), e2200425.

Zheng, L., Jiang, B., & Wu, Y. (2025). Maltodextrin production: Challenges and advances in enzymatic and metabolic synthesis for controlled polymerization degree. *Food Reviews International*, 1–19.

CHAPTER

8

Extraction, purification, and characterization of cyclodextrin glucanotransferases

Lucia Dhiantika Witasari

Universitas Gadjah Mada, Faculty of Agricultural Technology, Department of Food and Agricultural Product Technology, Yogyakarta, Indonesia

Introduction

Cyclodextrin glucanotransferases (CGTases; EC 2.4.1.19) are a class of enzymes that catalyze the conversion of starch and related linear α-1,4-glucans into cyclodextrins (CDs)—cyclic oligosaccharides composed of six, seven, or eight glucose units, known respectively as α-, β-, and γ-CDs. These cyclic molecules possess a hydrophobic central cavity and a hydrophilic outer surface, enabling them to form inclusion complexes with various guest molecules (Astray et al., 2009). CGTases belong to glycoside hydrolase family 13 (GH13) and are widely produced by microbial sources, particularly by *Bacillus* species. Their ability to perform intramolecular transglycosylation rather than hydrolysis sets them apart from most enzymes in the same family (Leemhuis et al., 2010a, 2010b).

The production of CDs using CGTases has gained industrial significance due to its efficiency, cost-effectiveness, and eco-friendliness compared to traditional chemical synthesis. In the pharmaceutical industry, CDs are utilized to enhance the solubility, stability, and bioavailability of poorly water-soluble drugs (Loftsson & Duchêne, 2007). In the food sector, they are employed to mask undesirable tastes or odors, stabilize sensitive ingredients, and encapsulate flavors or vitamins (Szente & Szejtli, 2004). CDs also have applications in cosmetics, agriculture, textiles, and environmental engineering, such as for removing organic pollutants through encapsulation mechanisms (Crini, 2014). The versatility of CGTases in generating such high-value products underscores their industrial relevance and the growing global demand for them.

This chapter provides an in-depth overview of CGTases, focusing on their biochemical properties, structural features, reaction mechanisms, and substrate specificity. It also explores current approaches to CGTase production, including recombinant expression and protein engineering approaches used to improve performance. Furthermore, the chapter highlights the diverse industrial applications of CDs derived from CGTase activity and discusses future perspectives for optimizing their use in sustainable and advanced bioprocessing systems.

Cyclodextrin Glucanotransferase. DOI: https://doi.org/10 1016/B978-0-443-41694-1.00010-1

Microbial sources of cyclodextrin glucanotransferases

The CGTase production technique is very similar to that of other enzyme manufacturing processes. CGTase production is usually initiated with shaking-flask culture and submerged fermentation in a selective medium containing essential elements for microbial growth. CGTases are predominantly produced by a variety of microbial species, with *Bacillus* spp. being the most common and extensively studied producers. Notable strains include *Bacillus circulans*, *Bacillus macerans*, and *Bacillus stearothermophilus*, which are known for their robust CGTase production and ability to generate different CD types (Mahat et al., 2004; Qi & Zimmermann, 2005; Wang et al., 2018). Other genera such as *Paenibacillus*, *Thermoanaerobacter*, and *Klebsiella* have also been identified as sources of CGTases with unique thermostability or product specificity traits (Kelly et al., 2009). The yield and properties of CGTases can vary significantly depending on genetic factors such as promoter strength, gene copy number, and regulatory elements involved in enzyme expression (Leemhuis et al., 2003). Additionally, environmental conditions—including pH, temperature, nutrient composition, and aeration—play crucial roles in modulating enzyme synthesis. For instance, slightly alkaline pH and moderate to high temperatures (typically 50°C–60°C) have been found to be favorable for CGTase expression in thermophilic strains (Martins & Hatti-Kaul, 2002). These parameters differ between organisms, making it challenging to determine the ideal fermentation conditions for optimal enzyme yield. Scientists have modified fermentation conditions such as pH, temperature, nutrient concentrations, and the composition of the production media (carbon and nitrogen sources), to optimize CGTase yield based on the bacteria utilized (Arce-Vázquez et al., 2016; Ravinder et al., 2014). However, ideal fermentation conditions for CGTase synthesis in diverse microbes range from 30°C to 65°C, pH 5.5 to 11, and incubation times of 24 to 120 hours, with a variety of substrates (Pardhi et al., 2025).

CGTases have been reported in a wide range of mesophilic, thermophilic, alkaliphilic, and extremophilic microorganisms, with distinct catalytic profiles and stability characteristics that determine their industrial suitability (Kelly et al., 2009; Qi & Zimmermann, 2005; Saini et al., 2022). Representative microbial sources of CGTases and their key characteristics are summarized in Table 8.1.

Extraction strategies for cyclodextrin glucanotransferases

The Fig. 8.1 presents an overview of major cell lysis strategies employed for the extraction of intracellular biomolecules, broadly categorized into mechanical, enzymatic, and chemical methods. Mechanical methods include physical forces to break open microbial or cellular structures. The key approaches shown in the figure include high-pressure homogenization, sonication, and bead milling, all of which efficiently rupture cells through shear, cavitation, or abrasive forces. These methods are widely used for large-scale and robust cell disruption applications. Enzymatic lysis utilizes cell wall-degrading enzymes, such as lysozyme and mutanolysin, which selectively hydrolyze peptidoglycan layers. This

Table 8.1 Microbial sources of cyclodextrin glucanotransferase (CGTase) and their characteristics.

Microorganism	Growth condition	CGTase product preference	Notable properties	References
Bacillus circulans	Mesophilic	β-CD dominant	Well-characterized enzyme; high productivity	Lawson et al. (1994)
Bacillus macerans	Mesophilic	α-/β-CD	Classical CGTase producer; early industrial use	Qi and Zimmermann (2005)
Bacillus stearothermophilus	Thermophilic	β-/γ-CD	High thermal stability	Avci and Donmez (2009)
Paenibacillus macerans	Mesophilic	β-CD	High substrate affinity	Kelly et al. (2009)
Bacillus pseudalcaliphilus	Alkaliphilic	β-CD	Active and stable at alkaline pH	Kitayska et al. (2011)
Thermoanaerobacter sp.	Thermophilic, anaerobic	γ-CD enriched	Anaerobic growth; thermostable CGTase	Martins and Hatti-Kaul (2002)
Klabsiella oxytoca	Mesophilic	β-CD	Extracellular CGTase secretion	Kelly et al. (2009)
Microbacterium terrae KNR9	Alkaliphilic	β-/γ-CD	High pH tolerance; novel source	Rajput et al. (2016)
Shouchella oshimensis P-106	Mesophilic	β-CD	Recently reported CGTase producer	Pardhi et al. (2025)
Extremophilic bacteria (metagenomic sources)	Variable	Variable	High stability under extreme conditions	Mesbah (2022)

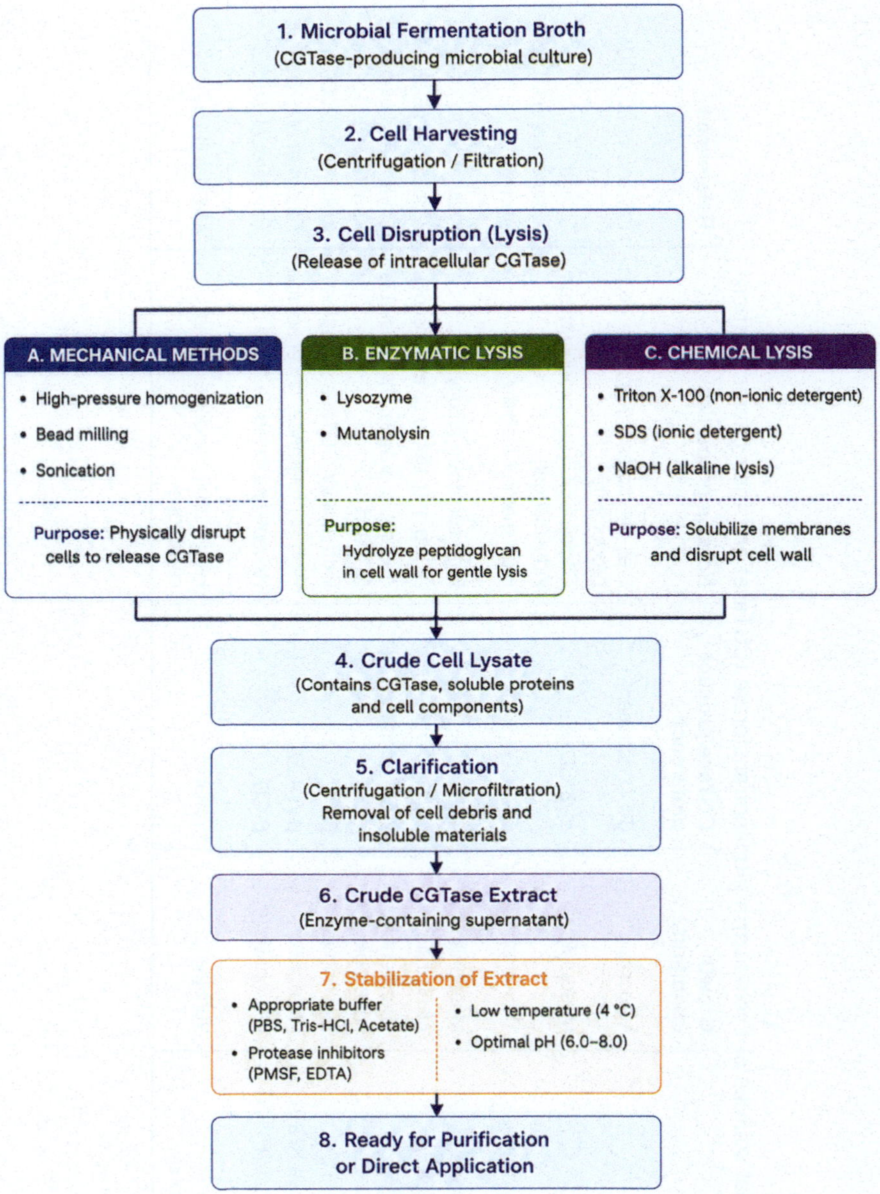

FIGURE 8.1

Overview of intracellular cyclodextrin glucanotransferase (CGTase) extraction. Following microbial harvesting, cells are lysed (mechanically, enzymatically, or chemically) to release the enzyme. The crude lysate is clarified via centrifugation/microfiltration and stabilized under cold, pH-controlled buffer with protease inhibitors prior to purification.

approach provides gentle, controlled lysis, making it suitable for sensitive proteins or applications that require high specificity. Chemical lysis involves the use of detergents or alkaline agents to solubilize cellular membranes. The figure lists sodium dodecyl sulfate (SDS), Triton X-100, and NaOH as common disruptants. These chemicals effectively permeabilize lipid bilayers and denature cellular components, enabling rapid and efficient extraction for downstream applications.

The efficient extraction of CGTases from microbial cultures is a crucial step in enzyme recovery that significantly affects yield, purity, and catalytic activity (Smith & Johnson, 2020). The first phase of this process involves cell lysis, which disrupts the microbial cell wall and releases intracellular CGTases into the surrounding solution. Various mechanical methods, such as high-pressure homogenization, bead milling, and sonication are commonly applied, particularly for robust producers like *Bacillus* spp (Lee et al., 2018). These methods are effective for large-scale operations but must be optimized to minimize heat generation and shear stress, which can denature the enzyme (Wang & Kim, 2019). In contrast, enzymatic lysis with cell wall-degrading enzymes such as lysozyme or mutanolysin provides a gentler alternative, particularly suitable for lab-scale extraction or for sensitive strains (Garcia et al., 2021). Chemical lysis—using agents such as Triton X-100, SDS, or NaOH—can also be employed, though the harshness of these reagents requires careful balancing to avoid loss of CGTase activity (Patel & Singh, 2017).

Once the cells are lysed, the resulting slurry is processed to obtain the crude extract, typically through centrifugation to remove cell debris and insoluble components (Zhao et al., 2016). The supernatant contains soluble proteins, including CGTase, and is considered the starting point for purification or direct application in some industrial processes (Kumar & Reddy, 2020). Key to maintaining enzyme functionality during this stage is temperature control—-performing lysis and extraction steps at 4°C significantly reduces the risk of thermal inactivation and proteolytic degradation (Fernandez & Lopez, 2015). Additionally, the buffer system used during extraction must be chosen based on the pH stability profile of the CGTase being recovered. Common buffers include phosphate-buffered saline, Tris-HCl, and sodium acetate, typically maintained at pH 6.0–8.0 to maintain enzyme integrity (Nguyen et al., 2019).

To further enhance enzyme stability and prevent degradation during extraction, it is advisable to supplement buffers with protease inhibitors such as phenylmethylsulfonyl fluoride or ethylenediaminetetraacetic acid (EDTA) (Choi & Park, 2018). The addition of chelating agents may also prevent metal ion-catalyzed oxidation or unwanted activation of metalloproteases (Singh & Mehta, 2017). In some cases, glycerol or other stabilizers are included to preserve enzyme structure during extended handling or freeze–thaw cycles (Martinez et al., 2021). Overall, the extraction strategy must be tailored to the microbial expression host, the physical and chemical properties of the target CGTase, and the intended downstream application—whether for research, diagnostics, or industrial-scale production. A well-optimized extraction protocol ensures high recovery of active CGTase and sets the stage for efficient purification and application (Brown & Davis, 2022).

In recent years, CGTase extraction strategies have progressed beyond conventional mechanical and chemical disruption toward process-integrated and sustainability-driven approaches. One notable development is the application of aqueous two-phase systems (ATPS), which enable simultaneous extraction and partial purification of CGTase under mild conditions. ATPS have been reported to improve enzyme recovery yield, reduce processing time, and preserve catalytic activity due to low interfacial tension and reduced protein denaturation (Bekavac et al., 2024). These systems are particularly attractive for large-scale CGTase recovery, as they significantly lower solvent consumption and energy input.

Another emerging strategy is membrane-assisted extraction, where microfiltration or ultrafiltration is directly coupled with fermentation. This approach allows continuous separation of CGTase while retaining microbial cells, thereby minimizing proteolytic degradation and enhancing volumetric productivity (Ogunbadejo & Al-Zuhair, 2024; Zhao et al., 2016). Such integration aligns with modern biorefinery concepts, in which enzyme recovery is embedded within upstream processing rather than treated solely as a downstream operation.

In parallel, advances in recombinant CGTase expression systems have reshaped extraction workflows. Engineered *Bacillus* strains capable of efficient extracellular secretion allow CGTase recovery directly from clarified fermentation broth, reducing reliance on cell disruption and simplifying extraction to centrifugation and membrane concentration steps (Leemhuis et al., 2010a, 2010b; Saini et al., 2022). These developments collectively demonstrate a shift toward extraction strategies that prioritize enzyme stability, scalability, and economic feasibility.

Purification techniques

The Fig. 8.2 illustrates a generalized workflow for protein purification, highlighting the sequential strategies commonly employed to enrich and isolate target proteins from complex biological mixtures. The process typically begins with ammonium sulfate precipitation, a classical salting out step used to concentrate proteins and remove bulk impurities. This is followed by ultrafiltration and dialysis, which help in buffer exchange, remove small molecules, and further concentrate the partially purified protein fraction.

The next stage involves chromatographic approaches, which provide higher resolution purification based on specific physicochemical properties of the target protein.

Ammonium sulfate precipitation

The process of precipitation involves the creation of submicron particles, which are then split up by shear forces and aggregate (Glatz & Fisher, 1986). Salts, polyelectrolytes, nonionic polymers, heat, pH adjustment, and organic solvents are only a few of the several types of precipitating agents. Ammonium sulfate is a common salt used for this purpose

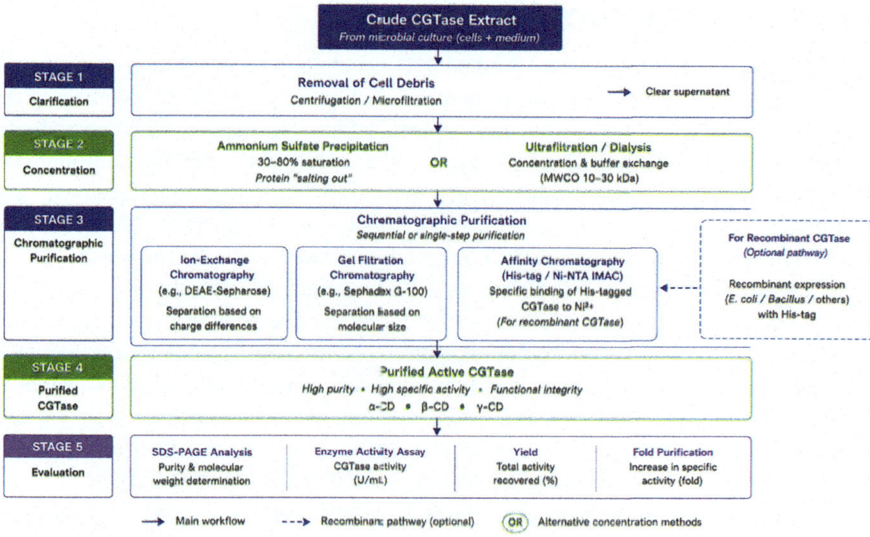

FIGURE 8.2

Purification strategies for cyclodextrin glucanotransferase (CGTase). The purification workflow begins with crude CGTase extract obtained from microbial culture, followed by clarification to remove insoluble materials. Enzyme concentration is achieved by either ammonium sulfate precipitation or ultrafiltration/dialysis before chromatographic purification. Chromatographic approaches include ion-exchange chromatography, gel filtration chromatography, and affinity chromatography for recombinant His-tagged CGTases. Purified active CGTase is subsequently evaluated by SDS–PAGE, enzyme activity assay, recovery yield, and fold purification to assess purity, catalytic activity, and purification efficiency.

through a method called salting out. The salting-out method is widely used for protein extraction and protein purification by adding high concentrations of salt to reduce the solubility of the target protein (Zhou et al., 2025). When ammonium sulfate is added to a solution, the salt concentration increases. This reduces the amount of water available to protect the hydrophobic parts of proteins, causing them to stick together and form a solid (precipitate). However, ammonium sulfate may contain metal ion impurities that can harm enzymes. Adding EDTA can help, as it binds to these metal ions and removes their negative effects (Sharaf & Al-Fadel, 2013; Wingfield, 2001).

The amount of ammonium sulfate needed to precipitate different proteins in a biological mixture varies with each protein, so it must be determined experimentally. As the ammonium sulfate saturation level increases, more protein can be collected from the solution. This happens because proteins with fewer hydrophilic areas tend to precipitate at lower saturation levels, whereas proteins with more hydrophilic areas require higher ammonium sulfate concentrations to precipitate (Nooralabettu, 2014).

Ultrafiltration and dialysis

Ultrafiltration is a commonly used method to concentrate protein solutions. It is simple to use, requires only basic equipment, and does not cause problems like damaging the protein structure or increasing the salt content, which can happen with precipitation methods. In ultrafiltration, small molecules like buffer salts can pass through the membrane and are removed, while larger molecules like proteins are retained. The membranes used in ultrafiltration can be made from materials like cellulose, nitrocellulose, or synthetic polymers. For proteins to work well, the membrane should have a hydrophilic surface. Ultrafiltration devices come in different sizes, suitable for volumes ranging from a few milliliters to several litres. The solution is pushed through the membrane using compressed air or, preferably, nitrogen gas. The ultrafiltrate is collected beneath the membrane. There are different ultrafiltration systems available. Instead of using air or gas pressure, some systems use centrifugal force or vacuum pressure to move the solution through the membrane. There are also filter units that can be attached to a syringe, allowing manual concentration (Bisswanger, 2019).

Dialysis is a special form of ultrafiltration. In this method, the protein solution is placed inside a semipermeable membrane (dialysis bag), which is then placed in a larger container filled with a buffer. Small molecules can pass through the dialysis membrane and mix with the outer solution, whereas larger molecules, such as proteins, cannot (Lazaro et al., 2024). Dialysis is often used to remove small unwanted substances, such as excess salts or other small molecules, from crude extracts or from solutions after ammonium sulfate precipitation. To fully remove these small molecules, the outer buffer must be changed several times (Andrew et al., 2002). Dialysis can also be used to concentrate protein. In this case, the dialysis bag is surrounded by a substance like polyethene glycol or Sephadex G-25, which absorbs water and small molecules from inside the bag. This reduces the liquid volume inside and increases the protein concentration (Bisswanger, 2019).

Chromatographic approaches

Ion-exchange chromatography

Ion-exchange chromatography is a method for separating components in a sample based on their affinity for charged sites on a solid support. This solid material is usually made from organic resins (Cummins et al., 2017). Since enzymes have both positive and negative charges, they can usually be separated using this technique. Specially designed resins help prevent enzymes from becoming damaged or losing their function during the process. Although the enzyme must remain dissolved in solution, the resin itself can withstand harsh chemicals, such as nonionic detergents and chaotropic agents, which help keep the enzyme soluble (Forman and Kennedy, 1997).

Proteins are made of amino acids, which contain different chemical groups that may carry positive, negative, or no charge at all. The ion-exchange resin contains fixed charged groups, such as carboxyl (negative) or quaternary ammonium (positive), that are chemically bonded to the resin. Each of these charged sites is paired with a counterion of the opposite charge, held only by ionic attraction. When a protein with the opposite charge comes close to the resin, it is attracted to the charged site, displaces the counterion, and binds to the resin. Proteins or molecules with no charge do not bind and can be washed away. The bound protein can later be released by adding another ion that binds more strongly to the resin than the protein does (Sahu & Jaiswani, 2018; Wheelwright, 1991).

Gel filtration chromatography

Gel filtration chromatography is a method for separating molecules by size. In this method, larger molecules move through the column faster than smaller ones. This happens because the material inside the column, called the stationary phase, has tiny pores. The ability of each molecule to enter these pores affects how quickly it travels (Fagain et al., 2016). Large molecules are too big to enter the pores, so they pass straight through and come out first. Medium-sized molecules can enter the pores to some extent and move more slowly. Small molecules can go deep into the pores, which slows them down even more. As the molecules move through the column, they randomly diffuse in and out of the pores, and this random movement takes time. The more a molecule enters the pores, the longer it takes to reach the end of the column. Therefore the smaller molecules, which diffuse farther into the matrix, take longer to migrate through the column (Wheelwright, 1991).

Affinity chromatography

Affinity chromatography is a technique that separates a target molecule, like an enzyme, from other substances by using a specific interaction between the enzyme and a substance (called a ligand) attached to a solid material. The process follows similar steps to ion-exchange chromatography, but the key difference is the type of binding group used. In affinity chromatography, the enzyme sticks to the ligand, while other unwanted molecules are washed away. After that, the enzyme is released, or eluted, from the column. Sometimes, the column can be reused without needing any regeneration (Wheelwright, 1991). For affinity chromatography steps to be successful, two conditions must be met. A specific ligand that can bind to the desired protein must be attached to a solid support; a means must be available to desorb the protein from the ligand; and a way must be provided to separate the protein from the ligand once purification is complete (Pina et al., 2013).

Purification of CGTase traditionally relies on multistep fractionation strategies that combine salt precipitation and chromatographic separation to achieve high purity and catalytic efficiency. Representative purification schemes illustrate the quantitative improvements achieved at each stage. For instance, extracellular CGTase from alkalophilic *Bacillus* sp. G1 was purified via sequential ammonium sulfate precipitation (30%–80% saturation), ion-exchange chromatography, and gel filtration (Sian et al., 2005). The enzyme was optimally recovered in the 60%–80% ammonium sulfate fraction, increasing specific activity from approximately 12 U/mg in the crude extract to 38 U/mg after dialysis, corresponding to approximately threefold purification with nearly 78% activity recovery. Similarly, CGTase from *Microbacterium terrae* KNR9 precipitated optimally at 70% saturation, yielding ~2.8-fold purification and ~75% recovery (Rajput et al., 2016). These data confirm that CGTases typically require moderate-to-high salt concentrations for efficient precipitation, consistent with their structural stability and surface hydrophobicity characteristics.

In ion-exchange chromatography, commonly performed using DEAE-Sepharose or DEAE-cellulose matrices, exploits the slightly acidic isoelectric point of CGTases (pI ~5.5–6.5). In the case of *Bacillus* sp. G1, application to a DEAE-Sepharose fast flow column equilibrated at pH 8.0 allowed enzyme elution at 0.25–0.35 M NaCl, increasing specific activity to ~95 U/mg and achieving nearly eightfold purification with ~52% overall yield (Sian et al., 2005). Further purification via size-exclusion chromatography (e.g., Sephadex G-100) yielded a 12–18-fold total purification and final specific activities of 120–150 U/mg, with overall yields typically ranging from 30% to 45%. Similar purification performance was reported for CGTase from *Bacillus pseudalcaliphilus*, achieving ~9.5-fold purification and final specific activity exceeding 100 U/mg (Kitayska et al., 2011).

While these classical methods remain effective, modern recombinant DNA technology has significantly streamlined CGTase purification through affinity tag strategies. Recombinant CGTases expressed in *Escherichia coli* or engineered *Bacillus* hosts are frequently fused with polyhistidine (His_6) tags at either the N- or C-terminus, enabling rapid purification via immobilized metal affinity chromatography (IMAC). In this approach, clarified cell lysate is applied directly onto Ni^{2+}-NTA or Co^{2+}-based resin, where the His-tag binds selectively to the immobilized metal ions. After washing to remove nonspecific proteins, recombinant CGTase is eluted using imidazole gradients (typically 100–250 mM).

In directed evolution studies of CGTase (Leemhuis et al., 2010a, 2010b), His-tagged variants purified via Ni-NTA chromatography achieved >90% purity in a single step, with specific activities comparable to or exceeding native enzymes. Purification folds of 10–20 were commonly achieved, with recovery yields between 60% and 85%, substantially higher than those of traditional multistep purification. In some engineered systems, recombinant CGTase reached specific activities above 180 U/mg following IMAC purification, demonstrating that affinity-based methods not only simplify downstream processing but also preserve catalytic integrity. Moreover, the use of extracellular secretion systems in recombinant *Bacillus* strains further reduces purification complexity, as enzyme recovery can proceed directly from clarified fermentation broth without cell disruption.

Affinity purification also enables incorporation of additional tags, such as maltose-binding protein or glutathione-S-transferase, which enhance solubility and folding during heterologous expression. After purification, protease cleavage sites engineered between the tag and CGTase allow removal of the affinity tag when required for structural or industrial applications. These strategies significantly reduce processing time and improve reproducibility, making recombinant affinity purification particularly attractive for large-scale enzyme production and protein engineering studies.

The efficiency of both classical and recombinant purification strategies is routinely confirmed by SDS-PAGE analysis. Purified CGTases consistently appear as a single band between 74 and 80 kDa. Native CGTase from *B. circulans* 251 exhibits a molecular mass of approximately 75 kDa (Lawson et al., 1994), whereas recombinant His-tagged variants typically show a slightly increased molecular mass (~76–82 kDa) due to the additional affinity tag. During purification, crude lysates display multiple protein bands; after IMAC purification, a dominant single band corresponding to CGTase becomes evident, confirming high purity. The disappearance of background proteins alongside increased specific activity validates purification efficiency.

Collectively, these examples demonstrate that while traditional precipitation and chromatographic methods remain reliable, affinity tag recombinant purification represents a more rapid, scalable, and efficient approach for CGTase isolation. The integration of IMAC with recombinant expression systems has substantially improved enzyme yield, purity, and reproducibility, thereby accelerating biochemical characterization, protein engineering, and industrial implementation.

Evaluation of purification efficiency

Sodium dodecyl sulfate-PAGE analysis

SDS-PAGE analysis confirms CGTase purity and estimates molecular weight (~75–80 kDa depending on source). Distinct bands corresponding to CGTase allow monitoring of purification progress and detection of contaminant proteins (Sian et al., 2005).

Enzyme activity assays

CGTase activity is commonly measured by quantifying CD production from starch substrates using colorimetric or chromatographic methods (Rojas et al.,

2019). Tracking activity through purification steps allows calculation of specific activity and guides optimization.

Yield and fold purification metrics

Yield reflects total recovered CGTase activity relative to crude extract, while fold purification indicates enrichment of enzymes per protein unit. These metrics provide a quantitative evaluation of purification strategy efficiency (Rajput et al., 2016).

Emerging purification technologies for cyclodextrin glucanotransferase

Although classical purification methods such as ammonium sulfate precipitation and column chromatography remain widely applied, recent studies emphasize process-intensified purification technologies that reduce the number of steps while maintaining enzyme activity. Membrane chromatography, which combines convective flow with ion-exchange functionality, enables faster processing and improved scalability compared to packed-bed systems (Cummins et al., 2017).

Similarly, expanded bed adsorption allows direct processing of unclarified fermentation broths, eliminating preliminary centrifugation steps and reducing overall processing time (Pina et al., 2013). This approach is particularly advantageous for CGTase-rich broths produced by *Bacillus* species.

More recently, affinity purification using magnetic nanoparticles has emerged as a promising alternative. Ligand-functionalized magnetic supports allow rapid CGTase binding and recovery under mild conditions, minimizing enzyme denaturation and facilitating reuse of the purification matrix (Bekavac et al., 2024; Grazu et al., 2005). These approaches reflect a paradigm shift from maximizing purity alone toward optimizing process efficiency, enzyme functionality, and cost-effectiveness.

Characterization of cyclodextrin glucanotransferase

Biochemical properties

Optimal pH and temperature

CGTases display optimal catalytic activity within a characteristic pH and temperature range that reflects the environmental niche of their source organism. Most CGTases function best between pH 5.0 and 8.0, with some thermophilic variants tolerating more extreme pH values (Kitayska et al., 2011). Temperature optima typically fall between

50°C and 70°C; thermostable CGTases retain activity even beyond 80°C, making them valuable for industrial processes involving heat (Avci & Donmez, 2009).

Thermal and pH stability

Beyond optima, the enzyme's stability under various conditions dictates practical utility. Stability assays often measure residual activity after incubation at elevated temperatures or nonoptimal pH values for set time intervals (e.g., 1–4 hours). Many *Bacillus* CGTases show half-lives of 2–6 hours at 60°C–70°C, and retain >80% activity between pH 5 and 9 (Berlina et al., 2021). Strategies such as enzyme immobilization or protein engineering can enhance these properties.

The biochemical properties of CGTases vary significantly depending on microbial origin, particularly with respect to optimal pH, temperature, and thermal stability, which directly influence their suitability for industrial CD production. Representative examples are summarized in Table 8.2 (Kitayska et al., 2011; Rajput et al., 2016; Sian et al., 2005). As shown in Table 8.2, CGTases from thermophilic and alkaliphilic microorganisms generally exhibit enhanced thermal stability and broader pH tolerance, making them particularly suitable for industrial CD production under harsh processing conditions.

Substrate specificity and product profile

Amylose and starch derivatives

CGTases act on α-1,4 glycosidic bonds of linear glucans such as amylose and modified starches, catalyzing the formation of CDs (Ji et al., 2020). Substrate branching degree influences enzyme efficiency, with higher specificity and turnover observed on less-branched amylose. Substrate modifications (e.g., gelatinized starch, maltodextrins) also affect product yields. Amylose is generally considered the preferred substrate because its linear structure readily fits into the CGTase substrate-binding groove, facilitating efficient cyclization. In contrast, highly branched substrates such as amylopectin require initial debranching or exhibit lower catalytic efficiency due to steric hindrance around α-1,6-glycosidic linkages. The physicochemical properties of starch substrates also influence CGTase activity and product distribution. Pretreatments such as gelatinization increase substrate accessibility by disrupting starch granule crystallinity, whereas partially hydrolyzed starches and maltodextrins provide shorter linear glucan chains that can alter cyclodextrin yield and the relative proportions of α-, β-, and γ-CDs. Factors including amylose content, molecular weight, degree of branching, and substrate concentration collectively determine enzyme affinity, catalytic turnover, and product specificity. Consequently, careful selection and modification of starch substrates are important strategies for optimizing cyclodextrin production in industrial processes, enabling improved reaction efficiency and selective production of desired cyclodextrin types.

Table 8.2 Optimal pH, temperature, and stability of selected cyclodextrin glucanotransferase.

Organism	Optimal pH	pH stability range	Optimal temp (℃)	Thermal stability ($t^{1/2}$ at optimal temp)	References
Bacillus firmus	6.0–7.0	5.0–9.0	60	~4 h at 60℃	Choi et al. (2019)
Thermophilic Bacillus sp.	6.5–7.5	5.5–9.5	70	~6 h at 70℃	Yang et al. (2021)
Bacillus subtilis	5.5–7.0	5.0–8.5	55	~3 h at 55℃	Moon et al. (2020)
Bacillus circulans	6.0–8.0	5.5–9.0	60	~4–5 h at 60℃	Sian et al. (2005), Lawson et al. (1994)
Bacillus pseudalcaliphilus	7.5–9.0	7.0–10.5	65	~5 h at 65℃	Kitayska et al. (2011)
Bacillus stearothermophilus	6.0–7.5	5.5–9.0	65–70	~6 h at 70℃	Avci and Donmez (2009)
Paenibacillus macerans	6.0–7.5	5.5–8.5	55–60	~3–4 h at 60℃	Kelly et al. (2009)
Microbacterium terrae KNR9	8.0–9.5	7.5–10.0	60	~4 h at 60℃	Rajput et al. (2016)
Thermoanaerobacter sp.	6.0–7.0	5.5–8.0	70	~6–7 h at 70℃	Martins and Hatti-Kaul (2002)
Shouchella oshimensis P-106	6.5–7.5	6.0–9.0	55–60	~3–4 h at 60℃	Pardhi et al. (2025)

Cyclodextrin types produced

CGTases catalyze the cyclization of linear α-1,4-glucans to produce three major cyclodextrins (CDs), which differ in the number of glucose units and cavity size:

- α-CD (6 glucose units)
- β-CD (7 glucose units)
- γ-CD (8 glucose units)

The relative proportions of these products depend primarily on the geometry of the enzyme active site, substrate-binding subsites, and the microbial origin of the CGTase (Kanai et al., 2004). Most Bacillus CGTases predominantly produce β-CD, whereas CGTases from thermophilic microorganisms often exhibit enhanced γ-CD production due to differences in substrate-binding loops and catalytic pocket architecture (Solanki et al., 2022). Because the cavity diameter increases from α-CD to γ-CD, each cyclodextrin exhibits distinct inclusion properties and industrial applications. α-CD is commonly used in food and pharmaceutical formulations, β-CD is the most widely produced and utilized for encapsulation and stabilization of bioactive compounds, while γ-CD, with its larger cavity and higher aqueous solubility, is particularly suitable for complexing larger molecules in pharmaceutical and nutraceutical applications.

Enzyme kinetics

Determination of K_m and V_{max}

The Michaelis constant (K_m) and maximum velocity (V_{max}) are derived by measuring initial reaction rates at increasing substrate concentrations, usually using soluble starch or amylose (Tincu et al., 2023). Typical K_m values for CGTases range from 0.5 to 5 mg/mL starch, reflecting moderate to high substrate affinity (Cheirsilp et al., 2009).

Lineweaver–Burk and Michaelis–Menten analysis

While Michaelis–Menten plots show hyperbolic saturation curves, Lineweaver–Burk plots linearize this data to facilitate parameter estimation. Modern practice favors nonlinear regression fitting due to bias introduced at low substrate concentrations (Cho & Lim, 2018). Accurate kinetic data support enzyme engineering and process optimization.

The kinetic properties of CGTases, including substrate affinity (K_m) and catalytic efficiency (V_{max}), vary considerably depending on the microbial source and reaction substrate. Representative kinetic parameters reported for CGTases acting on starch-based substrates are summarized in Table 8.3 (Rajput et al., 2016; Sian et al., 2005; Yang et al., 2021).

Structural analysis

Spectroscopic methods

Spectroscopic techniques provide rapid, non-destructive approaches for evaluating the structural integrity, purity, and conformational stability of cyclodextrin glucanotransferases (CGTases).

UV-Visible Spectroscopy is used for determining protein concentration and assessing enzyme purity. Proteins absorb ultraviolet light primarily due to the aromatic amino acid residues tryptophan, tyrosine, and phenylalanine, with a characteristic absorption maximum at approximately 280 nm. Measurement of absorbance at this wavelength enables rapid estimation of protein concentration using the Beer–Lambert law. The purity of enzyme preparations is commonly evaluated using the A280/A260 ratio, where contamination by nucleic acids results in increased absorbance at 260 nm. UV–Vis analysis serves as an effective quality-control tool throughout CGTases enzyme purification by monitoring protein recovery and identifying possible contaminants (Duncombe et al., 2021).

Fourier Transform Infrared (FTIR) Spectroscopy provides valuable information regarding the secondary structural organization of proteins by measuring the vibrational frequencies of peptide bonds. The amide I band (1600–1700 cm^{-1}), arising mainly from C=O stretching vibrations, is highly sensitive to protein secondary structure and is widely used to estimate the relative proportions of α-helices, β-sheets, β-turns, and random coils. The amide II band (1500–1600 cm^{-1}), originating from N–H bending and C–N stretching vibrations, provides complementary structural information. Changes in these spectral regions can indicate alterations in protein conformation resulting from thermal denaturation, pH variation, ligand binding, or protein engineering. For CGTases, FTIR analysis is particularly useful for evaluating structural stability after purification, immobilization, or chemical modification, as well as monitoring conformational changes induced by substrate or cyclodextrin binding (Miller et al., 2013). (Miller et al., 2013).

Circular Dichroism Spectroscopy is a highly sensitive technique for investigating protein secondary structure and conformational stability. In the far-UV region (190–250 nm), circular dichroism spectra provide quantitative estimates of α-helix, β-sheet, β-turn, and random-coil content by measuring the differential absorption of left- and right-circularly polarized light. Characteristic negative bands near 208 and 222 nm indicate α-helical structures, whereas β-sheet-rich proteins exhibit a negative band around 218 nm. Near-UV circular dichroism spectroscopy (250–320 nm) provides additional information regarding tertiary structure through signals originating from aromatic amino acid side chains and disulfide bonds. In CGTase studies, circular dichroism spectroscopy has been widely applied to compare native and engineered enzymes, assess structural integrity following purification or immobilization, and investigate the effects of mutations on protein stability and catalytic function (Linhares & Ramos, 2023).

Table 8.3 Kinetic parameters of representative cyclodextrin glucanotransferase.

Enzyme source	Substrate	K_m (mg/mL)	V_{max} (μ mol/min/mg)	References
Bacillus firmus	Soluble starch	1.2	150	Choi et al. (2019)
Thermophilic Bacillus sp.	Amylose	0.8	200	Yang et al. (2021)
Bacillus subtilis	Soluble starch	3.5	120	Kulshreshtha et al. (2020)
Bacillus circulans	Soluble starch	1.5	160	Sian et al. (2005)
Bacillus macerans	Amylose	2.1	140	Qi and Zimmermann (2005)
Bacillus stearothermophilus	Soluble starch	0.9	210	Avci and Donmez (2009)
Paenibacillus macerans	Soluble starch	2.4	135	Kelly et al. (2009)
Microbacterium terrae KNR9	Soluble starch	1.8	155	Rajput et al. (2016)
Thermoanaerobacter sp.	Amylose	0.6	230	Martins and Hatti-Kaul (2002)
Shouchella oshimensis P-106	Soluble starch	2.0	145	Pardhi et al. (2025)

Crystallography and molecular modeling

X-ray crystallography has elucidated high-resolution CGTase structures, detailing catalytic residues and substrate binding pockets, explaining specificity and product profiles (Lawson et al., 1994; Strokopytov et al., 1995). Molecular docking and dynamics simulations complement structural data, guiding mutagenesis for improved stability or altered CD selectivity (Gholampour et al., 2024).

High-resolution X-ray crystallographic studies have revealed that CGTases adopt a conserved $(\beta/\alpha)\backslash_8$ TIM (triosephosphate isomerase)-barrel fold characteristic of GH13 enzymes, with a catalytic Asp–Glu–Asp triad responsible for glycosidic bond cleavage and intramolecular transglycosylation leading to CD formation (Lawson et al., 1994; Strokopytov et al., 1995).

Structural analyzes have shown that CGTases possess a deep substrate binding groove that accommodates linear α-1,4-glucans and promotes intramolecular transglycosylation leading to CD formation. The catalytic triad, typically consisting of Asp, Glu, and Asp residues, functions through a double-displacement mechanism common to GH13 enzymes. Variations in loop regions surrounding the active site significantly influence CD size specificity, explaining why certain CGTases preferentially produce α-, β-, or γ-CDs. These structural insights provide a molecular basis for rational protein engineering aimed at improving CGTase stability, activity, and product selectivity.

Factors influencing cyclodextrin glucanotransferase activity and stability

Role of metal ions and cofactors

Metal ions significantly influence the catalytic performance and structural stability of CGTases, although they are not strictly required as catalytic cofactors. In many characterized CGTases from *Bacillus* species, divalent cations such as Ca^{2+} enhance thermal stability by binding to specific calcium-binding sites within the protein structure. These sites are typically located near domain interfaces, where Ca^{2+} contributes to structural rigidity and protects the enzyme against thermal denaturation and proteolytic degradation (Eijsink et al., 2011; Frommel & Hohne, 1981; Gohara & Cera, 2016).

For example, Ca^{2+} supplementation has been reported to increase the half-life of several bacterial CGTases at elevated temperatures, indicating a stabilizing effect on tertiary structure. Mg^{2+} and Mn^{2+} may also modulate catalytic efficiency depending on the enzyme source and reaction conditions.

In contrast, heavy metal ions such as Hg^{2+}, Pb^{2+}, and Cu^{2+} strongly inhibit CGTase activity. These ions may interact with catalytic residues or sulfhydryl groups, inducing conformational changes that impair substrate binding or disrupt active site geometry, ultimately leading to irreversible inactivation (Quyoom, 2014). The magnitude and direction of metal ion effects are often strain-dependent, highlighting the importance of optimizing ion composition for each CGTase used in industrial applications.

Presence of inhibitors or activators

CGTase activity is particularly sensitive to specific carbohydrates and reaction by-products that directly interact with its catalytic machinery. Because CGTases catalyze transglycosylation and cyclization reactions of starch, small saccharides such as glucose and maltose can act as competitive inhibitors by occupying subsites within the active cleft. Likewise, CDs, the reaction products may inhibit CGTase through product inhibition mechanisms, reducing catalytic turnover at high concentrations (Pinotsis et al., 2003).

Proteolytic contamination in fermentation broths may degrade CGTase, especially in mesophilic variants, thereby reducing overall stability during prolonged operation. Oxidative stress conditions can also modify sensitive residues involved in substrate binding.

Conversely, certain stabilizing additives have been shown to enhance CGTase performance. Polyols such as glycerol can stabilize CGTase by strengthening hydrophobic interactions and reducing conformational flexibility. Reducing agents like dithiothreitol may protect critical cysteine residues from oxidation, preserving catalytic competence (Grazu et al., 2005; Solanki et al., 2021). These modulators are particularly relevant during enzyme storage and repeated industrial use.

Effect of solvents, detergents, or surfactants

The catalytic efficiency of CGTase is strongly influenced by the physicochemical environment of the reaction medium. Low concentrations of nonionic surfactants such as Tween 20 or Triton X-100 have been reported to enhance CGTase activity by improving starch dispersion and increasing substrate accessibility to the active site (Holmberg, 2017; Yoon & Robyt, 2005). In some cases, surfactants reduce enzyme aggregation, thereby maintaining catalytic functionality.

However, ionic detergents such as SDS typically disrupt the secondary and tertiary structures of CGTase, leading to loss of activity due to protein unfolding.

Organic solvents also exert concentration-dependent effects. While high solvent concentrations generally destabilize CGTase by disturbing hydrophobic interactions, certain thermophilic CGTases retain partial activity in the presence of ethanol or methanol. This tolerance enables their application in biphasic or low-water systems aimed at modifying hydrophobic substrates (Bekavac et al., 2024; Salameh & Wiegel, 2010). Therefore solvent compatibility must be carefully evaluated for each CGTase variant intended for industrial biotransformation processes.

Protein engineering approaches to enhance stability

Protein engineering strategies have been extensively applied to improve CGTase stability under industrially relevant conditions. Site-directed mutagenesis targeting flexible loop regions, calcium-binding sites, or surface residues has successfully enhanced thermostability and pH tolerance in several CGTase variants. For instance, mutations that increase intramolecular hydrogen bonding or reduce conformational mobility can significantly prolong enzyme half-life at elevated temperatures without compromising cyclization activity.

Directed evolution approaches have also yielded CGTase mutants with improved catalytic efficiency or altered product specificity. In addition, fusion with stabilizing domains or immobilization on solid supports enhances operational stability and reusability, particularly in continuous production systems.

Collectively, these engineering approaches enable the rational design of CGTase variants with tailored stability profiles, supporting their expanded application in large-scale CD production.

Applications and industrial implications

Benefits of highly pure and stable cyclodextrin glucanotransferases

The availability of highly pure and stable CGTases is critical for advancing their industrial applications. Purified enzymes exhibit greater catalytic efficiency and reproducibility, enabling consistent production of CDs with defined properties (Solanki et al., 2022). Stability under harsh processing conditions, such as elevated temperatures, extreme pH, and presence of solvents is essential to maintain enzyme activity over extended operational periods, reducing costs related to enzyme replenishment (Ma et al., 2025). Moreover, stable CGTases minimize undesirable side reactions and contamination, which enhances product purity and downstream processing efficiency. These factors collectively contribute to improved process economics and sustainability in CD manufacturing.

Enhanced cyclodextrin production processes

The optimization of CGTase performance directly translates into enhanced CD production. Advances in enzyme engineering and purification have allowed for CGTases with improved substrate affinity, thermal tolerance, and product specificity, leading to higher yields of target CDs—particularly β- and γ-CDs under industrial conditions (Miao et al., 2024; Rabadiya et al., 2024). Process parameters such as enzyme dosage, substrate concentration, and reaction time are fine-tuned alongside enzyme properties to maximize productivity. Furthermore, immobilized CGTase systems have emerged, enabling enzyme reuse and continuous production modes that enhance operational stability and reduce costs (Mazzocato & Jacquier, 2024; Ogunbadejo & Al-Zuhair, 2024). Together, these improvements have made CD production more efficient, scalable, and environmentally friendly.

Integration into large-scale bioprocesses

Integrating CGTase-catalyzed reactions into large-scale bioprocesses requires careful consideration of enzyme stability, process economics, and downstream purification. The robustness of modern CGTases facilitates their use in high-volume bioreactors under continuous or fed-batch modes, often coupled with advanced substrate feeding strategies to maintain optimal enzymatic activity (Guajardo & Schrebler, 2024; Rocha et al., 2022). Additionally, coupling CGTase processes with membrane filtration and chromatography enhances CD recovery and purity, ensuring compliance with stringent pharmaceutical or food-grade standards. The scalability of CGTase-based bioprocesses has expanded CD availability for diverse applications, including drug delivery, food additives, and environmental remediation, underscoring their industrial significance (Mbanga et al., 2025).

Emerging trends in cyclodextrin glucanotransferase-based technologies

Recent technological trends are expanding the horizons of CGTase applications beyond traditional CD production. Protein engineering combined with machine learning is accelerating the design of CGTase variants with tailored specificity, stability, and catalytic profiles suited for novel substrates and reaction conditions (Leemhuis et al., 2010a, 2010b). Integration with nanotechnology and immobilization on novel supports enhances enzyme reusability and operational stability in harsh industrial environments (Ma et al., 2025). Moreover, hybrid biocatalytic systems coupling CGTase with other enzymes are being developed to synthesize complex oligosaccharides and functionalized CDs with applications in pharmaceuticals and cosmetics. The exploration of CGTases from extremophiles also opens opportunities for bioprocesses under unconventional conditions, such as high salinity or organic solvents, broadening the industrial applicability of these versatile enzymes (Mesbah, 2022).

Future directions

Challenges in cyclodextrin glucanotransferase research and production

Despite significant advancements in the study and industrial utilization of CGTases, several challenges remain. One major issue is achieving the balance between enzyme activity, stability, and specificity suitable for diverse industrial applications. Many naturally occurring CGTases exhibit limitations such as narrow pH and temperature ranges, susceptibility to inhibitors, or suboptimal product selectivity (Mahat et al., 2004). Additionally, large-scale production of CGTases with consistent quality and cost-effectiveness continues to be a bottleneck due to complex downstream purification processes and variable expression yields in microbial hosts (Saini et al., 2022). The presence of proteases and instability during storage and processing further complicate commercial viability. Addressing these challenges requires innovative approaches in enzyme engineering, fermentation optimization, and process integration.

Potential for genetic and protein engineering

Genetic and protein engineering hold tremendous potential to overcome the limitations of native CGTases. Advances in molecular biology tools, including site-directed mutagenesis, directed evolution, and CRISPR-based genome editing, allow precise modifications aimed at enhancing enzyme properties such as thermal stability, pH tolerance, and substrate specificity (Mao et al., 2024). Computational approaches, including molecular dynamics simulations and machine learning algorithms, facilitate rational design of variants with improved catalytic efficiency and novel functionalities. Fusion protein strategies and immobilization techniques can also be engineered genetically to improve enzyme recyclability and robustness under industrial conditions. Continued integration of these technologies promises to create tailor-made CGTases optimized for specific applications, significantly expanding their commercial potential (He et al., 2024; Jailani et al., 2023; Rybarczyk et al., 2025).

Exploration of extremophilic sources

The search for CGTases from extremophilic microorganisms is an exciting frontier that offers enzymes inherently adapted to harsh environmental conditions. Thermophiles, halophiles, acidophiles, and alkaliphiles produce enzymes with remarkable stability under extreme temperature, salinity, or pH, making them attractive candidates for industrial biocatalysis (Ashaolu et al., 2024). Mining metagenomic databases and environmental samples from extreme habitats provides access to a vast reservoir of novel CGTases with unique structural and functional attributes. These extremozymes can potentially function in unconventional reaction media such as organic solvents or high ionic strength solutions, opening new avenues for CD production and related biotransformations in challenging industrial contexts (Han et al., 2024; Mao et al., 2024).

Prospects for cyclodextrin glucanotransferase variants with novel properties

The convergence of protein engineering and extremophile exploration is likely to yield CGTase variants with unprecedented properties tailored to future industry needs. Potential novel features include broadened substrate specificity for diverse polysaccharides, enhanced regioselectivity for producing specific CD sizes, and improved resistance to chemical inhibitors and denaturants (Saini et al., 2022). Additionally, engineered CGTases with functionalities such as altered product profiles, reduced byproduct formation, or integrated multistep catalysis could revolutionize CD applications in pharmaceuticals, food, and environmental sectors. The development of smart enzyme systems capable of responding dynamically to process conditions via allosteric regulation or synthetic biology circuits represents a visionary frontier. Realizing these prospects will require multidisciplinary efforts combining biochemistry, molecular biology, computational design, and process engineering (Spiridon & Anghel, 2025).

Conclusion

The successful extraction, purification, and thorough characterization of CGTases are fundamental to unlocking their full biotechnological potential. Efficient extraction methods preserve enzyme activity and yield, while advanced purification techniques ensure high enzyme purity and stability—both critical for consistent industrial application. Detailed biochemical and structural characterization provides essential insights into enzyme functionality, guiding optimization for specific processes. Together, these foundational steps enable enhanced CD production with improved efficiency and product specificity. Looking forward, integrating protein engineering, exploration of extremophilic CGTases, and novel bioprocess technologies promises to overcome current limitations, driving innovation in enzyme design and application. This progress will broaden the scope of CGTase-based technologies, reinforcing their significance across pharmaceutical, food, and environmental industries.

References

Andrew, S. M., Titus, J. A., & Zumstein, L. (2002). Dialysis and concentration of protein solutions. *Current Protocols in Toxicology, 10*(1), 1–5. https://doi.org/10.1002/0471140856.txa03hs10.

Arce-Vázquez, M. B., Ponce-Alquicira, E., Delgado-Fornué, E., Pedroza-Islas, R., Díaz-Godínez, D., & Soriano-Santos, J. (2016). Integral use of amaranth starch to obtain cyclodextrin glycosyltransferase, by *Bacillus megaterium*, to produce β-cyclodextrin. *Frontiers in Microbiology, 7*, 1513. https://doi.org/10.3389/fmicb.2016.01513.

Ashaolu, T. J., Malik, T., Soni, R., Prieto, M. A., & Jafari, S. M. (2024). Extremophilic microorganisms as a source of emerging enzymes for the food industry: A review. *Food Science and Nutrition, 13*(1). https://doi.org/10.1002/fsn3.4540.

Astray, G., Gonzalez-Barreiro, C., Mejuto, J. C., Rial-Otero, R., & Simal-Gandara, J. (2009). A review on the use of cyclodextrins in foods. *Food Hydrocolloids, 23*(7), 1631–1640. https://doi.org/10.1016/j.foodhyd.2009.01.001.

Avci, A., & Donmez, S. (2009). A novel thermophilic anaerobic bacteria producing cyclodextrin glycosyltransferase. *Process Biochemistry, 44*(1), 36–42. https://doi.org/10.1016/j.procbio.2008.09.006.

Bekavac, N., Benkovic, M., Jurina, T., Valinger, D., Kljusuric, J. G., Tusek, A. J., & Salic, A. (2024). Advancements in aqueous two-phase systems for enzyme extraction, purification, and biotransformation. *Molecules (Basel, Switzerland), 29*(16), 1355. https://doi.org/10.3390/molecules29163776.

Berlina, Y. Y., Petrovskaya, L. E., Kryukova, E. A., Shingarova, L. N., Gapizov, S. S., Kryukova, M. V., Rivkina, E. M., Kirpichnikov, M. P., & Dolgikh, D. A. (2021). Engineering of thermal stability in a cold-active oligo-1,6-glucosidase from *Exiguobacterium sibiricum* with unusual amino acid content. *Biomolecules, 11*(8), 1229. https://doi.org/10.3390/biom11081229.

Bisswanger, H. (2019). *Practical enzymology* (3rd Edition). Weinheim: Wiley.

Brown, T., & Davis, R. (2022). Optimization of enzyme extraction protocols for industrial applications. *Journal of Biotechnology Advances, 45*(3), 123–135. https://doi.org/10.1016/j.jbiotecadv.2022.03.007.

Cheirsilp, B., Kitcha, S., & Maneerat, S. (2009). Kinetic characteristics of β-cyclodextrin production by cyclodextrin glycosyltransferase from newly isolated *Bacillus* sp. C26. *Electronic Journal of Biotechnology, 13*(4), 1–8.

Cho, Y. S., & Lim, H. S. (2018). Comparison of various estimation methods for the parameters of Michaelis-Menten equation based on in vitro elimination kinetic simulation data. *Translational and Clinical Pharmacology, 26*(1), 39–47. https://doi.org/10.12793/tcp.2018.26.1.39.

Choi, J., & Park, S. (2018). Use of protease inhibitors for enhanced enzyme stability during extraction. *Biochemical Engineering Journal, 134*, 12–19. https://doi.org/10.1016/j.bej.2018.01.014.

Choi, Y. J., Kim, J. W., & Lee, S. B. (2019). Biochemical characterization of CGTase from *Bacillus firmus*. *Journal of Molecular Catalysis B: Enzymatic, 152*, 12–20.

Crini, G. (2014). Review: A history of cyclodextrins. *Chemical Reviews, 114*(21), 10940–10975. https://doi.org/10.1021/cr500081p.

Cummins, P. M., Rochfort, K. D., & O'Connor, B. F. (2017). Ion-exchange chromatography: Basic principles and application. *Methods in Molecular Biology, 1485*, 209–223. https://doi.org/10.1007/978-1-4939-6412-3_11.

Duncombe, T. A., Ponti, A., Seebeck, F. P., & Dittrich, P. S. (2021). UV–Vis spectra-activated droplet sorting for label-free chemical identification and collection of droplets. *Analytical Chemistry, 93*(38), 13008–13013. https://doi.org/10.1021/acs.analchem.1c02822.

Eijsink, V. G. H., Matthews, B. W., & Vriend, G. (2011). The role of calcium ions in the stability and instability of a thermolysin-like protease. *Protein Science, 20*(8), 1346–1355. https://doi.org/10.1002/pro.670.

Fagain, C., Cummins, P. M., & O'Connor, B. F. (2016). Gel-filtration chromatography. *Methods in Molecular Biology, 1485*, 15–25. https://doi.org/10.1007/978-1-4939-6412-3_2.

Fernandez, L., & Lopez, M. (2015). Temperature effects on enzyme extraction and stability. *Enzyme and Microbial Technology, 67*, 34–41. https://doi.org/10.1016/j.enzmictec.2014.11.008.

Forman, H. J., & Kennedy, J. (1997). Effects of chaotropic agents versus detergents on dihydroorotate dehydrogenase. *Journal of Biology Chemistry, 252*(10), 3379–3381.

Frommel, C., & Hohne, W. E. (1981). Influence of calcium binding on the thermal stability of 'thermitase', a serine protease from *Thermoactinomyces vulgaris*. *Biochimica et Biophysica Acta (BBA) - Protein Structure, 670*(1), 25–31. https://doi.org/10.1016/0005-2795(81)90044-1.

Garcia, M., Rodriguez, P., & Sanchez, V. (2021). Enzymatic lysis methods for protein recovery from microbial cells. *Applied Microbiology and Biotechnology, 105*(9), 3675–3683. https://doi.org/10.1007/s00253-021-11358-6.

Gholampour, F., Hokmabady, L., & Ravari, F. (2024). Comprehensive molecular docking and molecular dynamics simulation study of unmodified and modified cyclodextrins: erlotinib inclusion complexes. *Molecular Simulation, 50*(17-18), 1–14. https://doi.org/10.1080/08927022.2024.2401617.

Glatz, C. E., & Fisher, R. R. (1986). Modeling of precipitation phenomena in protein recovery. *Separation, Recovery, and Purification in Biotechnology*.

Gohara, D. W., & Cera, E. D. (2016). Molecular mechanisms of enzyme activation by monovalent cations. *Journal of Biological Chemistry, 1*(1), 1–17. https://doi.org/10.1074/jbc.R116.737833.

Grazu, V., Abian, O., Mateo, C., Viera, F. B., Lafuente, R. F., & Guisan, J. M. (2005). Stabilization of enzymes by multipoint immobilization of thiolated proteins on new epoxy-thiol supports. *Biotechnology and Bioengineering, 90*(5), 597–605. https://doi.org/10.1002/bit.20452.

Guajardo, N., & Schrebler, R. A. (2024). Upstream and downstream bioprocessing in enzyme technology. *Pharmaceutics, 16*(1), 38. https://doi.org/10.3390/pharmaceutics16010038.

Han, J. R., Li, S., Li, W. J., & Dong, L. (2024). Mining microbial and metabolic dark matter in extreme environments: A roadmap for harnessing the power of multi-omics data. *Advanced Biotechnology, 2*(26), 1–18. https://doi.org/10.1007/s44307-024-00034-8.

He, Y., Wang, M., Ji, H., Cheng, Q., Liu, S., Huan, Y., Qian, T., & Yan, C. (2024). Molecular dynamics simulations for electrocatalytic CO_2 reduction: Bridging macroscopic experimental observations and microscopic explanatory mechanisms. *Advanced Functional Materials, 35*(3). https://doi.org/10.1002/adfm.202413703.

Holmberg, K. (2017). Interactions between surfactants and hydrolytic enzymes. *Colloid and Surface B: Biointerface, 168*, 169–177. https://doi.org/10.1016/j.colsurfb.2017.12.002.

Jailani, N., Jaafar, N. R., Rahman, R. A., & Illias, R. M. (2023). Robust cross-linked cyclodextrin glucanotransferase from *Bacillus lehensis* G1 aggregates using an improved cross-linker and a new co-aggregant for the production of cyclodextrins. *Enzyme and Microbial Technology, 169*. https://doi.org/10.1016/j.enzmictec.2023.110283.

Ji, H., Bai, Y., Li, X., Zheng, D., Shen, Y., & Jin, Z. (2020). Structural and property characterization of corn starch modified by cyclodextrin glycosyltransferase and specific cyclodextrinase. *Carbohydrate Polymers, 237*, 116–137. https://doi.org/10.1016/j.carbpol.2020.116137.

Kanai, R., Haga, K., Akiba, T., Yamane, K., & Harata, K. (2004). Role of Phe283 in enzymatic reaction of cyclodextrin glycosyltransferase from alkalophilic *Bacillus* sp.1011: Substrate binding and arrangement of the catalytic site. *Protein Science, 13*(2), 457–465. https://doi.org/10.1110/ps.03408504.

Kelly, R. M., Dijkhuizen, L., & Leemhuis, H. (2009). The evolution of cyclodextrin glycosyltransferase product specificity. *Applied Microbiology and Biotechnology, 84*(6), 1193–1200. https://doi.org/10.1007/s00253-009-2144-2.

Kitayska, T., Petrova, P., Ivanova, V., & Tonkova, A. I. (2011). Purification and properties of a new thermostable cyclodextrin glucanotransferase from *Bacillus pseudalcaliphilus*

8SB. *Applied Biochemistry and Biotechnology, 165*(5-6), 1285–1295. https://doi.org/10.1007/s12010-011-9346-4.

Kulshreshtha, N. M., Upadhyay, D., Shrivastava, D., Chauhan, A., & Singh, N. K. (2020). Production and optimization of cyclodextrin glucanotranferase from Bacillus sp. SS2. *Journal of microbiology, biotechnology and food sciences, 10*(2), 159–165.

Kumar, A., & Reddy, G. (2020). Recovery of CGTase from Bacillus strains for industrial use. *Industrial Microbiology, 11*(2), 100–110.

Lawson, C. L., Montfort, R., Strokopytov, B., Rozeboom, H. J., Kalk, K. H., Vries, G. E., Penninga, D., Dijkhuizen, L., & Dijkstra, B. W. (1994). Nucleotide sequence and X-ray structure of cyclodextrin glycosyltransferase from Bacillus circulans strain 251 in a maltose-dependent crystal form. *Journal of Molecular Biology, 236*(2), 590–600. https://doi.org/10.1006/jmbi.1994.1168.

Lazaro, L. G., Sabroso, C. M., Blanco, J. A., & Suarez, I. T. (2024). Assessment of in vitro release testing methods for colloidal drug carriers: The lack of standardized protocols. *Pharmaceutics, 16*(1), 103. https://doi.org/10.3390/pharmaceutics16010103.

Lee, H., Kim, S., & Park, J. (2018). Mechanical disruption techniques for microbial cell lysis in enzyme extraction. *Process Biochemistry, 68*, 107–115. https://doi.org/10.1016/j.procbio.2018.04.013.

Leemhuis, H., Kelly, R. M., & Dijkhuizen, L. (2010a). Directed evolution of cyclodextrin glycosyltransferase: selection and screening strategies. *Applied Microbiology and Biotechnology, 85*(3), 823–835. https://doi.org/10.1007/s00253-009-2282-6.

Leemhuis, H., Kelly, R. M., & Dijkhuizen, L. (2010b). Engineering of cyclodextrin glucanotransferases and the impact for biotechnological applications. *Applied Microbiology Biotechnology, 85*, 823–835. https://doi.org/10.1007/s00253-009-2221-3.

Leemhuis, H., Kralj, S., & Dijkhuizen, L. (2003). Transglycosylation reactions catalyzed by glucoside hydrolases: New developments and applications for enzymatic oligosaccharide synthesis. *Biotechnology and Genetic Engineering Reviews, 20*(1), 163–201. https://doi.org/10.1080/02648725.2003.10648001.

Linhares, L. A., & Ramos, C. H. I. (2023). Unlocking insights into folding, structure, and function of proteins through circular dichroism spectroscopy—A short review. *Applied Biosciences, 2*(4), 639–655. https://doi.org/10.3390/applbiosci2040040.

Loftsson, T., & Duchêne, D. (2007). Cyclodextrins and their pharmaceutical applications. *International Journal of Pharmaceutics, 329*(1–2), 1–11. https://doi.org/10.1016/j.ijpharm.2006.10.044.

Ma, X., Pronay, T. S., Gao, B., & Zhao, J. (2025). Nanoengineered enzyme immobilization: Toward biomedical, orthopedic, and biofuel applications. *ACS Publications, 1*(1), 1–17. https://doi.org/10.1021/acsomega.5c05589.

Mahat, M. K., Illias, R. M., Rahman, R. A., Rashid, N. A. A., Mahmood, N. A. N., Hassan, O., Aziz, S. A., & Kamaruddin, K. (2004). Production of cyclodextrin glucanotransferase (CGTase) from alkalophilic Bacillus sp. TS1-1: Media optimization using experimental design. *Enzyme and Microbial Technology, 35*(2004), 467–473. https://doi.org/10.1016/j.enzmictec.2004.07.008.

Mao, S., Jiang, J., Xiong, K., Chen, Y., Yao, Y., Liu, L., Liu, H., & Li, X. (2024). Enzyme engineering: Performance optimization, novel sources, and applications in the food industry. *Foods, 13*(23), 3846. https://doi.org/10.3390/foods13233846.

Martinez, D., Santos, R., & Lopez, A. (2021). Stabilizers for enzyme preservation during freeze-thaw cycles. *Journal of Enzyme Inhibition and Medicinal Chemistry, 36*(1), 92–98. https://doi.org/10.1080/14756366.2021.1878432.

Martins, R. F., & Hatti-Kaul, R. (2002). A novel cyclodextrin glycosyltransferase from an alkaliphilic bacterium. *Enzyme and Microbial Technology, 30*(1), 116–124. https://doi.org/10.1016/S0141-0229(01)00435-8.

Mazzocato, M. C., & Jacquier, J. C. (2024). Recent advances and perspectives on food-grade immobilisation systems for enzymes. *Foods, 13*. https://doi.org/10.3390/foods13132127.

Mbanga, B. G. F., Ngeva, Z. T., Badawy, W. M., Ebite, C., Onotu, O. P., Abogidi, C., Uzordinma, A. P., & Kaba, S. (2025). *Journal of Molecular Structure, 1328*, 1–18. https://doi.org/10.1016/j.molstruc.2025.141326.

Mesbah, N. M. (2022). Industrial biotechnology based on enzymes from extreme environments. *Front Bioeng Biotechnology, 10*, 1–16. https://doi.org/10.3389/fbioe.2022.870083.

Miao, B., Huang, D., Wang, T., Liu, H., Hao, Z., Yuan, H., & Jiang, Y. (2024). Enhancing trehalose production via *Bacillus* species G1 cyclodextrin glucanotransferase mutants: modifying disproportionation characteristics and thermal stability. *Frontiers in Microbiology, 15*. https://doi.org/10.3389/fmicb.2024.1500232.

Miller, L. M., Bourassa, M. W., & Smith, R. J. (2013). FTIR spectroscopic imaging of protein aggregation in living cells. *Biochimica et Biophysica Acta, 1828*(10), 2339–2346. https://doi.org/10.1016/j.bbamem.2013.01.014.

Moon, J. H., Lee, J. H., & Park, S. H. (2020). Characterization of cyclodextrin glucanotransferase from *Bacillus subtilis* for β-cyclodextrin production. *Biotechnology and Bioprocess Engineering, 25*, 421–430.

Nguyen, T., Tran, D., & Hoang, V. (2019). Buffer selection for enzyme extraction: A case study on CGTase. *Biotechnology Reports, 21*, e00318. https://doi.org/10.1016/j.btre.2019.e00318.

Nooralabettu, K. P. (2014). Optimisation of ammonium sulfate precipitation method to achieve high throughput concentration of crude alkaline phosphatase from brown shrimp (*Metapenaeus monoceros*) hepatopancreas. *International Journal of Analytical Bio-Science, 2*(1), 7–16.

Ogunbadejo, B. A., & Al-Zuhair, S. (2024). Enhanced performance of cyclodextrin glycosyltransferase by immobilization on an amine-induced macroporous metal organic framework. *LWT, 200*. https://doi.org/10.1016/j.lwt.2024.116221.

Pardhi, D. S., Rabadiya, K., Joshi, R. G., Raval, V. H., Panchal, R. R., & Rajput, K. N. (2025). Exploring the potential of *Shouchella oshimensis* P-106 for cyclodextrin glucanotransferase production. *Vidya - A Journal of Gujarat University, 4*(1), 1–9. https://doi.org/10.47413/03szp371.

Patel, R., & Singh, A. (2017). Chemical lysis for protein recovery: Balancing efficacy and enzyme activity. *Protein Expression and Purification, 133*, 1–9. https://doi.org/10.1016/j.pep.2017.01.002.

Pina, A. S., Lowe, C. R., & Roque, A. C. (2013). Challenges and opportunities in the purification of recombinant tagged proteins. *Biotechnology Advances, 32*(2), 366–381. https://doi.org/10.1016/j.biotechadv.2013.12.001.

Pinotsis, N., Leonidas, D. D., Chrysina, E. D., Oikonomakos, N. G., & Mavridis, I. M. (2003). The binding of β- and γ-cyclodextrins to glycogen phosphorylase b: Kinetic and crystallographic studies. *Protein Sciences, 12*(9), 1914–1924. https://doi.org/10.1110/ps.03149503.

Qi, Q., & Zimmermann, W. (2005). Cyclodextrin glucanotransferase: From gene to applications. *Applied Microbiology and Biotechnology, 66*(5), 475–485. https://doi.org/10.1007/s00253-004-1781-5.

Quyoom, S. (2014). An overview of inhibition of enzymatic activity by heavy metal ions. *Chemical Science Review and Letters, 3*(11), 381–387.

Rabadiya, K., Pardhi, D., Khushali, T., Patoliya, J., Rajput, K., & Joshi, R. (2024). A review on recent upgradation and strategies to enhance cyclodextrin glucanotransferase properties for its applications. *International Journal of Biological Macromolecules, 259*(2), 1–11. https://doi.org/10.1016/j.ijbiomac.2024.129315.

Rajput, K., Patel, K., & Trivedi, U. B. (2016). A novel cyclodextrin glucanotransferase from an alkaliphile Microbacterium terrae KNR 9: Purification and properties. *3 Biotechnology, 6*(2), 1–11. https://doi.org/10.1007/s13205-016-0495-6.

Ravinder, K., Prabhakar, T., Prashanthkumar, K., & Venuka, N. (2014). Optimization of process parameters of cyclodextrin glycosyltransferase isolated from novel mutated *Bacillus* sp. TPR71HNA6 by Taguchi orthogonal array method. *Research in Biotechnology, 5*(3), 10–20.

Rocha, R. A., Speight, R. E., & Scott, C. (2022). Engineering enzyme properties for improved biocatalytic processes in batch and continuous flow. *Organic Process Research & Development, 26*(7), 1914–1924. https://doi.org/10.1021/acs.oprd.1c00424.

Rojas, M. J., Fonseca, M. A., Lafuente, R. F., Giordano, R. L. C., & Tardioli, P. W. (2019). Recovery of starch from cassava bagasse for cyclodextrin production by sequential treatment with α-amylase and cyclodextrin glycosyltransferase. *Biocatalysis and Agricultural Biotechnology, 22*, 1–11. https://doi.org/10.1016/j.bcab.2019.101411.

Rybarczyk, A., Sultan, T., Hussain, N., Azam, H. M. H., Rafique, S., Zdarta, J., & Jesionowski, T. (2025). Fusion of enzymatic proteins: Enhancing biological activities and facilitating biological modifications. *Advances in Colloid and Interface Science, 340*. https://doi.org/10.1016/j.cis.2025.103473.

Sahu, P., & Jaiswani, R. (2018). Ion exchange resins: An approach for the development of advanced materials with industrial, pharmaceutical and clinical applications. *International Journal of Advance Research, Ideas, and Innovations in Technology, 4*(1), 465–481.

Saini, K., Pathak, V. M., Tyagi, A., & Gupta, R. (2022). Microbial cyclodextrin glycosyltransferases: Sources, production, and application in cyclodextrin synthesis. *Catalysis Research, 2*(3), 29. https://doi.org/10.21926/cr.2203029.

Salameh, M. A., & Wiegel, J. (2010). Effects of detergents on activity, thermostability and aggregation of two alkalithermophilic lipases from *Thermosyntropha lipolytica*. *The Open Biochemistry Journal, 4*(1), 22–28. https://doi.org/10.2174/1874091X01004010022.

Sharaf, E. F., & Al-Fadel, K. (2013). Green recycling of chicken feather and sheep wool using the partially purified alkaline protease from *Bacillus circulans* L. *Natural Science, 5*, 25–29. https://doi.org/10.4236/ns.2013.56A004.

Sian, H. K., Said, M., Hassan, O., Kamaruddin, K., Ismail, A. F., Rahman, R. A., Mahmood, N. A. B. N., & Illias, R. (2005). Purification and characterization of cyclodextrin glucanotransferase from alkalophilic *Bacillus* sp. G1. *Process Biochemistry, 40*(3-4), 1101–1111. https://doi.org/10.1016/j.procbio.2004.03.018.

Singh, N., & Mehta, R. (2017). Metal ion chelators in protease inhibition during protein extraction. *Analytical Biochemistry, 524*, 1–7. https://doi.org/10.1016/j.ab.2017.03.009.

Smith, J., & Johnson, M. (2020). Overview of CGTase extraction techniques. *Enzyme Research, 2020*, 123456. https://doi.org/10.1155/2020/123456.

Solanki, P., Awadhiya, P., & Banerjee, T. (2022). Conventional production optimization of cyclodextrin glucosyl transferase by a novel isolate of *Bacillus* sp. PBS1 from potato rhizosphere. *Journal of Microbiology, Biotechnology, and Food Sciences, 12*(1), 1–8. https://doi.org/10.55251/jmbfs.5130.

Solanki, P., Putatunda, C., Kumar, A., Bhatia, R., & Walia, A. (2021). Microbial proteases: Ubiquitous enzymes with innumerable uses. *3 Biotechnology, 11*(10), 428. https://doi.org/10.1007/s13205-021-02928-z.

Spiridon, I., & Anghel, N. (2025). Cyclodextrins as multifunctional platforms in drug delivery and beyond: Structural features, functional applications, and future trends. *Molecules (Basel, Switzerland), 30*(14). https://doi.org/10.3390/molecules30143044.

Strokopytov, B., Penninga, D., Rozeboom, H. J., Kalk, K. H., Dijkhuizen, L., & Dijkstra, B. W. (1995). X-ray structure of cyclodextrin glycosyltransferase complexed with acarbose. Implications for the catalytic mechanism of glycosidases. *Biochemistry, 34*(7), 2234–2240. https://doi.org/10.1021/bi00007a018.

Szente, L., & Szejtli, J. (2004). Cyclodextrins as food ingredients. *Trends in Food Science & Technology, 15*(3–4), 137–142. https://doi.org/10.1016/j.tifs.2003.09.019.

Tincu, C. E., Bouhadiba, B., Atanase, L. I., Stan, C. S., Popa, M., & Ochiuz, L. (2023). An accessible method to improve the stability and reusability of porcine pancreatic α-amylase via immobilization in gellan-based hydrogel particles obtained by ionic cross-linking with Mg2+ ions. *Molecules (Basel, Switzerland), 28*(12), 4695. https://doi.org/10.3390/molecules28124695.

Wang, H., Zhou, W., Li, H., & Bu, R. (2018). Optimization of the fermentation conditions for the mutant strain of β-cyclodextrin glycosyltransferase H167C to produce cyclodextrins. *3 Biotech, 8*(165). https://doi.org/10.1007/s13205-018-1182-6.

Wang, Y., & Kim, D. (2019). Heat and shear effects on enzyme activity during cell disruption. *Biochemical Engineering Journal, 148*, 43–50. https://doi.org/10.1016/j.bej.2019.04.011.

Wheelwright, S. M. (1991). *Protein purification: Design and scale up of downstream.* Munich: Hanser Publishers.

Wingfield, P. T. (2001). Protein precipitation using ammonium sulfate. *Current Protocols in Prorein Science, 13*(1), 1–10. https://doi.org/10.1002/0471140864.psa03fs13.

Yang, J., et al. (2021). Biochemical properties of CGTase from thermophilic *Bacillus* sp. *International Journal of Biological Macromolecules, 183*, 1024–1032. https://doi.org/10.1016/j.ijbiomac.2021.04.123.

Yoon, S. H., & Robyt, J. F. (2005). Activation and stabilization of 10 starch-degrading enzymes by Triton X-100, polyethylene glycols, and polyvinyl alcohols. *Enzyme and Microbial Technology, 37*(5), 556–562. https://doi.org/10.1016/j.enzmictec.2005.04.002.

Zhao, L., Chen, J., & Wu, H. (2016). Centrifugation and crude extract preparation in enzyme purification. *Journal of Microbiological Methods, 129*, 42–48. https://doi.org/10.1016/j.mimet.2016.06.007.

Zhou, B., Jiang, J., Huo, Y., Lin, J., Shang, C., & Wu, H. (2025). Extraction, purification and antioxidant activity of *Chlorella* polysaccharide. *Algal Research, 89*, 1–10. https://doi.org/10.1016/j.algal.2025.104077.

CHAPTER

9

Protein engineering of cyclodextrin glucanotransferases

Khushbu Rabadiya[1], Dimple Pardhi[1], Kiransinh N. Rajput[1] and Rushikesh Joshi[2]

[1]*Department of Microbiology & Biotechnology, University School of Sciences, Gujarat University, Ahmedabad, Gujarat, India,*

[2]*Department of Biochemistry & Forensic Science, University School of Sciences, Gujarat University, Ahmedabad, Gujarat, India*

Introduction

Proteins, particularly enzymes, are fundamental to the vast array of biochemical processes that sustain life. Their diverse catalytic capabilities make them indispensable tools in industrial biotechnology, pharmaceuticals, agriculture, food processing, and environmental remediation. However, native or wild-type enzymes often have limitations when used in industrial settings. These limitations may include poor stability under harsh reaction conditions, suboptimal catalytic efficiency, restricted substrate specificity, or unfavorable product profiles. To overcome these challenges and harness enzymes for specific applications, the field of protein engineering has emerged as a powerful and transformative scientific discipline.

Protein engineering is defined as the process of developing proteins with improved or novel functionalities by modifying their amino acid sequences. This engineering can be carried out through rational design, where modifications are based on structural and functional insights of the protein, or through directed evolution, which involves mimicking natural selection in the laboratory by introducing random mutations and selecting for desired traits. Both approaches have proven highly effective and are often used in combination to accelerate and optimize the engineering process. The schematics of the development process are shown in Fig. 9.1.

One of the cornerstones of rational protein engineering is site-directed mutagenesis, a technique that allows for the targeted modification of specific amino acids in a protein sequence. This method requires prior knowledge of the enzyme's structure-function relationship, often derived from crystallography, molecular modeling, or bioinformatics analysis. On the other hand, directed evolution bypasses the need for detailed structural information by using random mutagenesis and high-throughput screening to identify beneficial variants, making it

Cyclodextrin Glucanotransferase. DOI: https://doi.org/10.1016/B978-0-443-41694-1.00003-4

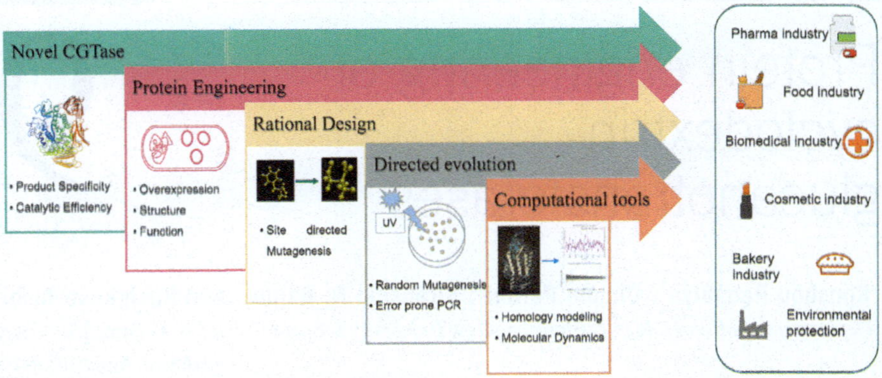

FIGURE 9.1

Design strategies of protein engineering and applications of cyclodextrin glycosyltransferase (*CGTase*).

especially useful when structural data is limited or unavailable (Kaul & Asano 2012)

In recent years, the integration of computational tools and molecular modeling has significantly enhanced the efficiency and precision of protein engineering. Techniques such as homology modeling, molecular docking, molecular dynamics (MD) simulations, and quantum mechanics/molecular mechanics (QM/MM) calculations have enabled researchers to predict the structural and functional impact of mutations with high accuracy (Xiong et al., 2021). These in silico approaches guide experimental designs, reduce time and resource expenditure, and increase the likelihood of success.

Structure of cyclodextrin glycosyltransferase

Cyclodextrin glycosyltransferase (CGTase) is a well-known starch-degrading enzyme classified within the glycoside hydrolase family 13. It facilitates both hydrolysis and transglycosylation reactions, including cyclization, coupling, and disproportionation. Based on the type of cyclodextrins (CDs) produced during the cyclization reaction, CGTases are categorized as α-, β-, or γ-CGTases, as well as combinations such as α/β- or β/γ-CGTases (Bosito et al., 2023). The enzyme's substrate-binding region comprises nine subsites: donor subsites from −7 to −1 and acceptor subsites at +1 and +2, with the active site positioned between the −1 and +1 subsites. Extensive molecular modification studies have been conducted targeting various regions of CGTase. It has been shown that subsites +1, +2, −3, −6, and −7 play important roles in the enzyme's cyclization function. With increasing understanding of CGTase structure and catalytic mechanisms, its potential for industrial use has gained considerable attention (Rabadiya et al., 2024). It has been proven that subsite −6 has an important role in transglycosylation reactions (Leemhuis et al., 2002). Nevertheless, to fully exploit this potential, it is critical to enhance CGTase production and functional properties to meet industrial demands for higher efficiency and product quality.

A particularly relevant application of protein engineering is seen in the optimization of CGTase, an industrially important enzyme involved in the conversion of starch into CDs. CDs are cyclic oligosaccharides with unique molecular structures that enable them to encapsulate hydrophobic molecules, making them valuable in drug delivery, food stabilization, and environmental detoxification (Esteso & Romero 2024). However, native CGTases often exhibit limitations in product specificity, stability, and efficiency. Through targeted engineering employing site-directed mutagenesis, directed evolution, and computational modeling, CGTases can be significantly improved to meet industrial demands (Han et al., 2014).

This chapter delves into the foundational concepts and advanced methodologies in protein engineering of CGTase. It explores the theoretical underpinnings of enzyme design, experimental strategies for mutagenesis, computational modeling techniques, and their synergistic application in improving CGTase. By examining these interconnected approaches, the chapter aims to provide a comprehensive understanding of how modern protein engineering can unlock the full potential of enzymes for industrial and biotechnological innovation.

Protein engineering approaches for cyclodextrin glycosyltransferase

Protein engineering is a dynamic field that integrates principles from biology, chemistry, and engineering to modify and optimize enzymes and other proteins for specific applications. This discipline plays a pivotal role in various industries, including pharmaceuticals, agriculture, and biofuels, by enhancing the efficiency and environmental sustainability of industrial processes through tailored enzymes (Sharma et al., 2021). The key objectives of enzyme engineering are enhancing catalytic activity and efficiency, altering substrate specificity, modifying product specificity, improving stability, increasing product yield, and eliminating allosteric regulation or product inhibition.

Directed evolution: mimicking natural selection

Directed evolution is a robust and versatile method for engineering enzymes by mimicking the principles of natural selection in a controlled laboratory setting. Unlike rational design, which depends on detailed structural and mechanistic knowledge, directed evolution operates without requiring prior information about the enzyme's structure or function. This makes it particularly effective for optimizing complex traits such as thermostability, activity under extreme pH conditions, and altered substrate or product specificity (Leemhuis et al., 2009).

The process typically involves three key steps: (1) generation of a diverse mutant library; (2) expression of mutant enzymes in a suitable host; and (3) screening or selection for desired enzymatic traits. Mutant libraries can be created using techniques such as error-prone polymerase chain reaction (epPCR), which introduces random point mutations throughout the gene, or DNA shuffling, which recombines segments from different parental genes to generate novel sequence combinations. Advanced techniques like saturation mutagenesis at hotspot residues further refine this approach by focusing mutational efforts on functionally important regions. Saturation mutagenesis targeting the acceptor subsite residue of Bacillus stearothermophilus NO_2 CGTase yielded a mutant that set a new benchmark for AA-2G production (Tao et al., 2018).

In CGTase engineering, directed evolution has been successfully employed to enhance catalytic efficiency and thermal stability. For example, mutant libraries generated via epPCR and subsequently screened for improved hydrolysis-to-transglycosylation ratios have yielded variants with superior CD production profiles. Similarly, evolution campaigns targeting the linker or domain interfaces have produced CGTases with enhanced resilience to heat and detergents, making them more suitable for industrial applications. For example, the enzyme activity of Geobacillus stearothermophilus CGTase was increased sevenfold by the directed evolution of the downstream box sequence using degenerate codons and a specific promoter sequence of PtacI in *E*scherichia *coli* BL21 (Deng et al., 2018). To minimize the undesired hydrolysis reaction while successfully maintaining the enzyme's cyclization activity, researchers employed a strategy of directed evolution. This was achieved by introducing random mutations throughout the cgt gene using epPCR, ultimately identifying mutations in two outer active site residues, Ser-77 and Trp-239, which reduced the enzyme's hydrolytic function up to 15-fold without compromising its ability to cyclize substrates (Kelly et al., 2008).

Random mutagenesis is a method to introduce arbitrary mutations to the target gene. Though there are many different techniques to achieve it, including chemical mutagenesis, UV irradiation, sequential mutagenesis, epPCR, and saturation mutagenesis, epPCR is widely used due to its simplicity and low cost. For example, random mutagenesis in the ABC transporter system of *E. coli* increased the secretion of CGTase, which is mediated by hemolysin (Low et al., 2010). The resulting mutations demonstrated progressive increases in CGTase secretion. Variants of CGTase with a higher product specificity for CD8 and a wider pH activity range were produced using a combination of epPCR and DNA shuffling. Compared to the wild-type CGTase, the mutant S77 showed the highest product specificity for γ-CD but a substantial decrease in γ-CGTase activity of only 4%. Paenibacillus pabuli US132 mutants were created using epPCR to create a powerful antistaling agent (Jemli, Jaoua, & Bejar 2016). A highly significant reduction in cyclization activity and specificity for starch dextrinization was found in one of the mutants. Another study used the directed evolution method to create three different mutants (I631T, I641T, and K647E) of B. stearothermophilus NO_2. The activity of K647E was raised to 1904 U/ml to produce 2-O-D-glucopyranosyl-L-ascorbic acid (Tao et al., 2020). To improve α-CD production, using epPCR mutant library was constructed. Among the positive mutants, Y167H showed higher potential for α-CD production (Song et al., 2014).

Furthermore, combination of directed evolution with computational tools, such as machine-learning-guided sequence analysis or in silico saturation mutagenesis, has enhanced the efficiency of library design and screening. This synergy accelerates the identification of beneficial mutations and reduces the number of experimental cycles required to obtain optimized enzymes.

DNA shuffling (molecular breeding) is a technique that involves the random recombination of DNA fragments from parent genes to generate mutant genes with unique attributes or combinations of desirable characteristics, such as thermostability and high activity. DNA shuffling can rapidly increase the size of a DNA library. It relies on homologous recombination or the similarity of DNA sequences, where fragmented DNA segments (e.g., 10–50 bp to over 1 kbp) are reassembled through polymerase chain reaction (PCR) without primers, followed by further amplification with primers. This method allows for the creation of hybrid or chimeric proteins with novel properties. For CGTase, domain shuffling has been successfully used to construct chimeric variants with tailored product specificity and enhanced thermal stability. For example, by shuffling domains between a thermostable G. stearothermophilus CGTase (GeoT) and a γ-CD-forming Bacillus sp. G825-6 CGTase variants were created that suppressed α-CD production and increased the yield of larger CDs (CD9-CD12) (Deng et al., 2018).

In summary, directed evolution is a powerful, empirical strategy that enables the continuous improvement of CGTase by exploring vast sequence space and selecting for enhanced performance. Its adaptability and success in generating industrially relevant enzyme variants make it an indispensable tool in modern protein engineering.

Rational design: structure-guided approaches

Rational design is a knowledge-driven approach in protein engineering that involves the intentional modification of specific amino acid residues in an enzyme based on structural, functional, and evolutionary data. Unlike directed evolution, which is largely empirical, rational design relies on a detailed understanding of enzyme architecture and catalytic mechanisms to guide targeted mutations that enhance enzyme performance (Teze et al., 2020). In the context of CGTase, rational design has been instrumental in engineering variants with improved product specificity, enhanced thermostability, and increased catalytic efficiency.

Site-directed mutagenesis: precision in protein engineering

Site-directed mutagenesis is a cornerstone technique in protein engineering that allows the targeted modification of specific amino acid residues within a protein's sequence. This precision approach is particularly valuable in the context of CGTase engineering, as it enables the manipulation of residues implicated in catalytic function, substrate binding, thermostability, and pH sensitivity (Wind et al., 1998). The method involves designing synthetic oligonucleotides that carry the desired mutation and using them to amplify the gene of interest via PCR. The modified gene is then cloned and expressed in a suitable host organism, commonly *E. coli*, to produce the mutant enzyme for functional evaluation.

In the case of CGTase, the active site typically includes residues such as Asp229, Glu257, and Asp328, which are crucial for glycosidic bond cleavage and transglycosylation activity. Substitutions at these positions have been shown to alter the ratio of α-, β-, and γ-CD products. For instance, replacing Glu257 with Gln has been reported to shift the product specificity towards γ-CD, which is of higher value in pharmaceutical formulations (Leemhuis et al., 2010). Additionally, site-directed mutagenesis has enabled the identification of distal residues—such as Ser182, Asn227, and Tyr195—whose structural roles in loop stabilization, domain interfaces, or surface exposure contribute to enzyme thermostability, folding efficiency, and solvent compatibility. These findings underscore the nonlocalized effects that remote residues can exert through allosteric modulation or by stabilizing conformational intermediates (Qi & Zimmermann 2005). (See Table 9.1) Fig. 9.1.

Site-directed mutagenesis thus provides a rational and efficient approach for CGTase engineering. Focusing on specific amino acid residues informed by structural and functional data, it enables precise modulation of enzyme properties to meet industrial requirements. When combined with computational modeling and experimental validation, this approach forms a powerful tool for tailoring CGTase for enhanced biocatalytic performance.

Functional improvements via site-directed mutagenesis

- Product specificity: In addition to producing CDs, CGTase also facilitates the enzymatic synthesis of 2-O-α-D-glucopyranosyl-L-ascorbic acid (AA-2G) via transglycosylation. Site-saturation mutagenesis was applied to CGTase from Paenibacillus macerans at three specific residues—Tyr195, Tyr260, and Gln265. Among the resulting variants, mutants Y195S, Y260R, and Q265K exhibited significantly higher AA-2G yields compared to the wild-type enzyme (Han et al., 2013). A single mutation at subsite -3, specifically the replacement of histidine with threonine (H43T mutation) in Bacillus sp. G1 CGTase significantly increased γ-CD production from 10% to approximately 39% when using tapioca starch as a substrate. This substantial increase is thought to result from reduced steric hindrance, which facilitates the formation of the larger γ-CD ring

Table 9.1 Key residues and their functional impact in cyclodextrin glycosyltransferase (CGTase) engineering (Kelly, Dijkhuizen, & Leemhuis 2009; Ara et al., 2021; Goh et al., 2012).

Position	Wild-type residue	Mutation(s)	Location/Subsite	Observed effect/function
183	Phe, Trp	Asn, Ser, Leu	Acceptor subsite +2	Cyclization, reduction of hydrolysis
232	Lys, Ala	Leu, Glu	Acceptor subsite +2	General activity
259	Phe, Tyr	Leu, Asn, Ile, Glu, His, Arg	Acceptor subsite +2	Cyclization, reduction of hydrolysis
194	Leu	Thr	Acceptor subsite +1	Cyclization
195	Tyr, Phe	All possibilities	Acceptor subsite +1	Cyclization
230	Ala	Val	Acceptor subsite +1	Transglycosylation specificity
233	His	Asp, Asn, Arg	Acceptor subsite +1	General activity significantly decreased α-cyclodextrin (CD) production
100	Tyr	Phe, Ser	Catalytic subsite -1	General activity
140	His	Asp, Asn, Arg	Catalytic subsite -1	General activity
227	Arg	Ala, Lys	Catalytic subsite -1	General activity
229	Asp	Ala, Asn	Catalytic nucleophile	Catalytic nucleophile, 4000–700,000-fold reduction in CD-forming activity
257	Glu	Ala, Gln	Acid–base catalyst	Acid–base catalyst, 4000–700,000-fold reduction in CD-forming activity
327	His	Asp, Asn, Arg	General activity	General activity
328	Asp	Asn, Ala	General activity, transition state stabilizer	General activity, transition state stabilizer, 4000–700,000-fold reduction in CD-forming activity
98	His	Asp	Donor subsite -2	General activity
47	Arg, Lys, His, Thr	Trp, Leu, Gln	Donor subsite -3	Cyclization
89	Tyr, Asp, Ser, Gln, Ala, Gly	Phe, Ser, Asp, Gly	Donor subsite -3	General activity, product specificity
94	Asn, Ser, Tyr, Glu	Ser, Gln	Donor subsite -3	Slightly decreased activities
196	Asp, Asn, Gly	His	Donor subsite -3	Reduced activity, more α-CD
371	Asp	Gly, Asn, Arg	Donor subsite -3	General activity, less α-CD

(Continued)

Table 9.1 Key residues and their functional impact in cyclodextrin glycosyltransferase (CGTase) engineering (Kelly, Dijkhuizen, & Leemhuis 2009; Ara et al., 2021; Goh et al., 2012). (*Continued*)

Position	Wild-type residue	Mutation(s)	Location/Subsite	Observed effect/function
167	Tyr	Phe	Donor subsite -6	Selection of long substrates
179	Gly	Leu	Donor subsite -6	Selection of long substrates
180	Gly, Asn	Leu	Donor subsite -6	Selection of long substrates
193	Asn	Leu, Gly	Donor subsite -6	Selection of long substrates
145	Ser, Met, Leu, Asp, Glu	Ala, Glu	Donor subsite -7	CD specificity
146	Ser, Glu, Val, Phe, Leu	Pro	Donor subsite -7	CD specificity
147	Thr, Asp, Asn	Trp	Donor subsite -7	CD specificity
21	Phe, Val	Leu	Other important residues	Limitation of hydrolysis
76	Ser, Thr	Pro	Other important residues	Hydrolysis
135	Asp	Asn, Ala	Other important residues	Influences the conformation of Glu257 and Arg227
326	Asp, Ser	Asp, Gln, Leu	Other important residues	Cyclization and general activity
43	His	Thr	Subsite -3	Increased γ-CD production (from 10% to ~39%)
182	Ser	Gly (S182G)	Adjacent to Ca-I site	4.3-fold increased thermostability (half-life 22 to 94 min at 60°C), 25% decreased β-cyclization activity, improved kcat/Km
182	Ser	Glu (S182E)	Adjacent to Ca-I site	Slightly increased thermostability (half-life 22 to 31 min at 60°C), 18%–30% decreased β/γ-cyclization activity
132	Asn	Arg (N132R)	Calcium binding site I	Completely disrupted catalytic activity
28	Asn	Arg (N28R)	Calcium binding site II	Increased β-CD (1.7-fold) and γ-CD (2.0-fold) cyclization activity, no effect on thermostability
197	Phe	Tyr (F197Y)	Acceptor subsite	Enhanced coupling activity (CspCGT13)
263	Gly	Ala (G263A)	Acceptor subsite	Enhanced coupling activity (CspCGT13)
266	Glu	Ala (E266A)	Acceptor subsite	Maintained enhanced coupling (CspCGT13)

(Kelly, Dijkhuizen, & Leemhuis 2009). Furthermore, the double and triple mutants demonstrated enhanced substrate specificity. Tyr195 has also been linked to variations in product specificity in Paenibacillus sp. 602–1. Subsites −7 and −6, located at the distal end of the donor-binding site, are known to form hydrogen bonds with glucose residues, potentially influencing the enzyme's preference for producing specific CD sizes. In another study, CGTase from P. macerans strain JFB05-01 was engineered using overlap extension PCR to create several mutants, including D147A, R146P, D147P, R146A/D147P, R146P/D147A, and R146P/D147P, aimed at evaluating the role of the −7 subsite in determining product specificity. Among these, the D147P mutant showed the highest β-CD yield, achieving 119% of the wild-type level (Wang, Duan, & Wu 2016). Overall, specific subsites within the CGTase substrate-binding region are closely associated with product specificity, and sequence analysis has revealed that conserved amino acids in donor-binding subsites near the catalytic site significantly influence the type of CDs produced. (see table 9.2) Fig 9.2.

- Substrate specificity: This study centers on enhancing the substrate selectivity of CGTase through molecular modification using protein engineering techniques. Mutant variants Y260R and Q265K of Paenibacillus macerans JFB05-01 CGTase exhibited significantly greater disproportionation activity compared to the wild-type enzyme. CD production from a novel substrate source, jackfruit seed bran (Artocarpus integrifolia L.), was achieved using CGTase derived from Bacillus licheniformis (Silva et al., 2019). In a separate effort, iterative saturation mutagenesis was employed to engineer the acceptor subsite of CGTase from B. stearothermophilus NO_2, aiming to improve specificity for AA-2G synthesis. Among the variants, the double mutant K228R/M230L showed enhanced substrate specificity and a substantial increase in AA-2G yield (Tao et al., 2018). Additionally, site-saturation mutagenesis was applied at the donor-binding region of this CGTase to further increase its specificity toward maltodextrin substrates.
- Thermostability: Thermostability is a crucial attribute for industrial enzymes like CGTase, as many starch processing applications require elevated operating temperatures. Enhancing an enzyme's ability to retain activity at high temperatures can significantly improve process efficiency and yield. A significant example is the S182G mutation in a Bacillus sp. G1 CGTase variant. This mutation, involving the substitution of serine at position 182 with glycine, is located adjacent to the calcium-binding site I and the active site cleft. It dramatically enhanced the enzyme's thermostability, increasing its activity half-life at 60°C from 22 minutes (for the parent CGTase) to 94 minutes, representing a 4.3-fold improvement. This observed improvement is attributed to structural changes, specifically the formation of a shorter α-helix and the alleviation of unfavorable steric strains due to the smaller glycine residue (Goh et al., 2012). Thermostability has also been improved through the replacement of structural domains; specifically, swapping the B domain of *Bacillus* sp. G1

Table 9.2 Improvement in the product specificity of cyclodextrin glycosyltransferase (CGTase) using site-directed mutagenesis.

Organism	Enzyme type	Site of modification	Functional outcome	References
B. circulans strain STB01	Β-CGTase	A315R, A315D, A315T, A315V, and A315I	A315R and A315D enhanced β-cyclodextrin (CD) specificity	Ban et al. (2015)
P. macerans strain JFB05-01	α-CGTase	R146A/D147P, R146P/D147A	Improved α-CD production	Wang, Duan, and Wu (2016)
B. circulans strain STB01	β-CGTase	D577G, D577A, D577V, D577L, and D577I	D577G and D577A elevated β-CD production	Li et al. (2016)
P. macerans	CGTase	F228M and F228V	Increase disproportionation activity by 120% and production of AA-2G by 84%	Chen et al. (2017)
Bacillus sp. N-227	Β-CGTase	Y127F, R254F, and D355R	Enhanced cyclization activity by 1.6–2.0-fold	Wang et al. (2017)
Bacillus sp. G-825-6	CGTase	Y183W, Y183R, D358R, Y183W/D358R, and Y183R/D358R	Y183R favored large-ring CD production with high specificity	Sonnendecker, Melzer, and Zimmermann (2019)
Paenibacillus barengoltzii	CGTase	A137V, A144V, L280A and M329I	A137V increased maltooligosaccharide production	Castillo et al. (2018)
Bacillus clarkii	γ-CGTase	T47H, F91N, F91L, F91W, Y186W	Y186W led to 94% higher γ-CD production than wild-type	Wang et al. (2020)
Paenibacillus macerans	α-CGTase	A156V/L174P, A156V/A166Y, A166Y/ L174P, and A156V/L174P/A166Y	Improved specificity toward large cyclic glucosyl saccharides	Han et al. (2020)

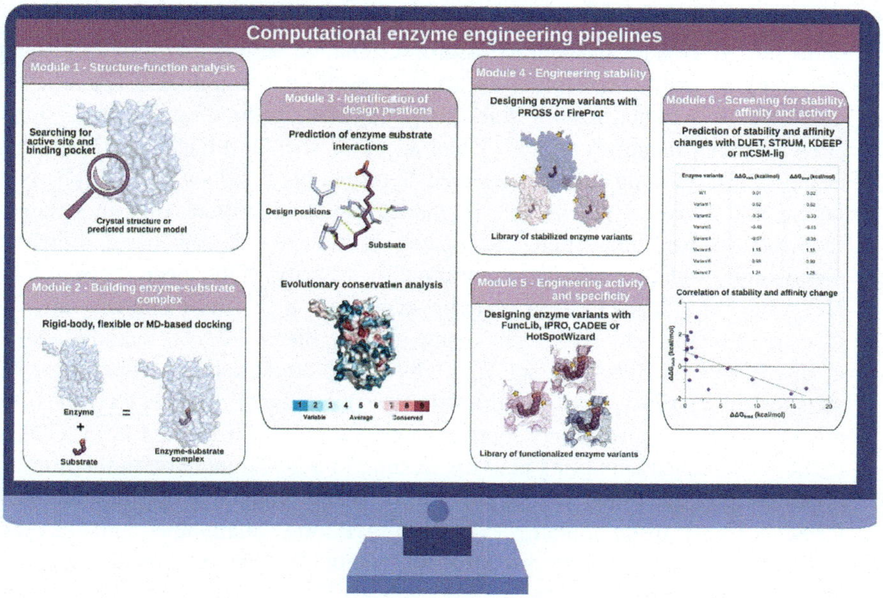

FIGURE 9.2

Computational enzyme engineering pipelines: a modular framework.

CGTase (H43T/Y87F) with that of a thermostable CGTase from B. stearothermophilus ET1 led to a 2.5-fold increase in half-life at 60°C compared to the wild-type (Goh et al., 2012). In another study, supplementation with Ca^{2+} ions enhanced the resistance of γ-CGTase from Bacillus sp. FJAT-44876 to heat-induced denaturation, further highlighting the importance of calcium in maintaining structural integrity under elevated temperatures (Li et al., 2023). The recombinant mutant N33K/S211G from Bacillus species G1 CGTase showed a 2.1-fold enhancement in transglycosylation activity as well as thermal stability and a prolonged half-life at optimum temperature over the wild type (Miao et al., 2024).

- Catalytic activity: Mutations can have a profound influence on the overall catalytic efficiency (kcat/Km) of CGTase. For example, despite a decrease in specific activity, the S182G mutant exhibited an improved catalytic efficiency (2 mL/mg·s) compared to the parent CGTase (1.82 mL/mg·s). The N28R mutant, as previously noted, significantly increased both β-CD and γ-CD cyclization activity. It is important to recognize that the core catalytic residues—Asp229 (nucleophile), Glu257 (acid–base catalyst), and Asp328 (substrate binding/transition state stabilization)—are highly sensitive to modification. Altering these residues can drastically reduce the enzyme's CD-forming activity, sometimes by factors of 4000–700,000-fold, underscoring their critical role in catalysis.

- Transglycosylation activity: CGTase facilitates the conversion of small molecules into glycosides through its disproportionation activity, thereby improving the stability and water solubility of the resulting compounds. The enzyme comprises five domains with nine identified substrate-binding subsites—donor sites spanning positions −7 to −1 and acceptor sites at +1 and +2 (Uitdehaag et al., 2000). Among these, conserved hydrophobic residues such as phenylalanine and tyrosine across the 12 subsites play a critical role in catalytic function. In particular, Leu277, located at subsite +2, has been shown to significantly enhance the disproportionation activity of B. stearothermophilus NO_2 CGTase. The L277M variant displayed a 2.4-fold higher ratio of disproportionation to hydrolysis compared to the wild-type enzyme (Kong et al., 2021). Studies on CspCGT13 have explored mutations near acceptor subsites to achieve more favorable interactions with alkyl glycosides for enhanced coupling reactions. For instance, the double mutant F197Y/G263A exhibited enhanced coupling activity, leading to a 2-fold increase in the primary coupling product when using γ-CD as a donor compared to the wild-type CspCGT13. A triple mutant, 197Y/G263A/E266A, maintained this increase (1.8-fold) in primary coupling product, although disproportionation activity remained similar to that of the double mutants. These mutations demonstrate the ability to precisely tailor CGTase for specific transglycosylation products beyond the classical CDs, expanding its biotechnological scope. Furthermore, a comprehensive list of positions and mutations affecting various aspects of CGTase activity, including cyclization, hydrolysis, and product specificity, has been compiled (e.g., Phe/Trp 183 and Phe/Tyr 259 in acceptor subsite +2 influence cyclization reduction; Asp/Asn/Gly 196 can lead to more α-CD; and Ser/Met/Leu/Asp/Glu 145, Ser/Glu/Val/Phe/Leu 146, and Thr/Asp/Asn 147 impact CD specificity) (Ara et al., 2021). Enhancing the transglycosylation capability of CGTase leads to the formation of more selective and functionally valuable glycosylated products.

Integration of rational design with site-directed mutagenesis

In CGTase, rational design has been used to modify residues around the catalytic triad Asp229, Glu257, and Asp328 to modulate the production ratio of α-, β-, and γ-CDs. For instance, altering residues lining the substrate tunnel or the acceptor subsites can shift the enzyme's regioselectivity, favoring the formation of a desired CD product. Mutations in flexible loop regions and surface-exposed domains have also been shown to enhance enzyme thermostability by promoting favorable intramolecular interactions and reducing entropy.

Computational tools play a central role in rational design. Software such as Rosetta, FoldX, HotSpot Wizard, and SwissSidechain can predict the thermodynamic impact of mutations, identify stabilizing residue substitutions, and visualize structural changes. These tools can simulate the local effects of amino acid replacements on hydrogen bonding, electrostatic interactions, and packing density, allowing researchers to prioritize the most promising variants for experimental validation.

Enzyme improvement through rational design is further supported by iterative rounds of modeling, mutagenesis, and functional screening. This cyclic process allows for the progressive refinement of enzyme properties. For example, studies have demonstrated that multipoint mutations designed via computational modeling and confirmed through site-directed mutagenesis can result in CGTase variants with significantly enhanced pH tolerance, improved resistance to denaturation, and a broadened substrate spectrum.

Moreover, rational design has facilitated the creation of CGTase variants suitable for specific industrial applications, such as food-grade enzymes with reduced allergenicity, enzymes with improved compatibility in detergent formulations, and variants capable of functioning in organic solvents. These advancements underscore the transformative potential of rational design in tailoring CGTase to meet diverse biotechnological needs.

In conclusion, rational design provides a powerful, targeted approach to enzyme improvement. When integrated with experimental validation and computational prediction, it enables the systematic optimization of CGTase for enhanced performance, stability, and application-specific functionality.

Computational tools for enzyme engineering

Computational enzyme engineering is a structure-guided approach that integrates molecular modeling, bioinformatics, and machine learning (ML) to rationally design and optimize enzymes for improved activity, stability, and specificity. The process begins with a detailed structure–function analysis to identify catalytic residues and the substrate-binding pocket, often informed by crystallographic data or homology models. Molecular docking techniques are then employed to simulate enzyme–substrate interactions, allowing researchers to visualize binding orientations and identify key residues involved in substrate recognition and catalysis (See Table 9.3). Based on these insights, designable positions are selected for mutagenesis, particularly residues within or near the active site or substrate tunnel. To enhance enzyme stability, algorithms such as PROSS and FireProt predict stabilizing mutations by integrating evolutionary conservation with energy-based models. For improving catalytic efficiency and substrate specificity, tools like FuncLib, IPRO, CADEE, and HotSpot Wizard generate and evaluate mutation libraries based on structural constraints and functional tolerance. The effects of designed mutations are subsequently assessed using predictive platforms such as DUET, STRUM, KDEEP, and mCSM-lig, which estimate changes in protein stability, ligand-binding affinity, and functional performance (Scherer et al., 2021).

This integrated computational pipeline streamlines enzyme engineering, reduces the experimental burden, and accelerates the development of highly efficient and tailored biocatalysts (see Fig. 9.2). Table 9.3.

Homology modelling and structural prediction

When experimental structures are unavailable, homology modeling is a crucial first step to generate a three-dimensional model of CGTase based on known structures of related proteins. For instance, the 3D structure of CGTase G1 was successfully determined through homology modeling using the CGTase from Bacillus stearothermophiles (1CYG) as a template, given its high sequence identity and similarity. This process typically involves automated sequence alignment, initial model building using software like Modeler, and subsequent energy minimization to refine the structure. Model quality is then rigorously assessed using programs such as ERRAT and Verify 3D, which evaluate geometric accuracy and sequence-structure compatibility, respectively (Goh et al., 2009). Such models provide a foundational understanding of the enzyme's architecture, including the active site and substrate-binding subsites, which are critical for guiding rational design efforts. Similarly, research has examined the influence of different cross-linking agents on CGTase G1 CLEA development for CD synthesis, integrating molecular modeling with experimental optimization (Jailani et al., 2022). Some studies also show the relationship between ionic interaction and thermal stability of the protein (Goh et al., 2008).

Molecular docking

Molecular docking is a computational technique used to predict the position and orientation—known as the binding mode—of a ligand (a small molecule) when it binds to a protein receptor or enzyme. This technique is crucial for virtual screening in drug discovery and for gaining a detailed understanding of protein-ligand interactions at an atomic level. Tools like AutoDock, GOLD, and Glide simulate the binding of CD precursors or acceptors within CGTase's catalytic pocket, providing key information about substrate orientation, binding affinity, and reaction feasibility. Docking results inform rational design strategies by pinpointing residues that influence substrate specificity and turnover rate. Molecular docking was employed to study the binding conformation of fisetin, a flavonoid with low water solubility, in the active site of CGTase from Paenibacillus sp. RB01 (Lorthongpanich et al., 2022). Mutational studies and structural analysis of CGTase CspCGT13 focused on modifying its acceptor subsites to achieve more favorable interactions with alkyl glycosides, aiming to improve the synthesis of these nonionic surface-

Table 9.3 The most commonly used software in protein design engineering.

Software	Primary function	Website
Discovery Studio	Comprehensive suite for structure-based design, docking, simulation, and analysis	http://softwaretopic.in/former.com/discovery-studio-4-0-server/
AutoDock Vina	Fast, free docking program using empirical scoring functions for ligand–protein binding	https://www.cgl.ucsf.edu/chimera/data/downloads/1.7/docs/ContributedSoftware/vina/vina.html (kazlab.umn.edu)
AutoDock	Classic, versatile docking package used for small-molecule and macromolecule binding	http://autodock.scripps.edu/
Chimera/ChimeraX	Structural analysis, molecular docking preparation	cgl.ucsf.edu/chimera/
MOE-Dock (in MOE)	Commercial docking tool within MOE suite for molecular modeling and virtual screening	https://www.chemcomp.com/Products.htm
Rosetta	De novo design, folding, docking, and multisite mutation optimization	https://www.rosettacommons.org/
STRUM	Rapid scanning of stabilizing/destabilizing mutations	zhanglab.ccmb.med.umich.edu/STRUM/
NAMD	Highly scalable molecular dynamics (MD) engine for large biomolecular systems	http://www.ks.uiuc.edu/Research/namd/
FuncLib	Enzyme activity and specificity enhancement	https://funclib.weizmann.ac.il/step/fl_terms/
Amber	Widely used MD suite with force fields and energy calculations	http://ambermd.org/
GROMACS	High-performance MD software with parallel scaling and community support	http://www.gromacs.org/
Sybyl	Commercial MD, docking, and cheminformatics platform	https://en.freedownloadmanager.org/Windows-PC/SYBYL-X.html
YASARA	Commercial MD analysis and modeling package	http://www.yasara.org/
Consensus Finder	Web tool for identifying consensus residues from MSAs to guide stabilizing mutations	https://kazlab.umn.edu/
FireProt	Predicts stabilizing mutations using energy and evolutionary information	loschmidt.chemi.muni.cz/fireprot/
HotSpot Wizard	Identification of mutation-prone sites, Protein function modification	hotspotwizard.org
I-Mutant 2.0	SVM-based server predicting protein stability changes upon mutations ($\Delta\Delta G$)	https://folding.biofold.org/i-mutant/i-mutant2.0.html

(Continued)

Table 9.3 The most commonly used software in protein design engineering. (*Continued*)

Software	Primary function	Website
FoldX	Rapid estimation of mutation-induced stability changes	www.foldxsuite.crg.eu
POVME	Calculates volumes of binding pockets for drug design and enzyme engineering	https://github.com/POVME/POVME
MSDsite	Predicts potential substrate–protein binding sites on structures	https://www.ebi.ac.uk/pdbesite/pdbemotif/
POCASA	Detects and characterizes active pockets in protein structures	http://g6altair.sci.hokudai.ac.jp/g6/Research/POCASA_e.html
CaverDock	Simulates ligand passage through enzyme tunnels or channels	https://loschmidt.chemi.muni.cz/caverdock/

active agents. This computational insight can then directly guide targeted mutagenesis experiments, as demonstrated by the engineering of CGTase CspCGT13 for improved alkyl glycoside synthesis (Ara et al., 2021). Molecular docking thus transforms enzyme engineering from a purely empirical process to a more rational, hypothesis-driven one, enabling the design of enzymes with tailored substrate specificities and product profiles, which is crucial for developing novel biocatalytic processes.

Molecular dynamics simulations

MD simulations are employed to study the dynamic behavior of CGTase, providing insights into conformational changes, stability, and interactions with substrates or inhibitors over time. These simulations are particularly valuable for investigating the relationship between enzyme thermostability and structural features like salt bridges. For example, in silico mutations and MD simulations have been used to explore how introducing new salt bridges into thermally unstable regions of Bacillus clarkii 7364 γ-CGTasecould improve its thermostability. Studies have monitored the dynamic motions of these salt bridges, revealing that those in thermally unstable regions tend to tighten at elevated temperatures, thereby helping to maintain the stable conformation of the surrounding spatial neighborhood (Hu et al., 2024). This computational strategy has demonstrated its viability in improving the thermal stability of mesophilic CGTases by strategically introducing stablilizing salt-bridge interactions. MD studies also aid in evaluating the stability of engineered variants by comparing root mean square deviation, root mean square fluctuation, and hydrogen-bonding patterns relative to wild-type enzymes. Enhanced loop rigidity, increased intramolecular hydrogen bonds, or reduced solvent-accessible surface area are often indicative of improved stability and functional robustness.

Application to cyclodextrin glycosyltransferase: exploring thermal unfolding and conformational changes

MD simulations have been extensively used to better understand the thermostability mechanisms of proteins in general, and specifically for enzymes like CGTase.

A notable study involved MD simulations of the thermal unfolding reaction of CGTase from Bacillus macerans. These simulations were carried out for 1.5 nanoseconds each at five different temperatures, providing a dynamic view of how the enzyme responds to increasing thermal stress. The simulations revealed that the distortion of the enzyme was initiated simultaneously in the N-terminal and domain D. The thermal unfolding of the outer domains of CGTase was observed to be faster than that of the catalytic core domain. Importantly, the catalytic center, residing within the $(\alpha/\beta)_8$ TIM-barrel, remained well protected at simulation temperatures up to 600K, indicating its inherent stability. The unfolding of the eight β-sheets followed a random ordered mechanism, with specific β-sheets (8, 1, and 6) unfolding more rapidly than others. The study also meticulously analyzed the influence of nonbonded interactions, such as hydrogen bonds and salt bridges, on protein stability. The results clearly indicated that the protein's stability is not evenly distributed over the whole structure, suggesting localized regions of vulnerability or robustness (Fu et al., 2009). This detailed dynamic analysis provides invaluable insight into the structure-stability relationship of CGTase. It contributes significantly to the understanding of protein structural properties, the role of nonbonded interactions in stabilizing secondary peptide structures and promoting folding, and ultimately, how these factors influence enzyme action.

Force fields and software in molecular dynamics simulations

Software: A variety of sophisticated software packages are available for performing MD simulations. Tools such as GROMACS and NAMD are widely recognized for their capabilities. Contemporary software can even simulate entire organelles, viruses, and cellular environments, modeling interactions between thousands of molecules simultaneously. OpenMM is another notable software that integrates ML potentials, enabling PyTorch-based models to compute forces and energy in simulations, which can significantly improve accuracy with minimal cost increase.

Force fields: The choice of force field is paramount for the accuracy of MD simulations. Widely used force fields for biomolecular simulations include CHARMM, AMBER, and OPLS. There is a continuous and dynamic effort to improve these force fields, often leveraging data-driven approaches and ML algorithms to refine parameters and capture complex interactions. Specialized force fields have also been developed for specific types of systems, such as membrane environments or carbohydrates, allowing for more accurate and realistic simulations in true biological contexts.

Quantum mechanics/molecular mechanics approaches

QM/MM simulations combine quantum mechanical calculations for the reactive site with molecular mechanical modeling of the surrounding protein environment. This hybrid method is particularly useful for studying the catalytic mechanism of CGTase, including glycosidic bond cleavage and transition-state stabilization. Software tools such as ORCA and Gaussian, when integrated with CHARMM or AMBER, allow for high-precision investigation of enzyme catalysis at the electronic level.

In silico mutagenesis and energy calculations

Computational tools enable in silico mutagenesis, allowing researchers to predict the effects of specific amino acid substitutions on CGTase properties before experimental validation. This approach can quickly identify promising candidate sequences from large theoretical pools based on physical principles. For instance, the S182G mutation in Bacillus sp. G1 CGTase, which significantly enhanced thermostability, was supported by in silico modeling suggesting that the change led to a shorter α-helix and alleviated unfavorable steric strains. Similarly, computational simulations have been used to investigate the relationship between thermostability and salt bridges in Bacillus macerans CGTase.

ML and artificial intelligence (AI)-based tools: Recent advances in AI and ML have introduced powerful data-driven approaches to protein engineering. ML algorithms trained on large datasets of protein sequences and structures can predict beneficial mutations and functional outcomes with increasing accuracy. Tools such as AlphaFold2, ProteinMPNN, and DeepMutScan are increasingly being adopted to complement traditional structure-based methods.

In summary, computational tools and MD simulations serve as indispensable components of modern CGTase engineering. They offer predictive power, mechanistic understanding, and design precision that significantly accelerate the development of optimized enzyme variants tailored for industrial applications.

Emerging trends and future directions in cyclodextrin glycosyltransferase protein engineering

The field of protein engineering is experiencing rapid transformation, driven by advancements in computational power, AI, and synthetic biology. These emerging trends are poised to revolutionize the engineering of enzymes like CGTase, expanding their capabilities and applications far beyond current limits.

Advanced computational and artificial intelligence/machine learning approaches

AI and ML are increasingly being integrated into protein engineering workflows, enabling the design of novel proteins with desired properties. These technologies facilitate the analysis of vast datasets and the prediction of protein structure and function with unprecedented accuracy. AI is moving protein engineering from an empirical, trial-and-error process to a more predictive and generative science. This allows for the design of "unnatural" enzymes—proteins with functionalities or stabilities that have no direct equivalents in nature. This fundamental shift means that, instead of relying on slow iterative experimental cycles, researchers can computationally design CGTase variants with highly specific properties, such as enhanced stability at extreme temperatures or tailored product specificity, thereby significantly accelerating the development of novel biocatalysts for complex industrial challenges. This computational power allows for the rapid screening and rejection of millions of possible combinations, a process that would otherwise take decades of trial and error in a laboratory setting.

Integration with synthetic biology

Synthetic biology, a rapidly evolving field, involves the design and construction of new biological systems, such as genetic circuits, to achieve specific functions. In the context of enzyme engineering, synthetic biology enables the creation of novel enzymes with desired properties and is crucial for the design and optimization of metabolic pathways. This discipline allows for the selection and assembly of enzymes to produce a desired product, adhering to principles such as modularity, standardization, and optimization of enzyme activity, stability, and specificity.

Synthetic biology enables the design of entire metabolic pathways and cellular factories, moving beyond single-enzyme optimization to integrated bioproduction systems. This means that CGTase engineering can be part of a larger, rationally designed biological system in which the enzyme's production, activity, and interaction with other cellular components are precisely controlled. This holistic approach allows for the creation of highly efficient and sustainable biomanufacturing platforms, where

CGTase can be optimized not just in isolation, but within the context of a complete cellular environment, leading to more predictable and scalable industrial outcomes.

Challenges and opportunities in industrial scale-up

Protein engineering of CGTase is not limited to CD production; it extends to the synthesis of a wide array of novel glycosylated products. For example, CGTase is increasingly engineered to improve its intermolecular transglycosylation ability for synthesizing glucosides such as alkyl glucosides, ascorbic acid 2-glucoside, sophoricoside glucoside, and particularly glucosylated steviol glycosides. These glycosylated compounds often exhibit enhanced water solubility and bioavailability, making them valuable in pharmaceuticals, food, and cosmetics. For instance, hesperidin glucoside, produced by engineered CGTase, has promising applications in skincare due to its antioxidant and antiinflammatory properties (Chen et al., 2022).

Conclusions

CGTase is an enzyme of profound biological significance and immense industrial value. Its natural role in enabling microorganisms to monopolize starch substrates by converting them into CDs for energy underscores its inherent catalytic efficiency and versatility. This natural proficiency—encompassing cyclization, coupling, disproportionation, and hydrolysis—provides a robust foundation for its diverse applications in the pharmaceutical, food, cosmetic, and chemical industries, primarily through the production of CDs.

However, the inherent limitations of wild-type CGTase, such as suboptimal product specificity, low conversion rates, and insufficient stability under industrial conditions, necessitate targeted protein engineering. These limitations are interconnected, collectively impacting the economic viability and market potential of CGTase-catalyzed processes. To overcome these hurdles, protein engineering aims to tailor CGTase variants with enhanced properties, including improved yields, specific CD ratios, increased thermostability, and optimized pH profiles.

The methodologies employed in CGTase protein engineering are sophisticated and increasingly integrated. Rational design, guided by detailed structural and mechanistic understanding, allows for precise, site-directed modifications to achieve desired changes in product specificity and stability. Directed evolution, conversely, offers a powerful, knowledge-independent approach, mimicking natural selection to discover novel functionalities and improve existing ones, especially when detailed structural information is scarce. Semirational design, a synergistic blend of these two approaches, optimizes the engineering process by focusing randomization on promising regions, resulting in smaller, higher-quality mutant libraries and more efficient discovery of tailored enzymes.

Computational modeling and simulation, including homology modeling, MD, and in silico mutagenesis, provide critical insights into CGTase structure–function relationships and enable the prediction of mutation effects. The advent of AI and ML has profoundly transformed computational protein design, mitigating its historical limitations by providing highly accurate predictive and generative models for protein structure and function, thereby accelerating the design cycle and enabling the development of "unnatural" enzymes.

Looking forward, the integration of advanced computational tools, particularly AI and ML, with synthetic biology approaches represents a paradigm shift in CGTase engineering. These technologies enable the design of "unnatural" enzymes with unprecedented precision and the construction of entire biological systems for efficient bioproduction. While challenges in high-throughput screening and industrial scale-up persist, continued advancements in these fields promise to unlock the full potential of CGTase. This will lead to the development of highly efficient, robust, and versatile biocatalysts capable of synthesizing a broader spectrum of high-value glycosylated products, solidifying CGTase's role as a cornerstone enzyme in the bioeconomy.

References

Ara, K. Z., Gulshan, J. A., Linares-Pastén, J., Jönsson, M., Viloria-Cols, S., Ulvenlund, P., Adlercreutz, & Karlsson, E. N. (2021). Engineering CGTase to improve synthesis of alkyl glycosides. *Glycobiology, 31*(5), 603–612. https://doi.org/10.1093/glycob/cwaa109.

Ban, X., Zhengbiao Gu, C., Li, M., Huang, L., Cheng, Y., Hong, & Li, Z. (2015). Mutations at calcium binding site III in cyclodextrin glycosyltransferase improve β-cyclodextrin specificity. *International Journal of Biological Macromolecules, 76*, 224–229. https://doi.org/10.1016/j.ijbiomac.2015.02.036.

Bosito, E. F. M., Montecillo, A. D., Franco, R. A. G., Sabino, N. G., Bertuso, A. L. G., Mendoza, B. C., & Lantican, N. B. (2023). Alkalihalobacillus lehensis M136, a novel alkaliphilic, cyclodextrin glucanotransferase (CGTase)-producing isolate from manleluag hyperalkaline spring in Pangasinan, Philippines. *Philippine Journal of Science, 152*(1), 109–124.

Castillo, J. D. L. M., Caminata Landriel, S., Sànchez Costa, M., Taboga, O. A., Berenguer, J., Hidalgo, A., Ferrarotti, S. A., & Costa, H. (2018). A single mutation in cyclodextrin glycosyltransferase from paenibacillus barengoltzii changes cyclodextrin and maltooligosaccharides production. *Protein Engineering, Design and Selection, 31*(10), 399–407. https://doi.org/10.1093/protein/gzy034.

Chen, H., Liu, Y., Ren, X., Wang, J., Zhu, L., Lu, Y., & Chen, X. (2022). Engineering of cyclodextrin glycosyltransferase through a size/polarity guided triple-code strategy with enhanced a-glycosyl hesperidin synthesis ability. *Applied and Environmental Microbiology, 88*(17), 1–13 https://doi.org/10.1128/aem.01027-22.

Chen, S., Xiong, Y., Su, L., Wang, L., & Wu, J. (2017). Position 228 in paenibacillus macerans cyclodextrin glycosyltransferase is critical for 2-O-D-glucopyranosyl-L-ascorbic acid synthesis. *Journal of Biotechnology, 247*, 18–24. https://doi.org/10.1016/j.jbiotec.2017.02.011.

Deng, C., Li, J., Shin, H. D., Du, G., Chen, J., & Liu, L. (2018). Efficient expression of cyclodextrin glycosyltransferase from geobacillus stearothermophilus in *Escherichia coli* by promoter engineering and downstream box evolution. *Journal of Biotechnology, 266*, 77–83. https://doi.org/10.1016/j.jbiotec.2017.12.009.

Esteso, M. A., & Romero, C. M. (2024). Cyclodextrins: Properties and applications. *International Journal of Molecular Sciences, 25*(8), 1–5. https://doi.org/10.3390/ijms25084547.

Fu, Y., Ding, Y., Chen, Z., & Xu, W. (2009). Exploring unfolding pathway of CGTase: Insight from molecular dynamics simulation. *Riv Biol Sep-Dec, 102*(3), 347–364.

Goh, K., Nor Muhammad, M., Osman, M., Raja Noor Zaliha Raja Abdul Rahman, H., & Illias, R. M. (2009). A predominant β-CGTase G1 engineered to elucidate the relationship between protein structure and product specificity. *Journal of Molecular Catalysis B: Enzymatic, 57*(1–4), 270–277. https://doi.org/10.1016/j.molcatb.2008.09.016.

Goh, K., Nor, M., Mahadi, M., Hassan, O., Noor Zaliha, R., Abdul Rahman, R., & Illias, R. M. (2008). Molecular modeling of a predominant β-CGTase G1 and analysis of ionic interaction in CGTase. *Biotechnology (Reading, Mass.), 7*(3), 418–429. https://doi.org/10.3923/biotech.2008.418.429.

Goh, P., Hong, Illias, R. M., & Goh, K. M. (2012). Domain replacement to elucidate the role of B domain in CGTase thermostability and activity. *Process Biochemistry, 47*(12), 2123–2130. https://doi.org/10.1016/j.procbio.2012.07.033.

Goh, P., Hong, Illias, R. M., & Goh, K. M. (2012). Rational mutagenesis of cyclodextrin glucanotransferase at the calcium binding regions for enhancement of thermostability. *International Journal of Molecular Sciences, 13*(5), 5307–5323. https://doi.org/10.3390/ijms13055307.

Han, R., Li, J., dong Shin, H., Chen, R. R., Du, G., Liu, L., & Chen, J. (2014). Recent advances in discovery, heterologous expression, and molecular engineering of cyclodextrin glycosyltransferase for versatile applications. *Biotechnology Advances, 32*(2), 415–428. https://doi.org/10.1016/j.biotechadv.2013.12.004.

Han, R., Liu, L., Dong, H., Shin, Chen, R. R., Li, J., Du, G., & Chen, J. (2013). Systems engineering of tyrosine 195, tyrosine 260, and glutamine 265 in cyclodextrin glycosyltransferase from paenibacillus macerans to enhance maltodextrin specificity for 2-O-d-glucopyranosyl-lascorbic acid synthesis. *Applied and Environmental Microbiology, 79*(2), 672–677. https://doi.org/10.1128/AEM.02883-12.

Han, R., Ni, Jie, Dong, J., Xu, G., & Ni, Y. (2020). Engineering of cyclodextrin glycosyltransferase reveals PH-regulated mechanism of enhanced long-chain glycosylated sophoricoside specificity. *Applied and Environmental Microbiology, 86*(7), 1–14.

Hu, G., Luo, J., Bao, T., Hu, X., Yang, J., & Zhang, H. (2024). Rational design of disulfide bonds to improve the thermostability and γ-cyclodextrin production capacity of γ-cyclodextrin glycosyltransferase. *Process Biochemistry, 144*(June), 248–255. https://doi.org/10.1016/j.procbio.2024.06.008.

Jailani, N., Jaafar, N. R., Suhaimi, S., Mackeen, M. M., Bakar, F. D. A., & Illias, R. M. (2022). Cross-linked cyclodextrin glucanotransferase aggregates from *Bacillus lehensis* G1 for cyclodextrin production: Molecular modeling, developmental, physicochemical, kinetic and thermodynamic properties. *International Journal of Biological Macromolecules, 213*, 516–533. https://doi.org/10.1016/j.ijbiomac.2022.05.170.

Jemli, S., Jaoua, M., & Bejar, S. (2016). US132 cyclodextrin glucanotransferase engineering by random mutagenesis for an anti-staling purpose. *Molecular Biotechnology, 58*(8–9), 551–557. https://doi.org/10.1007/s12033-016-9952-z.

Kaul, P., & Asano, Y. (2012). Strategies for discovery and improvement of enzyme function: State of the art and opportunities. *Microbial Biotechnology, 5*(1), 18–33. https://doi.org/10.1111/j.1751-7915.2011.00280.x.

Kelly, R. M., Dijkhuizen, L., & Leemhuis, H. (2009). The evolution of cyclodextrin glucanotransferase product specificity. *Applied Microbiology and Biotechnology, 84*(1), 119–133. https://doi.org/10.1007/s00253-009-1988-6.

Kelly, R. M., Leemhuis, H., Rozeboom, H. J., Van Oosterwijk, N., Dijkstra, B. W., & Dijkhuizen, L. (2008). Elimination of competing hydrolysis and coupling side reactions of a cyclodextrin glucanotransferase by directed evolution. *Biochemical Journal, 413*(3), 517–525. https://doi.org/10.1042/BJ20080353.

Kong, Demin, Wang, L., Su, L., & Ju, J. (2021). Effect of Leu 277 on disproportionation and hydrolysis activity in *Bacillus stearothermophilus* NO_2 cyclodextrin glucanotransferase. *Applied and Environmental Microbiology, 87*(12), 1–13 https://doi.org/10.1128/AEM.03151-20.

Leemhuis, H., Kelly, R. M., & Dijkhuizen, L. (2009). Directed evolution of enzymes: Library screening strategies. *IUBMB Life, 61*(3), 222–228. https://doi.org/10.1002/iub.165.

Leemhuis, H., Kelly, R. M., & Dijkhuizen, L. (2010). Engineering of cyclodextrin glucanotransferases and the impact for biotechnological applications. *Applied Microbiology and Biotechnology, 85*(4), 823–835. https://doi.org/10.1007/s00253-009-2221-3.

Leemhuis, H., Uitdehaag, J. C. M., Rozeboom, H. J., Dijkstra, B. W., & Dijkhuizen, L. (2002). The remote substrate binding subsite -6 in cyclodextrin-glycosyltransferase controls the transferase activity of the enzyme via an induced-fit mechanism. *Journal of Biological Chemistry, 277*(2), 1113–1119. https://doi.org/10.1074/jbc.M106667200.

Li, X., Zheng, D., Wu, J., Jin, Z., Svensson, B., & Bai, Y. (2023). Increasing γ-CD conversion rates by improving thermostability of *Bacillus* Sp. FJAT-44876 γ-CGTase. *Food Bioscience, 51*(October 2022), 102204. https://doi.org/10.1016/j.fbio.2022.102204.

Li, Z., Huang, M., Gu, Z., Holler, T. P., Cheng, L., Hong, Y., & Li, C. (2016). Asp577 mutations enhance the catalytic efficiency of cyclodextrin glycosyltransferase from *Bacillus circulans*. *International Journal of Biological Macromolecules, 83*, 111–116. https://doi.org/10.1016/j.ijbiomac.2015.11.042.

Lorthongpanich, N., Mahalapbutr, P., Rungrotmongkol, T., Charoenwongpaiboon, T., & Prousoontorn, M. H. (2022). Fisetin glycosides synthesized by cyclodextrin glycosyltransferase from *Paenibacillus* Sp. RB01: Characterization, molecular docking, and antioxidant activity. *PeerJ, 10*(May), e13467. https://doi.org/10.7717/peerj.13467.

Low, K. O., Mahadi, N. M., Rahim, R. A., Rabu, A., Bakar, F. D. A., Murad, A. M. A., & Illias, R. M. (2010). Enhanced secretory production of hemolysin-mediated cyclodextrin glucanotransferase in *Escherichia coli* by random mutagenesis of the ABC transporter system. *Journal of Biotechnology, 150*(4), 453–459. https://doi.org/10.1016/j.jbiotec.2010.10.001.

Miao, B., Huang, D., Wang, T., Liu, H., Hao, Z., Yuan, H., & Jiang, Y. (2024). Enhancing trehalose production via *Bacillus* species G1 cyclodextrin glucanotransferase mutants:

Modifying disproportionation characteristics and thermal stability. *Frontiers in Microbiology, 15*(November). https://doi.org/10.3389/fmicb.2024.1500232.

Qi, Q., & Zimmermann, W. (2005). Cyclodextrin glucanotransferase: From gene to applications. *Applied Microbiology and Biotechnology, 66*(5), 475–485. https://doi.org/10.1007/s00253-004-1781-5.

Rabadiya, K., Pardhi, D., Thaker, K., Patoliya, J., Rajput, K., & Joshi, R. (2024). A review on recent upgradation and strategies to enhance cyclodextrin glucanotransferase properties for its applications. *International Journal of Biological Macromolecules*. https://doi.org/10.1016/j.ijbiomac.2024.129315 Elsevier B.V.

Scherer, M., Fleishman, S. J., Jones, P. R., Dandekar, T., & Bencurova, E. (2021). Computational enzyme engineering pipelines for optimized production of renewable chemicals. *Frontiers in Bioengineering and Biotechnology, 9*(June), 1–13. https://doi.org/10.3389/fbioe.2021.673005.

Sharma, A., Gupta, G., Ahmad, T., Mansoor, S., & Kaur, B. (2021). Enzyme engineering: Current trends and future perspectives. *Food Reviews International, 37*(2), 121–154. https://doi.org/10.1080/87559129.2019.1695835.

Silva, B. L., Marques, G. L., Reis, N. S., Maldonado, R. R., Santos, R. L. S. R., & Aguiar-Oliveira, E. (2019). Enzymatic production of Β-cyclodextrin from jackfruit seeds (*Artocarpus Intergrifolia* L.). *Brazilian Journal of Chemical Engineering, 36*(4), 1393–1402. https://doi.org/10.1590/0104-6632.20190364s20180343.

Song, B., Yue, Y., Xie, T., Qian, S., & Chao, Y. (2014). Mutation of Tyrosine167Histidine at remote substrate binding subsite -6 in α-cyclodextrin glycosyltransferase enhancinga α-cyclodextrin specificity by directed evolution. *Molecular Biotechnology, 56*, 232–239.

Sonnendecker, C., Melzer, S., & Zimmermann, W. (2019). Engineered cyclodextrin glucanotransferases from *Bacillus* sp. G-825-6 produce large-ring cyclodextrins with high specificity. *MicrobiologyOpen, 8*(6). https://doi.org/10.1002/mbo3.757.

Tao, X., Su, L., Wang, L., Chen, X., & Wu, J. (2020). Improved production of cyclodextrin glycosyltransferase from *Bacillus stearothermophilus* NO_2 in *Escherichia coli* via directed evolution. *Applied Microbiology and Biotechnology, 104*(1), 173–185. https://doi.org/10.1007/s00253-019-10249-8.

Tao, X., Wang, T., Su, L., & Wu, J. (2018). Enhanced 2- O-α- d -glucopyranosyl- l -ascorbic acid synthesis through iterative saturation mutagenesis of acceptor subsite residues in bacillus stearothermophilus NO_2 cyclodextrin glycosyltransferase. *Journal of Agricultural and Food Chemistry, 66*(34), 9052–9060. https://doi.org/10.1021/acs.jafc.8b03080.

Teze, D., Jiao, Z., Wiemann, M., Kazi, Z. G., Lupo, R., Rønne, M. E., Carlström, G., et al. (2020). Rational enzyme design without structural knowledge: A sequence-based approach for efficient generation of glycosylation catalysts. *ChemRxiv, no. 1*. https://doi.org/10.26434/CHEMRXIV.11538708.V1.

Uitdehaag, J. C. M., Van Alebeek, G. J. W. M., Van Der Veen, B. A., Dijkhuizen, L., & Dijkstra, B. W. (2000). Structures of maltohexaose and maltoheptaose bound at the donor sites of cyclodextrin glycosyltransferase give insight into the mechanisms of transglycosylation activity and cyclodextrin size specificity. *Biochemistry, 39*(26), 7772–7780. https://doi.org/10.1021/BI000340X.

Wang, H., Zhou, W., Li, H., Rie, B., & Piao, C. (2017). Improved activity of β-cyclodextrin glycosyltransferase from *Bacillus* sp. N-227 via mutagenesis of the conserved residues. *3 Biotech, 7*(2), 1–10. https://doi.org/10.1007/s13205-017-0725-6.

Wang, L., Duan, X., & Wu, J. (2016). Enhancing the α-cyclodextrin specificity of cyclodextrin glycosyltransferase from *Paenibacillus macerans* by mutagenesis masking subsite -7. *Applied and Environmental Microbiology, 82*(8), 2247–2255. https://doi.org/10.1128/AEM.03535-15.

Wang, L., Xia, Y., Su, L., & Wu, J. (2020). Modification of *Bacillus clarkii* Γ-cyclodextrin glycosyltransferase and addition of complexing agents to increase γ-cyclodextrin production. *Journal of Agricultural and Food Chemistry, 68*, 12079–12085. https://doi.org/10.1021/acs.jafc.0c05408.

Wind, R. D., Uitdehaag, J. C. M., Buitelaar, R. M., Dijkstra, B. W., & Dijkhuizen, L. (1998). Engineering of cyclodextrin product specificity and pH optima of the thermostable cyclodextrin glycosyltransferase from *Thermoanaerobacterium thermosulfurigenes* EM1. *Journal of Biological Chemistry, 273*(10), 5771–5779. https://doi.org/10.1074/jbc.273.10.5771.

Xiong, W., Liu, B., Shen, Y., Jing, K., & Savage, T. R. (2021). Protein engineering design from directed evolution to de novo synthesis. *Biochemical Engineering Journal, 174*(June), 108096. https://doi.org/10.1016/j.bej.2021.108096.

CHAPTER

Immobilization of cyclodextrin glucanotransferases

10

Babatunde A. Ogunbadejo and Sulaiman Al-Zuhair
United Arab Emirates University, Al Ain, United Arab Emirates

Introduction

Enzyme immobilization is a critical technique that can be used to overcome several challenges limiting the use of enzymes as industrial catalysts, including enzyme instability, high production costs, and reusability problems. Immobilization can enhance enzymes' activity, stability, specificity, and other characteristics while rendering them insoluble, thus enabling recycling and easing reusability, making them appropriate for use in industrial processes (Sheldon, 2007). Unfortunately, despite the benefits that enzyme immobilization on solid supports promotes, there may be drawbacks because of the altered protein structure, restricted access to the active region, or problematic mass transfer. Therefore the ideal conditions and procedures for immobilization should be determined for a particular enzyme.

This chapter delves into the fundamentals and science of cyclodextrin glucanotransferases (CGTase) immobilization. Key challenges and advancements in the field will be discussed, highlighting how novel approaches like nanotechnology and bioengineering are revolutionizing immobilization practices. The chapter concludes by discussing the benefits of CGTase immobilization, as well as envisioning future directions for integrating immobilized CGTases into large-scale sustainable processes. Through this discussion, the chapter aims to provide a comprehensive understanding of the current state and potential of CGTase immobilization in advancing industrial biotechnology.

Techniques for enzyme immobilization

Enzyme immobilization is the process of securely attaching enzymes to solid supports to increase their use in a variety of applications. The enzymes mostly become more stable by this process, thereby prolonging their survival in different environments (Guisan, 2006). Additionally, it facilitates enzyme reusability, which can drastically reduce production costs (Saifuddin et al., 2013). These advantages

Cyclodextrin Glucanotransferase. DOI: https://doi.org/10.1016/B978-0-443-41694-1.00016-2

make immobilized enzymes highly in demand in various industries, including food manufacturing, environmental remediation, medicines, and textile production.

Enzyme immobilization techniques may be broadly divided into two categories: physical and chemical methods (Mohamad et al., 2015). In the physical methods, enzymes are attached to supports via weaker, noncovalent interactions (Brena and González-Pombo, 2013; Costa et al., 2005). Adsorption, in which enzymes adhere to surfaces by ionic bonds or van der Waals forces, is one example (1). Entrapment is another technique in which enzymes are held inside membranes or gels, including polyacrylamide gels or alginate beads. Although the physical techniques are simpler and less expensive, the weakness of the enzyme-support interactions may eventually lead to loss of effectiveness due to enzyme leaking.

The second category is the chemical technique, which involves the enzyme and solid supports creating robust covalent bonds (Brena and González-Pombo, 2013; Chiou & Wu, 2004; Mohamad et al., 2015). Covalent binding and cross-linking are some of the chemical techniques that can be employed. Covalent binding involves attaching the enzyme to the support surface via certain groups on the enzyme surface, like amino or carboxyl groups, to produce enzyme-support complexes (Gimenez et al., 2019). Cross-linking uses chemicals like glutaraldehyde to form a bridge between the enzyme molecules in order to produce extremely robust enzymes that are resistant to leaching (Rojas et al., 2019). Although these chemical processes may require more intricate and expensive processes, they often result in more stable immobilized enzymes that are appropriate for harsh conditions.

Each of these methods offers advantages and disadvantages. Physical methods are usually simpler and faster, ideal for quick applications, but the resulting biocatalyst may not withstand extreme conditions. Chemical techniques provide stronger attachment, making enzymes more resistant to physical and chemical stress. However, they can sometimes lead to a reduction in enzyme activity if not carefully controlled. Selecting the right method depends on the specific enzyme, the intended use, and the conditions it will face during operation. Effective immobilization can increase enzyme efficiency by restricting their movement, leading to higher reaction rates. It enhances stability by protecting enzymes from denaturation caused by temperature, pH changes, or other environmental factors. Reusability through immobilization means enzymes can be used repeatedly in industrial processes, lowering overall costs and reducing waste. This makes enzyme immobilization an important tool in developing sustainable and cost-effective technologies across many industries.

The need for immobilization

Enzyme immobilization aids in addressing key issues that have been identified with free or native enzymes, such as enzyme handling, storing, and reuse. It has grown as an important research area because of the need to overcome these issues in industrial operations where enzymes are utilized to accelerate chemical reactions.

Reusing enzymes by immobilization can significantly save costs while increasing production efficiency. Enzyme reuse is more attractive by the fact that enzyme costs represent a significant portion of the overall expenses; as such, maintaining low enzyme costs might increase a process' profitability when producing high-value goods.

Previously, enzymes were largely preserved by lyophilization, also known as freeze-drying, by eliminating water present in them to enhance their storage. The structure of the enzymes might be harmed by this action, resulting in less active enzymes in reactions as a result of such damage (Sheldon, 2007). The enzymes may also become less stable and function more slowly as a result of structural deformation, which makes process control more difficult. Immobilizing enzymes on supports can solve many of these problems. When enzymes are attached to a solid support, they are more accessible and easier to handle. This approach distributes enzymes over a larger surface area, exposing more active parts of the enzyme to the substrate molecules. As a result, the reaction rate can increase, and enzyme efficiency improves. Immobilization also facilitates easy separation of enzymes from the reaction mixture. This makes it easier to reuse enzymes multiple times without losing activity.

The choice of support material is essential for successful enzyme immobilization. An ideal support must be inert, meaning it won't react chemically with the enzyme or interfere with its activity. It should resist microbial growth to prevent contamination. The support must also not hinder substrate molecules from reaching the enzyme's active site. Diffusional issues can slow down reactions or cause uneven enzyme activity. Supports like porous ceramics, synthetic polymers, or natural materials such as silica are often used because they meet these criteria. Proper support materials improve enzyme stability, extend their usable life, and help maintain high activity levels over time. In summary, enzyme immobilization is a key strategy to enhance enzyme utility in various industrial processes. It protects enzymes from structural damage, makes them easier to manage, and allows for repeated use. The right support material and method of immobilization can significantly influence the success of immobilization, ensuring high stability and activity while minimizing costs and operational problems. The interplay of the desired properties of enzymes and supports is shown in Fig. 10.1.

Surface adsorption

The most basic and popular method for immobilizing enzymes is physical adsorption, also known as surface adsorption. In this process, enzymes adhere to solid supports via weak, noncovalent bonds, such as ionic bonds, hydrogen bonds, van der Waals forces, and hydrophobic interactions (Schöffer et al., 2013). The natural structure and function of the enzyme are mostly preserved, and the procedure is simple and doesn't call for intricate chemical processes. In this immobilization method, the physical interaction between the enzyme and the

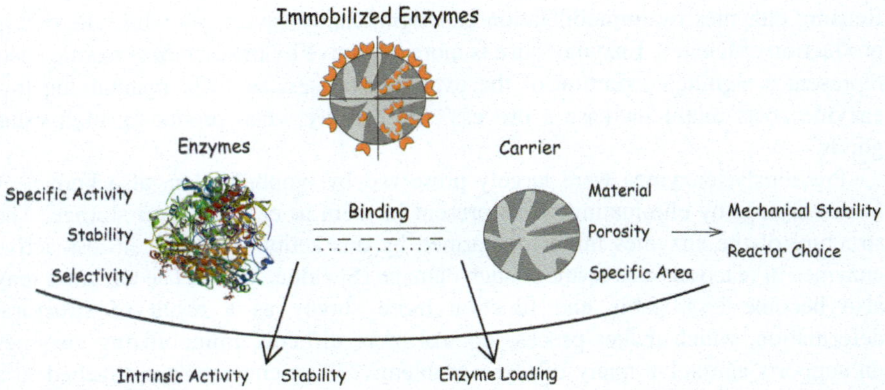

FIGURE 10.1

Properties of enzymes and supports for biocatalysts (Bolivar and Eisl, 2016).

support is what causes surface adsorption (Sheldon, 2007). It is accomplished by soaking the support for a certain amount of time—often referred to as the incubation time—in an enzyme-buffered solution. An alternative is to allow the enzyme solution to dry on the support surface before removing any detached enzymes. Physical adsorption is appealing due to its reversible nature, which allows the immobilized enzymes to be detached from the support when enzymatic activity declines under moderate circumstances. However, when exposed to industrial environments, the weak forces holding the enzymes leave them vulnerable to leaching off the support (Sobral et al., 2002; Tardioli et al., 2006).

Several researchers have delved into this method for CGTase immobilization in order to discover its mechanism for different types of surfaces. Jamil et al. (2018) showed that electrostatic interaction played a crucial role during CGTase immobilization on polyvinylidene difluoride (PVDF) hollow fiber membrane. PVDF is a polymer with several useful properties such as a high surface-to-volume ratio, good mechanical strength, operational durability, and high chemical resistance, and is widely used in ultrafiltration, pervaporation, and microfiltration (Algieri et al., 2017). These properties have made PVDF to find use as support for several enzymes, including CGTase. The structure of the PVDF membrane contains repeating monomer units of -CF_2-CH_2- with positively charged hydrogen atoms and negatively charged fluoride atoms (Fontananova et al., 2015). The CGTase surface possesses a net negative charge, thereby establishing electrostatic interaction with the positively charged hydrogen atoms on the PVDF surface (Jamil et al., 2018). The immobilization of CGTase onto PVDF supports resulted in an immobilization yield reaching up to 80% when using a 10% (v/v) CGTase concentration (Jamil et al., 2018). This means that most of the enzyme molecules attached successfully to the support material at this concentration. When the enzyme concentration increased beyond this point, the immobilization yield started to decline. This decline likely occurs because all the available attachment sites on the PVDF surface become saturated at 10%, leaving no room for additional enzymes to bind. Adding more enzymes does not increase the number of enzymes attached; instead, it causes the enzymes to crowd together despite the limited space. This overcrowding could lead to overlapping or clustering of enzyme molecules, which might hinder their activity or cause some molecules to detach over time. It could also lead to denaturation, limited active site accessibility, and loss of activity (Prousoontorn & Pantatan, 2007; Talbert & Goddard, 2012). Temperature also played an important role in improving how many CGTase molecules attached to the support. As the temperature increased from 20°C to 25°C, the immobilization yield rose sharply from 25% to a very high 95.6%. This suggests that at 25°C, the molecules of CGTase vibrate more quickly and actively (Dwevedi, 2016; Mateo et al., 2007). These faster vibrations help to form stronger bonds between the enzyme and the PVDF surface. The energy from this movement makes it easier for enzyme molecules to find and connect with the support, increasing the overall yield (Blanco et al., 1989). This temperature effect highlights the importance of molecular motion in enzyme attachment. When CGTase molecules vibrate more, they are more likely to interact effectively with the support surface. The faster movement increases the chances of forming stable links. However, other studies have found different optimal temperatures for immobilizing CGTase on different supports, indicating that the best temperature depends on the material used. For example, some supports may require higher or lower temperatures for maximum enzyme binding (Ivanova, 2010; Sobral and Rodrigues, 2003). Each type of support interacts differently with the enzyme, which can change how temperature impacts the process.

The findings underline that there is a point at which adding more enzyme no longer benefits the process. Instead, it can cause overcrowding and reduce the number of active enzymes attached. Similarly, increasing temperature within an optimal range enhances the chances of forming strong, stable attachments. These results help guide the best conditions for fixing CGTase onto PVDF and similar supports in lab and industrial settings. Proper control of these factors ensures the highest possible activity of the immobilized enzymes, which can improve processes like starch modification or synthesis of valuable compounds.

Covalent binding

In contrast to surface attachment, covalent binding of CGTase to the support offers a more robust biocatalyst. In this method, the functional groups present on CGTase and the carrier surface play a major role in the formation of covalent bonds (Ferrarotti et al., 2006). Although it increases the enzyme's stability and produces stronger binding, properties highly desired for industrial applications, the primary drawback is that the chemical linkages may alter the activity of the connected enzyme (Sobral et al., 2002; Sulaiman et al., 2015).

Covalent binding of CGTase typically involves the amino (–NH_2) and carboxyl groups (–COOH) present in the residues. These groups react with the chemically activated support material to create a covalent bond (Batista-Viera et al., 1994; Grazú et al., 2005). The surface of the support is chemically modified to introduce desired reactive groups onto it. Some of the widely used activation methods are via glutaraldehyde, for introducing aldehyde groups, capable of bonding with the amino groups on the CGTase, epoxy activation, and carbodiimide activation, used for bonding carboxyl groups to amino groups on the enzyme. To achieve covalent binding, the enzyme and activated support are mixed under carefully regulated pH and temperature. The amino groups (-NH_2) on lysine residues present in CGTase are frequently used for attachment by CGTase (Schöffer et al., 2017). In order to avoid nonspecific adsorption or adverse reactions, inert molecules (such as glycine or ethanolamine) are used to block the unreacted sites on the support after immobilization. Cellulose nanofiber (CNF) was used to establish covalent binding has immobilized CGTase by covalent binding (Sulaiman et al., 2017). It has been demonstrated that the surface area of the support significantly influences the effectiveness of immobilized CGTase, with the CNF with the smallest diameter exhibiting the maximum activity and enzyme binding.

The immobilization of a thermophilic CGTase from *Thermoanaerobacter* sp. by covalent binding to Eupergit C has also been investigated (Martín et al., 2003). Eupergit C is made by copolymerization of methacrylamide, N, N' -methylene-bis-(methacrylamide), allyl glycidyl ether, and glycidyl methacrylate to form macroporous beads having a diameter between 100–250 mm (Katchalski-Katzir & Kraemer, 2000). This support is an appropriate carrier for covalently immobilizing enzymes for use in industrial settings. It has good characteristics and is commercially available in several forms. It can bind proteins via its oxirane, sulfhydryl, and carboxyl groups depending on the immobilization conditions (Katchalski-Katzir & Kraemer, 2000). At neutral and alkaline pH, the oxirane groups bind with the amino groups of the protein molecules to produce stable covalent bonds. Also, Eupergit C can bind to enzyme molecules using the sulfhydryl groups and through the carboxyl groups in the acidic, neutral, and alkaline regions. The wide pH range in which Eupergit C can covalently bind enzyme molecules ensures that the enzyme remains stable without significant loss in activity (Katchalski-Katzir & Kraemer, 2000). The experimental studies on using Eupergit C for CGTase immobilization showed the catalytic efficiency of the immobilized CGTase peaked at 4.4%, representing 70% of the added protein using a 10:1000 ratio (w/w) CGTase to Eupergit C. Above this, a decrease in recovered activity was observed, which was attributed to diffusional resistance faced by the substrate and insolubilization process (Hernaiz & Crout, 2000). The specific activity of CGTase showed an eightfold reduction after immobilization on Eupergit C. The observed reduction may be attributed to several factors, ranging from steric hindrances within the pores of Eupergit C, internal diffusional limitations, modifications of the CGTase conformation and/or the active site, and the nature of the microenvironment within the Eupergit C, which differs significantly from the CGTase native environment (Clark, 1994).

A macroreticular hydrophilic carrier resin (FE-4611), which was synthesized by polymerization of glycidyl methacrylate, with ethylene glycol as cross-linking agent and introduction of primary amine to act as the active group, has also been used as an immobilization support for CGTase (Sakai, Yamamoto, et al., 1991). The carrier resin (FE-4611), with an average pore size of 650 Å and specific surface area of $30\,m^2/g$ (for pores larger than 500 Å), is hydrophilic in sharp contrast to conventional styrene-based ion exchange resins that are highly hydrophobic and prone to organic fouling due to the high molecular weight substrate used in the process (Sakai, Yoshida, et al., 1991). The hydrophobicity of the conventional resins led to loss of CGTase activity. The hydrophilic resin (FE-4611), on the other hand, is subjected to little or no organic fouling, which qualifies it as an adequate support. The optimum pH of the immobilized CGTase ranged from 6.0 to 8.0 compared to the native, with an optimum pH of 6.0. The study also showed that substrate flow rate could have a drastic effect on the product mixture (Sakai, Yamamoto, et al., 1991).

Another material with high affinity for proteins is chitosan. It has been shown in various studies to be an excellent support for enzyme immobilization due to its cheapness, as it is derived majorly from chitin, a byproduct from the fish industry, and also it is readily available (Spagna et al., 2001). The immobilization of CGTase on chitosan can facilitate the recovery of free CGTase activity and maintain adequate pH. CGTase from *Bacillus firmus* has been immobilized on chitosan via covalent bonding with hexamethylenediamine (HEMDA) and γ-aminopropyl-triethoxysilane (γ-APTS) at various test conditions. The chitosan was first hydrated by several washings with 10% volume of acetone and ethanol each, followed by distilled water to obtain a filtrate with neutral pH (Sobral et al., 2002). The results after immobilization by covalently binding of CGTase with γ-APTS showed that no activity was observed for the immobilized enzyme. This could be due to γ-APTS attacking the chitin, which is the precursor of chitosan (Bon and Freire, 1984). This shows that using γ-APTS isn't appropriate for CGTase immobilization on chitosan. A better result was obtained using HEMDA at 7°C without any agitation, with protein fixation attaining 89.7% and specific activity at 0.263 μmol β-CD/min/mg of protein (Sobral et al., 2002).

In another related study, silica surface was functionalized with different chemical groups in order to develop site-directed immobilization protocols using several amino acid residues present on the CGTase surface (Schöffer et al., 2017). Mesoporous silicates have found wide use in this regard owing to its inert and rigid nature, alongside its porous structure, which provides a high surface area for better loading of enzymes without introducing mass transfer limitations (Carlsson et al., 2014; Zucca & Sanjust, 2014). Silica was functionalized by silanization using organic solvents such as 3-mercaptopropyltrimethoxysilane (MPTMS) and 3-aminopropyltrimethoxysilane (APTMS) to produce thiol-functionalized silica (Si-SH) and amino-functionalized silica (Si-NH), respectively. The amino groups on Si-NH can be activated by glutaraldehyde so as to react with the ε-amino groups from lysine residues on the CGTase to form a covalent binding, while Si-SH can form a disulfide covalent bond with the native cysteine residue for site-directed immobilization after it has been acted upon by reducing agents such as dithiothreitol (DTT) or 2-mercaptoethylamine to release the thiol groups (Grazú et al., 2003; Grazú et al., 2005; Migneault et al., 2004; Rueda et al., 2016; Schöffer et al., 2017). Using a moderate protein concentration of 10 mg/g of support to avoid protein crowding, which could lead to protein-protein interactions, CGTase was immobilized on silica activated by a combination of APTMS and glutaraldehyde, and using MPTMS. Both supports showed very high immobilization yields of 85.37% (±0.95) on Si-SH and 96.56% (±0.45) on Si-NH, but the efficiencies of the biocatalysts were rather low, 11.91% (±0.72) and 2.86% (±0.37) and for Si-SH-CGTase and Si-NH-G-CGTase. The observed higher immobilization yield using the amino groups on the lysine residue could be attributed to the ease of accessing it compared to the cysteine residue. The low immobilization efficiency could be due to inappropriate orientation of the active sites postimmobilization, thereby affecting the enzyme-substrate interaction. It is clear that there is a difference in the results obtained from the two forms of immobilization of CGTase on silica due to the location of the residue that was employed in each case (Schöffer et al., 2017). A look into the optimum conditions for both biocatalysts showed that they both showed the best activity at pH 5.5, although Si-NH had a wider pH range compared to Si-SH. For temperature, Si-NH gave better activity at high temperatures compared to Si-SH, an indication that the enzyme's structures are better preserved when immobilized via amino groups compared to the thiol groups (Schöffer et al., 2017).

Encapsulation and entrapment

Encapsulation is the process of enclosing the enzyme in a semipermeable matrix or membrane, which prevents enzyme loss while permitting substrate and product flow. In order to shield CGTase from adverse operating conditions like high temperatures or extremely high pH, the immobilization process takes place in a restricted microenvironment (Ogunbadejo & Al-Zuhair, 2021). Enzyme encapsulation is an immobilization technique in which the enzyme is contained inside a porous material. The entrapped enzyme's ability to diffuse out is limited, but it is not really linked to the support. Enzyme entrapment occurs quickly, under moderate circumstances, and there is less chance of denaturing because the enzyme is not chemically reacting with the support. However, because the substrate may have trouble reaching all of the active sites, this approach has limits with regard to mass transfer. Similar to entrapment, encapsulation involves the suspension of cells and enzymes in solutions, but in a regulated setting. By enclosing delicate enzymes and organisms in tiny vesicles with porous membranes, the encapsulation approach is designed to protect them. There are very limited works on CGTase using this immobilization type.

Cross-linking and aggregation

One of the more recent developments in immobilization techniques is the cross-linked enzyme aggregates (CLEAs), commonly referred to as carrier-free immobilization because they do not require the use of a support (Zhang et al., 2019). In the case of enzyme immobilization on supports, the enzyme is attached at a point to the support surface, but in cross-linking, the enzyme gains greater stability because it is stabilized by links in a 3D structure; therefore cross-linking enzymes typically exhibit superior mechanical stability, shear stress resistance, and temperature tolerance compared to other immobilized enzymes. Mostly, the enzyme aggregates are formed using bifunctional (e.g., glutaraldehyde) or polyfunctional (e.g., aldehyde–pectin, aldehyde–dextran, aldehyde–starch) chemical agents (Sheldon, 2011). Also, it can be achieved with the help of salts, nonionic polymers, or organic solvents. Compared to previous immobilization techniques, this approach has a number of benefits, including simplicity, aggregates' operational and thermal stability, and the ease with which it may be used on many enzymes simultaneously.

The impact of the cross-linking agent on CGTase CLEA activity has been examined (Rojas et al., 2019). The method was not particularly effective, though, and even at the ideal concentration of the cross-linking agent glutaraldehyde, the immobilized enzyme's activity recovery was less than 10%, while the aggregate's activity loss was more than 80%. Diffusional resistance to the large starch substrate entering the CLEA's structure or incorrect enzyme cross-linking, which results in a higher loss of enzymes and a lower activity recovery, were the two possible explanations for the observed poor activity recovery. The use of CLEA can be used to minimize conformational changes in CGTase after immobilization for product specificity. Zhang et al. (2019) utilized glutaraldehyde for cross-linking recombinant CGTase at 85°C. The CLEA-CGTase acted upon potato starch solution for 420 min and achieved 100% β-CD production from soluble starch after 420 min.

CGTase from *Bacillus megaterium* has been immobilized on various supports by using glutaraldehyde activation, a chemical process that helps attach enzymes to surfaces more firmly (Migneault et al., 2004; Panajotova et al., 2025). Researchers tested several different materials as supports for this process, including DEAE-cellulose, chitosan, CM-cellulose, polymers activated with glutaraldehyde, pumice stone, and perlite. These materials were chosen because they offer different properties, such as charge, porosity, and surface chemistry, which can influence how well the enzyme binds and remains active. The goal was to find the best support that could hold the enzyme securely and maintain its activity over time. The results of the immobilization process showed some clear differences among the supports. Notably, no enzymatic activity was detected when chitosan, CM-cellulose, or chitosan treated with glutaraldehyde were used as supports. This suggests that these materials did not favor the attachment or the stability of CGTase, possibly due to their surface chemistry or how they interact with the enzyme. In contrast, significant success was achieved with DEAE-cellulose, a type of ion-exchange resin. This support allowed the enzyme to attach effectively, with over

40% of the enzyme activity retained after immobilization. This level of activity indicates that DEAE-cellulose provides a suitable environment for CGTase to stay attached and functional. When glutaraldehyde was added to the system along with DEAE-cellulose, the immobilization yield increased further, reaching up to 51.25%. The use of glutaraldehyde helped to form stronger links between the enzyme and the support surface, essentially "gluing" the enzyme more securely. This enhancement shows that glutaraldehyde plays a crucial role in improving enzyme immobilization efficiency. It acts as a cross-linker, creating covalent bonds that improve stability and prevent enzyme leaching during use. The study also explored how different concentrations of glutaraldehyde affect the immobilization process. It was observed that increasing the concentration of glutaraldehyde up to 2% resulted in improved enzyme activity and stability. This suggests that at this level, glutaraldehyde effectively enhances the binding without causing damage to the enzyme or the support. However, when the concentration exceeded 2%, the activity of CGTase experienced a decline. Higher levels of glutaraldehyde can sometimes cause excessive cross-linking, which may restrict the enzyme's flexibility or block its active sites, thereby reducing activity (Migneault et al., 2004; Ogunbadejo & Al-Zuhair, 2023). This indicates that there is an optimal concentration of glutaraldehyde that maximizes immobilization success while preserving enzyme function. These findings highlight that the support material and the chemical conditions used during immobilization are crucial factors. DEAE-cellulose combined with an appropriate concentration of glutaraldehyde offers the best results among the supports compared in terms of enzyme attachment and activity retention. Understanding these parameters helps in designing better processes for CGTase reuse in industrial applications (Panajotova et al., 2025).

Challenges facing immobilization

Despite its advantages, immobilizing enzymes has drawbacks as well. These include the possibility of denaturation, altered kinetic properties, mass transfer restrictions, decreased catalytic efficiency with insoluble substrates, and possible decreases in enzyme activity brought on by conformational changes or uncontrolled orientation. However, these problems do not impact all immobilization methods (Robescu & Bavaro, 2025).

Support materials for cyclodextrin glucanotransferases immobilization

The type of support material used during the immobilization of enzymes significantly influences the performance and stability of the biocatalyst. The interaction between the enzyme and its support can alter the enzyme's chemical, biochemical, and mechanical properties. This means that the choice of support isn't just about holding the enzyme in place (Sobral and Rodrigues, 2003). It also affects how well the enzyme performs its catalytic function, how long it lasts, and how easy it is to recover and reuse. The support acts like a host that determines the enzyme's environment—affecting factors such as activity, stability, and resistance to harsh conditions. Therefore selecting the right support is a critical step in designing effective immobilized enzyme systems.

An ideal support should possess several key properties. It must be chemically inert, meaning it does not react with or deactivate the enzyme. It should be hydrophilic, or water-attracting, to facilitate interaction with aqueous reaction media. The support must also resist microbial attack, which could degrade the support and compromise enzyme activity (Ogunbadejo & Al-Zuhair, 2021). Biocompatibility is important too, especially when enzymes are used in medical or food applications, to ensure safety. Cost is another major factor; supports should be inexpensive to keep overall process expenses manageable. While no single support fulfills all these needs perfectly for every enzyme and application, the selected support should have features such as reactive functional groups to bind enzymes securely, easy regeneration options for reuse, and a nontoxic nature to avoid introducing contaminants (Datta et al., 2017). Choosing the right support is inherently complex. It involves considering multiple factors, including the specific enzyme involved, the conditions under which the reaction occurs, and the properties of the reaction media. The physical and chemical traits of the support, like pore size distribution and surface chemistry, are also vital. For example, the support pore size must be compatible with the enzyme's size to prevent leakage or inaccessibility. Surface chemistry influences how well the enzyme can be immobilized through covalent bonds or physical adsorption. These factors collectively determine how long the enzyme remains active, how efficiently it catalyzes reactions, and how easy it is to handle during repeated cycles. An iterative cycle that can be followed in developing a robust immobilized CGTase system on any support is depicted in Fig. 10.2

Supports can be classified into different categories based on their chemical composition, physical nature, and application history. Chemically, supports are divided into organic and inorganic types. Organic supports often include materials like cellulose, agarose, or synthetic polymers, which are popular in various industries. Inorganic supports include materials such as silica, metal oxides, or glass, valued for their stability and resistance to extreme conditions (Sobral and Rodrigues, 2003). In terms of physical nature, supports are categorized as natural or synthetic. Natural supports, like lignocellulosic materials, are inexpensive and readily available. Synthetic supports, such as polystyrene or polyacrylamide beads, offer better control over surface properties and pore size.

Another classification considers how commonly the support has been used in enzyme immobilization. Conventional supports include silica gels and activated carbons, which have a long history of successful application. Metal-organic frameworks (MOFs) are a newer class of supports with highly ordered structures, offering high surface areas and tunable pore sizes. Nanomaterials, such as nanoparticles or nanotubes, have gained interest due to their unique properties like large surface area and high reactivity (Ogunbadejo, & Al-Zuhair, 2021, 2023;

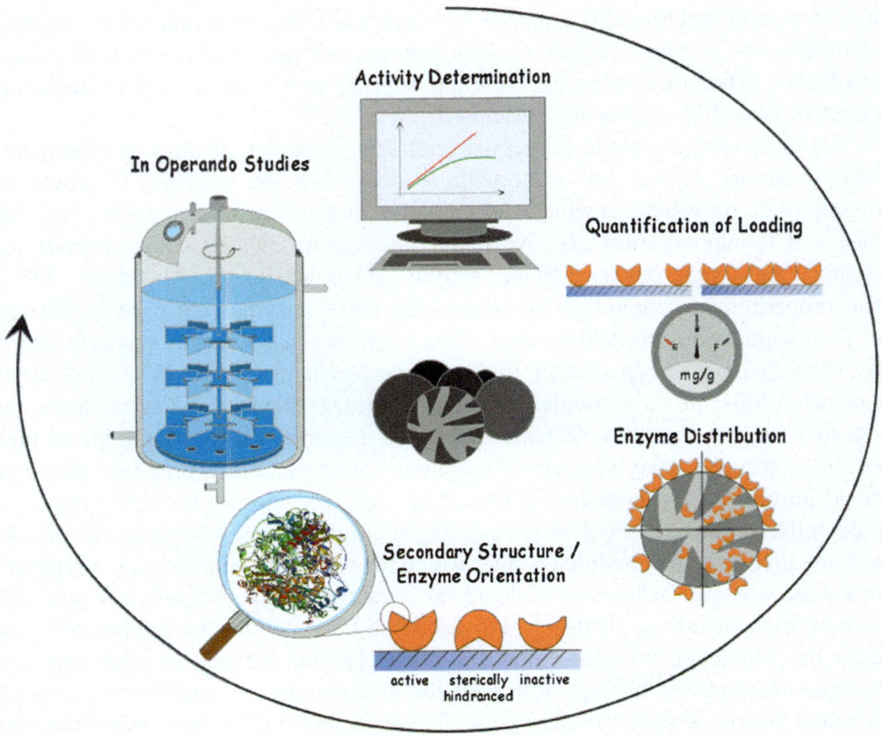

FIGURE 10.2

Cycle for optimizing immobilized enzyme (Bolivar and Eisl, 2016).

Ogunbadejo et al., 2024). Focusing on immobilization of CGTase, the support materials used vary widely depending on factors like application needs and enzyme stability requirements. These materials are chosen based on their proven effectiveness, ease of functionalization, and availability. The subsequent sections describe the different types of support materials most commonly used for immobilizing CGTase, highlighting their advantages and limitations in various industrial or research settings.

Conventional supports

Some of the supports that have been used for CGTase immobilization include porous glass, alumina, purified sand, alginate, and polystyrene. The attachment of biologically active molecules to porous glass beads was developed from its resistance to microbial attack and ease of regeneration, which can be achieved by heating to remove all organic material. In order to determine the best immobilization method for CGTase on porous glass, several actions were performed on the surface of the porous glass (Steighardt & Kleine, 1993). First, the porous glass beads (Coming CPG 10, 200 mesh, 2000 Å pore diameter) were dried by heating at 500°C for 6 h (Robinson et al., 1971), followed by the introduction of 2.0 U/g of CGTase to the dried porous glass and the immobilization by adsorption was carried out at 25°C for 2 h. the result was compared to covalent binding of CGTase onto glutaraldehyde-activated spherical porous glass (trisoperl), which had been treated with 3-aminopropyltriethoxysilane (APTES). Covalent coupling of CGTase to trisoperl performed better, showing about 25% CGTase activity after immobilization compared to 7% via absorption.

In the search for cheap and readily available inert support that won't require the activation step, which is necessary during CGTase immobilization onto inorganic carriers, purified sea sand was used (Iyer et al., 2003). Purified sea sand was synthesized by neutralizing aqueous alkali metal silicate with acid, followed by calcination in order to eliminate trace metals. Its high mechanical stability and absence of mineral contaminants like barium and aluminum make purified sea sand to find application in column packing for chromatography and pharmaceutical applications, respectively. Another important factor that makes it suitable as a CGTase immobilization support is its low cost compared to other matrices. Maximum adsorption of CGTase on purified sea sand was attained after a 4 h incubation period, and it was observed that CGTase adsorption increased with pH, reaching a peak at pH 7.0, and then declined as pH rose. Conversely, the results in the study showed that adsorption decreased with increasing temperature between 5°C–20°C, due to the weakening of dipolar interactions between the silanol groups on the purified sea sand and the enzyme. The optimum CGTase adsorption of 4.88 U was attained, representing 98% of the starting CGTase activity (Iyer et al., 2003).

The synthetic adsorption resin made from a polystyrene/divinylbenzene matrix, sold under the trade name DIAION HP-20, is produced by Mitsubishi Chemical Industrial Company in Tokyo, Japan (Kato & Horikoshi, 1984). This resin is known for its large surface area and its ability to have a controlled pore size distribution. These features make it an ideal support material for enzymes like CGTase. The resin's large surface area means it can attach more enzyme molecules, which can improve overall enzyme activity. Its controlled pore size allows for better immobilization and easier access for substrates to reach the active sites of the enzyme. Because of these qualities, DIAION HP-20 has become a popular choice in enzyme immobilization processes. DIAION HP-20 offers several advantages when used with CGTase. One of these benefits is that the enzyme can attach directly to the resin, simplifying the process of immobilization. This direct attachment not only saves time but also ensures that the enzyme remains stable and active. When the enzyme begins to lose activity over time, the resin can be easily reactivated. This makes it cost-effective, as the immobilized enzyme can be reused multiple times without needing complete replacement. Additionally, the resin's optimal pH of around 9.0 plays a crucial role in enzyme stability and activity. It helps prevent the retrogradation of starch, a process where starch molecules re-aggregate and become less soluble. This pH level, therefore supports longer operation times and maintains the efficiency of starch processing through CGTase's activity. Research has shown that CGTase derived from alkalophilic *Bacillus* sp. No. 38-2 (ATCC 21783) can be immobilized on DIAION HP-20 successfully (Kato & Horikoshi, 1984). During immobilization, the enzyme binds firmly to the resin, which results in notable stability improvements. Tests demonstrated that the immobilized CGTase remained stable at temperatures up to 65°C. It also maintained its activity at a pH of 8.0 during an incubation period of 20 min. These conditions point to a significant enhancement in enzyme resilience compared to its free form, which often loses activity more quickly under high temperatures or nonideal pH levels. After immobilization, the acid-CGTase activity disappears or becomes undetectable. This suggests that the enzyme's properties changed once it attaches to the resin, possibly due to structural modifications or substrate accessibility. This change does not hinder the enzyme's use; rather, it indicates that the immobilized form might favor specific reactions or stabilize certain enzyme functions better than the soluble form. Such characteristics make DIAION HP-20 an attractive support material in industrial applications. It simplifies enzyme reuse, extends operational life, and enhances enzyme stability. These benefits are especially valuable in large-scale starch processing, where consistent enzyme performance is critical. The stability at higher temperatures and specific pH ranges can lead to increased process efficiency. Choosing the right support, like DIAION HP-20, improves overall productivity and reduces costs. Its ability to allow for easy enzyme reactivation and long-term stability makes it suitable for continuous industrial operations involving CGTase.

There has also been growing interest in using natural supports as immobilization matrices. One of the cheap, readily available, and renewable derivatives of plants is the vegetable sponge. The loofa sponge, made from matured dried fruit of *Luffa cylindrica*, possessing an open fibrous network with interconnecting voids, can provide rapid contact between immobilized cells and the surrounding aqueous medium. It is nontoxic, easy to use, and is very stable during repeated usage (Iqbal et al., 2005; Lim et al., 2021). The loofa sponge was produced by cutting the loofa into disks about 2.5 cm in diameter and 2–3 mm thick, soaked in boiling water for 30 mins, properly cleaned with tap water, and put in distilled water, which was replenished three to four times under 24 h. The loofa disks were then oven-dried at 70°C before it was used as immobilization matrix (Iqbal et al., 2005). Another natural material with a large amount of free hydroxyl groups on the surface is Curdlan, which is a high molecular weight polymer of glucose, made by pure culture fermentation from a nonpathogenic and nontoxicogenic strain of *Agrobacterium biobar* or *Agrobacterium radiobacter*. The presence of enormous hydroxyl groups has made it a suitable natural candidate for enzyme immobilization since the free hydroxyl groups can be activated to make them accessible for enzymes (Nishinari and Hirashima, 1998; Saudagar & Singhal, 2004).

Another material that has been utilized for CGTase immobilization is polyethylene films. Prior to the introduction of CGTase onto the surface, carboxylic acid was grafted to the surface using the radiation-induced graft copolymerization method, which is well-known for functionalizing different polymers (Choi et al., 2002).

Metal-organic frameworks

The use of MOFs as immobilization support for enzymes has gathered lots of interest recently due to the superior properties they possess compared to the conventional supports (Li et al., 2016). MOFs offer higher surface area, ability to tune the pore network, excellent thermal and chemical stability, and proven biocompatibility with most enzymes (Cohen et al., 2012). Some of the benefits that have been identified postimmobilization of enzymes on MOFs include improved thermal and pH stability, ease of enzyme recovery for multiple reuse cycles, provide a protective microenvironment for the enzymes, and also enhance enzyme efficiency. One of the promising classes of MOFs in biocatalysis is the Materials Institute Lavoisier materials (MIL-101), due to their good stability under different operating conditions (Gkaniatsou et al., 2020; Lebedev et al., 2005; Zare and Bordbar, 2018). MIL-101 is synthesized through the coordination of trivalent cations (e.g., Cr^{3+} or Al^{3+}) with carboxylic acid groups (e.g., using terephthalic acid) to produce materials with pore sizes around 2.9 to 3.4 nm (Zou et al., 2022).

The adsorption of CGTase on MIL-101 (Cr) was examined to evaluate how effectively the support could immobilize the enzyme. Results showed that the CGTase was able to reach a maximum binding capacity of 31.6 mg/g of support material. This rate of enzyme uptake significantly exceeded that of zeolite, a common support used in such applications, by more than 24%. The higher adsorption capacity of MIL-101 (Cr) mainly comes from its larger surface area, which provides more space for the enzyme to attach. Additionally, the surface of MIL-101 (Cr) contains many functional groups known to work well with enzymes. The most important among these are carboxylic groups (-COOH), which are highly compatible with enzyme molecules (Li et al., 2022; Ogunbadejo & Al-Zuhair, 2023). This compatibility makes MIL-101 (Cr) an excellent host for enzyme immobilization, as the presence of these groups enhances the binding process. Another factor that likely contributed to the enzyme's high uptake on MIL-101 (Cr) involves the electrostatic forces between the support and the enzyme. To explore this, zeta potential analysis was performed. Zeta potential measures the electrical charge on particle surfaces, which influences how particles interact. The enzyme CGTase displayed a slightly negative charge with a zeta potential of about -0.06 mV—almost neutral. In contrast, MIL-101 (Cr) had a positive charge, with a zeta potential of approximately 0.53 mV (Ogunbadejo and Al-Zuhair, 2023). This difference means that electrostatic attraction exists between the enzyme and the support, encouraging the enzyme to stick strongly to the surface. Such attractions can also trigger changes in the enzyme's structure.

To understand how the enzyme's structure might change, the secondary structure of CGTase was analyzed by examining the amide I band using Fourier transform infrared spectroscopy. This band reveals details about protein folding, including the presence of beta-sheets. For the free CGTase in phosphate buffer, beta-sheets were the dominant structure, making up about 56% of the enzyme's secondary structure. After immobilization on MIL-101 (Cr), the proportion of beta-sheets increased significantly, reaching 84.1%. This suggests that when CGTase binds to MIL-101, it tends to form aggregates on the surface (Fu et al., 1999). These aggregates promote interactions between enzyme molecules, which can influence how the enzyme behaves. The increased beta-sheet content points to a more ordered and possibly more stable protein structure upon attachment. The combination of a high surface area, enzyme-compatible carboxylic groups, and electrostatic attraction explains how MIL-101 (Cr) achieves such a high enzyme loading capacity. These features create an ideal environment for CGTase to attach strongly and maintain its structure. The shift towards more beta-sheets indicates that the enzyme adapts its shape, possibly resulting in improved stability or activity.

To improve how starch molecules reach the active sites of the enzyme CGTase, different researchers have looked for ways to reduce internal diffusion issues. These issues occur when molecules get trapped inside the support material, slowing down the reaction. One promising solution is the use of two-dimensional (2D) MOF materials. These are highly porous structures that can provide clear pathways for molecules to access enzymes easily. By using 2D MOFs, researchers aim to create a more open environment that allows starch to interact freely with CGTase, increasing the efficiency of the process. A specific example of this approach involved the use of Ca-TMA, a 2D MOF. This material was made by mixing an ethanolic solution of trimesic acid with a calcium nitrate solution. The process results in a layered, flat structure that can host enzymes like CGTase. The design of Ca-TMA makes it well-suited for enzyme immobilization because the thin layers provide ample space for the enzyme and the substrate to interact efficiently. This reduces the chances of molecules getting trapped or slowed down inside the support. When Ca-TMA was used to immobilize CGTase, it showed excellent results. The enzyme retained a high level of activity even after being fixed to the support with minor changes to the secondary structures of the CGTase. This indicates that Ca-TMA does not cause significant damage or change to the enzyme's structure. Detailed analysis of the enzyme's conformational structure, especially focusing on the amide I region (a part of the protein structure that reveals folding and stability), confirmed that the enzyme largely maintained its secondary structures (Ogunbadejo et al., 2024). This is important because enzymes depend on their structure to function properly, and support materials that preserve this structure will improve the enzyme's performance.

The benefit of using Ca-TMA and similar 2D MOF supports lies in their ability to hold the enzyme in a stable and active form while allowing easy access for substrate molecules. These materials are designed to minimize internal diffusional barriers, making the enzymatic process faster and more efficient. This approach could lead to improved industrial applications where enzyme stability and activity are critical, such as in starch processing or biofuel production. Overall, the use of 2D MOF materials like Ca-TMA represents a significant step forward in enzyme immobilization technology. It combines high enzyme activity with structural support that maintains the enzyme's natural shape. This balance could open new pathways for creating more effective biocatalysts in various industrial processes.

Nanomaterials/nanoparticles

Recent progress in nanostructured materials has led to a drop in costs for products such as nanocomposites, nanoparticles, nanotubes, and nanofibers (Wang, 2006). This makes them more practical for use in various applications, especially as supports for enzyme immobilization. These nanoscale materials are valued for their outstanding properties, which include a very high surface area to volume ratio, allowing more enzymes to attach (Mateo et al., 2007). They also support effective enzyme loading and pose less resistance to the movement of molecules through them. The tiny pores within nanomaterials create a confining environment that can trap enzymes, while the constant, random motion of nanoparticles—known as Brownian motion—helps keep enzymes in close contact with the support (Ibrahim et al., 2014). Their ability to self-assemble into structured shapes further enhances their usefulness. Because of these qualities, nanomaterials are playing an increasingly important role in enzyme immobilization, especially in processes where enzyme reuse and stability are crucial.

To make the separation of nanomaterials from the solution easier, researchers have also turned to magnetic nanoparticles. These materials respond quickly to magnetic fields, allowing their separation from solutions without complex processes. Among these, magnetic nanoparticles made of magnetite (Fe_3O_4) have gained attention. Magnetite is favored because it exhibits superparamagnetism, meaning it can be magnetized when needed but shows no residual magnetism when the field is removed. This property helps prevent clumping and makes handling easier. Additionally, magnetite is nontoxic, small in size, and has high field reversibility—features that make it suitable for enzyme applications. Researchers often functionalize magnetite nanoparticles to improve their affinity for enzymes. One common approach involves coating the magnetite with a layer of silica that has been chemically treated with amino groups, producing $Fe_3O_4@NH_2\text{-}SiO_2$ nanospheres. To create these structures, Fe_3O_4 nanoparticles are dispersed into a mixture of ethanol, water, and concentrated ammonia, followed by the addition of tetraethyl orthosilicate, which forms a silica shell around the particles. To introduce amino groups on the surface, a silane compound, for example APTES, is added to the solution (Ibrahim et al., 2013). This process results in a silica layer rich in silanol groups, which serve as points for further chemical bonding. Next, these amino-functionalized nanospheres are activated using glutaraldehyde. Glutaraldehyde acts as a link, binding to the amino groups and creating reactive sites that can attach enzymes such as CGTase. The process has been optimized by studying several factors, and the experimental results have shown that the best results occur when 5% glutaraldehyde solution is used, and the immobilization process is carried out for about 2 h (Ibrahim et al., 2013). Under these conditions, the enzyme immobilization yield reached approximately 93.4%, with a loading efficiency of around 94.5%. This means most of the enzymes present in the solution successfully attached to the support. Further studies explored how different initial amounts of CGTase affected the process. As the amount of free enzyme increased, the activity on the support also increased, reaching an optimal load at about 470 mg. Beyond this point, enzyme activity started to decline. This drop was likely caused by the enzyme molecules crowding the support's surface, leading to steric hindrance. Also, unfavorable interactions between enzyme molecules could reduce overall activity. As a result, the amount of enzyme loaded onto the $Fe_3O_4@NH_2\text{-}SiO_2$ support reached approximately 47 μg/mg of support. These advancements highlight the potential of nanomaterials, especially magnetic ones, to improve enzyme immobilization. The careful modification of nanostructures can significantly boost enzyme loading, activity, and ease of recovery. This approach supports the development of more efficient, reusable enzyme systems that can be used in various industrial processes.

In another related study, Abdelnasser et al. (Ibrahim et al., 2014) focused on anchoring CGTase onto specially designed double-shell silica nanospheres. These nanospheres have a core-shell structure composed of magnetic and silica layers that can be functionalized for better enzyme attachment. The synthesis process started with creating magnetite nanoparticles, which act as the core. Next, a dense silica layer was coated onto these particles to provide stability and a surface suitable for further modification. After this, another layer of mesoporous silica was added. The mesoporous silica's tiny pores increase the surface area and allow for more efficient enzyme immobilization. Before attaching the CGTase enzyme, an activation step was performed on these nanospheres. This step involved preparing the surface so that the enzyme could anchor effectively. The enzyme was then attached through two main methods: physical adsorption, where enzymes stick loosely to the surface, and covalent binding, which involves forming stronger chemical bonds. Results clearly showed that covalent binding offered many advantages. It achieved a higher immobilization yield, meaning more enzyme molecules were successfully attached, and a greater loading efficiency, indicating that a larger percentage of the enzyme was retained on the support. Specifically, the nanomaterial designated as mag@d-SiO_2@m-SiO_2-NH_2, which features amino groups attached within the silica gel using anionic surfactant (APTES), showed outstanding performance. When amino groups were anchored using anionic surfactant, the immobilization yield reached 98.1%, and the loading efficiency was 96.2%. These values were significantly higher than when amino groups were introduced with cationic surfactants. The enhanced performance with anionic surfactants was primarily due to the thicker silica shell created during synthesis. This thicker layer supported more amino groups, which in turn provided more sites for enzyme attachment. This clearly shows that the type of surfactant used plays a crucial role. The anionic surfactant created a more favorable environment for enzyme binding, thanks to the larger shell thickness. It also involved a costructure directing agent that promoted electrostatic interaction. This agent helped attract the negatively charged silica layer to the surfactant molecules, strengthening the attachment points of the enzyme. As a result, the nanomaterials could hold more enzymes on their surface. The capacity of these nanospheres to hold enzymes was also notable. The activated nanomaterials could carry up to 58 μg protein/mg carrier (Ibrahim et al., 2014). This is among the highest values reported in scientific studies for enzyme immobilization on silica supports. Such high loading capacity indicates these nanospheres are highly efficient supports for enzymes like CGTase. The success of this nanomaterial largely depends on the way it was functionalized. In this case, the amino groups were introduced during the synthesis phase, not added afterward. This approach allowed for a denser and more uniform distribution of amino groups on the surface. The greater density of free amino groups provided more opportunities for enzyme molecules to attach. This method results in stronger and more stable enzyme-support interactions, which can improve the overall activity and durability of the immobilized enzyme. This nanomaterial demonstrates that in situ functionalizatio-n—adding functional groups during synthesis—can significantly improve enzyme immobilization. Its design leads to better enzyme loading, higher retention rates, and potentially greater stability (Deng et al., 2005). These factors make it a promising platform for applications that require robust enzyme activity, such as industrial biocatalysis or biosensors. The findings from this study highlight the importance of carefully choosing surfactants and synthesis strategies to optimize enzyme-support systems.

Methods to enhance cyclodextrin glucanotransferases stability

Protein orientation

Studies controlling the enzyme orientation to enable site-specific immobilization on the support surface have demonstrated that certain attachment sites can optimize the stability of immobilized enzymes by minimizing undesired enzyme–surface interactions as well as the stearic effects of the enzyme active sites after immobilization (Schöffer et al., 2017; Wu et al., 2015). Epoxy supports have proved to be very convenient to stabilize enzymes via multipoint immobilization, with the first step involving adsorption of the protein, followed by formation of covalent bonds with the epoxy groups. It has been established that thiol-functionalized epoxy supports offer site-specific covalent immobilization of enzymes (Grazú et al., 2005). This idea was employed for reversible immobilization of CGTase onto thiosulfinate-agarose, a thiol-containing support which forms disulfide bonds between the thiol-reactive groups on the support and the thiol groups present in the enzyme (Viera et al., 2012). Prior to immobilization, the enzyme was first reduced using DTT due to the lack of reactivity or activity between the enzyme and the support. It was discovered that the optimal condition for enzyme reduction was 30 min incubation period, using 100 mM DTT in 100 mM phosphate buffer (pH 8.0), which led to a two-fold increment in the thiol groups on the enzyme. During the immobilization of the reduced CGTase onto the thiolsulfinate-agarose, different conditions were used so as to work out the optimal conditions needed. The thiolation step improved the immobilization yield to 90% under the optimized conditions (Viera et al., 2012).

Protein engineering

Another method of enhancing CGTase activity and stability is through protein engineering. Several methods have been investigated for protein engineering of CGTase, which include directed evolution, site-directed mutagenesis, and a mix of site-directed mutagenesis with X-ray crystallography (Leemhuis et al., 2010; Wind et al., 1998). Directed evolution is a powerful method that can be used in altering the properties of enzymes, such as resistance to denaturing, tightness of binding, or reaction specificity (Ghadessy et al., 2004; Kazlauskas, 2005). The availability of mechanistic and structural information about CGTase has allowed the generation of saturated sets of mutant amino acids for initial evolutionary studies via semirational mutagenesis.

Chemical modification

The numerous applications of CGTase have stimulated interest in improving its stability and functionality. One of the approaches that can be used to enhance CGTase activity, selectivity, and specificity is by chemical modification. Chemical modification serves as a versatile process that allows for easy preparation of modified biomolecules in large quantities from commercially available materials (Alcalde et al., 1999).

Succinylation, a method of transferring a succinyl group to the lysine residue of an enzyme, has been studied on CGTase from *Thermoanaerobacter* sp. 501 (Alcalde et al., 2001). The modification of the amino group on CGTase was done using succinic anhydride in order to understand if it significantly affects CGTase properties. An excess molar amount of succinic anhydride was used in the reaction because the reaction with CGTase occurs competitively with its hydrolysis to succinic acid. The amount of amino groups modified by succinic acid out of the 25 amino groups (24 lysines and the N-terminal α-amino group -Ala1-) present in CGTase from *Thermoanaerobacter* sp. 501 was determined using MALDI-TOF mass spectrometry. It was discovered that 11 succinic anhydride molecules were incorporated into CGTase as each succinylation site added 100.1 Da to the mass of the native CGTase, which indicates that 11 amino groups in CGTase were modified into stable derivatives, causing a shift in the isoelectric point of CGTase from pI 6.3 (neutral) to 5.2. Succinylation of CGTase led to a 1.3-fold decrease in specific activity for cyclization, 2.4 and 2.1 times lower activities for coupling of α- and β-cyclodextrin (CD), respectively, and a slightly reduced saccharification activity, while the disproportionation reaction showed higher activity postmodification. Further probe into the behavior of the modified CGTase during disproportionation reaction revealed that the chain length of the malto-oligosaccharide acceptor affected the disproportionation activity, with an increase in the degree of polymerization of the acceptor leading to a decrease in disproportionation activity, which can be attributed to the steric effect at the acceptor binding site (Alcalde et al., 2001).

In another study on chemical modification, the carboxylic residues (43 Asp and 17 Glu) present in CGTase from *Thermoanaerobacter* sp. were altered using glycine ethyl ester in the presence of acarbose in order to study the effect of changes in the hydrogen bonds during enzyme-substrate interaction on product specificity. Amino acid analysis was used to estimate the amount of glycine residues incorporated per mole of CGTase, and the degree of substitution was determined to be 15%, meaning that nine negatively charged carboxylic groups were transformed into neutral and stable Asp-Gly or Glu-Gly moieties. Analysis of the product mixture using the modified CGTase revealed that the initial production of β-CD and γ-CD was reduced compared to the native CGTase, while there were no noticeable changes in the hydrolytic and disproportionation activities. The study also showed that chemical modification of CGTase affected the ratio of each type of CD in the product mixture.

The effect of modifying the amino groups of CGTase from *Thermoanaerobacter* using acetic anhydride on its specificity has also been analyzed (Alcalde et al., 1999). Prior to the chemical modification, CGTase was purified by affinity chromatography, followed by the reaction with acetic anhydride, which converts the positively charged ε-amino groups of Lys and the N-terminal α-group to zero charge. As short-chain acid anhydrides show low selectivity during acylation of amino groups, modification of other functional groups such as thio groups of Cys, hydroxylic groups of Tyr, and amino groups of His could also occur (Geoghegan, 1996; Means & Feeney, 1990). The degree of substitution was determined fluorimetrically to be 85% using the fluorescamine method, which involves binding fluorescamine to the free amino groups on the native CGTase. The high degree of acylation, that is 85% obtained, indicated that the properties of the modified amino groups on CGTase should be homogeneous. The acetylated CGTase showed higher saccharifying activity of 127 U/mg compared to 82 U/mg of the native CGTase due to the high presence of short-chain gluco-oligosaccharides in the product mixture, indicating that the modified CGTase behavior had shifted towards α-amylase activity (Alcalde et al., 1999). Surprisingly, chemical modification led to a decrease in cyclization, coupling, and disproportionation activity. The loss in cyclization activity after acetylation of the residues on CGTase could be attributed to the changes made to the subsites in the active center, leading to a reduction in interactions between the glucose units and the subsites (Wind et al., 1998). The results showed that CGTase, which is a transglycosylase, can be effectively converted into an efficient hydrolase.

Application of ultrasound

One promising strategy to improve the activity and stability of the CGTase enzyme is the application of ultrasonication. This technique uses sound waves at frequencies higher than what humans can hear, typically above 20 kHz, to influence enzyme performance. When applied under certain conditions, ultrasonication can induce changes in enzyme structure through a process known as acoustic cavitation. This phenomenon occurs when sound waves cause the formation, growth, and collapse of tiny bubbles in a liquid. The intense energy released during bubble collapse generates localized high temperatures and pressure, which can temporarily alter the enzyme's shape and conformation (Hou et al., 2019; Pacheco et al., 2023). These structural modifications often lead to enhanced enzyme stability or increased activity, making ultrasonication a useful tool in enzyme engineering and process optimization. Studies have explored how ultrasonication affects CGTase activity and stability under various experimental settings. Jeferson et al. started by identifying the enzyme's optimal temperature by varying parameters such as ultrasonication time, temperature, and pH levels to observe how these factors influence the enzyme's performance. The experimental process followed is summarized in Fig. 10.3 which outlines the key steps taken to evaluate the effects of ultrasonication, including initial temperature testing, subsequent adjustments of sonication duration, and pH shifts before measuring enzyme activity. This detailed approach helps isolate the impacts of each variable, ensuring an accurate assessment. To quantify the effects on CGTase, the relative enzymatic activity, which is a ratio that compares the activity of CGTase treated with ultrasonication to that of untreated enzyme, both measured under the same conditions, and the storage

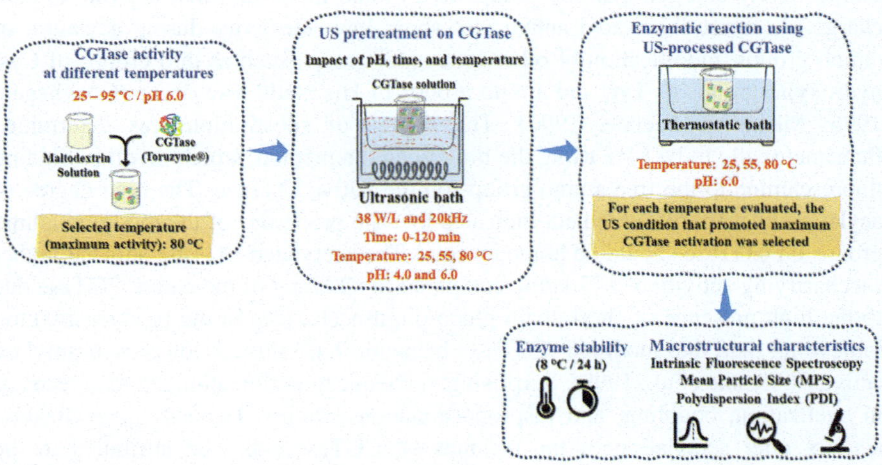

FIGURE 10.3

Methodology for evaluating ultrasound effect on cyclodextrin glucanotransferases (*CGTase*) (Cunha et al., 2024).

stability over time after sonication were used. Results from these experiments show notable improvements at lower, suboptimum temperatures. For instance, when ultrasonication was applied at 25°C, which is below the enzyme's usual optimal temperature, the activity increased significantly. The highest increment was observed with a sonication time of about 30 min, resulting in an activity increase of 93% at 25°C and 68% at 55°C. At the optimal temperature of 80°C, the increase was 31% with a sonication time of 60 min. This pattern suggests that ultrasonication is particularly effective at enhancing activity at temperatures below the enzyme's optimum, likely because the enzyme's natural catalysis is less efficient under those conditions (Cunha and Pacheco, 2024). The explanation for these differences lies in the enzyme's energy requirements. At suboptimal temperatures, enzyme molecules have less energy and are less likely to interact effectively with substrates (Patist & Bates, 2008). Ultrasonication appears to expose more of the enzyme's active sites, the regions responsible for catalysis, making it easier for the enzyme to bind and react with substrates. This effect results in an apparent boost in activity. At higher, optimal temperatures, the enzyme already operates at high efficiency. Modifications caused by ultrasonication may have less impact because the enzyme's activity is already near its maximum. In such conditions, structural changes from sonication might even be slightly disruptive, although the overall effect remains positive. These findings highlight ultrasonication as a valuable tool in improving CGTase activity. It offers a way to boost enzyme activity, especially when operating under conditions where the enzyme is naturally less active. By adjusting sonication time and temperature, scientists can tailor treatments to maximize stability and performance.

Benefits of immobilization for industrial processes

The search for more efficient and scalable methods to use enzymes in industrial processes has become very important. Industries need ways to reduce costs and improve processes. Enzymes can be expensive and slow to handle. Finding ways to make them more practical for large-scale use is essential. One of the best methods to address these challenges is enzyme immobilization. This technique involves attaching enzymes to a solid support or structure. It transforms free enzymes into immobilized enzymes, making them easier to manage during production.

Enzyme immobilization has many benefits. For example, it simplifies the handling process. Instead of dealing with enzymes floating freely in a mixture, they remain attached to a solid, making them easier to separate from the product. Reusing enzymes becomes possible, which helps cut costs. A single batch of enzymes can be used multiple times, lowering the need for fresh enzymes in each run. This reusability is vital in industries like pharmaceuticals, food, and biofuels. Immobilized enzymes also give better control over chemical reactions. Producers can more easily regulate factors such as temperature, pH, and reaction time. This control improves the quality and consistency of the final product. In addition, enzyme immobilization increases both operational stability—meaning enzymes last longer during processing—and storage stability, so enzymes stay effective for longer periods when stored.

This method also simplifies large-scale production. Immobilized enzymes make it easier to scale up processes from small lab settings to industrial levels. They can operate continuously in reactors, increasing production capacity. This shift allows industries to meet high demand without sacrificing quality or efficiency. Another crucial advantage is the easy separation of enzymes from the final product. When enzymes are immobilized, they remain bound to a support material, making their removal straightforward. This process reduces protein contamination, which is important for products like food or pharmaceuticals, where purity is critical. It also helps meet safety and regulatory standards more easily. For example, in the production of high-fructose corn sirup, immobilized enzymes make the process cleaner and faster. They help convert starch into sugars more efficiently and with less waste. Similarly, immobilized CGTase is used in the synthesis of CDs, which are valuable in the food and pharmaceutical industries for improving the solubility and stability of active compounds. In summary, immobilizing CGTase offers multiple advantages that help industries produce products more efficiently and cost-effectively. By making enzyme handling easier, enabling reuse, and controlling reactions better, this technique supports large-scale, sustainable, and high-quality production.

References

Alcalde, M., Plou, F. J., Andersen, C., Martín, M. T., Pedersen, S., & Ballesteros, A. (1999). Chemical modification of lysine side chains of cyclodextrin glycosyltransferase from Thermoanaerobacter causes a shift from cyclodextrin glycosyltransferase to α-amylase specificity. *FEBS Letters, 445*(2-3), 333–337. https://doi.org/10.1016/s0014-5793(99)00134-9.

Alcalde, M., Plou, F. J., Martín, M. T., Valdés, I., Méndez, E., & Ballesteros, A. (2001). Succinylation of cyclodextrin glycosyltransferase from Thermoanaerobacter sp. 501 enhances its transferase activity using starch as donor. *Journal of Biotechnology, 86*(1), 71–80. https://doi.org/10.1016/s0168-1656(00)00422-3.

Algieri, C., Donato, L., & Giorno, L. (2017). Tyrosinase immobilized on a hydrophobic membrane. *Biotechnology and Applied Biochemistry, 64*(1), 92–99. https://doi.org/10.1002/bab.1462.

Batista-Viera, F., Manta, C., & Carlsson, J. (1994). Solid-phase thiolsulfinates for the reversible immobilization of thiols. *Applied Biochemistry and Biotechnology, 44*(1), 1–14. https://doi.org/10.1007/bf02921846.

Blanco, R. M., Calvete, J. J., & Guisán, J. M. (1989). Immobilization-stabilization of enzymes; variables that control the intensity of the trypsin (amine)-agarose (aldehyde) multipoint attachment. *Enzyme and Microbial Technology, 11*(6), 353–359. https://doi.org/10.1016/0141-0229(89)90019-7.

Bolivar, J. M., Eisl, I., & Nidetzky, B. (2016). Advanced characterization of immobilized enzymes as heterogeneous biocatalysts. *Catalysis Today, 259*, 66–80. https://doi.org/10.1016/j.cattod.2015.05.004, http://www.sciencedirect.com/science/journal/09205861.

Bon, E., Freire, D. G., Mendes, M. F., & Soares, V. F. (1984). Immobilization of glucoamylase on inexpensive supports: A comparative evaluation. In *Biotechnology and Bioengineering Symposium*. John Wiley & Sons, 447–455 14.

Brena, B., González-Pombo, P., & Batista-Viera, F. (2013). *Immobilization of enzymes: A literature survey*. Springer Science and Business Media LLC, 15–31. 10.1007/978-1-62703-550-7_2 1051.

Carlsson, N., Gustafsson, H., Thörn, C., Olsson, L., Holmberg, K., & Åkerman, B. (2014). Enzymes immobilized in mesoporous silica: A physical–chemical perspective. *Advances in Colloid and Interface Science, 205*, 339–360. https://doi.org/10.1016/j.cis.2013.08.010.

Chiou, S. H., & Wu, W. T. (2004). Immobilization of *Candida rugosa* lipase on chitosan with activation of the hydroxyl groups. *Biomaterials, 25*(2), 197–204. https://doi.org/10.1016/S0142-9612(03)00482-4, http://www.journals.elsevier.com/biomaterials/.

Choi, S. H., Kim, M. S., Ryoo, J. J., Lee, K. P., Shin, H. D., Kim, S. H., & Lee, Y. H. (2002). Immobilization of a cyclodextrin glucanotransferase (CGTase) onto polyethylene film with a carboxylic acid group and production of cyclodextrins from corn starch using CGTase-immobilized PE film. *Journal of Applied Polymer Science, 85*(11), 2451–2457. https://doi.org/10.1002/app.10879.

Clark, D. S. (1994). Can immobilization be exploited to modify enzyme activity? *Trends in Biotechnology, 12*(11), 439–443. https://doi.org/10.1016/0167-7799(94)90018-3.

Cohen, S. M. (2012). Postsynthetic methods for the functionalization of metal-organic frameworks. *Chemical Reviews, 112*(2), 970–1000. https://doi.org/10.1021/cr200179u.

Costa, S., Azevedo, H.S., & Reis, R.L. (2005). Enzyme immobilization in biodegradable polymers for biomedical applications.

Cunha, J. S., Pacheco, F. C., Martins, C. C. N., Pacheco, A. F. C., Tribst, A. A. L., & Junior, B. R. C. L. (2024). Use of ultrasound to improve the activity of cyclodextrin glycosyltransferase in the producing of β-cyclodextrins: Impact on enzyme activity, stability and insights into changes on enzyme macrostructure. *Food Research International, 191*.

Datta, R., Anand, S., Moulick, A., Baraniya, D., Pathan, S. I., Rejsek, K., Vranova, V., Sharma, M., Sharma, D., Kelkar, A., & Formanek, P. (2017). How enzymes are adsorbed on soil solid phase and factors limiting its activity: A Review. *International Agrophysics, 31*(2), 287–302. https://doi.org/10.1515/intag-2016-0049.

Deng, H., Li, X., Peng, Q., Wang, X., Chen, J., & Li, Y. (2005). Monodisperse magnetic single-crystal ferrite microspheres. *Angewandte Chemie - International Edition, 44*(18), 2782–2785. https://doi.org/10.1002/anie.200462551.

Dwevedi, A. (2016). *Basics of enzyme immobilization*. Springer Nature, 21–44. https://doi.org/10.1007/978-3-319-41418-8_2.

Ferrarotti, S. A., Bolivar, J. M., Mateo, C., Wilson, L., Guisan, J. M., & Fernandez-Lafuente, R. (2006). Immobilization and stabilization of a cyclodextrin glycosyltransferase by covalent attachment on highly activated glyoxyl-agarose supports. *Biotechnology Progress, 22*(4), 1140–1145. https://doi.org/10.1021/bp0600740.

Fontananova, E., Bahattab, M. A., Aljlil, S. A., Alowairdy, M., Rinaldi, G., Vuono, D., Nagy, J. B., Drioli, E., & Di Profio, G. (2015). From hydrophobic to hydrophilic polyvinylidenefluoride (PVDF) membranes by gaining new insight into material's properties. *RSC Advances, 5*(69), 56219–56231. https://doi.org/10.1039/C5RA08388E.

Fu, K., Griebenow, K., Hsieh, L., Klibanov, A. M., & Langer, R. (1999). FTIR characterization of the secondary structure of proteins encapsulated within PLGA microspheres1An article of related interest has been published by Yang et al. in J. Pharm. Sci., 88(2), Feb. 1999, accepted Nov. 1998.1. *Journal of Controlled Release, 58*(3), 357–366. https://doi.org/10.1016/s0168-3659(98)00192-8.

Geoghegan, K. F. (1996). Modification of amino groups. *Current Protocols in Protein Science, 4*(1). https://doi.org/10.1002/0471140864.ps1502s04.

Ghadessy, F. J., Ramsay, N., Boudsocq, F., Loakes, D., Brown, A., Iwai, S., Vaisman, A., Woodgate, R., & Holliger, P. (2004). Generic expansion of the substrate spectrum of a DNA polymerase by directed evolution. *Nature Biotechnology, 22*(6), 755–759. https://doi.org/10.1038/nbt974.

Gimenez, G. G., Silva, R. M., Francisco, C. P., Rando, Fd. S., Dantas, J. H., De Souza, H. M., & Matioli, G. (2019). Immobilization of commercial cyclomaltodextrin glucanotransferase into controlled pore silica by the anchorage method and covalent bonding. *Process Biochemistry, 85*, 68–77. https://doi.org/10.1016/j.procbio.2019.07.002, www.elsevier.com/inca/publications/store/4/2/2/8/5/7.

Gkaniatsou, E., Ricoux, R., Kariyawasam, K., Stenger, I., Fan, B., Ayoub, N., et al. (2020). Encapsulation of microperoxidase-8 in MIL-101 (Cr)-X nanoparticles: Influence of metal–organic framework functionalization on enzymatic immobilization and catalytic activity. *ACS Appllied Nano Materials, 3*(4), 3233–3243.

Grazú, V., Abian, O., Mateo, C., Batista-Viera, F., Fernández-Lafuente, R., & Guisán, J. M. (2005). Stabilization of enzymes by multipoint immobilization of thiolated proteins on new epoxy-thiol supports. *Biotechnology and Bioengineering, 90*(5), 597–605. https://doi.org/10.1002/bit.20452.

Grazú, V., Ovsejevi, K., Cuadra, K., Betancor, L., Manta, C., & Batista-Viera, F. (2003). Solid-phase reducing agents as alternative for reducing disulfide bonds in proteins. *Applied Biochemistry and Biotechnology, 110*(1), 23–32. https://doi.org/10.1385/ABAB:110:1:23.

Guisan, J. M. (2006). *Immobilization of enzymes and cells*. Springer, 22.

Hernaiz, M. J., & Crout, D. H. G. (2000). Immobilization/stabilization on Eupergit C of the β-galactosidase from *B. circulans* and an α-galactosidase from *Aspergillus oryzae*. *Enzyme and Microbial Technology, 27*(1-2), 26–32. https://doi.org/10.1016/S0141-0229(00)00150-2.

Hou, F., Ma, X., Fan, L., Wang, D., Wang, W., Ding, T., Ye, X., & Liu, D. (2019). Activation and conformational changes of chitinase induced by ultrasound. *Food Chemistry, 285*, 355–362. https://doi.org/10.1016/j.foodchem.2019.01.180.

Ibrahim, A. S. S., Al-Salamah, A. A., El-Toni, A. M., El-Taye, M. A., & Elbadawij, Y. B. (2014). Cyclodextrin glucanotransferase immobilization onto functionalized magnetic double mesoporous core-shell silica nanospheres. *Electronic Journal of Biotechnology, 17*(2), 55–64. https://doi.org/10.1016/j.ejbt.2014.01.001, http://www.ejbiotechnology.info/index.php/ejbiotechnology/article/downloadFile/2014.01.001/1831.

Ibrahim, A. S. S., Al-Salamah, A. A., El-Toni, A. M., El-Tayeb, M. A., & Elbadawi, Y. B. (2013). Immobilization of cyclodextrin glucanotransferase on aminopropyl-functionalized silica-coated superparamagnetic nanoparticles. *Electronic Journal of Biotechnology, 16*(6). https://doi.org/10.2225/vol16-issue6-fulltext-8, http://www.ejbiotechnology.info/index.php/ejbiotechnology/article/viewFile/v16n6-8/1802.

Iqbal, M., Saeed, A., Edyvean, R. G. J., O'sullivan, B., & Styring, P. (2005). Production of fungal biomass immobilized loofa sponge (FBILS)-discs for the removal of heavy metal ions and chlorinated compounds from aqueous solution. *Biotechnology Letters, 27*(17), 1319–1323. https://doi.org/10.1007/s10529-005-0477-y.

Ivanova, V. (2010). Immobilization of cyclodextrin glucanotransferase from paenibacillus macerans atcc 8244 on magnetic carriers and production of cyclodextrins. *Biotechnology and Biotechnological Equipment, 24*, 516–528. https://doi.org/10.1080/13102818.2010.10817893.

Iyer, J. L., Shetty, P., & Pai, J. S. (2003). Immobilisation of cyclodextrin glucanotransferase from Bacillus circulans ATCC 21783 on purified seasand. *Journal of Industrial Microbiology and Biotechnology, 30*(1), 47–51. https://doi.org/10.1007/s10295-002-0009-x, www.springerlink.com/app/home/journal.asp.

Jamil, N., Che Man, R., Suhaimi, S., Md Shaarani, S., Iffah Mohd Arshad, Z., Kholijah Abdul Mudalip, S., & Zubaidah Sulaiman, S. (2018). Effect of enzyme concentration and temperature on the immobilization of cyclodextrin glucanotransferase (CGTase) on hollow fiber membrane. *Materials Today: Proceedings, 5*(10), 22036–22042. https://doi.org/10.1016/j.matpr.2018.07.065, https://www.sciencedirect.com/journal/materials-today-proceedings.

Katchalski-Katzir, E., & Kraemer, D. M. (2000). Eupergit® C, a carrier for immobilization of enzymes of industrial potential. *Journal of Molecular Catalysis - B Enzymatic, 10*(1-3), 157–176. https://doi.org/10.1016/S1381-1177(00)00124-7.

Kato, T., & Horikoshi, K. (1984). Immobilized cyclomaltodextrin glucanotransferase of an alkalophilic *Bacillus* sp. No. 38-2. *Biotechnology and Bioengineering, 26*(6), 595–598. https://doi.org/10.1002/bit.260260606.

Kazlauskas, R. J. (2005). Enhancing catalytic promiscuity for biocatalysis. *Current Opinion in Chemical Biology, 9*(2), 195–201. https://doi.org/10.1016/j.cbpa.2005.02.008, http://www.elsevier.com/locate/cbi.

Lebedev, O. I., Millange, F., Serre, C., Van Tendeloo, G., & Férey, G. (2005). First direct imaging of giant pores of the metal–organic framework MIL-101. *Chemistry of Materials: A Publication of the American Chemical Society, 17*(26), 6525–6527.

Leemhuis, H., Kelly, R. M., & Dijkhuizen, L. (2010). Engineering of cyclodextrin glucanotransferases and the impact for biotechnological applications. *Applied Microbiology and Biotechnology, 85*(4), 823–835. https://doi.org/10.1007/s00253-009-2221-3.

Li, P., Modica, J. A., Howarth, A. J., Vargas L., E., Moghadam, P. Z., Snurr, R. Q., Mrksich, M., Hupp, J. T., & Farha, O. K. (2016). Toward design rules for enzyme immobilization in hierarchical mesoporous metal-organic frameworks. *Chem, 1*(1), 154–169. https://doi.org/10.1016/j.chempr.2016.05.001, http://www.cell.com/chem/home.

Li, R., Liu, S., Zhou, X., Liu, H., Zhou, H., Wang, C., Liu, Y., & Zhang, X. (2022). Efficient immobilization of catalase on mesoporous MIL-101 (Cr) and its catalytic activity assay. *Enzyme and Microbial Technology, 156*, 110005. https://doi.org/10.1016/j.enzmictec.2022.110005.

Lim, C. H., Rasti, B., Sulistyo, J., & Hamid, M. A. (2021). Comprehensive study on transglycosylation of CGTase from various sources. *Heliyon, 7*(2). https://doi.org/10.1016/j.heliyon.2021.e06305, http://www.journals.elsevier.com/heliyon/.

Martín, M. T., Plou, F. J., Alcalde, M., & Ballesteros, A. (2003). Immobilization on Eupergit C of cyclodextrin glucosyltransferase (CGTase) and properties of the immobilized biocatalyst. *Journal of Molecular Catalysis B: Enzymatic, 21*(4-6), 299–308. https://doi.org/10.1016/S1381-1177(02)00264-3.

Mateo, C., Palomo, J. M., Fernandez-Lorente, G., Guisan, J. M., & Fernandez-Lafuente, R. (2007). Improvement of enzyme activity, stability and selectivity via immobilization techniques. *Enzyme and Microbial Technology, 40*(6), 1451–1463. https://doi.org/10.1016/j.enzmictec.2007.01.018.

Means, G. E., & Feeney, R. E. (1990). Chemical modifications of proteins: History and applications. *Bioconjugate Chemistry, 1*(1), 2–12. https://doi.org/10.1021/bc00001a001.

Migneault, I., Dartiguenave, C., Bertrand, M. J., & Waldron, K. C. (2004). Glutaraldehyde: Behavior in aqueous solution, reaction with proteins, and application to enzyme crosslinking. *Biotechniques, 37*(5), 790–802. https://doi.org/10.2144/04375rv01, https://www.future-science.com/journal/btn.

Mohamad, N. R., Marzuki, N. H. C., Buang, N. A., Huyop, F., & Wahab, R. A. (2015). An overview of technologies for immobilization of enzymes and surface analysis techniques for immobilized enzymes. *Biotechnology and Biotechnological Equipment, 29*(2), 205–220. https://doi.org/10.1080/13102818.2015.1008192, http://www.tandfonline.com/doi/pdf/10.1080/13102818.2015.1008192.

Nishinari, K., Hirashima, M., Takaya, T., & Miyoshi, E. (1998). *Rheological and DSC studies of aqueous dispersions and gels of curdlan*. Elsevier BV, 26–33. https://doi.org/10.1533/9781845698362.1.26.

Ogunbadejo, B. A., Aljahoushi, K. A., Alzamly, A., Greish, Y. E., & Al-Zuhair, S. (2024). Immobilization of cyclodextrin glycosyltransferase onto three dimensional- hydrophobic and two dimensional- hydrophilic supports: A comparative study. *Biotechnology Journal, 19*(1). https://doi.org/10.1002/biot.202300195, http://onlinelibrary.wiley.com/journal/10.1002/(ISSN)1860-7314.

Ogunbadejo, B. A., & Al-Zuhair, S. (2023). Bioconversion of starch to Cyclodextrin using Cyclodextrin glycosyltransferase immobilized on metal organic framework. *Biocatalysis and Agricultural Biotechnology, 53*. https://doi.org/10.1016/j.bcab.2023.102878, http://www.journals.elsevier.com/biocatalysis-and-agricultural-biotechnology/.

Ogunbadejo, B., & Al-Zuhair, S. (2021). MOFs as potential matrices in cyclodextrin glycosyltransferase immobilization. *Molecules (Basel, Switzerland), 26*(3), 680. https://doi.org/10.3390/molecules26030680.

Pacheco, A. F. C., De Souza, L. B., Paiva, P. H. C., Lelis, C. A., Vieira, E. N. R., Tribst, A. A. L., & Leite Júnior, B. Rd. C. (2023). Impact of ultrasound on pumpkin seed protein concentrate hydrolysis: Effects on Alcalase, protein, and assisted reaction. *Applied Food Research, 3*(1). https://doi.org/10.1016/j.afres.2023.100281, https://www.journals.elsevier.com/applied-food-research.

Panajotova, H., Zhekova, B., Dobrev, G., Dobreva, V., Ivanov, G., Mihalev, K., Dobrev, G., Kostov, G., & Balev, D. (2025). Conditions for immobilization of cyclodextrin glucanotransferase from *Bacillus megaterium* on DEAE-cellulose activated with glutaraldehyde. *BIO Web of Conferences, 170*, 02002. https://doi.org/10.1051/bioconf/202517002002.

Patist, A., & Bates, D. (2008). Ultrasonic innovations in the food industry: From the laboratory to commercial production. *Innovative Food Science & Emerging Technologies, 9*(2), 147–154. https://doi.org/10.1016/j.ifset.2007.07.004.

Prousoontorn, M. H., & Pantatan, S. (2007). Production of 2-O-α-glucopyranosyl l-ascorbic acid from ascorbic acid and β-cyclodextrin using immobilized cyclodextrin glycosyltransferase. *Journal of Inclusion Phenomena and Macrocyclic Chemistry, 57*(1-4), 39–46. https://doi.org/10.1007/s10847-006-9163-5.

Robescu, M. S., & Bavaro, T. (2025). A comprehensive guide to enzyme immobilization: All you need to know. *Molecules (Basel, Switzerland), 30*(4), 939. https://doi.org/10.3390/molecules30040939.

Robinson, P. J., Dunnill, P., & Lilly, M. D. (1971). Porous glass as a solid support for immobilisation or affinity chromatography of enzymes. *Biochimica et Biophysica Acta (BBA) - Enzymology, 242*(3), 659–661. https://doi.org/10.1016/0005-2744(71)90160-4.

Rojas, M. J., Amaral-Fonseca, M., Zanin, G. M., Fernandez-Lafuente, R., Giordano, R. D. L. C., & Tardioli, P. W. (2019). Preparation of crosslinked enzyme aggregates of a thermostable cyclodextrin glucosyltransferase from *Thermoanaerobacter* sp. Critical effect of the crosslinking agent. MDPI, Brazil. *Catalysts, 9*(2). https://doi.org/10.3390/catal9020120, https://www.mdpi.com/2073-4344/9/2/120/pdf.

Rueda, N., Dos Santos, J. C. S., Ortiz, C., Torres, R., Barbosa, O., Rodrigues, R. C., Berenguer-Murcia, Á., & Fernandez-Lafuente, R. (2016). Chemical modification in the design of immobilized enzyme biocatalysts: Drawbacks and opportunities. *Chemical Record (New York, N.Y.), 16*(3), 1436–1455. https://doi.org/10.1002/tcr.201600007, http://onlinelibrary.wiley.com/journal/10.1002/(ISSN)1528-0691.

Saifuddin, N., Raziah, A. Z., Junizah, A. R., & Urriolabeitia, E. P. (2013). Carbon nanotubes: A review on structure and their interaction with proteins. *Journal of Chemistry, 2013*(1). https://doi.org/10.1155/2013/676815.

Sakai, S., Yoshida, S., Mikuni, K., & Hara, K. (1991). Continuous conversion of maltose to glucose with immobilized glucoamylase. *Journal of the Japanese Society of Starch Science, 38*(1), 37–39. https://doi.org/10.5458/jag1972.38.37.

Sakai, S., Yamamoto, N., Yoshida, S., Mikuni, K., Ishigami, H., & Hara, K. (1991). Continuous production of glucosyl-cyclodextrins using immobilized cyclomaltodextrin glucanotransferase. *Agricultural and Biological Chemistry, 55*(1), 45–51. https://doi.org/10.1271/bbb1961.55.45.

Saudagar, P. S., & Singhal, R. S. (2004). Curdlan as a support matrix for immobilization of enzyme. *Carbohydrate Polymers, 56*(4), 483–488. https://doi.org/10.1016/j.carbpol.2004.03.015.

Schöffer, Jd. N., Matte, C. R., Charqueiro, D. S., de Menezes, E. W., Costa, T. M. H., Benvenutti, E. V., Rodrigues, R. C., & Hertz, P. F. (2017). Directed immobilization of CGTase: The effect of the enzyme orientation on the enzyme activity and its use in packed-bed reactor for continuous production of cyclodextrins. *Process Biochemistry, 58*, 120–127. https://doi.org/10.1016/j.procbio.2017.04.041, www.elsevier.com/inca/publications/store/4/2/2/8/5/7.

Schöffer, J.da N., Klein, M. P., Rodrigues, R. C., & Hertz, P. F. (2013). Continuous production of β-cyclodextrin from starch by highly stable cyclodextrin glycosyltransferase immobilized on chitosan. *Carbohydrate Polymers, 98*(2), 1311–1316. https://doi.org/10.1016/j.carbpol.2013.07.044.

Sheldon, R. A. (2011). Cross-linked enzyme aggregates as industrial biocatalysts. *Organic Process Research and Development, 15*(1), 213–223. https://doi.org/10.1021/op100289f.

Sheldon, R. A. (2007). Enzyme immobilization: The quest for optimum performance. *Advanced Synthesis and Catalysis, 349*(8-9), 1289–1307. https://doi.org/10.1002/adsc.200700082.

Sobral, K. C. A., Rodrigues, R. M. O., De Oliveira, R. D., De Moraes, F. F., & Zanin, G. M. (2002). Immobilization of cyclodextringlycosyltransferase (CGTase) from Bacillus firmus in commercial chitosan. *Journal of Inclusion Phenomena, 44*(1-4), 383–386. https://doi.org/10.1023/A:1023004927302.

Sobral, K. A., Rodrigues, R. O., Oliveira, R. D., Olivo, J. E., De Moraes, F. F., & Zanin, G. M. (2003). *Evaluation of Supports and Methods for Immobilization of Enzyme Cyclodextringlycosyltransferase*. Springer Science and Business Media LLC, 809–819. https://doi.org/10.1007/978-1-4612-0057-4_68 105–108.

Spagna, G., Barbagallo, R. N., Casarini, D., & Pifferi, P. G. (2001). A novel chitosan derivative to immobilize α-L-rhamnopyranosidase from *Aspergillus niger* for application in beverage technologies. *Enzyme and Microbial Technology, 28*(4-5), 427–438. https://doi.org/10.1016/S0141-0229(00)00340-9.

Steighardt, J., & Kleine, R. (1993). Production and immobilization of a proteinase-reduced cyclodextrin glycosyltransferase preparation. *Applied Microbiology and Biotechnology, 39*(1), 63–68. https://doi.org/10.1007/BF00166850.

Sulaiman, S., Mokhtar, M. N., Naim, M. N., Baharuddin, A. S., & Sulaiman, A. (2015). A Review: Potential usage of cellulose nanofibers (CNF) for enzyme immobilization via covalent interactions. *Applied Biochemistry and Biotechnology, 175*(4), 1817–1842. https://doi.org/10.1007/s12010-014-1417-x, http://www.springer.com/humana+press/journal/12010.

Sulaiman, S., Cieh, N. L., Mokhtar, M. N., Naim, M. N., & Kamal, S. M. M. (2017). Covalent immobilization of cyclodextrin glucanotransferase on kenaf cellulose nanofiber and its application in ultrafiltration membrane system. *Process Biochemistry, 55*, 85–95. https://doi.org/10.1016/j.procbio.2017.01.025, www.elsevier.com/inca/publications/store/4/2/2/8/5/7.

Talbert, J. N., & Goddard, J. M. (2012). Enzymes on material surfaces. *Colloids and Surfaces B: Biointerfaces, 93*, 8–19. https://doi.org/10.1016/j.colsurfb.2012.01.003.

Tardioli, P. W., Zanin, G. M., & De Moraes, F. F. (2006). Characterization of Thermoanaerobacter cyclomaltodextrin glucanotransferase immobilized on glyoxyl-agarose. *Enzyme and Microbial Technology, 39*(6), 1270–1278. https://doi.org/10.1016/j.enzmictec.2006.03.011.

Viera, S. E., Batista-Viera, F., & Ovsejevi, K. (2012). Development and characterization of a solid-phase biocatalyst based on cyclodextrin glucantransferase reversibly immobilized onto thiolsulfinate- agarose. *Applied Biochemistry and Biotechnology, 167*(1), 164–176. https://doi.org/10.1007/s12010-012-9686-8.

Wang, P. (2006). Nanoscale biocatalyst systems. *Current Opinion in Biotechnology, 17*(6), 574–579. https://doi.org/10.1016/j.copbio.2006.10.009.

Wind, R. D., Uitdehaag, J. C. M., Buitelaar, R. M., Dijkstra, B. W., & Dijkhuizen, L. (1998). Engineering of cyclodextrin product specificity and pH optima of the thermostable cyclodextrin glycosyltransferase from *Thermoanaerobacterium thermosulfurigenes* EM1. *Journal of Biological Chemistry, 273*(10), 5771–5779. https://doi.org/10.1074/jbc.273.10.5771.

Wu, J. C. Y., Hutchings, C. H., Lindsay, M. J., Werner, C. J., & Bundy, B. C. (2015). Enhanced enzyme stability through site-directed covalent immobilization. *Journal of Biotechnology, 193*, 83–90. https://doi.org/10.1016/j.jbiotec.2014.10.039, www.elsevier.com/locate/jbiotec.

Zare, A., Bordbar, A. K., Jafarian, F., & Tangestaninejad, S. (2018). Candida rugosa lipase immobilization on various chemically modified *Chromium terephthalate* MIL-101. *Journal of Molecular Liquids, 254*, 137–144.

Zhang, J., Li, M., & Mao, H. (2019). Cross-linked enzyme aggregates of recombinant cyclodextrin glycosyltransferase for high-purity β-cyclodextrin production. *Journal of Chemical Technology and Biotechnology, 94*(5), 1528–1533. https://doi.org/10.1002/jctb.5912, http://onlinelibrary.wiley.com/journal/10.1002/(ISSN)1097-4660.

Zou, M., Dong, M., & Zhao, T. (2022). Advances in metal-organic frameworks MIL-101(Cr). *International Journal of Molecular Sciences, 23*(16), 9396. https://doi.org/10.3390/ijms23169396.

Zucca, P., & Sanjust, E. (2014). Inorganic materials as supports for covalent enzyme immobilization: Methods and mechanisms. *Molecules (Basel, Switzerland), 19*(9), 14139–14194. https://doi.org/10.3390/molecules190914139.

SECTION

Cyclodextrins and their Applications IV

CHAPTER

Structure and properties of cyclodextrins

11

Grégorio Crini[1], Marc Fourmentin[2], Dario Lacalamita[3], Miriana Kfoury[4], Chiara Mongiovì[5], Chong Liu[6], Sophie Fourmentin[4] and Nadia Morin-Crini[1]

[1]*Université Marie et Louis Pasteur, Laboratoire Chrono-environnement, Besançon, France,*
[2]*Université Littoral Côte d'Opale, UR 4493, LPCA, Laboratoire de Physico-Chimie de l'Atmosphère, Dunkerque, France,*
[3]*Università degli Studi di Bari Aldo Moro, Dipartimento di Chimica, Bari, Italy,*
[4]*Université Littoral Côte d'Opale, UR 4492, UCEIV, Unité de Chimie Environnementale et Interactions sur le Vivant, Dunkerque, France,*
[5]*Istituto per i Processi Chimico-Fisici, CNR, Bari, Italy,*
[6]*Department of Chemical & Materials Engineering, University of Auckland, New Zealand*

Cyclodextrins—an introduction

Cyclodextrins (CDs), or cycloamyloses, are naturally occurring amphiphilic cage molecules, discovered accidentally in 1891 by the French chemist and pharmacist Antoine Villiers, who named them *cellulosines* (Crini, 2014). CD are cyclic oligosaccharides obtained by enzymatic degradation of starch (Fig. 11.1). The most studied commercial CDs are α-, β-, and γ-CD molecules, containing six, seven, and eight glucopyranose units in their cyclic structure, respectively. Each glucose unit is linked *via* α-1,4-glycosidic bonds, forming a cyclic structure (Fig. 11.1). Their primary characteristic is their remarkable ability to form host-guest inclusion complexes, a property that is highly valued in numerous and diverse industrial sectors. Indeed, CDs and their derivatives have led to major theoretical and practical advances in chemistry, biology, biochemistry, health sciences, and agriculture.

Although CDs have been known for 134 years, their widespread use only began in the 1980s (Szejtli, 1982, 1984; Atwood et al. 1984). Initially costly, the three main CDs—α-CD, β-CD, and γ-CD—were long regarded as mere laboratory curiosities. Indeed, their main obstacle was not only their price but also their presumed toxicity and the lack of sufficient knowledge of these substances. In addition, very few researchers were convinced of the industrial potential of CDs. It was not until 1979 that they began to be produced on an industrial scale (Kainuma, 1984; Loftsson & Duchêne, 2007; Szejtli, 1982).

In the early 1980s, the emergence of various inclusion compounds on the market contributed to the expansion of CD applications. Several factors were responsible

Cyclodextrin Glucanotransferase. DOI: https://doi.org/10.1016/B978-0-443-41694-1.00013-7

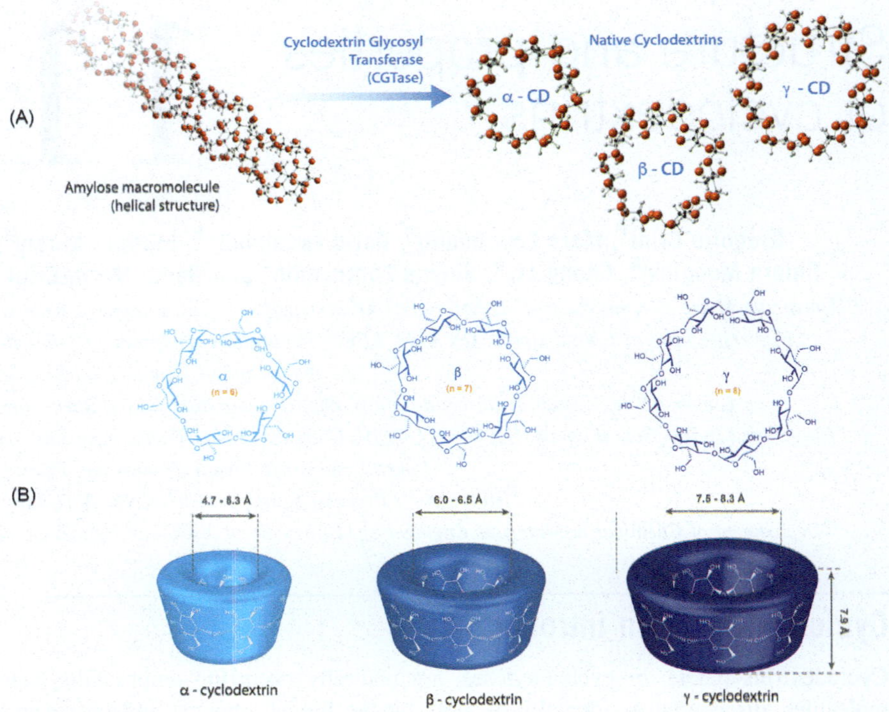

FIGURE 11.1

(A) Schematic representation of the production and (B) structure of the three main cycloamyloses or native cyclodextrins (*CDs*) from the enzymatic degradation of starch (amylose macromolecule) by cyclodextrin glycosyltransferase (*CGTase*): cyclohexaamylose or α-cyclodextrin (α-CD, number of glucopyranose units (n) = 6), cycloheptaamylose or β-cyclodextrin (β-CD, n = 7), and cyclooctaamylose or γ-cyclodextrin (γ-CD, n = 8).

for this development: exponential advancements in scientific knowledge regarding CD structure and properties, the gradual establishment of industrial-scale production leading to significantly reduced costs, progress in assessing their nontoxicity, biocompatibility, and biodegradability, and above all, a deeper understanding of molecular encapsulation, which revealed significant innovation and development potential (Szejtli, 1982, 1984, 1991). Their introduction into the food and pharmaceutical industries, for instance, led to remarkable advancements and spectacular industrial progress, resulting in the first commercially available drugs and ready-to-eat meals containing CDs, which quickly gained success.

As early as 1976, Japan authorized the use and commercialization of CDs, and their inclusion derivatives in the food industry, extending this approval to the pharmaceutical sector in 1983. A year later, the first medications incorporating CDs were marketed by Japanese manufacturers, such as the complex between β-CD and a prostaglandin (a lipid-soluble hormone secreted by immune system cells and involved in various physiological reactions), sold under the name PROSTARMON®, and the β-CD/piroxicam complex (a nonsteroidal antiinflammatory drug) (Frömming & Szejtli, 1994; Duchêne et al. 2009; Sharma & Baldi, 2016).

In Europe, Hungary was the first country to authorize the use of CDs in the food industry in 1983 (Szejtli, 1984), followed 4 years later by other European nations (Belgium, Denmark, Spain, France, and Luxembourg), albeit exclusively as flavor carriers. Amongst the list of prestigious researchers who have contributed to the development of CDs at the industrial scale, Professor József Szejtli, a Hungarian carbohydrate chemist known as the "Godfather of Cyclodextrins" or "Mr Cyclodextrin," played a fundamental role as an eminent scientist and a visionary (Crini, 2014; Crini, Fenyvesi, et al., 2021).

Beginning in 1988, an Italian company commercialized the CD/piroxicam complex under various brand names (CYCLADOL®, BREXIN®, PIRODEX®, OXIDEX®). Pharmaceutical formulations quickly proliferated in the market, including TAKETIAM® or TEXODIL® (α-CD/cefotiam complex) as an antibacterial agent for treating microbial infections, SPORANOX® (HP-β-CD/itraconazole complex) as an antifungal, and ALOPEXY® (γ-CD/minoxidil inclusion complex) for combating hair loss.

Since then, the inclusion complexation property of CDs has been extensively exploited, and they have become integral to various industrial sectors, ranging from pharmaceuticals and food processing to biomedicine, radiology, cosmetics, hygiene, agriculture, agrochemistry, chemistry, catalysis, chromatography, biotechnology, textiles, packaging, nanotechnology, and environmental applications, for example, soil and water treatment (Crini et al., 2018, 2020; Crini and Fourmentin, 2020; Fenyvesi et al., 2019; Fourmentin et al., 2018a, 2018b; Liu et al., 2020, 2024; Morin-Crini et al., 2021;). Most of these applications rely on the ability of CDs to form inclusion complexes.

After a brief overview of some general aspects of CDs, and their unique structure, this chapter describes their properties, with a particular focus on their ability to form inclusion complexes. This chapter also examines the formation of these complexes by detailing the forces involved and the factors influencing their stability, emphasizing the characteristics of this complexation.

Cyclodextrin description

Nomenclature

During experiments on the degradation of carbohydrates under the action of ferments, Professor Villiers was the first to observe in 1891 the formation of "unwanted crystals with particular properties," that is, the formation of CDs. These substances were initially called "cellulosines" by Villiers because he assumed that the novel crystalline substances, obtained from digests of *B. amylobacter*, were "a kind of cellulose." However, at the beginning of the 1900s, numerous researchers remained unconvinced by Villiers' conclusions, pointing out the fact that "Villiers used impure cultures but his digests contained sufficient *B. macerans* to account for the small amount of crystalline dextrin obtained" (Crini and Fourmentin, 2020).

The first great scientist who left its mark on the history of these oligosaccharides was Professor Franz Schardinger, an Austrian chemist and bacteriologist. Schardinger is considered the "Founding Father" of CD. In 1903, he discovered that a type of extremely heat-resistant microorganism, named *Rottebacillus I*, was able to dissolve starch and form crystalline by-products, remarkably similar to "cellulosines" reported by Villiers. Indeed, while studying the digestibility of warm meals, Schardinger was the first to distinguish between two types of "*krystallisiertes dextrins*" by organic solvent precipitation, which he called crystallized dextrin A and crystallized dextrin B, and several months later, crystallized polysaccharides. Schardinger identified the crystalline carbohydrate in Villiers' publications as dextrin B. These two substances were then considered as the intermediate decomposition products of starch.

Professor Schardinger is recognized as the first to have laid the foundations of CD chemistry, describing their fundamental properties, suggesting their ability to form complexes, isolating the strain of bacteria responsible for dextrin formation, that is, *B. macerans*, and hypothesizing that the crystallized substances were cyclic "polysaccharides" (Crini, 2014, 2020b).

A decade later, in the mid-1910s, the German chemist and biochemist Hans Pringsheim used the name "*krystallisiertes polyamylosen*," that is, crystallized polyamyloses, distinguishing two series, the α-series of dextrins containing 2n D-glucose units per molecule and the β-series containing 3n D-glucose units per molecule. This nomenclature was similar to that introduced by Schardinger. Pringsheim demonstrated that polyamyloses could form adducts with various solvents. He was the first to propose a hypothetical cyclic structure of polyamyloses, which was only demonstrated 20 years later with the work of Professor Robert Rundle and his doctoral student Dexter French on dextrin crystallography.

In 1923, Professor Paul Karrer, a Swiss organic chemist best known for his research on vitamins, was the first to propose that dextrins are composed of maltose units only joined by α-(1→4) glucosidic linkages. However, just such as Schardinger and Pringsheim, Karrer failed to elucidate the cyclic structure of the dextrins (Crini and Fourmentin, 2020; Crini, 2014, 2020a, 2021).

In the 1920s, as a tribute to the pioneering work of Professor Franz Schardinger, the German chemist Karl Johann Freudenberg, a student of the renowned Professor Emil Fischer, called them "Schardinger dextrins" and referred to these compounds as α-dextrin and β-dextrin. Dextrins/CDs were called "Schardinger dextrins" in his honor, almost up to the 1970s, or also sometimes simply as dextrins (Crini, 2014).

Schardinger dextrins were also named "cycloamyloses" by the American chemist Dexter French in 1942, "cycloglucanes" by Freudenberg in 1943, and finally "cyclodextrins" in 1949 by the young German student Friedrich Cramer, a pupil of Freudenberg. The model of "cycloamyloses," proposed by Professor

French, recognized as one of the leading authorities on the chemistry and biochemistry of dextrins and starch, was constructed from glucopyranose units in the boat conformation. Until the 1980s, the term "*cycloamyloses*" was also commonly used, but it was gradually replaced by "*cyclodextrins*," the nomenclature we use today (Crini and Fourmentin, 2020; Crini, 2014).

The term "cyclodextrins" is indeed currently the most used. A Greek letter indicates the number of glucose units that make up the CDs: α for 6, β for 7, γ for 8, δ for 9, etc. CDs with five or fewer glucose units do not exist due to steric constraints, while those containing nine or more glucose units are difficult to synthesize and purify, with less favorable complexation properties.

Among the CDs, the most well-known and studied are α-CD (cyclohexaamylose, cyclomaltohexaose, α-cycloamylose), β-CD (cycloheptaamylose, cyclomaltoheptaose, β-cycloamylose), and γ-CD (cyclooctaamylose, cyclomaltooctaose, γ-cycloamylose) (Fig. 11.1). Their CAS numbers are as follows: 10016-20-3 for α-CD, 7585-39-9 for β-CD, and 17465-86-0 for γ-CD. Each of these three CDs is referred to as a native CD. β-CD is the most widely produced, least expensive, and most used commercial derivative. Its IUPAC nomenclature is as follows: 5,10,15,20,25,30,35-heptakis (hydroxymethyl)-2,4,7,9,12,14,17,19,22,24,27,29,32,34-tetradecaoxaoctacyclo[31.2.2. 2.3,6.28,11.213,16.218,21.223,262.28,31]nonatetracontane-36,37,38,39,40,41,42,43,44, 45,46,47,48,49-tetradecole.

To distinguish the three native CDs, that is, α-CD, β-CD, and γ-CD, from those containing a larger number of glucose units, the term "large-cycle cyclodextrins" is used (Larsen, 2002). These molecules have also been designated using a Greek prefix. However, this designation is a source of confusion because it does not describe the size of the macrocycle. For this reason, large-cycle CDs are more simply referred to as cycloamyloses (cyclomaltooses) and by the abbreviation CD_n, where n denotes the number of glucose units in the macrocycle. For example, cyclomaltoheneicosaose or Π-CD contains 21 glucose units, and its abbreviation is CD_{21}. Nowadays, more than 150-membered cyclomaltoses are known, but the largest isolated with a demonstrated structure contains 39 units.

Production

In the mid-1950s, French was the first to demonstrate that CD could be produced at an industrial scale by a relatively simple technology based on the fermentation of starch. At that time, cycloamyloses were still laboratory curiosities, difficult to produce and purify, and also extremely expensive. In fact, native CDs are derived from the enzymatic degradation of starch, specifically from amylose extracted from potato starch, which provides the highest production yields among various starch sources. The enzyme responsible for this transformation is cyclodextrin glycosyltransferase (CGTase), produced by various *Bacillus* species. This process disrupts the helical structure of the amylose macromolecule, resulting in the formation of a cyclic structure. French proposed a mechanism of action based on a *trans*-glucosylation reaction to interpret the formation of dextrins by *B. macerans*, known as the glycosidic exchange concept. The enzyme performed three *trans*-glucosylation reactions involving cyclization, coupling, and disproportionation, and also a hydrolysis reaction (Crini, 2014; Crini, Fenyvesi, et al., 2021).

The primary CGTase-producing *Bacillus* species include *B. macerans*, *Klebsiella pneumoniae*, *B. stearothermophilus*, *B. megaterium*, *B. circulans*, and *Alkalophilic bacterium* No. 38-2. Depending on their origin, these enzymes act specifically on starch, yielding defined proportions of α-CD, β-CD, and γ-CD, with one of these always being predominant. *B. macerans* and, more selectively, *K. pneumoniae* are used for α-CD production, while *A. bacterium* No. 38-2 is employed for β-CD synthesis. Industrial processes focus on enzymological and thermodynamic factors, as well as the addition of selective precipitating agents to preferentially obtain one of the three CDs.

The first protocol describing the synthesis of Schardinger dextrins with high purity was proposed by Freudenberg in 1935. For more than 10 years, this protocol was studied and modified, and in 1947, Freudenberg's group described the first scheme for the isolation of pure fractions of dextrins (Crini, 2014). However, the homogeneity and chemical purity of the produced CDs were low. Two years later, French proposed a new protocol for the preparation, separation, and purification of cycloamyloses on a larger-than-laboratory scale, which was much simpler than that of Freudenberg. In the mid-1950s, the industrially produced CDs exceeded 95%. At the same time, Cramer also investigated the enzymatic production of CDs, introducing a new solvent technology. French's and Cramer's protocols served as the basis for those developed at the end of the 1970s, when industrial production of CDs began, in particular Professor Szejtli's patents (Crini, Fenyvesi, et al., 2021; Szente & Fenyvesi, 2016; Szente, 2004). The synthesis of native CDs, whose production protocols are now well established, has also enabled the development of modified CD derivatives. Among these, the most well-known is hydroxypropyl-β-cyclodextrin (HP-β-CD), whose production began in the early 1980s (Szente & Szejtli, 1999).

Types of cyclodextrins

At the beginning of the last century, Schardinger recognized only dextrin-α and dextrin-β, while Freudenberg obtained γ-dextrin in 1948, although it had previously been regarded by him as a cyclic hepta-saccharide (Crini, 2014). Two years later, both the groups of Freudenberg and French elucidated the structure of γ-dextrin. In the late 1950s, French and coworkers established precise molecular weights, exact chemical structures, dimensions, and types of bonding in the three cycloamyloses.

Structure

In the early 1950s, the simultaneous work of Freudenberg and French, followed by that of Cramer, showed that the ring shape of CDs was a consequence of the C-1 conformation of the glucopyranose subunits and their α-1,4-type glycosidic linkages. As a consequence of this conformation of the glucopyranose units, all the secondary hydroxyl groups were located on one of the two edges of the ring, while all the primary hydroxyl groups were located on the other edge (Fig. 11.1). These intrinsic characteristics of CDs were not clearly demonstrated until the mid-1960s, notably with the work of Professor Benito Casu, an Italian chemist and biochemist, considered a pioneer in the dissemination of the knowledge of CDs and glycosaminoglycan research (Crini & Torri, 2017; Torri et al., 2022). At that time, the first infrared and nuclear magnetic resonance spectroscopic equipment enabled spectacular advances in the molecular structural characterization of oligosaccharides and polysaccharides.

Between 1942 and 1965, French was the first to contribute greatly to the preparation, characterization, and molecular structural knowledge of cycloamyloses. Very quickly, he became a pioneer in the understanding of their structure, publishing an impressive number of results on cycloamyloses, which are still used as references today. In 1950, French, studying the periodate oxidation of the three cycloamyloses, demonstrated that all cycloamyloses had a cyclic structure in which each *D*-glucose unit was linked to the next by an α-*D*-(1→4)-glycosidic bond, and that the interior of each cavity was apolar. On the side where the secondary hydroxyl groups are situated, the diameter of the cavity is larger than on the side with the primary hydroxyl groups, since the free rotation of the latter reduces the effective diameter of the cavity. The lining of the internal cavity is formed by hydrogen atoms and glucosidic oxygen-bridge atoms; therefore, this surface is slightly apolar. Cycloamyloses were initially regarded by the French as cylinders, then conical cylinders, and finally as "capsules."

However, Freudenberg and Cramer pointed out the fact that a CD molecule should be regarded rather as a truncated cone, a "torus," than a cylinder or a capsule. These researchers were also the first to propose a detailed study on the hydrophobic and hydrophilic regions of an α-dextrin "torus." In his doctoral work published in 1949, the young Cramer reported conclusions similar to those of French. The term "cyclodextrins" was then introduced to designate the cyclic structure of oligosaccharides composed of 6 to 13 glucose units (glucopyranose units) linked by α-(1→4) glycosidic bonds in a $^4C_1$ chair conformation. Structurally, CD molecules are in the form of a conical cylinder or a torus, a consequence of the chair conformation of the glucopyranose residues and their glycosidic bonds. The core of this particular structure can then trap or encapsulate other substances.

The three main conclusions of French, Freudenberg, and Cramer were the following: (1) all the glucose residues of CDs were in the $^4C_1$ chair conformation; (2) the cavity was lined by the hydrogen atoms and the glycosidic oxygen bridges, respectively; and (3) the nonbonding electron pairs of the glycosidic oxygen bridges were directed towards the inside of the cavity, producing a high electron density. These conclusions were demonstrated in the mid-1960s with the work of Casu (Crini & Torri, 2017; Torri et al., 2022) and finally accepted in the late 1970s with that of Szejtli and his famous book (Szejtli, 1982), which is still considered a reference in the scientific community (Crini, Fenyvesi, et al., 2021; Szente & Fenyvesi, 2016; Szente, 2004).

Nowadays, CDs/cycloamyloses are considered as nonreducing oligosaccharides containing six, seven, or eight units linked by α-*D*-(1→4) bonds, having both hydrophobic and hydrophilic regions. The truncated toroid structure consists, on the outside, of primary hydroxyl groups carried by the C-6 carbons bordering the small diameter (one hydroxyl per glucose on the narrowest side) and secondary hydroxyl groups carried by the C-2 and C-3 carbons bordering the large diameter (hydroxyls located opposite each other on the wide part of the conical cylinder), and in the internal part of hydrogen atoms, carbon, and ether-oxide bonds. The nonbonded electron pairs of the glycosidic oxygen bridges are oriented towards the interior of the cavity, producing a high electron density, which allows CDs to have basic properties. The dimensions of the truncated conical shape vary depending on the number of glucose units. From this particular amphiphilic structure arise all the particularities of this molecule and its properties: the high electron density due to the glycosidic oxygens transforms the interior of the CD into a hydrophobic cavity having an apolar character (Lewis's base) and a highly hydrophilic character for the external part (Fourmentin et al., 2018a).

Properties

Form—CDs are white crystalline powders with no odor and a slightly sweet taste (Szejtli, 1982, 1996, 1998).

Thermal behavior—The dehydration of β-cyclodextrin occurs in two stages: first, a loss of seven water molecules located outside the molecule, then departure of four water molecules fixed inside the cavity, the stable hydrate at room temperature being the undecahydrate (Sandilya et al., 2020; Szejtli, 1998).

Solubility—The hydrophilic surface of CDs provides relatively good water solubility. Among the three native CDs, β-CD is the least soluble in water. Its solubility is 1.85 g/100 mL at 25°C and increases rapidly with temperature (at 45°C, the solubility is 3.1 g/100 mL). This is the reason why β-CD is the easiest to purify by recrystallization. The solubility of CDs varies depending on the nature and concentration of the solvent. β-CD exhibits excellent solubility in certain organic solvents (dimethylformamide, dimethyl sulfoxide, pyridine, *N*,*N*-dimethylacetamide). The difference in water solubility between CDs is explained by the formation of hydrogen bonds that link the secondary OH groups in pairs. These intramolecular bonds stabilize the molecule, making its structure rigid, and thus preventing hydration of the molecule. The hydrogen layer bonds are not observed in α- and γ-CDs, because the size of the cycles is either too small (α-CD) or too large (γ-CD) to observe this phenomenon. In addition, γ-CD has a noncoplanar, more flexible structure, and is therefore the most soluble of the three native CDs. In order to increase the solubility of β-CD, its structure is chemically modified either by grafting neutral (2-hydroxypropyl, etc.) or ionic (sulfate, sulfobutyl, etc.) solubilizing groups directly onto the molecule, or by breaking the hydrogen bonds between the OH-2 and OH-3 of adjacent units by grafting methyl groups (Szejtli, 1982).

Reactivity—The primary hydroxyls at position six on the macrocycle are more reactive than the secondary hydroxyls, and they are more basic. The hydroxyls at position two are the most acidic, and those at position three on the ring are the most sterically hindered. The pK_a of the secondary hydroxyl groups of β-CD is 12.2, which is low compared to noncyclic dextrins. Aqueous solutions of CDs have a significant viscosity at high concentrations and depending on the temperature, which can pose problems at high concentrations (Musuc, 2024).

Stability—CDs are nonreducing oligosaccharides, relatively stable and little subject to hydrolysis. These molecules are completely resistant to the usual enzymatic hydrolysis of starches, in particular β-amylase, since they do not have terminal groups susceptible to being attacked by this enzyme. However, α-amylase is capable of hydrolyzing CDs but, in general, at a low rate (Rekharsky and Inoue, 1998).

Chirality—CDs are chiral compounds (optically active) with five asymmetric carbon atoms per glucose unit; they will form a diastereoisomeric inclusion complex with a racemic mixture of a chiral guest molecule; a nonsymmetric guest molecule can take various orientations to enter the CD cavity: this is called bi- or multimodal (discrimination phenomenon widely used in chromatography) (Szejtli, 1998).

Toxicology—Since the 1980s, numerous toxicological tests have been carried out on CDs and their derivatives, particularly orally, and have shown the nontoxic nature of these substances. In humans, the daily intake should not exceed 1 to 2 g per day. The three native CDs are authorized in the agri-food industry as food additives (E457, E459, and E458 for α-, β-, and γ-CD, respectively). These products hardly diffuse through biological membranes and are not absorbed during intestinal transit. They slowly hydrolyze in the body to release their six, seven, or eight glucose units. They are neither embryotoxic, nor teratogenic, nor mutagenic, nor carcinogenic. Their metabolism is identical to that of potato starch. More generally, α-, β-, and γ-CDs are considered harmless to all living organisms. However, during oral administration, undesirable effects, in particular gastric, may appear. The high affinity of CD for lipids and their ability to extract lipid substances from the intestinal membrane can cause poorly tolerated local irritations and stimulate intestinal and gastric secretions, and therefore the appearance of diarrhea, the complexation of bile salts, and the increase in the level of lipids in the blood (creatinine levels) (Frömming & Szejtli, 1994; Szejtli, 1996; Duchêne et al. 2009; Musuc, 2024).

Biodegradability—CDs are considered environmentally friendly and degrade in the natural environment, with no harmful effects on environmental compartments and living organisms. However, modified molecules degrade less easily than native molecules: 82% of β-CD is biodegraded in 28 days compared to only 6% for HP-β-CD (according to data sheets) (Szjetli 1982, Szejtli, 1996).

Inclusion complex formation—The most remarkable property of CDs is their ability to give inclusion complexes (Fig. 11.2), which has the consequence of profoundly modifying the intrinsic properties of the guest molecule. The macrocyclic structure, the apolar internal space, and the external hydrophilicity confer on these substances this particular complexation capacity (Szejtli, 1982, 1998; Maazaoui and Abderrahim, 2015; Jacob & Nair, 2018; Jansook et al., 2018).

Cyclodextrins—inclusion complexes

An inclusion complex, a concept first put forward by Pringsheim in 1924, taken up by Freudenberg in 1935, studied by French in 1942, and definitively established by Cramer between 1952 and 1957 (Crini, 2014, 2020a; Crini, Fenyvesi, et al., 2021), is a molecular association between one or more molecules, one of which is the host (the receptor) and the other is the guest (the substrate) (Crini and Fourmentin, 2020; Crini, 2014; Crini, Fenyvesi, et al., 2021). The guest molecule is then totally or partially encapsulated, the receptor playing the role of the host molecule. This is called a host/guest type complex (Fig. 11.2A). The interactions between the host molecule and the guest molecule are weak, which allows for easy and gentle dissociation, a very interesting property for the transport and release of a target

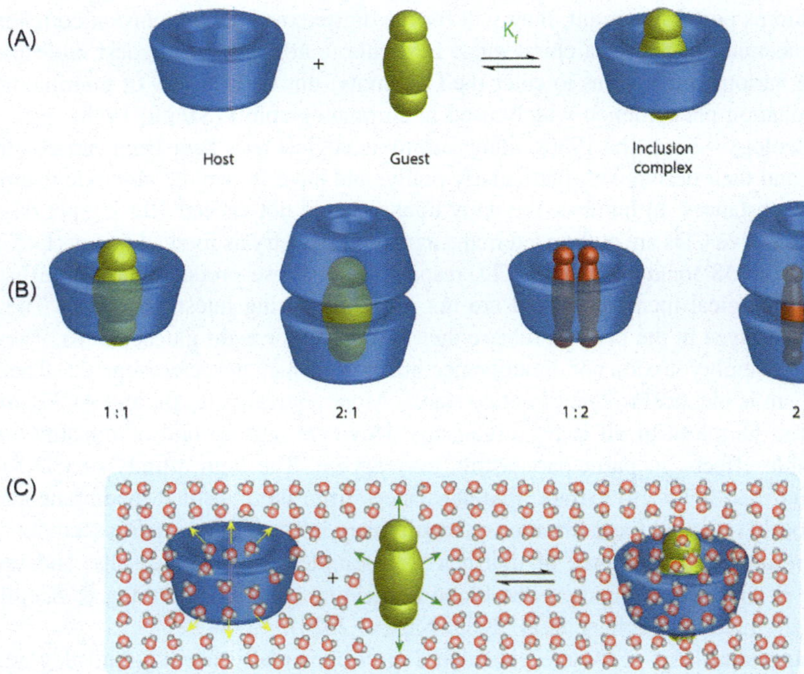

FIGURE 11.2

(A) Inclusion phenomena between a cyclodextrin (*CD*) molecule (the host) and an organic molecule (the guest) to form inclusion complexes (K_f is the formation constant); (B) association between CD and organic molecules leading to host/guest inclusion complexes of different stoichiometry; and (C) schematic representation of the formation of an inclusion complex between a guest and a CD molecule; the force causing the formation of the complex is the energetic stabilization of the system by the replacement in the cavity of water molecules by hydrophobic substrates, water molecules being repulsed by the hydrophobic potential guest (the main and remarkable conclusion of Szejtli's work).

molecule. Molecules other than CDs, such as crown ethers, clathrates, intercalates, zeolites, porphyrins, and cyclophanes, allow for such molecular associations (Frömming & Szejtli, 1994; Szejtli & Osa, 1996; Szejtli, 1982, 1996, 1998).

In 1956, Cramer was the first to demonstrate that the interior of the CD cavity could be considered as a lipophilic microenvironment in which a nonpolar hydrophobic molecule could be placed, without any covalent bond between the host molecule and the guest molecule. Indeed, the guest is kept inside the cavity by noncovalent forces, for example, hydrogen bonds, hydrophobic interactions, and van der Waals forces, and therefore weak, which makes the reversibility of the system possible. In the nomenclature proposed by Cramer, if the guest is located inside the CD cavity, it is then called an inclusion complex. In this case, complex formation requires a good match between the size of the host molecule and that of the CD. If the guest interacts with the host outside the cavity, it is called an association complex. In all cases, the external part of the host must be compatible with the solvent used to avoid problems of insolubility or aggregation (Loftsson et al., 2023).

Depending on the nature of the CD and the guest, an inclusion complex has its own properties, such as its stoichiometry, its formation constant, and its spatial structure. Most of the studies described 1:1 inclusion complexes: one CD molecule corresponded to one guest molecule (Fig. 11.2A). The substances concerned were very diverse: molecules (aliphatic or aromatic hydrocarbons, acids, amines, phenolic substances, organometallic structures), macromolecules (synthetic polymers, polypeptides, etc.), and ions (minerals including metals). Another important point was indeed raised by Cramer on the fact that any inclusion complex is a dissociation-association equilibrium. At the same time, the same conclusions were reported by French, but the paternity of these conclusions is attributed to Cramer and Freudenberg (Crini, 2014).

Complexes preparation

The complexation process can occur both in the solid state and in solution (Szejtli, 1982, 1984, 1996, 1998). Various methods for preparing these complexes have been reported in the literature, primarily depending on the properties of the substrate. In general, water is the preferred solvent, as the formation of complexes in a homogeneous aqueous solution is rapid (Fourmentin et al., 2018a, 2018b; Nitin et al., 2024). However, it should be noted that highly water-soluble substances tend to form weak complexes. Complexation can also occur in the presence of polar organic solvents (*e.g.*, DMSO, DMF), although the range of suitable solvents is limited, and in deep eutectic solvents, as demonstrated recently (Moufawad et al., 2019). Many solvents, such as pyridine and toluene, form stable complexes with CDs and thus compete with the intended substrate. Moreover, the stability of complexes in aqueous solution decreases significantly upon the addition of an organic solvent.

Several methods are available for preparing these complexes, including solution-based techniques (coprecipitation, lyophilization), suspension methods, kneading (mechanical mixing), and dry grinding (Nitin et al., 2024). For example, in the coprecipitation method, the substrate, dissolved in a solvent such as acetone, ethanol, ether, or chloroform, is added to a cold aqueous solution of CD. The mixture is maintained for several hours, after which water is removed by cold drying to yield the solid-phase complex. Alternatively, spray drying or filtration can be used to obtain a precipitate. This technique is simple and relatively fast, though in some cases, complex formation may take several days.

In the lyophilization method, the guest molecule is first dissolved in water, and CD is then added under stirring, adjusting the molar ratios accordingly. The resulting solution is filtered, lyophilized, and washed with a solvent before the residue is dried under vacuum. In some cases, the last two steps can be omitted. This method is particularly suitable for water-soluble guest molecules or thermolabile compounds, yielding a powdered sample with high inclusion efficiency.

For suspension-based methods, the substrate is mixed with a CD suspension in water. The addition of the substrate increases the viscosity of the mixture, and its concentration must be optimized to allow vigorous mixing of both reactants. The complex is then recovered by filtration.

The kneading (mechanical mixing) method involves blending CD with a small amount of water (without complete dissolution) in which the substrate has been predissolved. During mixing, viscosity increases significantly, forming a paste that is subsequently dried, ground, and washed. Complex formation occurs because the CD/water complex is thermodynamically less favorable than the CD/substrate complex, which also exhibits different crystalline structures. If the resulting complex is soluble or decomposes upon drying, lyophilization is performed. Kneading is widely used in industrial applications.

In the dry grinding method (also known as solid-phase complexation), the two reactants are coground without the addition of solvent or water. In some cases, a volatile oil, such as lemon oil, can act as a grinding aid for CD. Other complexation techniques involve the use of supercritical CO_2.

The kinetics of complex formation depend on the preparation method. In homogeneous solutions, complexation occurs very rapidly, whereas in heterogeneous systems, the formation rate is limited by the dissolution rate of CD. In many applications, the formed complex must be easily dissociable to release the guest molecule, such as a drug or an aroma. This can be achieved through boiling, solvent extraction, acid hydrolysis (which opens the CD ring), or enzymatic hydrolysis.

The simplest method for obtaining inclusion complexes is to bring the guest molecule into contact with an aqueous solution of CD. However, the resulting inclusion complexes represent a relatively simplified explanation of a much more complex phenomenon. CD not only form host-guest inclusion complexes but also noninclusion complexes, a concept introduced by Cramer in 1956. The hydroxyl groups on the external surface can form hydrogen bonds with other molecules, enabling CD—similar to other dextrins, noncyclic oligosaccharides, and polysaccharides—to form molecular structures with water-insoluble lipophilic substances. Additionally, CD can form aggregates capable of solubilizing hydrophobic molecules in a micelle-like manner (Loftsson et al., 2023).

CD aggregates also exhibit complexing abilities, which have led to the exploration of host-guest architectures resembling nanotubes. However, this remains a subject of ongoing debate, as results vary significantly depending on the analytical technique used. Before discussing the properties of an inclusion complex, its formation must be thoroughly characterized and demonstrated.

Evidence of inclusion complexes

Many characterization techniques are available to study an inclusion complex. Indeed, any analytical method that allows the highlight of characteristic changes (change in solubility, retention in chromatography, UV-visible absorbance or fluorescence, chemical shift of protons in NMR, etc.) can be used for the study of complexation. However, depending on the form of the complex, either in the solid state or in the liquid state, certain methods are more suitable. For example, to study inclusion complexes in the solid state, techniques such as X-ray diffraction, thermo-analytical methods (changes in melting, boiling, decomposition or oxidation temperature, polymorphic transition), thin-layer chromatography, infrared and Raman spectroscopy, or solid-state NMR (CPMAS, HRMAS) are preferred. In solution, the complexes are characterized by solubility diagrams, the study of their fluorescence or circular dichroism, by potentiometry, electrochemistry (polarography) or microcalorimetry, and by spectroscopy techniques (UV-Visible, NMR, mass, spin). However, the state of the art in the literature shows that UV-visible and NMR spectroscopy are the most used techniques because they are relatively simple while giving both qualitative (evidence of inclusion) and quantitative (determination of stoichiometry and formation constant) information (Nitin et al., 2024; Szejtli & Osa, 1996; Szejtli, 1982, 1998).

Factors affecting the stability of inclusion complexes

The main forces involved in the formation of an inclusion complex with CDs include van der Waals forces, electrostatic interactions, hydrophobic interactions, and hydrogen bonding (Szejtli, 1982). In all cases, the stabilization energies of inclusion complexes are relatively low, as they arise from noncovalent interactions (Fourmentin et al., 2018b). The synergy of these interactions leads to a dynamic equilibrium between the encapsulated guest (G) and the CD. Furthermore, the dissociation constants of CD:G complexes are highly dependent on the solvent. Complexation is stronger in solvents with higher polarity (Fig. 11.3).

The stability of a complex strongly depends on the geometric factor. Nevertheless, other factors influence the complexation of molecules by CD. Three types of factors can be distinguished (Szejtli, 1982):

1. The nature of the guest molecule, in addition, to its size and shape, its hydrophobicity, water solubility, and ionization state must be considered.
2. The characteristics of CD (native and modified).

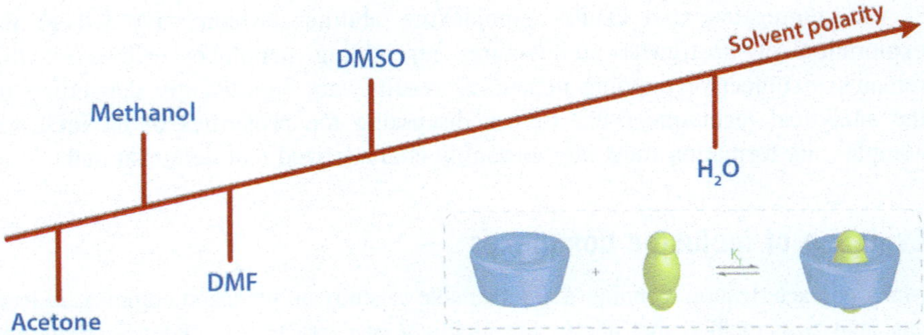

FIGURE 11.3

Stability of inclusion complexes between a cyclodextrin and a guest molecule as a function of solvent polarity (K_f is the formation constant).

3. The context of the solution (pH, ionic strength, temperature, and presence of a cosolvent).

The ionic strength of the solution, particularly its variation, is a key factor in the stability of CD complexes. For instance, in the case of neutral substrates, an increase in ionic strength reduces the solubility of guest molecules in the aqueous phase, thereby increasing the formation constant. Generally, higher temperatures enhance the solubility of complexes but may also destabilize them. Moreover, heat sensitivity varies from one complex to another: some decompose at temperatures as low as 50°C–60°C, while others remain stable at higher temperatures, particularly insoluble complexes (Szejtli, 1982).

The polarity and charge of the guest molecule, which depend on pH, are also crucial factors in the formation of inclusion complexes, as they influence the underlying mechanisms. CDs preferentially complex molecules that are less polar than water, whereas highly hydrophilic, hydrated, or ionized molecules exhibit little to no complexation. A frequently cited example is sodium benzenesulfonate, which has two functional groups capable of entering the α-CD cavity: the benzene ring and the sulfonate group. The insertion of the benzene ring into the cavity is more stable than the orientation of the polar sulfonate group. This can be explained by the fact that the CD cavity is less polar than the surrounding aqueous medium; therefore the more polar a guest molecule is, the less likely it is to enter the cavity, resulting in lower complex stability. However, there are notable exceptions. For instance, sodium *p*-nitrophenolate forms a complex that is 13 times more stable than *p*-nitrophenol, whereas the opposite is observed for benzoic acid and its corresponding anion.

Stoichiometry

The stoichiometry of an inclusion complex between host (cyclodextrin, denoted as CD) and guest (denoted as G) molecules is defined by the ratio a/b, where a and b are the respective stoichiometric coefficients of the host and guest in the chemical equilibrium of (11.1).

The most common inclusion complexes are of the 1:1 (host:guest) type ((11.2)). However, when a molecule is too large to be fully enclosed within a single CD cavity, its other end can also be encapsulated by another CD molecule. This can lead to the formation of 2:1, 2:2, 3:1, 3:2, or even 4:5 complexes. Fig. 11.2B illustrates the structural arrangements of host/guest stoichiometries, including 1:1, 2:1, 1:2, and 2:2 complexes. Additionally, 1:2 complexes or ternary complexes involving three components are also possible (Szejtli, 1982). The extreme case of complexation stoichiometry is found in rotaxanes, which are supramolecular assemblies consisting of multiple CD threaded onto a long aliphatic chain (macromolecule).

$$aCD + bI \rightleftarrows CD_a/I_b \tag{11.1}$$

$$CD + I \rightleftarrows CD/I \tag{11.2}$$

In the case of an amphiphilic molecule, the most apolar (hydrophobic) part preferentially enters the CD cavity to maximize the overlap of hydrophobic regions, while the more polar (hydrophilic) part remains outside the cavity. This ensures better contact between the solvent and the external hydroxyl groups, allowing for potential stabilization of the complex through hydrogen bonding (Szejtli, 1982).

The topology of CD complexes is therefore highly variable, as it depends on the shape of both the host and the guest, as well as the reaction medium. Various interactions can be established accordingly:

1. Inclusion can be a simple association, as described in Cramer's nomenclature.
2. Inclusion can be complete, meaning that it involves one molecule of each partner (*e.g.*, benzaldehyde in β-CD).
3. Inclusion can be partial, involving only the most lipophilic part of the molecule (*e.g.*, prostaglandin $F_{2\alpha}$ in β-CD).
4. Inclusion can be axial when the guest molecule is long and slender (*e.g.*, 1,4-butanediol in β-CD).
5. Inclusion can adopt a 2:1 sandwich-type arrangement when the guest is small but bulky (*e.g.*, ferrocene in α-CD).

The stoichiometry of inclusion complexes can be determined by the continuous variations known as Job's method. This method, which is applicable in UV-visible spectrophotometry, fluorescence, and NMR, consists of observing a chosen parameter (absorbance in UV, fluorescence intensity, or variation of chemical shift in NMR) in the free and complexed state. The variation of the parameter is proportional to the concentration of the complexed molecule, and therefore a function of the molar fraction of the guest molecule, which allows the determination of the stoichiometry by a simple experimental plot. There are other methods to determine the stoichiometry of complexes, such as electrospray mass spectrometry.

Formation constant

If the stoichiometry of the inclusion complex is known, it is then possible to compare the relative strength of several complexes by determining an equilibrium constant, the formation constant, K_f. Some authors use the term association constant (K_{ass}) or even stability constant (K_s) (Fourmentin et al., 2018b).

The formation constant is described in Eqs. (11.3) and (11.4), where CD, I and, CD/I represent the molar concentrations of CD, guest molecule, and inclusion complex, respectively. The unit of K_f is therefore 1/M. Eq. (11.3) is generally valid in dilute solution and with a 1:1 stoichiometry, whereas Eq. (11.4) applies even if 1:2, 2:1, and 2:2 stoichiometries can be observed at higher concentrations, or with specific substrates. The higher the value of K_f, the more stable the complex will be. K_f is the inverse of the dissociation constant of the equilibrium K_d (*i.e.*, $K_f = 1/K_d$).

$$K_f = \frac{[CD_a/I_b]}{[CD]^a\,[I]^b} \tag{11.3}$$

$$K_f = \frac{[CD/I]}{[CD]\,[I]} \tag{11.4}$$

Several factors influence the value of K_f, such as the geometry and polarity of the substrate. When the guest molecule is ionizable, the constant is also pH dependent. Moreover, in general, K_f decreases with increasing temperature. The results obtained are explained by the fact that the enthalpy of formation ΔH is less than zero (the most frequent case).

The most used experimental techniques to determine the formation constant are UV-visible spectrophotometry (measurement of the absorbance of the guest molecule at a characteristic wavelength as a function of the CD concentration, the guest concentration being fixed), and NMR spectroscopy (measurement of the chemical shift of the protons of the CD as a function of the guest molecule concentration, the CD concentration being fixed), and static headspace coupled to gas chromatography (measurement of reduction in peak area of volatile guests as a function of the CD concentration, the guest concentration being fixed) (Fourmentin et al., 2018a, 2018b).

Complexation phenomena

In the late 1950s, French and later Cramer showed that the complexation phenomenon was the result of a multitude of interactions between the three components of the system: CD, guest molecule, and solvent, which led to the most stable thermodynamic state. However, their work showed that the mechanism of formation of the inclusion complex is governed essentially by a determining factor, the geometric factor, defining a geometric compatibility. Indeed, from this point of view, the inclusion depends on the relative size of the CD cavity compared to the size of the guest molecule: if the guest is too large, it will not be able to penetrate inside the CD cavity. On the other hand, if it is too small, it will have few interactions with the CD. The steric effect therefore plays an important role in the complexation phenomenon. Also, α-CD more easily complexes small molecules such as aliphatic chains, β-CD can complex aromatic compounds or heterocycles, and γ-CD accommodates macromolecules (macrocycles, steroids, etc.). For example, naphthalene is too bulky for α-CD but forms a stable complex with β-CD, while anthracene only enters γ-CD (Crini, French, et al., 2021; Crini, 2014; Szejtli, 1982).

Five elementary steps have been proposed to explain the formation of a CD/guest inclusion complexe, hypothesizing that organic substrates dissolved in water exhibit a preference for hydrophobic environments (Frömming & Szejtli, 1994; Szejtli, 1982):

1. In aqueous solution, the cavity of a CD is occupied by water molecules that are in an unfavorable energy state, due to polar–apolar interactions, and are therefore easily replaced by appropriate molecules, less polar than water. The guest substrate then approaches the CD molecule, and the water molecules begin to escape from the cavity to acquire an energy level corresponding to that of the gaseous state. van der Waals interactions and the number of hydrogen bonds decrease, while the degrees of freedom of translation and rotation of the released water molecules increase.
2. The guest molecule is then released from the water layer that coats it and acquires a different state. The water layer disperses and rearranges.
3. The guest molecule, considered to be in the state of a perfect gas, penetrates inside the cavity, and the complex is stabilized by van der Waals interactions and/or by hydrogen bonds.
4. The expelled water molecules rearrange and create hydrogen bonds between them, which stabilize the complex.
5. The structure of the water is restored around the part of the guest molecule that remains in contact with the solvent and integrated into the water layer, hydrating the CD.

In 1976, the German biochemist and crystallographer Wolfram Saenger provided two additional clarifications to explain the formation of an inclusion complex: (1) CD molecules absorb the energy of water molecules retained in the cavity and then take on a relaxed conformation (in this state, water molecules can be easily substituted by another potential substrate) and (2) the substrate associates with the outer surface of the CD and only penetrates the cavity after absorbing the activation energy (transition from the higher-energy conformation of the water-CD complex to the lower-energy conformation of the CD-guest molecule complex). Saenger introduced the concept of releasing strain energy within the CD molecule during the formation of a complex. At the same time, the American chemist Myron Lee Bender published the same conclusions (Crini, 2014). However, the conclusions of these two remarkable scientific authorities, essentially based on the formation of hydrophobic interactions, have aroused many controversies and debates mainly concerning the nature of the binding forces.

An apolar binding force is normally characterized by a very favorable entropy change. However, it turns out that the formation of an inclusion complex is accompanied by a favorable enthalpy change and an unfavorable entropy change. The main force governing inclusion, in the case of complexation in aqueous solution, is the replacement of the high-energy water molecules contained in the cavity by the guest. The complexation is thus accompanied by a desolvation of the guest molecule and the CD cavity. These water molecules are thus transferred into the solution while the guest interacts with the interior of the CD. The release of water molecules leads to an increase in the entropy of the system ($\Delta S° > 0$). However, the hydrophobic effects compensate for the unfavorable effects (desolvation of the substrates) and lead to a negative enthalpy change ($\Delta H° < 0$). The entropy change has several origins. The first contribution to this entropy change is that the water, initially surrounding the apolar guest in an ordered manner, becomes disorganized. The second is that there is, at the same time, a reduction in the degrees of freedom of translation and rotation due to the host/guest complexation. These two opposite effects are almost equal. Also, the overall entropy change can be very small or even zero. The formation of an inclusion complex is associated with an energetically favorable interaction between a relatively apolar guest molecule and an imperfectly solvated hydrophobic cavity (Szejtli, 1982).

In 1978, Bender was the first researcher to observe and interpret the fact that complexation is indeed accompanied by a gain in enthalpy and a loss of entropy. The more the guest molecule penetrates the CD cavity, the greater the change in enthalpy, and the greater the stability of the complex, a phenomenon that is all the more true as the molecule is apolar. Bender then proposed four interpretations (Crini, 2014):

1. The establishment of weak van der Waals type bonds (interactions of dipole–dipole nature) between the host and the guest.
2. The formation of hydrogen bonds between the guest molecule and the hydroxyls of the CD.
3. The release of water molecules with high energy during substitution by the guest molecule.
4. The release of strain energy within the CD molecule during complex formation.

These interpretations will be taken up and formally summarized, reformulated, and demonstrated in 1982 by Szejtli, and illustrated by a famous scheme (Fig. 11.2C). This schematic illustration is still a reference in academic and industrial communities interested in CDs, in which he explained the enthalpy gain by the spontaneous arrival of the guest, expelling the active water molecules retained in the apolar cavity of the CD in aqueous solution (Szejtli & Osa, 1996; Szejtli, 1982, 1984, 1996, 1997, 1998; Szente, 2004). In the formation of the complex, van der Waals bonds and hydrogen bonds have a major role, strongly involved in the energetic stabilization of the system to create the apolar–apolar association. However, Szejtli assumed that these bonds were too weak to confer on their own sufficient stability of the complex and also highlighted steric interactions (Crini, Fenyvesi, et al., 2021).

His three main conclusions were: (1) the guest molecule, less polar than water, directly replaced the water molecules in the cavity; (2) the CD molecules absorbed the energy of the water molecules retained in the cavity; and (3) the organic guest dissolved in water entered in the cavity because it had a preference for hydrophobic environment (Crini, Fenyvesi, et al., 2021). The complexation phenomenon resulted from a multitude of interactions between the three components of the system CD-substrate-solvent leading to a state that was more thermodynamically stable overall.

Despite the numerous published results demonstrating the existence of this set of interpretations that we have just commented on, there is still a debate on the mechanisms and forces that act to achieve complexation. Some researchers conclude that hydrophobic and van der Waals interactions predominate during complex formation, others emphasize steric effects and hydrogen bonds, and still others question the main conclusion of Professor Szejtli's work by stating that the water present inside the CD cavity plays little role in the complexation phenomenon.

To summarize all that we have just described, it is important to remember that the formation of an inclusion complex is the result of a set of intermolecular interactions (noncovalent forces) and that each type of interaction plays a role in the complexation (Szejtli, 1982, 1984, 1996, 1998).

Conclusions

In this chapter, we have described the structure and properties of CDs, naturally occurring cage-like molecules, with a particular emphasis on their remarkable ability to form host-guest inclusion complexes. This molecular encapsulation results from various intermolecular interactions, including van der Waals forces, electrostatic interactions, hydrophobic interactions, and hydrogen bonding, each of which plays a role in the complexation process. The stability of these complexes depends primarily on the ability of the guest molecule to fit into the CD cavity and its polarity. The industrial development of CDs, which began in the 1980s, has highlighted their complexing properties. Indeed, CD-based encapsulation is widely used in various fields, including chemistry, biology, analytical science, pharmaceuticals, medicine, and the food industry, all of which are discussed in this book.

Acknowledgments

GC, DL, CM, and NMC would like to thank the *Université Marie et Louis Pasteur* (Besançon, France), the University of Bari (Bari, Italy), and the *Fonds européen de développement régional* (FEDER) for funding this work under the CHRYSALIDE II 2025–2027 project, which allowed a Franco-Italian partnership to be established between the universities of Besancon and Bari, and to welcome DL and CM as visiting researchers in France.

References

Atwood, JL, Davies, JED, & MacNicol, DD (1984). *Inclusion compounds*. London: Oxford University Press.

Crini, G. (2014). Review: A history of cyclodextrins. *Chemical Reviews, 114*, 10940–10975. https://doi.org/10.1021/cr500081p.

Crini, G. (2020a). The contribution of Franz Schardinger to cyclodextrins: A tribute on the occasion of the centenary of his death. *Journal of Inclusion Phenomena and Macrocyclic Chemistry, 97*, 19–28. https://doi.org/10.1007/s10847-020-00990-3.

Crini, G. (2020b). Twenty years of dextrin research: A tribute to Professor Hans Pringsheim (1876–1940). *Journal of Inclusion Phenomena and Macrocyclic Chemistry, 98*, 11–27. https://doi.org/10.1007/s10847-020-01013-x.

Crini, G. (2021). The contribution of Professor Karrer (1889–1971) to dextrins. *Journal of Inclusion Phenomena and Macrocyclic Chemistry, 99*, 155–167. https://doi.org/10.1007/s10847-021-01049-7.

Crini, G., Fenyvesi, É., & Szente, L. (2021). Outstanding contribution of Professor József Szejtli to cyclodextrin applications in foods, cosmetics, drugs, chromatography and biotechnology: A review. *Environmental Chemistry Letters, 19*, 2619–2641. https://doi.org/10.1007/s10311-020-01170-y.

Crini, G., Fourmentin, S., Fenyvesi, E., Torri, G., Fourmentin, M., & Morin-Crini, N. (2018). Cyclodextrins – From molecules to applications. *Environmental Chemistry Letters, 16*, 1316–1375. https://doi.org/10.1007/s10311-018-0763-2.

Crini, G., French, A. D., Kainuma, K., Jane, J. L., & Szente, L. (2021). Contributions of Dexter French (1918–1981) to cycloamylose/cyclodextrin and starch science. *Carbohydrate Polymers, 257*, 117620. https://doi.org/10.1016/j.carbpol.2021.117620.

Crini, G., Fourmentin, S., & Lichtfouse, E. (2020). *The history of cyclodextrins. Environmental chemistry for a sustainable world 52*. Springer Nature ISBN: 978-3-030–49307-3.

Crini, G., & Torri, G. (2017). In memoriam Benito Casu (1927–2016). *Carbohydrate Research, 134*, 227–228. https://doi.org/10.1016/j.carres.2017.06.001.

Cyclodextrins. In J. Szejtli, & T. Osa (Vol. Eds.), *Comprehensive supramolecular chemistry: 3*Pergamon Oxford, UK.

Duchêne, D, Bochot, A, & Loftsson, T (2009). Cyclodextrins and their use in pharmacy and cosmetology. *STP Pharma Pratiques, 19*, 15–27.

Fenyvesi, É., Puskás, I., & Szente, L. (2019). Applications of steroid drugs entrapped in cyclodextrins. *Environmental Chemistry Letters, 17*, 375–391. https://doi.org/10.1007/s10311-018-0807-7.

Fourmentin, S., Crini, G., & Lichtfouse, E. (2018a). *Cyclodextrins applications in medicine, food, environment and liquid crystals. Environmental chemistry for a sustainable world.* Switzerland: Springer Nature ISBN: 978-3-319–76162-6.

Fourmentin, S., Crini, G., & Lichtfouse, E. (2018b). *Cyclodextrins fundamentals, reactivity and analysis. environmental chemistry for a sustainable world*. Switzerland: Springer Nature ISBN: 978-3-319-76159-6.

Frömming, K. H., & Szejtli, J. (1994). *Cyclodextrins in pharmacy. Topics in inclusion science* (5). Kluwer Academic Publishers. https://doi.org/10.1007/978-94-015–8277-3.

Jacob, S., & Nair, A. B. (2018). Cyclodextrin complexes: Perspective from drug delivery and formulation. *Drug Development Research, 79*(5), 201–217. https://doi.org/10.1002/ddr.21452.

Kainuma, K. (1984). Starch oligosaccharides: Linear, branched, and cyclic. In R. L. Whistler, J. N. BeMiller, & E. F. Paschall (Eds.). *Starch - chemistry and technology* (pp. 125–152).

Academic Press Ltd. chapter V https://doi.org/10.1016/B978-0-12–746270-7.50011-2 2nd edition.

Jansook, P., Ogawa, N., & Loftsson, T. (2018). Cyclodextrins: Structure, physicochemical properties and pharmaceutical applications. *International Journal of Pharmaceutics, 535*(1–2), 272–284. https://doi.org/10.1016/j.ijpharm.2017.11.018.

Larsen, K. L. (2002). Large cyclodextrins. *Journal of Inclusion Phenomena, 43*, 1–13. https://doi.org/10.1023/A:1020494503684.

Liu, C., Crini, G., Wilson, L. D., Balasubramanian, P., & Li, F. Y. (2024). Removal of contaminants present in water and wastewater by cyclodextrin-based adsorbents: A bibliometric review from 1993 to 2022. *Environmental Pollution, 348*, 123815. https://doi.org/10.1016/j.envpol.2024.123815.

Liu, Q. M., Zhou, Y., Lu, J., & Zhou, Y. B. (2020). Novel cyclodextrin-based adsorbents for removing pollutants from wastewater: A critical review. *Chemosphere, 241*, 125043. https://doi.org/10.1016/j.chemosphere.2019.125043.

Loftsson, T., & Duchêne, D. (2007). Cyclodextrins and their pharmaceutical applications. *International Journal of Pharmaceutics, 329*(1–2), 1–11. https://doi.org/10.1016/j.ijpharm.2006.10.044.

Loftsson, T., Sigurdsson, H. H., & Jansook, P. (2023). Anomalous properties of cyclodextrins and their complexes in aqueous solutions. *Materials, 16*(6), 2223. https://doi.org/10.3390/ma16062223.

Maazaoui, R., & Abderrahim, R. (2015). Applications of cyclodextrins: formation of inclusion complexes and their characterization. *International Journal of Advanced Research, 3*(2), 1030–1044.

Morin-Crini, N., Fourmentin, S., Fenyvesi, É., Lichtfouse, E., Torri, G., Fourmentin, M., & Crini, G. (2021). 130 years of cyclodextrin discovery for health, food, agriculture and the industry: A review. *Environmental Chemistry Letters, 19*, 2581–2617. https://doi.org/10.1007/s10311-020-01156-w.

Moufawad, T., Moura, L., Ferreira, M., Bricout, H., Tilloy, S., Monflier, E., Costa Gomes, M., Landy, D., & Fourmentin, S. (2019). First evidence of cyclodextrin inclusion complexes in a deep eutectic solvent. *ACS Sustainable Chemistry & Engineering, 7*(6), 6345–6351. https://doi.org/10.1021/acssuschemeng.9b00044.

Musuc, A. M. (2024). Cyclodextrins: Advances in chemistry, toxicology, and multifaceted applications. *Molecules, 29*(22), 5319. https://doi.org/10.3390/molecules29225319.

Nitin, M. G., Priya, A. S., Nitesh, D. P., Rakhi, S., Ashish, K. K., & Ganesh, D. S. (2024). Methods of preparation and characterization of cyclodextrin encapsulated inclusion complex: Review. *International Journal of Scientific Research in Science and Technology, 11*(3), 563–571. https://doi.org/10.32628/ijsrst24113129.

Rekharsky, M. V., & Inoue, Y. (1998). Complexation thermodynamics of cyclodextrins. *Chemical Reviews, 98*(5), 1875–1918. 10.1021/cr970015o.

Sandilya, A. A., Natarajan, U., & Priya, M. H. (2020). Molecular view into the cyclodextrin cavity: Structure and hydration. *ACS Omega, 5*(40), 25655–25667. https://doi.org/10.1021/acsomega.0c02760.

Sharma, N., & Baldi, A. (2016). Exploring versatile applications of cyclodextrins: An overview. *Drug Delivery, 23*(3), 739–757. https://doi.org/10.3109/10717544.2014.938839.

Szejtli, J. (1982). *Cyclodextrins and their inclusion complexes*. Akadémiai Kiadó ISBN: 963-05-2850-9.

Szejtli, J. (1984). Industrial applications of cyclodextrins. In J. L. Atwood, J. E. D. Davies, & D. D. MacNicol (Eds.). *Inclusion compounds* (pp. 331–390). Academic Press 3.

Szejtli, J. (1991). The use of cyclodextrins in biotechnological operations. In D. Duchêne (Ed.). *New trends in cyclodextrins and derivatives* (pp. 595–626). Éditions de santé chapter 17.

Szejtli, J. (1996). Chemistry, physical and biological properties of cyclodextrins. In J. L. Atwood, E. D. Davies, D. D. MacNicol, & F. Vögtle (Vol. Eds.), *Comprehensive supramolecular chemistry: 3*, (pp. 5–40). Pergamon Oxford, UK cyclodextrins. Szejtli J and Osa T (eds).

Szejtli, J. (1997). Utilization of cyclodextrins in industrial products and processes. *Journal of Materials Chemistry, 7*, 575–587. https://doi.org/10.1039/A605235E.

Szejtli, J. (1998). Introduction and general overview of cyclodextrin chemistry. *Chemical Reviews, 98*, 1743–1753.

Szente, L. (2004). Prof. Dr. Szejtli József. 1933–2004. *Cyclodextrin News, 18*, 268–269.

Szente, L., & Fenyvesi, É. (2016). József Szejtli and his important findings in the pharmaceutical application of cyclodextrins. *Magyar Kémiai Folyóirat, 122*, 60–64.

Szente, L., & Szejtli, J. (1999). Highly soluble cyclodextrin derivatives: Chemistry, properties, and trends in development. *Advanced Drug Delivery Reviews, 36*, 17–28. https://doi.org/10.1016/S0169–409X(98)00092-1.

Torri, G., Naggi, A., Lichtfouse, E., & Crini, G. (2022). Professor Casu's contribution to cyclodextrins, the remarkable cage-shaped molecules: A review. *Environmental Chemistry Letters, 20*, 2085–2095. https://doi.org/10.1007/s10311-022-01417-w.

CHAPTER

Current status of cyclodextrin production

12

Dimple S. Pardhi[1], Khushbu Rabadiya[1], Rushikesh Joshi[2] and Kiransinh N. Rajput[1]

[1]Department of Microbiology & Biotechnology, University School of Sciences, Gujarat University, Ahmedabad, Gujarat, India,

[2]Department of Biochemistry & Forensic Science, University School of Sciences, Gujarat University, Ahmedabad, Gujarat, India

Introduction

Cyclodextrin (CD) production is an industrially important process because of the unique molecular structure of CDs, which can form non-covalent inclusion complexes with a variety of hydrophobic molecules and help develop a wide range of industrial products. The complexation behavior of CDs contributes to an improvement of the solubility, stability, and controlled release of the guest compounds, which is why they are highly utilized in the pharmaceutical, food, and cosmetic industries (Pardhi et al., 2023). In the pharmaceutical industry, CDs are used to increase drug bioavailability, mask undesired tastes, reduce irritation, and protect active pharmaceutical ingredients (APIs) from degradation. Furthermore, their use in food and cosmetic products for flavor encapsulation and stabilization of volatile or sensitive components depicts the industrial need for CDs.

Apart from this, CD production is gaining significance due to its developing role in textile, environmental, and chemical processing industries. The inclusion complex of CDs is known to offer an innovative solution for the removal of organic pollutants and coloring agents, such as pesticides, dyes, and other hazardous compounds from wastewater. In textile applications, CD-based materials are used to impart functional finishes such as controlled release of fragrances, antimicrobial agents, and bioactive compounds (Del Valle, 2004). Notably, the large-scale manufacturing of CDs using cyclodextrin glucanotransferase (CGTase) is an eco-friendly approach as it employs renewable starch substrates, mild reaction conditions, and low amounts of toxic solvents. These factors underscore the importance of CD production as a sustainable, high-value product for the industry.

Cyclodextrin Glucanotransferase. DOI: https://doi.org/10.1016/B978-0-443-41694-1.00009-5

Industrial demand trends for cyclodextrins: α, β, γ

Among the naturally occurring CDs, β-cyclodextrin (β-CD) accounts for the largest share of global industrial demand due to its favorable cavity size, cost-effectiveness, and broad applicability across pharmaceutical, food, cosmetic, and chemical industries. β-CD is extensively used to improve the solubility and stability of poorly water-soluble drugs, as well as for flavor and odor encapsulation in food products (Loftsson & Duchêne, 2007). Its widespread regulatory acceptance and ease of production using CGTase-based enzymatic processes have further strengthened its dominant industrial position.

In contrast, α-cyclodextrin (α-CD) shows growing but relatively moderate demand, primarily driven by its use in the food and nutraceutical sectors. Owing to its smaller cavity size and resistance to digestion, α-CD is increasingly used as a dietary fiber, fat-binding agent, and functional food ingredient. Its recognized physiological benefits, including cholesterol-lowering effects, have contributed to rising demand in health-oriented food formulations and dietary supplements, particularly in developed markets (Astray et al., 2009).

The demand for γ-cyclodextrin (γ-CD), although currently lower than that of β-CD, is increasing steadily due to its larger cavity size and superior solubility. These properties make γ-CD especially suitable for encapsulating bulky molecules in advanced pharmaceutical formulations, functional foods, and speciality chemical applications (Li et al., 2007). Regulatory approval of γ-CD for food use and ongoing research into high-value drug delivery systems are expected to further expand its industrial utilization, particularly in niche and high-performance applications.

Market drivers: pharmaceutical, food, cosmetic, and environmental sectors

The status of CD production is strongly influenced by sustained growth in the pharmaceutical sector, which remains the primary market driver. Increasing emphasis on formulation strategies for poorly water-soluble compounds, along with the expansion of advanced drug delivery systems, has led to a steady rise in industrial demand for CDs. This trend has encouraged manufacturers to optimize enzymatic production processes, improve yield efficiency, and diversify CD types to meet sector-specific performance requirements (Muniz et al., 2024).

Simultaneously, the food and cosmetic industries have emerged as significant contributors to CD market growth. Rising consumer preference for clean-label, natural, and functionally enhanced products has accelerated the adoption of CDs as multifunctional ingredients (Mangolim et al., 2014). This shift has led to increased production volumes, particularly for food-grade and cosmetic-grade CDs, and has stimulated interest in scalable, cost-effective manufacturing technologies that comply with regulatory and safety standards (Astray et al., 2009).

Additionally, growing awareness of environmental sustainability has positioned CDs as attractive materials in eco-oriented industrial processes. The demand for greener alternatives in pollution control, waste treatment, and sustainable materials has indirectly driven CD production, especially those synthesized through environmentally gentle enzymatic paths (Morillo et al., 2020). Collectively, these market drivers reflect a transition from niche utilization toward large-scale, diversified industrial production of CDs, reinforcing their strategic importance in modern bioprocessing industries.

Industrial scenario of cyclodextrin production

Enzymatic cyclodextrin production

Industrial production of CDs is primarily an enzymatic conversion process that transforms starch raw materials into cyclic oligosaccharides using CGTase (Crini, 2014). The overall production flow can be understood in stages, such as cultivation and preparation of the enzyme, starch pretreatment, enzymatic conversion, and downstream separation and purification. These stages are common to most industrial setups, regardless of scale.

In commercial processes, CGTase is produced by submerged fermentation using starch or dextrin as a carbon source, and the enzyme is then applied to pretreated starch substrates such as corn, potato, or tapioca starch. Starch pretreatment typically involves gelatinization at elevated temperatures to disrupt crystalline granules, followed by partial liquefaction using α-amylase to reduce viscosity and enhance substrate accessibility, while maintaining sufficient chain length for efficient cyclization. Under controlled pH (generally 6.0–10.5) and temperature conditions (40°C–70°C depending on enzyme origin), CGTase catalyzes intramolecular transglycosylation reactions converting linear α-1,4-glucans into cyclic oligosaccharides, namely α-, β-, and γ-CD; however, the product distribution depends strongly on the microbial source of CGTase, with *Bacillus circulans* typically favoring β-CD production, whereas enzymes from *Paenibacillus macerans* may yield relatively higher proportions of γ-CD (Pardhi et al., 2023).

Reaction equilibrium can be shifted toward a desired CD by incorporating complexing agents that selectively precipitate one CD form, thereby enhancing overall yield. Following enzymatic conversion, downstream processing constitutes a major economic step and involves selective crystallization (particularly efficient for β-CD due to its lower solubility), membrane filtration, solvent extraction, and recrystallization to obtain food- or pharmaceutical-grade purity. Industrial-scale production is carried out by companies such as Wacker Chemie AG and Roquette, which use optimized fermentation, enzymatic conversion, and purification to manufacture high-quality CDs for use in pharmaceuticals, food technology, and biotechnology. Recent technological advancements include the use of immobilized CGTase systems for enzyme reuse, recombinant expression for improved productivity and altered specificity, and continuous membrane-based bioprocesses, all of which contribute to enhanced process efficiency, reduced production costs, and improved sustainability in large-scale CD manufacturing (Biwer et al., 2002; Del Valle, 2004; Szejtli, 1998).

The generalized steps for industrial CD production are as follows:

1. **Starch liquefaction and gelatinization:** Industrial CD production begins with starch (typically corn, potato, or tapioca starch) dispersed in water or a suitable buffer solution and gelatinized by heating. Partial liquefaction is often carried out to improve substrate accessibility for the enzyme. This step ensures uniform starch availability for enzymatic conversion.
2. **Enzymatic conversion using CGTase:** The gelatinized starch slurry is treated with CGTase under controlled pH and temperature conditions. CGTase catalyzes intramolecular transglycosylation reactions, converting linear starch chains into cyclic oligosaccharides, that is, α-, β-, and γ-CDs. The enzyme source and reaction conditions strongly influence CD composition and yield, making this step critical for commercial viability.
3. **Enzyme inactivation and solid removal**: After completion of the reaction, the enzyme is inactivated (commonly by heat treatment), and insoluble materials are removed by filtration or centrifugation to obtain a clarified CD-containing solution.
4. **CD separation and purification:** Individual CDs are separated using selective precipitation (commonly for β-CD), complexation, membrane filtration, or chromatographic techniques. This stage determines product purity and is one of the most cost-intensive steps in industrial production.
5. **Crystallization and drying:** Purified CDs are crystallized, washed to remove impurities, and dried using spray drying or vacuum drying. The final product is milled and standardized to meet pharmaceutical, food, or industrial specifications.

The efficiency, specificity, and stability of the enzyme are critical parameters for industrial-scale production of CDs. Industrial CD production primarily depends on CGTase from *Bacillus* spp., while a broader range of species are used in research and niche production. The product profile of CDs (α, β, or γ) can be tailored by selecting specific CGTases or engineering enzyme variants with altered substrate binding domains (Pardhi et al., 2025a). Moreover, the yield and selectivity of the process can be enhanced through reaction optimization or the use of auxiliary agents like organic solvents and complexing agents. Optimizing enzymatic production is essential to meet the growing global demand for CDs.

At the industrial level, CD manufacturing relies on specialized enzyme-producing industries that supply CGTase as well as on vertically integrated CD producers that implement or tailor CGTase internally for optimized production. Globally recognized enzyme suppliers such as Novozymes A/S, Amano Enzyme Inc., and Hayashibara Co., Ltd. produce standardized CGTase preparations derived from bacterial sources, which are distributed to CD manufacturers for enzymatic starch conversion and consistent performance in large-scale processes. CGTase from *Thermoanaerobacter* sp., sold commercially as Toruzyme by Novozymes, and CGTase from *P. macerans*, marketed by Amano Enzyme, are well-known enzyme products used in CD synthesis. In addition, major CD producers like Wacker Chemie AG, Roquette Frères, and Ensuiko Sugar Refining Co., Ltd. often develop or customize their own CGTase systems within integrated production operations to achieve desired product distributions (α, β, γ) and commercial efficiency, reflecting a mix of external supply and in-house enzyme adaptation that underpins industrial CD production.

Modified cyclodextrin production

In the industrial production of CDs, a clear distinction must be made between the biological construction of the macrocycle and its subsequent chemical functionalization: while native CDs are derived exclusively through the enzymatic biotransformation of starch, their modified counterparts are synthesized via traditional chemical methods to tailor their solubility and complexation properties. Even though total chemical synthesis of native CDs has been demonstrated at the laboratory scale, it is not commercially feasible. Therefore, industrial production relies exclusively on the enzymatic conversion of starch using CGTase.

Although the production of native CDs occurs efficiently by enzymatic degradation of starch, the production of modified CDs requires specific chemical intervention to improve their physiochemical characteristics. The main aim of chemical synthesis of CDs is to overcome the limitations and drawbacks of native CDs, such as the poor aqueous solubility of β-CD. By replacing the primary and/or secondary alcoholic groups with functional groups such as hydroxypropyl, methyl, and sulfobutyl ether, chemists can significantly improve the microenvironment around the cavity and the overall solubility of the macrocycle. This can be achieved by nucleophilic substitution reactions in alkaline conditions, in which the reactivity of C-6 hydroxyl groups can be exploited compared to C-2 and C-3 positions, as discussed in Szejtli (1998).

The general procedure for the synthesis of these derivatives is as follows: The synthesis of these derivatives is generally performed by following a selective functionalization approach (Fig. 12.1). The challenge associated with this approach is that a given molecule of CD has several identical hydroxyl groups (21 for β-CD), and hence the challenge is to attain a particular degree of substitution. The general approach to the synthesis of these derivatives is to dissolve the native CD in a solvent such as dimethylformamide or water, then add an etherifying or esterifying reagent along with a catalyst. For instance, the synthesis of a widely used CD derivative, hydroxypropyl-β-cyclodextrin (HP-β-CD), is achieved by reacting β-CD with propylene oxide in a basic environment. The final product is not a single compound but a mixture of isomers with varying degrees of substitution, which is then purified by dialysis to achieve uniform complexation behavior.

Beyond basic substitution, sophisticated chemical techniques have enabled the synthesis of "bridged" or "capped" CDs, which can provide a more rigid, hydrophobic microenvironment for targeted drug molecules. These are synthesized using "capping reagents" that react with two different hydroxyl groups on the same ring, thus effectively narrowing the entrance to the CD cavity. In addition to this, the synthesis of

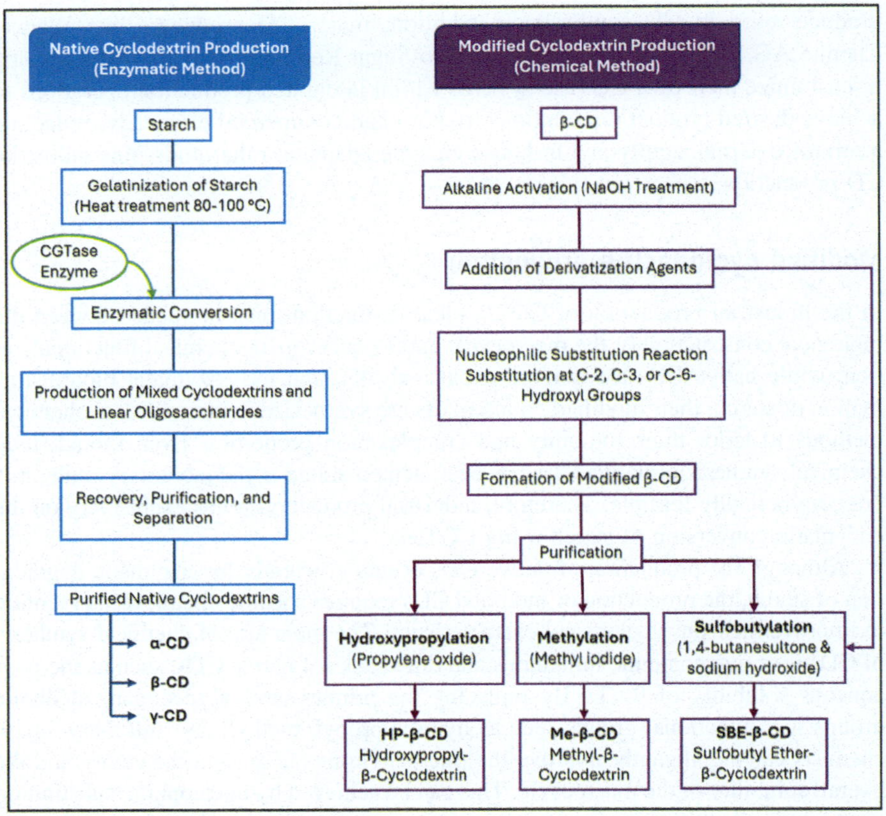

FIGURE 12.1

Process of native and modified cyclodextrin (*CD*) production.

"click-cyclodextrins" has also gained significant momentum, using copper-catalyzed azide-alkyne cycloaddition to conjugate complex biological ligands with the CD framework. Such a degree of structural modification cannot be achieved through enzymatic means, thus making chemical synthesis the basis of supramolecular chemistry.

Most importantly, after completing either simple or complex modifications, the products need to be purified and quality-controlled. After the completion of the derivatization reactions, purification steps such as neutralization, dialysis, precipitation, filtration, and spray drying are carried out. Although the total chemical synthesis of CDs is highly complex and economically infeasible, modifying β-CD, produced via an enzymatic route, has been industrially viable. This has ensured the extension of the applications of chemically modified CDs.

The differences in CD production strategies (Table 12.1) significantly influence market segmentation and value distribution.

Table 12.1 Comparative analysis of cyclodextrin (*CD*) production by enzymatic and chemical methods.

Parameter	Native CD (enzymatic method)	Modified CD (chemical method)
Starting material	Starch (renewable source)	β-CD (enzymatically produced)
Key catalyst/ reagent	Cyclodextrin glycosyltransferase (CGTase)	Chemical reagents (e.g., propylene oxide, methylating agents, sulfobutylating agents)
Type of reaction	Enzymatic transglycosylation and cyclization	Hydroxyl group substitution (etherification/alkylation)
Reaction conditions	Mild, aqueous, eco-friendly	Alkaline medium, controlled temperature, solvent-based
Main products	α-CD, β-CD, γ-CD	HP-β-CD, ME-β-CD, SE-β-CD
Structural nature	Native cyclic oligosaccharide	Substituted derivatives with altered physicochemical properties
Degree of substitution (DS)	Not applicable	Critical parameter; determines solubility and performance
Solubility	β-CD has limited water solubility	Significantly enhanced aqueous solubility
Production complexity	Relatively simple and optimized	Multistep reaction and purification
Purification process	Crystallization and separation	Neutralization, dialysis, spray drying
Industrial scalability	Highly scalable	Scalable but cost-intensive
Environmental impact	Green biocatalytic process	Chemical waste management is required
Regulatory acceptance	Widely accepted in food and pharma	Certain derivatives approved for injectable use
Cost of production	Lower cost	Higher cost due to reagents and processing
Market share (volume)	Dominates global production volume	Lower production volume
Market share (revenue)	Moderate revenue	High revenue contribution per unit
Application sector	Food, agriculture, cosmetics, basic pharma	Advanced pharmaceuticals, parenteral formulations, biotechnology
Patent & innovation activity	Limited innovation (mature technology)	High patent activity and R&D focus
Value addition level	Basic functional excipient	High-value drug delivery enhancer
Market growth driver	Bulk demand and cost efficiency	Pharmaceutical innovation and solubility enhancement

Raw materials and substrates

Industrial production of CDs relies on starch and starch-derived substrates as primary raw materials because CGTase specifically acts on starch chains to form cyclic oligosaccharides. The most widely used substrate in commercial CD manufacture is corn starch, which is readily available in large quantities, cost-effective, and has a high amylopectin content that favors CD formation (Indrajit & Singh, 2024). Corn starch is often liquefied to reduce viscosity and improve enzyme accessibility before enzymatic conversion by CGTase in industrial processes; it remains the dominant source due to its global agricultural production and processing infrastructure. Soluble starch is commonly used in laboratory studies due to its pregelatinized nature and high enzymatic accessibility, leading to high CD yields. However, its relatively high cost makes it less suitable for industrial-scale applications (Wadetwar et al., 2006).

In addition to corn starch, a variety of agricultural starches are used and investigated as substrates, depending on regional availability and specific production goals. The effective substrates with CGTase efficiency linked to starch morphology and crystallinity include potato starch, cassava (tapioca) starch, pea starch, amaranth starch, rice starch, wheat starch, and sago starch that support CD formation (Arce-Vázquez and Ponce-Alquicira, 2016; Benavent-Gil et al., 2021; Johnson et al., 2010; Laga, 2008; Pishtiyski et al., 2008; Rajput et al., 2016; Vongpichayapaiboon et al., 2016; Wu et al., 2022). For instance, research comparing starch substrates reported that potato and sago starches both yield significant amounts of β-CD, with potato starch often outperforming other grains under similar conditions, likely due to their granule structures and amylopectin contents (Pardhi et al., 2025b).

Moreover, native starches such as tapioca (cassava) and corn starch are preferred at the industrial level, as they offer a more favorable balance between yield, cost, and large-scale availability (Pardhi et al., 2025c). Tapioca and cassava starches have gained industrial relevance due to their low processing cost and efficient conversion, while corn starch remains the most widely adopted substrate because of its consistent supply and scalability. Emerging sources such as amaranth starch have also demonstrated high CD yields and may become viable alternatives in the future. Overall, an ideal industrial substrate must combine high enzymatic accessibility with economic viability and reliable availability to sustain commercial CD production. The comparative data on properties, yield, and cost (USD/kg) of starch substrates used for CD production are presented in Table 12.2.

In industrial CD production, gelatinized starch is generally preferred, as it provides greater enzymatic accessibility for most commercially used CGTases. Although raw starch (RS) has attracted interest because certain thermostable CGTases can act directly on ungelatinized substrates, its application remains limited by low enzyme accessibility, particularly in high-amylose or highly crystalline starches. Wadetwar et al. (2006) reported that the conversion of RS to β-CD is generally restricted unless a prior gelatinization or liquefaction step is employed.

Table 12.2 Properties, cost of different starch substrates and CD yield.

Starch source	Properties	Approx. cost (USD/kg)	CD yield	Industrial significance
Tapioca (cassava)	Low viscosity; high amylopectin content	0.73-0.88	High (up to 87.4%)	Supports efficient CGTase cyclization; widely available; favourable yield-cost balance
Corn (maize)	~23% amylose; moderately branched structure	0.21-0.26	Moderate (24-35%)	Most cost-effective and scalable substrate; slower reaction rate
Amaranth	~80% amylopectin; highly branched starch	0.83-1.04	High (up to ~40%)	High susceptibility to CGTase; promising high-yield substrate; niche availability
Sago	Limited DP (9-12) oligosaccharides; high viscosity	0.62-0.83	Low	Regionally available; unfavourable structure and viscosity reduce CD yield
Rice flour	High viscosity; moderate amylopectin; protein and lipid content	0.42-0.52	Low-moderate	Easily sourced; requires pre-treatment; increased processing cost
Wheat flour	Contains proteins and lipids; non-starch components	0.26-0.36	Low	Common in food industry; CGTase inhibition unless pre-treated

*Reference: Lee & Kim, 1991; Wadetwar et al., 2006; Pérez-Luna et al., 2016.

Therefore, the selection between gelatinized and RS as a substrate must be guided by CGTase compatibility, process requirements, and overall production constraints, making substrate form an important consideration in industrial CD manufacturing.

Downstream processing and yield

At the industrial scale, downstream processing represents a critical cost-determining step in CD manufacturing, often accounting for a significant proportion of total production expenses. Following enzymatic conversion, the reaction mixture typically contains CDs along with residual starch, linear and branched oligosaccharides, unreacted dextrins, and enzyme residues. Industrial recovery strategies therefore focus on selective separation of CDs using methods such as complexation-based precipitation, membrane filtration, adsorption, and crystallization, depending on the type of CD produced (Gastón et al., 2015; Kantzow et al., 2015). Among these, selective precipitation using complexing agents (e.g., organic solvents or guest molecules) has been widely adopted for β-CD because of its relatively low solubility compared with α- and γ-CDs.

CD yield at the industrial level is influenced not only by CGTase activity but also by downstream losses during separation and purification. Typical industrial yields for β-CD range from 30%–50% when using a starch substrate, whereas α- and γ-CDs generally exhibit lower recoverable yields due to their higher solubility and more complex purification requirements. To improve overall recovery, industries employ ultrafiltration and nanofiltration systems (Mimi Sakinah et al., 2014) to remove residual enzymes and low-molecular-weight sugars prior to crystallization, thereby enhancing product concentration and purity. Continuous processing strategies and recycling of unconverted dextrins are also commonly integrated to improve material efficiency and economic feasibility (Fenelon et al., 2015).

In recent years, industrial trends have shifted toward process integration and yield optimization, aiming to reduce solvent usage, energy consumption, and waste generation (Fenelon et al., 2018). Advances in enzyme specificity, reaction control, and downstream integration have enabled improved CD selectivity, reduced the formation of unwanted by-products, and simplified purification. Additionally, the adoption of membrane-based separation and solvent recovery systems has contributed to higher effective yields and lower operational costs (Schöffer et al., 2013). As a result, downstream processing is no longer viewed as a separate step but as an integral component of yield optimization and commercial viability in CD production.

Challenges in industrial cyclodextrin production

The large-scale production of CDs from laboratory to industrial scale faces many technical and practical issues. One of the main limitations is the low efficiency and stability of CGTase during industrial processes. The activity of the enzyme can be affected by high substrate concentrations, high temperatures, and long reaction times, resulting in low conversion rates and the formation of undesirable compounds, such as linear dextrins and oligosaccharides (Del Valle, 2004; Fenelon et al., 2018). The repeated use of the enzyme in industrial production processes requires strict control of pH, temperature, and ionic strength.

Another major challenge is the variability associated with the substrate. Although starch is the main substrate, its type, size, and crystallinity vary according to the source and processing technique. This can result in variations in CD yield and downstream processing efficiency. High-amylose or high-crystallinity starch is more difficult to convert with enzymes, requiring either pretreatment or the use of special CGTases, which are expensive (Lee & Park, 1992; Wadetwar et al., 2006).

Industrial-scale downstream processing is also a major bottleneck. The separation and purification of CDs from starch, linear oligosaccharides, and enzymes require large amounts of solvent or the use of complex separation techniques such as ultrafiltration, crystallization, or extractive bioconversion (Gastón et al., 2015; Kantzow et al., 2015). The recovery of α- and γ-CDs is particularly difficult due to their high solubility, which requires multistep purification processes, thereby increasing the production time and cost.

Moreover, economic and environmental factors also affect the feasibility of industries. The high energy requirements involved in the starch gelatinization process, as well as the use of solvents in the subsequent recovery steps, increase the cost of production and create sustainability issues. Another issue is the disposal of waste materials, especially starch and oligosaccharides, which must be handled in accordance with environmental laws (Astray et al., 2009; Szejtli, 1998).

Global cyclodextrin production statistics

Since 2025, the market for CDs in India has been growing at a substantial rate due to their increasing use in different sectors. As of 2024, the global market for CDs was valued at around USD 1414.66 billion, and it is expected to grow to USD 2852.16 billion by 2035, with a compound annual growth rate (CAGR) of 6.58% (Market Research Future, 2026). In 2023, North America accounted for more than 40% of the global market. Europe and the Asia-Pacific regions are also major contributors to the global market. The Asia-Pacific region is expected to experience substantial market growth due to rising demand in countries such as China, India, and Japan (Market Research Future, 2026). The global demand for CDs in different sectors is shown in Fig. 12.2.

Amongst the various applications, the pharmaceutical industry is the largest consumer of CDs, as they are used to improve the solubility, stability, and bioavailability of drugs (Global Market Insights, 2024). The pharmaceutical industry is expected to register a CAGR of 6.2% over the forecast period (Allied Market Research, 2024).

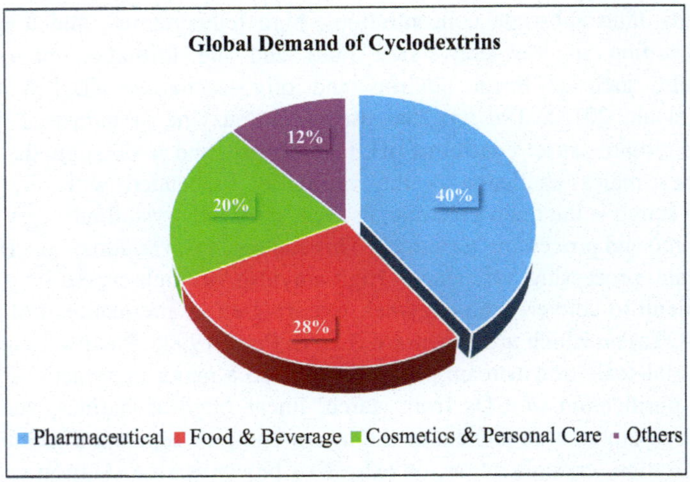

FIGURE 12.2

Worldwide demand for cyclodextrins in different fields.

CDs are finding applications in the food industry for the purpose of flavor encapsulation, odor masking, and stabilization of sensitive ingredients (Market Research Future, 2026). The food and beverage industry is expected to register a CAGR of 4.0% from 2024 to 2032 and reach USD 85.5 million by 2032 (Global Market Insights, 2024). The cosmetic industry uses CDs for the stabilization of volatile compounds, fragrance release, and improving the delivery of active agents. This market is expected to grow substantially due to increasing demand for sophisticated skincare and personal care products. They are used in various industrial processes, including textiles, agriculture, and environmental engineering. The industrial applications market is expected to register moderate growth, with growing applications in different industrial processes (Market Research Future, 2026). To meet the demand for CDs, the industry is actively looking for ways to produce them. The major manufacturers of CDs worldwide are listed in Table 12.3.

Although overall production data for the industrial sector are not often made public due to commercial confidentiality, market research studies do offer important information on CD production levels, market trends, and production costs. Recent studies have shown that the overall consumption of CDs in 2023 was estimated to be in the range of 850–900 tons, with β-CD contributing over 40% to the overall production because of its relatively more efficient production process and widespread industrial applications (Business Research Insights, 2023; 360 Research Reports, 2025). On the other hand, α- and γ-CDs are produced in lower quantities due to lower enzymatic selectivity and higher purification costs, although they remain valuable for specific pharmaceutical and nutraceutical applications.

Market predictions indicate that the CD market will continue to grow at a steady pace through 2026. The market value of CD is expected to increase from

Table 12.3 Worldwide industries manufacturing cyclodextrins (*CDs*).

Company	Country	Year	Specialization[s]/brands[b] and use[u]
CycloLab Cyclodextrin R&D Ltd.	Hungary	1991	[b]R&D, SBECD, analytical services [u]Pharma R&D, injectable formulations
Tokyo Chemical Industry Co., Ltd.	Japan	1970s	[s]Research chemicals, including CDs [b]R&D, pharma, academic
Ensuiko Sugar Refining Co., Ltd.	Japan	1980s	[s]Bulk CD production [u]Food, pharma
Sigma-Aldrich (Merck KGaA)	USA/ Germany	1980s	[s]Lab-grade and specialty CDs [u]Laboratory, diagnostics, research
Wacker Chemie AG	Germany	1988	[b]CAVAMAX® (α-, β-, γ-CD), CAVASOL® (modified CDs) [u]Pharma, food, cosmetics, industrial
Roquette Frères	France	1990s	[b]KLEPTOSE® (β-CD) [u]Pharma, food, nutraceuticals
Ashland Global Holdings Inc.	USA	2000s	[s]Modified CDs, solubility enhancers [u]Drug delivery, stability
Zibo Qianhui Biological Tech Co.	China	2005	[s]Bulk CD and derivatives [u]Pharma, food, agrochemicals
Shandong Xinda Fine Chemical Co.	China	2006	[s]Industrial CDs, sulfated & hydroxypropyl derivatives [u]Pharma, food, textiles
Wellona Pharma	India	2013	[s]Export-quality β-CD and API intermediates [u]Pharmaceuticals, contract manufacturing

USD 243 million in 2025 to USD 257–260 million in 2026, registering a CAGR of around 6% (Global Growth Insights, 2026). Geographically, North America remains a major market for CD due to the well-established pharmaceutical and biotech industries, while Europe remains a stable market due to the region's acceptance and established excipient markets. However, the Asia-Pacific market is expected to grow the fastest due to the development of healthcare infrastructure, the food processing industry, and a low-cost manufacturing base (Global Growth Insights, 2026).

In addition to the increasing demand, the trend of production costs is another important consideration that impacts the production of industrial CDs. Production costs are highly dependent on the price of starch, the availability of the CGTase enzyme, energy costs, and the complexity of downstream processing. It has been reported that fluctuations in corn and potato starch prices have led to a 12%–18% increase in the cost of β-CD production in recent years (PW Consulting, 2026).

Additionally, the production of CD is estimated to be 30%–40% more costly than traditional excipient production, mainly because of the use of enzymes and the complexity of purification (24 Chemical Research, 2025).

In summary, the comparison of production, demand growth until 2026, and cost development reveals a sector in transition from a bulk β-CD market to higher-value CDs and their derivatives. Although cost is a hindrance, demand growth and technological development are anticipated to facilitate the sustainable production of CDs on an industrial scale.

Industrial production of CDs is highly biased toward the production of β-CD, which contributes the major share of the total production due to its high conversion efficiency in the CGTase-catalyzed starch cyclization reaction and relatively lower cost of purification. On the other hand, α- and γ-CDs are produced in relatively lower amounts, as their production requires CGTases with higher selectivity. As a result, the production of β-CD is given prime importance on an industrial scale, while α- and γ-CDs are usually produced on a speciality or high-value scale, primarily for pharmaceutical and nutraceutical applications.

Although production scale is not always publicly disclosed, based on industry disclosures and patent documents (Table 12.4), it is evident that β-CD production occurs at multiton to kiloton scales. On the other hand, the production of γ-CD and chemically modified CDs occurs on a smaller scale due to their specific uses and higher production costs. The difference in the production scale of CDs highlights the challenges of large-scale selective CD synthesis.

The global production of CDs is mainly localized in regions close to starch resources, with well-established enzyme technology and suitable industrial infrastructure. The Asia-Pacific region, especially China and Japan, is a major hub for bulk CD production, thanks to low production costs and abundant carbohydrate resources. The European region is more inclined to produce high-purity CDs, thanks to its well-established regulatory compliance infrastructure and advanced downstream processing technologies. This regional differentiation is more related to their industrial policies than to market demand.

At the industrial scale, the production of CD is also intimately related to the availability and activity of CGTase enzymes. Some companies choose to produce CGTase in-house, while others source it from enzyme suppliers. Industrial-scale production is optimized for the stability and reusability of the enzyme, as the cost of the enzyme is a major factor in the overall production costs. This integration of production also highlights the significance of biocatalyst optimization in industrial production efficiency.

One of the challenges in evaluating the global production statistics of CD is the lack of publicly available information on the production data. This is because most of the companies producing CDs consider their production data as confidential information. This is due to the companies' competitive nature. Therefore the evaluation of global CD production is primarily based on capacity and market indicators. This is also the case when evaluating the global production of CDs using enzyme-based industrial processes.

Table 12.4 Patents related to cyclodextrin (*CD*) production and applications.

Patent number	Country/patent office	Type of CD involved	Application area	Reference (Applicant)
EP 0017242 A1	European Patent Office (EPO)	α-CD, β-CD	Enzymatic production of CDs	Sanraku-Ocean Co., Ltd (1980)
US 4,477,568	United States Patent Office (USPTO)	α-CD, β-CD, γ-CD	Industrial manufacture of CDs	Hokse et al. (1984)
EP 0614971 A3	European Patent Office (EPO)	β-CD	Purification of β-CDs	Amano Pharmaceutical Co., Ltd. (1994)
US 6,660,804 B1	United States Patent Office (USPTO)	α-, β-, γ-CD derivatives	CD polymers and inclusion complexes	Université des Sciences et Technologies de Lille (2003)
US 5,241,059	United States Patent Office (USPTO)	CD derivatives	Modified CDs for industrial applications	Toppan Printing Co., Ltd. (1993)
EP 0301847 A1	European Patent Office (EPO)	α-, β-, γ-CD	CD inclusion complexes (detergent & formulations)	Microgenics Corporation (1989)
US 2017/ 0096647 A1	United States Patent Office (USPTO)	CGTase-derived CDs	Enzyme engineering for CD production	Green Biologics Ltd. (2017)

Modified cyclodextrin market and role

The modified CD market, often referred to as the CD derivatives segment, is emerging as a specialized and fast-growing portion of the broader CD industry. In 2024, the global CD derivatives market was valued at approximately USD 124.82 million and is projected to reach around USD 195.3 million by 2033, with a CAGR of about 5.1% over the forecast period. This growth is underpinned by increasing adoption across pharmaceuticals, food and beverage, and cosmetics, where derivatives such as HP-β-CD and methylated CDs are preferred for their enhanced solubility and complexation properties compared to native forms. The pharmaceutical sector alone accounts for more than 42% of total global demand for these derivatives, underscoring their central role in advanced drug formulations that improve the bioavailability and stability of APIs (Global Growth Insights, 2025).

A major driver of the modified CD market is the need to address formulation challenges posed by poorly water-soluble drugs. Traditional CDs have limited aqueous solubility, which restricts their inclusion complex performance; however, chemically modified CDs overcome this limitation, offering improved dissolution profiles and enabling the formulation of otherwise difficult drugs. This has led to significant integration of derivatives in oral, topical, and parenteral drug products, aligning with the pharmaceutical industry's emphasis on enhanced delivery systems and targeted therapies. In addition to pharmaceuticals, modified CDs are increasingly used in the food and beverage industry (approximately 29% share) for flavor encapsulation and in cosmetics (nearly 15% share) for ingredient stabilization and controlled fragrance release (Global Growth Insights, 2025).

Geographically, the market exhibits balanced global demand, with Europe holding roughly 33%, Asia-Pacific about 28%, and North America around 27% of the derivatives market in 2024, reflecting diversified adoption across mature and emerging economies. Growth prospects in the Asia-Pacific are particularly strong due to expanding pharmaceutical manufacturing and increased research activities, while North America and Europe continue to lead with high R&D investments and supportive regulatory environments for excipient innovation. Alongside commercial drivers, sustainability trends such as demand for biodegradable excipients derived from renewable resources are also shaping the competitive landscape, encouraging development of green production methods for modified CD derivatives (Global Growth Insights, 2025; PW Consulting, 2024).

Research and patent trends in cyclodextrins

The global market for CDs is expected to grow significantly across various sectors, driven by their diverse applications and the increasing need for improved product formulations. There has been a substantial increase in publications on CDs over the past 5 years (Fig. 12.3), indicating growing interest in their applications across fields such as pharmaceuticals, environmental studies, food technology, and

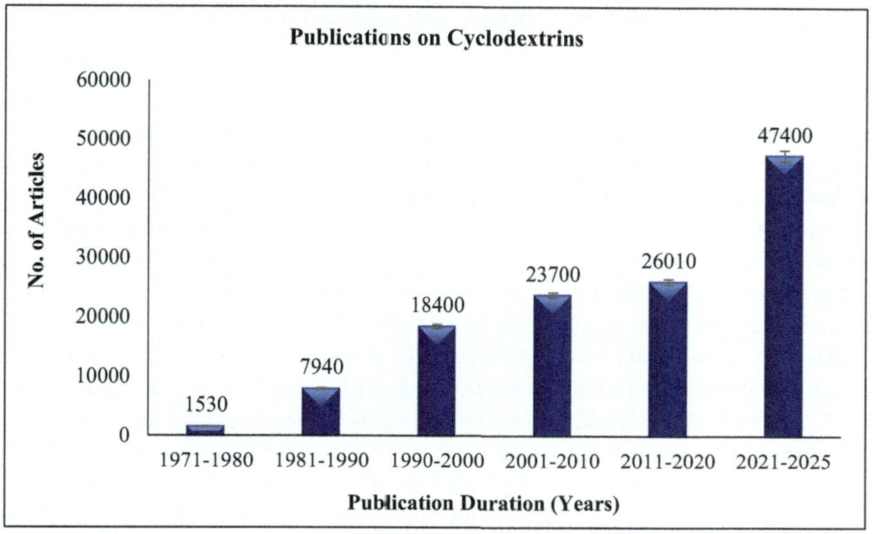

FIGURE 12.3

Evolution of publications on cyclodextrins since 1971.

nanotechnology. The bibliometric study of Scopus and Web of Science databases reveals that the key trends are:

- Advanced drug delivery systems using CD-based nanocarriers.
- Environmental remediation applications (e.g., removal of pollutants and heavy metals).
- CD-functionalized materials and polymers.
- Growing interdisciplinary research integrating CDs with materials science and biotechnology.

The publication of CDs over the last 55 years is shown in Fig. 12.3. The publication data have been collected from Scopus, Web of Science, and Google Scholar. It is evident that the number of publications on CDs has been steadily increasing over the years. Between 2011 and 2020, a total of 26,010 publications were recorded. It is worth noting that from 2021 to 2025, the number of publications exceeded 47,400. The main contributors to CD research are China, the USA, Japan, and India, and there is a growing trend of collaboration between academia and industry. Some selected patents related to CD production and applications are mentioned in Table 12.4.

Table 12.4 shows some important patents for making and using CDs. These patents show how scientists and companies are working to make CDs more efficiently and find new ways to use them in medicine, food, and industry. Overall, the patents indicate that CDs are becoming increasingly important and widely used across various fields.

Recent advances and future perspectives

One of the most important factors that will influence the future of CD production is the cost-value relationship. Although native β-CD is still relatively cheap, thanks to the availability of production methods and high demand, the production costs of α- and γ-CDs, as well as modified CDs, are much higher. This is due to lower enzymatic selectivity and additional chemical modification steps. Nevertheless, the high added value of CDs in pharmaceutical, food, cosmetic, and biotechnological applications makes them economically justifiable, especially for high-purity and application-specific CDs. Future industrial production will therefore aim to increase enzyme efficiency and product selectivity and to reduce purification costs.

The role of regulatory aspects, especially generally recognized as safe (GRAS) status, is very important for the development of CD applications. The native CDs, α-, β-, and γ-CD, have already gained GRAS status for application in food preparations in several parts of the world, including the USA and the European Union. But the regulatory assessment for modified CDs (HP-β-CD, sulfobutyl ether-β-CD, etc.) is more stringent, especially for applications in the pharmaceutical and food industries. With the increasing need for the development of sophisticated CD formulations, there will be a need for the standardization of regulations and thorough toxicological analysis.

One of the most promising areas for future development in CDs is the synthesis of new CD derivatives to improve solubility, binding specificity, and functionality. On the other hand, trends in biotechnology and nanotechnology are also opening new areas of application for CDs beyond traditional inclusion complexes. CDs are now being incorporated into nanosponges, polymeric matrices, and nanomaterials for applications in drug delivery, environmental remediation, and smart packaging. The convergence of CD chemistry with the latest advances in biotechnology and nanotechnology is likely to open new areas of industrial applicability for CDs in the coming years.

Conclusion

The production of CDs has become a mature yet ever-evolving technology, driven by growing demand for these compounds in the pharmaceutical, food, cosmetic, and environmental industries. There has been substantial progress in understanding substrate selection, CGTase-mediated transformations, and downstream processing technologies for the industrial production of CDs with improved purity and yield. The preponderance of β-CD production is driven by its economic viability and versatility, while the rising interest in α-, γ-, and modified CDs indicates a trend toward the production of high-value, application-specific CDs. However, issues such as production costs, enzyme specificity, and purification efficiency remain important in industrial-scale production.

Future R&D and industrial development are anticipated to include the discovery and development of high-performance CGTases, optimization of substrate metabolism, and development of cost-effective downstream processing technologies. Moreover, the development of new CD derivatives and their application in advanced biotechnological and nanotechnological applications will further increase the industrial relevance. Meeting the regulatory and sustainability challenges will also be crucial for future development. In summary, innovation in the enzymatic, process, and application areas will shape the future of CD production and development.

References

24 Chemical Research. (2025). Cyclodextrin market, global outlook and forecast 2025–2032. https://www.24chemicalresearch.com/reports/298739/global-cyclodextrin-forecast-market-2025-2032-889.

360 Research Reports. (2025). Cyclodextrin market size and share: Value distribution and production usage in 2023. https://www.360researchreports.com/press-release/cyclodextrin-market-15625.

Amano Pharmaceutical Co., Ltd. (1994). *β-Cyclodextrin purification process* (European Patent No. EP 0614971 A3). European Patent Office.

Allied Market Research. (2024). Cyclodextrin market by type (alpha-cyclodextrin, beta-cyclodextrin, gamma-cyclodextrin), by end-user industry (food, pharmaceutical, cosmetics and personal care, chemical, textile, others): Global opportunity analysis and industry forecast, 2023–2032. https://www.alliedmarketresearch.com/cyclodextrin-market-A107969.

Arce-Vázquez, M. B., Ponce-Alquicira, E., Delgado-Fornué, E., Pedroza-Islas, R., Díaz-Godínez, G., & Soriano-Santos, J. (2016). Integral use of amaranth starch to obtain cyclodextrin glycosyltransferase by *Bacillus megaterium* to produce β-cyclodextrin. *Frontiers in Microbiology, 7*, 1513.

Astray, G., Gonzalez-Barreiro, C., Mejuto, J. C., Rial-Otero, R., & Simal-Gándara, J. (2009). A review on the use of cyclodextrins in foods. *Food Hydrocolloids, 23*(7), 1631–1640. https://doi.org/10.1016/j.foodhyd.2009.01.001.

Benavent-Gil, Y., Rosell, C. M., & Gilbert, E. P. (2021). Understanding CGTase action through the relationship between starch structure and cyclodextrin formation. *Food Hydrocolloids, 112*, 106316. https://doi.org/10.1016/j.foodhyd.2020.106316.

Biwer, A., Antranikian, G., & Heinzle, E. (2002). Enzymatic production of cyclodextrins. *Applied Microbiology and Biotechnology, 59*, 609–617.

Business Research Insights. (2023). Cyclodextrin market size, share and global forecast: Production volume estimate for 2023. Business Research Insights. https://www.businessresearchinsights.com/market-reports/cyclodextrin-market-127345.

Crini, G. (2014). Review: a history of cyclodextrins. *Chemical Reviews, 114*(21), 10940–10975. https://doi.org/10.1021/cr500081p.

Del Valle, E. M. M. (2004). Cyclodextrins and their uses: A review. *Process Biochemistry, 39*(9), 1033–1046. https://doi.org/10.1016/S0032-9592(03)00258-9.

Fenelon, V. C., Aguiar, M. F., Miyoshi, J. H., Martinez, C. O., & Matioli, G. (2015). Ultrafiltration system for cyclodextrin production in repetitive batches by CGTase from *Bacillus firmus* strain 37. *Bioprocess and Biosystems Engineering, 38*(7), 1291–1301. https://doi.org/10.1007/s00449-015-1369-8.

Fenelon, V. C., Miyoshi, J. H., Mangolim, C. S., Noce, A. S., Koga, L. N., & Matioli, G. (2018). Different strategies for cyclodextrin production: Ultrafiltration systems, CGTase immobilization and use of a complexing agent. *Carbohydrate Polymers, 192*, 19–27. https://doi.org/10.1016/j.carbpol.2018.03.035.

Gastón, R. J., Costa, H., & Ferrarotti, S. A. (2015). Continuous production of cyclodextrins in an ultrafiltration membrane reactor catalyzed by cyclodextrin glycosyltransferase from *Bacillus circulans* DF 9R. *Biotechnology Progress, 31*(3), 695–699. https://doi.org/10.1002/btpr.2053.

Global Growth Insights. (2026). Cyclodextrin market size, share, growth and industry analysis, by types (alpha-cyclodextrin, beta-cyclodextrin, gamma-cyclodextrin), by covered applications (food industries, beverages, pharmaceutical industries, cosmetic industries, chemical industries, others) and regional insights and forecast to 2035. https://www.globalgrowthinsights.com/market-reports/cyclodextrin-market-100564.

Global Growth Insights. (2025). *Cyclodextrin derivatives market size & global outlook 2033*. https://www.globalgrowthinsights.com/market-reports/cyclodextrin-derivatives-market-114376.

Global Market Insights. (2024). Cyclodextrin market size & share, growth forecasts 2024–2032. https://www.gminsights.com/industry-analysis/global-cyclodextrin-market.

Green Biologics Ltd. (2017). *Cyclodextrin glucanotransferase* (U.S. Patent Application No. US 2017/0096647 A1). United States Patent and Trademark Office.

Hokse, H., Kaper, F. S., & Wijpkema, J. T. (1984). *Process for the manufacture of cyclodextrin* (U.S. Patent No. 4,477,568). United States Patent and Trademark Office.

Indrajit, W. S., & Singh, A. (2024). Exploring the interplay: Amylose to amylopectin ratio and formation of resistant starch. *International Journal of Advanced Biochemistry Research, 8*(6), 531–536.

Johnson, R., Moorthy, S. N., & Padmaja, G. (2010). Production of high fructose syrup from cassava and sweet potato flours and their blends with cereal flours. *Food Science and Technology International, 16*(3), 251–258. https://doi.org/10.1177/1082013210366770.

Kantzow, C., Mayer, A., & Weuster-Botz, D. (2015). Continuous gas fermentation by *Acetobacterium woodii* in a submerged membrane reactor with full cell retention. *Journal of Biotechnology, 212*, 11–18. https://doi.org/10.1016/j.jbiotec.2015.07.020.

Laga, A. (2008). The effect of tapioca hydrolysate concentration and minimal acceptor on cyclodextrin production. *Jurnal Teknologi dan Industri Pangan, 19*(2), 149–150. https://journal.ipb.ac.id/jtip/article/view/348.

Lee, Y. D., & Kim, H. S. (1991). Enzymatic production of cyclodextrins from unliquefied corn starch in an attrition bioreactor. Biotechnology and Bioengineering, 37(9), 795-801. https://doi.org/10.1002/bit.260370902.

Lee, Y. H., & Park, D. C. (1992). Direct synthesis of cyclodextrin in a heterogeneous enzyme reaction system containing insoluble extruded starch. In S. Furusaki, I. Endo, & R. Matsuno (Eds.). *Biochemical Engineering for 2001*Tokyo: Springerhttps://doi.org/10.1007/978-4-431-68180-9_33.

Li, Z., Wang, M., Wang, F., Gu, Z., Du, G., Wu, J., & Chen, J. (2007). Gamma-cyclodextrin: A review on enzymatic production and applications. *Applied Microbiology and Biotechnology, 77*(2), 245–255. https://doi.org/10.1007/s00253-007-1166-7.

Loftsson, T., & Duchêne, D. (2007). Cyclodextrins and their pharmaceutical applications. *International Journal of Pharmaceutics, 329*(1–2), 1–11. https://doi.org/10.1016/j.ijpharm.2006.10.044.

Mangolim, C. S., Moriwaki, C., Nogueira, A. C., Sato, F., Baesso, M. L., Neto, A. M., & Matioli, G. (2014). Curcumin–β-cyclodextrin inclusion complex: Stability, solubility, characterisation by FT-IR, FT-Raman, X-ray diffraction and photoacoustic spectroscopy, and food application. *Food Chemistry, 153*, 361–370. https://doi.org/10.1016/j.foodchem.2013.12.067.

Market Research Future. (2026). Cyclodextrin market research report by type (alpha-cyclodextrin, beta-cyclodextrin, gamma-cyclodextrin), by application (pharmaceutical formulations, cosmetics and personal care products, food and beverage industry, industrial applications), by source (corn, wheat, potato), by grade (pharmaceutical grade, food grade, industrial grade) and by regional (North America, Europe, South America, Asia Pacific, Middle East and Africa) – Forecast to 2035. https://www.marketresearchfuture.com/reports/cyclodextrin-market-24250.

Microgenics Corporation. (1989). *Cyclodextrin inclusion complexes and separation process* (European Patent No. EP 0301847 A1). European Patent Office.

Mimi Sakinah, A. M., Ismail, A. F., Illias, R. M., Zularisam, A. W., Hassan, O., & Matsuura, T. (2014). Effect of substrate and enzyme concentration on cyclodextrin production in a hollow fibre membrane reactor system. *Separation and Purification Technology, 124*, 61–67. https://doi.org/10.1016/j.seppur.2014.01.005.

Morillo, E., Madrid, F., Lara-Moreno, A., & Villaverde, J. (2020). Soil bioremediation by cyclodextrins: A review. *International Journal of Pharmaceutics, 591*, 119943. https://doi.org/10.1016/j.ijpharm.2020.119943.

Muniz, I. C. B., Santos, J. B., de Oliveira, R. M., Santos, F. G., de Souza Junior, F. C., Oyama, L., Fontan, R. C. I., & Bonomo, R. C. F. (2024). Current advances in obtaining novel cyclodextrin glycosyltransferases for optimizing the synthesis of cyclodextrins. *Process Biochemistry, 145*, 195–209. https://doi.org/10.1016/j.procbio.2024.07.008.

Pardhi, D. S., Rabadiya, K., Joshi, R. G., Raval, V. H., Panchal, R. R., & Rajput, K. N. (2025a). Screening and optimization of cyclodextrin glucanotransferase production by an alkaliphile *Paenibacillus daejeonensis* P-83. *The Microbe*. https://doi.org/10.1016/j.microb.2025.100358.

Pardhi, D. S., Rabadiya, K., Joshi, R. G., Panchal, R. R., Raval, V. H., & Rajput, K. N. (2025b). Purification and characterization of cyclodextrin glucanotransferase by *Paenibacillus daejeonensis* P-83. *Next Bioengineering, 1*, 100002. https://doi.org/10.1016/j.nxbio.2025.100002.

Pardhi, D., Rabadiya, K., Joshi, R., Raval, V., Panchal, R., & Rajput, K. (2025c). Exploring the potential of Shouchella oshimensis P-106 for cyclodextrin glucanotransferase production. *Vidya - A Journal of Gujarat University, 4*(1), 1–9. https://doi.org/10.47413/03szp371.

Pardhi, D. S., Rabadiya, K. J., Panchal, R. R., Raval, V. H., Joshi, R. G., & Rajput, K. N. (2023). Cyclodextrin glucanotransferase: Fundamentals and biotechnological implications. Microbial surfactants: A journey from fundamentals to recent advances. *Applied Microbiology and Biotechnology*. https://doi.org/10.1007/s00253-023-12708-9.

Pérez-Luna, A., Olvera, C., Gonzalez, A. Martinez, A., & Aguilar, C. N. (2016). Integral use of amaranth starch to obtain cyclodextrin glycosyltransferase and cyclodextrins using Bacillus megaterium in submerged fermentation. Frontiers in Microbiology, 7, 1513. https://doi.org/10.3389/fmicb.2016.01513.

Pishtiyski, I., Popova, V., & Zhekova, B. (2008). Characterization of cyclodextrin glucanotransferase produced by *Bacillus megaterium. Applied Biochemistry and Biotechnology, 144*, 263–272. https://doi.org/10.1007/s12010-007-8009-y.

PW Consulting. (2024). *Worldwide cyclodextrin derivatives market research report 2026, forecast to 2032*. https://pmarketresearch.com/worldwide-cyclodextrin-derivatives-market-research-2024-by-type-application-participants-and-countries-forecast-to-2030/.

PW Consulting. (2026). β-cyclodextrin API market. https://pmarketresearch.com/hc/β-cyclodextrin-api-market/.

Rajput, K. N., Patel, K. C., & Trivedi, U. B. (2016). Screening and selection of medium components for cyclodextrin glucanotransferase production by new alkaliphile *Microbacterium terrae* KNR 9 using Plackett–Burman design. *Biotechnology Research International, 2016*, 3584807. https://doi.org/10.1155/2016/3584807.

Sanraku-Ocean Co., Ltd. (1980). *Process for producing cyclodextrins* (European Patent No. EP 0017242 A1). European Patent Office.

Schöffer, J. D. N., Klein, M. P., Rodrigues, R. C., & Hertz, P. F. (2013). Continuous production of β-cyclodextrin from starch by highly stable cyclodextrin glycosyltransferase immobilized on chitosan. *Carbohydrate Polymers, 98*(2), 1311–1316. https://doi.org/10.1016/j.carbpol.2013.07.044.

Szejtli, J. (1998). Introduction and general overview of cyclodextrin chemistry. *Chemical Reviews, 98*(5), 1743–1754. https://doi.org/10.1021/cr970022c.

Toppan Printing Co., Ltd. (1993). *Cyclodextrin derivatives* (U.S. Patent No. 5,241,059). United States Patent and Trademark Office.

Université des Sciences et Technologies de Lille. (2003). *Cyclodextrin polymers and/or cyclodextrin derivatives with complexing and ion-exchange properties* (U.S. Patent No. 6,660,804 B1). United States Patent and Trademark Office.

Vongpichayapaiboon, T., Pongsawasdi, P., & Krusong, K. (2016). Optimization of large-ring cyclodextrin production from starch by amylomaltase from *Corynebacterium glutamicum* and effect of organic solvent on product size. *Journal of Applied Microbiology, 120*(4), 912–920. https://doi.org/10.1111/jam.13087.

Wadetwar, R. N., Upadhye, K. P., Bakhle, S. A., Deshpande, S. H., & Nagulwar, V. P. (2006). Production of β-cyclodextrin: Effect of pH, time, and additives. *Indian Journal of Pharmaceutical Sciences, 68*(4), 520–523.

Wu, H., Li, X., Ji, H., Svensson, B., & Bai, Y. (2022). Improved production of gamma-cyclodextrin from high-concentrated starch using enzyme pretreatment under swelling condition. *Carbohydrate Polymers, 284*, 119124.

CHAPTER

Large-ring cyclodextrins

13

Sakshi Patel and Niha Mohan Kulshreshtha

Department of Biotechnology, B. Lal Institute of Biotechnology, Jaipur, Rajasthan, India

Introduction

Cyclodextrins (CDs) are circular oligosaccharides that consist of units of D-glucopyranose connected by α-(1,4) glycosidic bonds to form a typical truncated-cone shape. Three naturally found types, namely, α-, β-, and γ-CD, contain six, seven, and eight glucose units, respectively, and have found extensive application in pharmaceuticals, food sciences, and in the environment due to their ability to form inclusion complexes with hydrophobic guest molecules (Esteso & Romero, 2024). CDs can selectively enclose these guest molecules and alter their solubility, stability, and bioavailability, making them invaluable for numerous scientific and industrial enterprises. Although α-, β-, and γ-CDs are the most commonly studied and widely used industrially, another category of CD, namely, large-ring cyclodextrins (LR-CDs), is rapidly gaining attention. LR-CDs are CDs containing more than eight glucose units in the ring, often between 9 and 12 or more. Examples include δ-CD (9 units), ε-CD (10 units), and others of even larger ring sizes. LR-CDs were only identified recently, in the 1960s, when French et al. presented evidence of the existence of 9–13 units long galactopyranose (French et al., 1965; Pulley & French, 1961). Nevertheless, the initial studies of these molecules were constrained by the inability of these molecules to synthesize in large amounts, as they were highly soluble in water and were too flexible (Endo, 2011). These molecules were not revisited until the discovery of a method for preparing of LR-CD mixtures (Kobayashi et al., 1986), and the crystal structure of δ-CD (Fujiwara et al., 1990) was determined. The increased cavity size of LR-CDs, which enables them to trap larger guest molecules, makes them appropriate for inclusion complexes involving macromolecules or bulkier molecular structures (Sonnendecker et al., 2019). All these characteristics make LR-CDs a promising component of future supramolecular systems, with potential to offer new possibilities in drug delivery, materials science, environmental remediation, and molecular recognition. This has opened new avenues for the applications of LR-CDs (Crini et al., 2018).

Cyclodextrin Glucanotransferase. DOI: https://doi.org/10.1016/B978-0-443-41694-1.00008-3

Structure and physicochemical properties

CDs are a family of structurally homologous natural products synthesized in the process of bacterial breakdown of cellulose. The α-, β-, and γ-CDs are cyclic oligosaccharides with a bucket-like structure comprising (α-1–4)-linked α-D-glucopyranose units with a central hydrophobic cavity and an outer hydrophilic surface (Fig. 13.1A-C). These CDs are truncated cones (rather than perfect cylinders) because of the chair conformation of the glucopyranose units. The major hydroxyl groups of the sugar molecules are located at the smaller portion of the cone, whereas the minor hydroxyl groups are located at the larger portion of the cone. The hydroxyl groups are facing the cone surface. The skeletal carbons and ethereal oxygen of the glucose molecules line the cavity center, giving it a lipophilic nature. LR-CDs are built in a similar manner as α-, β-, and γ-CDs, but are a little elliptical, distorted boat-like (Fig. 13.1D-E) with a narrow groove-like cavity.

The availability of crystal structures of LR-CDs is limited by the low crystallinity due to (1) the high flexibility of the molecules, which makes the formation of intra and interchain hydrogen bonds difficult, and (2) the low nucleation rate of sugars, which is the preceding step to crystal growth (Endo & Ueda, 2004). In the CD10 and CD14 structures, the anticonformation (band flip) was observed between adjacent glucopyranose units at two sites instead of the usual syn conformation. The CDs with even larger rings adopt a more folded conformation (Fig. 13.1F) due to

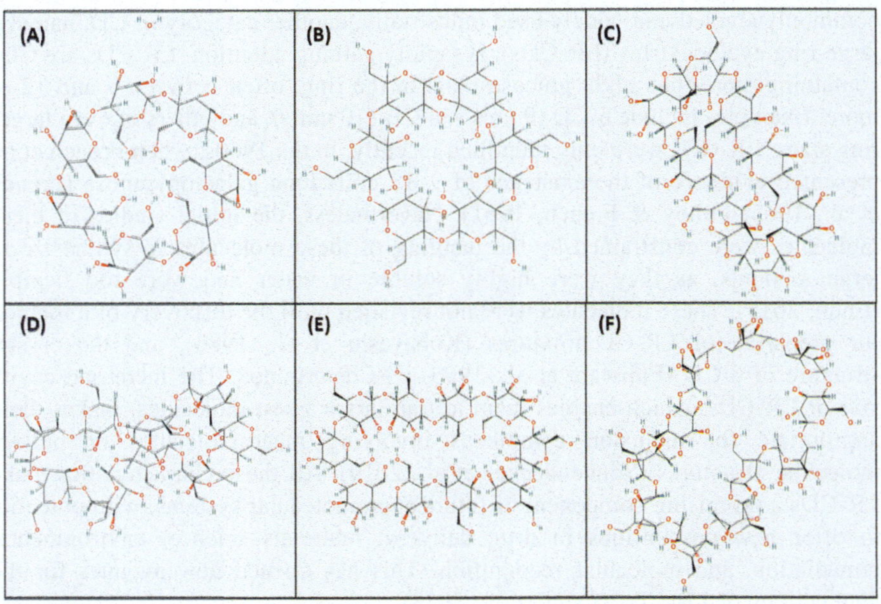

FIGURE 13.1

Structure of cyclodextrins with different degrees of polymerization.

steric hindrance arising from significantly greater ring strain (Ivanov & Jaime, 2004). The structure of CD26 displayed binding of two antiparallel V-amylose helices with band flips (Endo & Ueda, 2004).

While CD complexes and single CD molecules are viewed as homogeneous molecular dispersions in aqueous systems, CD water solutions have historically been considered true solutions. It has recently been demonstrated that CDs and CD complexes self-associate to form aggregate- or micelle-like structures. In addition, polymers have been found to interact with such systems, and the resultant aggregates can solubilize drugs by noninclusion complex formation (Saokham et al., 2018). LR-CDs have higher solubility than the most commonly used α-CDs and have a superior ability to form inclusion complexes with less water-soluble molecules. This property has been attributed to the presence of a larger number of hydrophobic cavities in LR-CDs. The aqueous solubility of LR-CDs, except for CD9, CD10, and CD14, is greater than 100 g/100 mL due to their high flexibility. The acid hydrolysis properties of CD10 to CD21 were found to be similar (Table 13.1).

Synthesis of large-ring cyclodextrins

Synthesis of LR-CDs is both enzymatically and chemically challenging. The primary enzymes that are used to convert starch-derived substances into mixtures of LR-CDs are cyclodextrin glucanotransferases (CGTases), which are low-yielding and provide low selectivity for the product. More precise chemical and template-directed methods offer scaffolding of rings assisted by a molecular scaffold, but are tedious and require multistep purification. The additional challenges in the production of LR-CDs are linked to the enzyme price, reaction optimization, purification efficiency, and retention of ring-size selectivity on the practical scale (Fig. 13.2).

Enzymatic synthesis

Intramolecular cyclization of starch substrates for α-1,4 transglycosylation to form CDs is catalyzed by CGTase. It predominantly produces α-, β-, and γ-CDs, but under conditions of lack of regulation, LR-CDs (nine glucose units) are also produced as secondary products (Pardhi et al., 2023). Two forms of reactions catalyzed by the CGTases—coupling and cyclization— have to be balanced in order to raise the LR-CD production yield (Endo & Ueda, 2004). A coupling reaction can be inhibited by removing acceptors and controlling the temperature appropriately. CD (nine units) is synthesized on a multigram scale using enzymatic dynamic combinatorial chemistry and a dianionic boron-based template. The purity can be above 95%, and the yield is about 7% after high-performance liquid chromatography (HPLC) purification (Hansen et al., 2025). Enzymatic conditions need to be optimized to favor bigger rings in CGTase. Three primary factors influence the relative proportion of LR-CDs.

Table 13.1 Physicochemical properties of large-ring cyclodextrins.

Cyclodextrin (CD)	Glucopyranose units	Molecular weight	Solubility in water (g/100 mL)	Specific rotation [α]25D	Surface tension (mN/m)	Half-life of ring opening (h)
α CD	6	972	14.5	+147.8	72	33
β CD	7	1135	1.85	+161.1	73	29
γ CD	8	1297	23.2	+175.9	73	15
δ CD (CD9)	9	1459	8.19	+187.5	72	4.2
ε CD (CD10)	10	1621	2.82	+204.9	72	3.2
ζ CD (CD11)	11	1783	>150	+200.8	72	3.4
η CD (CD12)	12	1946	>150	+197.3	72	3.7
θ CD (CD13)	13	2107	>150	+198.1	72	3.7
ι CD (CD14)	14	2270	2.30	+199.7	73	3.6
κ CD (CD15)	15	2432	>120	+203.9	73	2.9
λ CD (CD16)	16	2594	>120	+204.2	73	2.5
μ CD (CD17)	17	2756	>120	+201.0	72	2.5
ν CD (CD18)	18	2919	>100	+204.0	73	3.0
ξ CD (CD19)	19	3081	>100	+201.0	73	3.4
σ CD (CD20)	20	3243	>100	+199.7	73	3.4
π CD (CD21)	21	3405	>100	+205.3	73	3.2

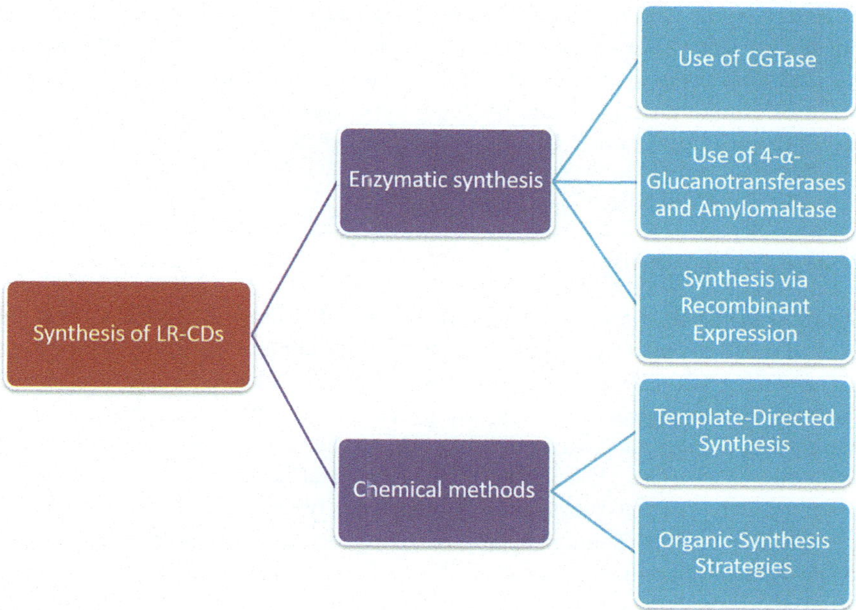

FIGURE 13.2

Strategies for synthesis of large-ring cyclodextrins (*LR-CDs*).

(1) Substrate Concentration, (2) pH, and (3) temperature (Endo et al., 2002). A lower starch level is favorable for forming LR-CD because it reduces competition from smaller rings. The ideal pH for LR-CD production is 5–6. Changes in the optimal value can reduce enzyme efficiency and alter the product distribution. The best temperature is 50ºC–60ºC, which ensures that starch will gelatinize without CGTase denaturation (Krusong et al., 2022).

Maltodextrin glucanotransferase (CGTase) and 4-α-glucanotransferases, which are often referred to as amylomaltases, are involved in the synthesis of LR-CDs. The enzymes of *Corynebacterium glutamicum* are of special interest because they are capable of forming LR-CDs of extremely high degrees of polymerization (between 16 and more than 100 glucose units). These enzymes have intramolecular transglycosylation reactions, which allow the formation of much larger macrocycles than those made by CGTase. More recent protein engineering technologies, particularly site-directed mutagenesis of key residues, including P228Y, A413F, and G417F, have led to a radical increase in the ability to control ring-size and selectivity in LR-CD homologs. Engineered amylases are more catalytically active, yield fewer side products, and are more stable; hence, they have great promise as biocatalysts for producing LR-CDs at scale. They are superior to other CDs in producing large macrocycles with a higher level of uniformity, which makes them useful in industrial biotechnology and the formation of CDs with specific free arm lengths (Ngawiset et al., 2023).

Heterologous expression of CGTase with specificity can be used to produce more CD using the microorganism of choice. *Bacillus pseudofirmus and Paenibacillus macerans* are used for heterologous expression, and modified CGTases are more catalytic and produce higher amounts of β- and γ-CDs, which have a direct impact on LR-CD pathways (Guedes et al., 2023).

Chemical methods

Chemical synthesis can be template-directed or direct organic synthesis. Template-directed synthesis provides a powerful strategy for controlling enzymatic pathways and accessing LR-CDs that are otherwise difficult to obtain. In enzyme-mediated dynamic systems, glycosyltransferases generate a complex, transient mixture of interconverting linear and cyclic α-1,4-D-glucans that normally evolves toward small malto-oligosaccharides. The introduction of artificial template molecules stabilizes this kinetically trapped subsystem by selectively binding and amplifying specific macrocyclic products. Through appropriate template selection, the enzymatic system can be directed not only toward native CDs (α, β, and γ) with high selectivity, but also toward the formation of nonnatural LR-CDs containing 9 or 10 glucose units. These templates shift the balance between kinetic and thermodynamic control, enabling out-of-equilibrium self-assembly and prolonging the lifetime of the dynamic combinatorial library. As a result, template-directed synthesis provides a robust, tunable approach to the selective production of LR-CDs, thereby expanding the structural diversity and functional potential of CD-based materials (Larsen & Beeren, 2019). Hansen et al. (2025) employed a superchaotropic dodecaborate template ($B_{12}Cl_{12}^{2-}$) to selectively direct the enzymatic synthesis of δ-CD from a dynamic combinatorial library of interconverting CDs. This simple, scalable, and template-recyclable strategy enables access to large quantities of δ-CD, opening the door for systematic studies of its properties and potential applications (Hansen et al., 2025). Chemical techniques create predefined persubstituted LRs by gradually glycosylating protected glucose units. Although laborious, this enables solubility augmentation and structural design. For example, such linear-to-cyclic pathways are used to construct hydroxypropyl and methyl derivatives of LR-CDs.

Challenges in large-ring cyclodextrin synthesis

The major challenge in LR-CD synthesis is the low yield. Enzymatic synthesis biases toward smaller CDs because of thermodynamically favorable conditions and enzyme kinetics, frequently leading to less than 10% LR-CD yields. Even under optimized CGTase and templating conditions, product yields are modest (~7%).

Another issue is the production of a large number of byproducts. The product of enzymatic synthesis is a mixture of reduced CDs, linear glucose chains, and analogs. The purification of LR-CDs is therefore tedious and involves multiple time-consuming steps, such as HPLC, size-exclusion chromatography, and a solvent extraction process. These processes are not easy to scale. With chemical methods, additional impurities can be added, such as protecting groups, coupling factors, and template impurities, and they requires a lot of purification to achieve pure macrocycles (Krusong et al., 2022).

Structural and physicochemical characteristics

LR-CDs, with nine or more α-(1–4)-linked groups of glucopyranoses, are defined by unique structural and physicochemical properties that distinguish them from the conventional α-, β-, and γ-cyclodextrins. Their macrocyclic structures are much larger and offer higher flexibility of conformational changes, increased volumes of their cavity, and modified patterns of hydrogen bonding. These properties affect their solubility, thermal properties, inclusion capacity, and host–guest properties. Consequently, LR-CDs are specialized in terms of their structural adaptations and functional characteristics and are of great importance to the advanced supramolecular, pharmaceutical, and material science applications.

Structural features

The LR-CDs maintain the truncated-cone geometry of native CDs, but contain greater conformational flexibility. Every glucopyranose subunit again takes the $^4C_1$ chair conformation, and the two ends, primary (C6-OH) and secondary (C2- and C3-OH), form the narrow and wider ends, respectively. Changing the C6 hydroxyl orientations (axial or equatorial) in combination with intramolecular hydrogen-bond networks in LR-CDs provides dynamic belt-like structure motifs, such as smaller CDs yet more flexible. This conformational plasticity allows LR-CDs to take up a range of conformations, from almost planar toroids to rings with a shape of a saddle-cone, based on the size of the ring and guest occupancy. Their inclusion properties and host–guest selectivity are directly affected by conformation, which has also been demonstrated with the help of combined spectroscopic and modeling studies (Poulson et al., 2022).

Conversion of γ-CD (eight units) to LR-CD (≥nine units) causes the internal cavity sizes to increase from 7.5–8.3 Å to >10 Å, and this is critical to guest dynamics and complex stability. This has been demonstrated through molecular dynamics simulations of the 4-terpineol/LR-CD complexes. Thus LR-CD structures are widened and anisotropic; some of them are found to be flattened or elliptic. Further, the hydration of the molecules and bound guest molecules also affects the final structure. In contrast, native CDs are characterized by well-defined cavities, as obtained with X-ray diffraction (Cac et al., 2022).

Characterization techniques

The primary characterization techniques for LR-CDs include nuclear magnetic resonance (NMR) spectroscopy, matrix-assisted laser desorption/ionization time-of-flight (MALDI-TOF), and X-ray crystallography. These techniques provide complementary information, shedding light on the exact structure and binding properties of the LR-CD molecules. Due to the flexible nature of LR-CDs, the high-resolution NMR (^{1}H, ^{13}C, correlation spectroscopy, nuclear overhauser effect spectroscopy [NOESY], and rotating frame overhauser enhancement spectroscopy) is necessary to elucidate the exact structures. Cavity hydrophobicity and guest proximity are observed as changes in the signal of internal protons (H3 and H5). NOESY cross-peptides between glucose units are necessary in the study of LR-CD conformation as they confirm the closure of rings and spatial orientations. The most important rim structure (equatorial vs axial orientation) is disclosed by the presence or absence of diastereotopic splitting at C6, which conveys information on the flexibility of the ring structures (Crini et al., 2018).

Taking into account the low specificity of enzymatic production, the analysis of the MALDI-TOF analysis allows distinguishing between LR-CD homologs (DP = 9–22) effortlessly. The immense mass precision of the approach (±0.5 Da) allows verification of the composition and purity by analyzing the cooccurrence of small CDs (DP = 6–8) (Musuc, 2024). X-ray crystal structures of LR-CD inclusion complexes can be used to validate cavity geometries and guest-binding topologies. For example, the structure of the CD 17-tocopherol inclusion complex has been studied by X-ray crystallography. The data generated using X-ray crystallography can be used as the basis for molecular modeling (molecular dynamic simulations), where computed binding energies and cavity volumes can be used to reveal dynamic conformations at 300K, frequencies of breathing of pores, and host–guest affinities (Kerdpol et al., 2021; Osaki et al., 2020). These experiments provide a wealth of information on the predicted behavior of the LR-CDs with guest molecules.

Physicochemical properties

Native CDs are soluble to different extents: α (14.5 g/L), β (1.85 g/L), and γ (23.2 g/L), in accordance with their cavity rigidity and intramolecular hydrogen-bonding arrangement. Due to their higher flexibility of conformation and hydrogen-bond networks, LR-CDs are more soluble in aqueous solutions than γ CDs (Table 13.1). They are also soluble in slightly polar organic solvents, which makes it easier to encapsulate hydrophobic guests and allows the creation of larger compositions (Ji et al., 2024).

Thermogravimetric analysis (TGA) and differential scanning calorimetry (DSC) are used to indicate the crystallinity and melting behavior of LR-CDs. Abrupt melting points indicate crystalline organization, which is typical of native CDs. On the other hand, LR-CDs are most likely to have more slow DSC transitions, which are indicative of semicrystalline or amorphous structures. TGA studies of LR-CD complexed with tocopherol reveal that host–guest interactions can lead to improved thermal integrity, dehydration processes (below 100ºC), internal water removal (above 100ºC to 105ºC), and macrocycle breakage (above 300ºC) (Cao et al., 2020). Unlike natural CDs that disintegrate in a solution, solutions below pH 3 or above pH 12, as the glycosidic bonds are broken, LR-CDs are less rigid and have a variety of alkaline pores. They have similar stability profiles despite the fact that ring diameter and guest inclusion can provide some resistance. LR-CDs can be structurally stable up to approximately 300°C, after which they partially degrade (Musuc, 2024).

Host–guest chemistry and inclusion complexes

LR-CDs with nine or more α-1,4-linked glucose residues exhibit more complex host–guest chemistry than α-, β-, and γ-CDs. Their cavity flexibility, in conjunction with their enhanced hydrophobic cavity size, gives them an advantage, especially for the inclusion complexation of large, structurally heterogeneous guest molecules. Table 13.2 provides the characteristics of LR-CD inclusion complexes.

Mechanism of inclusion

The most important stabilizing forces in LR-CDs inclusion complexes are hydrophobic interactions and van der Waals interactions. Upon guest entry, energetically unfavorable water molecules held in the hydrcphobic cavity are expelled, allowing complex formation. The majority of LR-CDs are 1:1 inclusion complexes, in which a single guest molecule is fully accommodated within the expanded cavity. But guests of higher-order of stoichiometry can be induced by larger (or amphiphilic) guests. For example, LR-CDs with DP = 14 can form 1:2 host–guest complexes with dyes such as Nile red, with one dye molecule occupying two adjacent macrocyclic cavities. It proves the flexible and adaptable character of LR-CD host–guest architecture (Cao et al., 2020).

Table 13.2 Characteristics of large-ring cyclodextrin (*LR-CD*) inclusion complexes.

Characteristics	LR-CD inclusion complex
Stabilizing elements	Van de- Waals forces and hydophobic contacts
Guest molecule size	One guest molecule is fully contained
Host–guest ratio	Mostly 1:1, higher order possible
Guest size and polarity	Hydrophobicity and size are essential
Ring size and functional groups	Increases solubility, hydrophobic interaction, and hydrogen bonding

Factors affecting binding

Hydrophobicity, molecular dimensions, and overall polarity of the guest molecule are critical factors in determining the binding affinity of the LR-CDs. With their much larger cavities, they can accommodate larger hydrophobic guests, such as terpenoids, steroidal drug molecules, carotenoids, long-chain fatty acids, and large organic dyes. Thus the molecular volume relates well to the size of the internal cavity. Smaller or more polar molecules can still be drawn into the cavity, but they tend not to replace the energetically stabilized water molecules that are already there within LR-CDs, and this causes weaker complex formation as a result of enthalpic penalties. Cavity flexibility and capacity increase with the increase in ring-size (CD $_9$–CD_{12}), which allows stronger van der Waals interactions and accommodates larger guest molecules. Also, a guest polarity can affect binding orientation. Nonpolar molecules are usually centrally oriented, whereas amphiphilic guests can anchor through polar groups on the rim of the CD. Therefore the overall stability and functionality of complexation are affected by the molecular complementarity between the guest structure and the LR-CD cavity.

The size of the rings and the presence of functional groups have a strong impact on the binding behavior of LR-CDs. Larger rings offer greater volumetric capacity and flexibility, making it possible to accommodate larger or more structurally challenging guests with increased stability. Host–guest interactions are further functionalized by chemical functionalization, methylation, hydroxypropylation, sulfation, or grafting with aromatic linkers. These alterations enhance solubility in aqueous solutions, strengthen hydrophobic interactions, and create opportunities for directional hydrogen bonding at the rim of the CD. Cavity polarity and steric environment can also be modulated by functional groups, thereby increasing guest selectivity, binding affinity, and loading efficiency. This targeted derivatisation enables LR-CDs to be versatile in specialized applications in pharmaceuticals, the environment, and materials (Yu & Chen, 2019) (Table 13.1).

Comparative binding studies

The relative affinities of LR-CDs to α-, β-, and γ-CDs are always found to be higher toward large hydrophobic guests. For instance, inclusion complexes in supramolecular hydrogels leverage the flexibility of the LR-CD domain to achieve guest affinities and loading capacities that are not possible with smaller CDs. The greater cavity and induced-fit mechanism provide deeper guest encapsulation and improved thermal stability, both essential in drug delivery and material applications (Musuc, 2024).

The presence of LR-CDs enables significant inclusion of water-insoluble dyes due to their larger cavities and greater structural flexibility. These LR-CDs are able to encapsulate large-sized chromophores better than natural α-, β-, or γ-CDs, enhancing dye dispersibility, thermal stability, and photostability in aqueous solutions. LR-CD complexes are also useful in preventing dye oxidation, aggregation, and photobleaching, and are thus useful in textiles, food coloring, fluorescent probes, and optical materials. Moreover, the reversible host–guest property of LR-CDs enables controlled dye release and, therefore they find applications in smart coatings and sensing (Tian et al., 2020).

The CD-based inclusion complexes in supramolecular nanocarrier systems are particularly important, as hydrophobic sequestration is an effective way to increase drug solubility, stability, and controlled delivery (Jin et al., 2019). Even greater benefits are provided by LR-CDs, whose open spaces offer more room to encapsulate very large or highly hydrophobic drugs, such as anticancer agents, steroids, and poorly soluble antibiotics. They have a flexible macrocyclic group that enables greater insertion of the guest and high affinity for binding with their tertiary group in the body, leading to increased bioavailability and less degradation of sensitive drug molecules. LR-CD complexes are also useful for sustained and targeted delivery and can be incorporated into more sophisticated therapeutic systems, such as nanogels, micelles, and responsive drug-delivery systems (Jin et al., 2019).

The interactions that occur between the host/guest molecules with the multivalent CDs, which are bound onto the polymer and the complementary guest molecules, result in the formation of a supramolecular hydrogel, in which reversible and noncovalent binding determines the gelation. The increased cavity size and flexibility of LR-CDs allow the formation of more robust and dynamic hydrogel networks that can accommodate larger hydrophobic crosslinkers. These hydrogels have adjustable mechanical strength, self-healing properties, and sensitivity to stimuli such as pH, temperature, or enzymatic activity. Hydrogels based on LR-CDs have been of interest in biomedical engineering for controlled drug delivery, injectable tissue-engineering scaffolds, and local biologic delivery. Their flexible structure allows them to tightly regulate gel permeability and therapeutic loading capacity (Liu et al., 2018).

Applications of large-ring cyclodextrins

The LR-CDs, a macrocyclic oligosaccharides consisting of nine or more units of glucose, have become one of the most versatile supramolecular hosts due to their much-expanded hydrophobic cavity and increased aqueous solubility, as well as their favorable biocompatibility. Their macrocyclic structure is flexible, which enables them to trap a wide range of guest molecules, including large hydrophobic molecules that cannot be incorporated into smaller CDs. This has made LR-CDs prominent in various scientific and industrial uses. They have special structural merits that allow them to be useful in drug therapy, food and dietary supplement formulation, environmental detoxification, and advanced materials science and enable the application of controlled release, stabilization, pollutant removal, and smart supramolecular assemblies (Fig. 13.3).

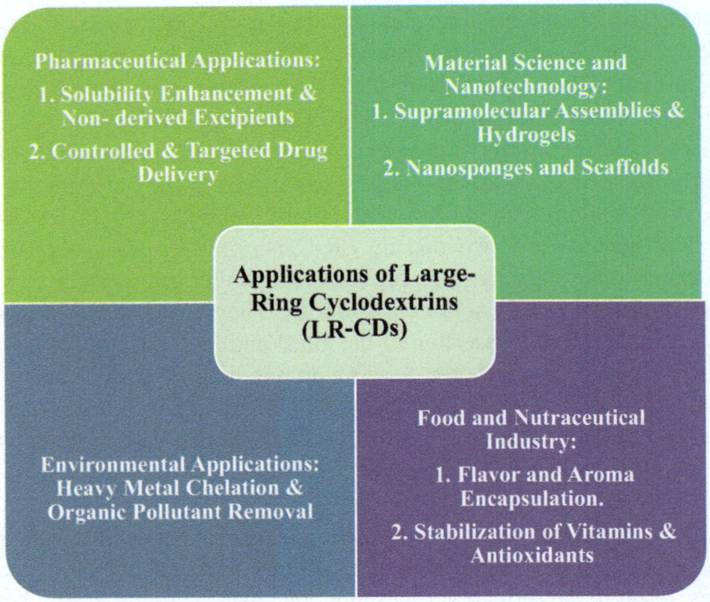

FIGURE 13.3

Applications of large-ring cyclodextrins (*LR-CDs*).

Pharmaceutical applications

In the pharmaceutical industry, LR-CDs have been used for solubility enhancement, enabling their use as excipients. The capacity of LR-CDs to improve the aqueous solubility of poorly water-soluble drugs is higher by virtue of their larger and more flexible cavities. Ismail et al. (2021) found that domperidone complexed with an enzymatically synthesized CD_{33} increased the solubility threefold, whereas an isolated LR-CD fraction enhanced the solubility 2.7-fold better than native or small-ring CDs. These results highlight LR-CDs as highly efficient pharmaceutical excipients that can enhance drug dissolution, stability, and overall bioavailability and could be used as valuable assets in the creation of modern, more effective therapeutic formulations (Ismail et al., 2021).

Another area in which LR-CDs are particularly useful is controlled and targeted drug delivery. The use of LR-CDs as versatile targets in controlled and targeted drug therapies has been enabled by their increased cage size, diverse geometries, and ability to bind to their hosts. The capacity to entrap large hydrophobic reagents, nucleic acids, and multicomponent therapeutic agents, combined with the possibility of controlling the drug release characteristics, such as sustained delivery, timed delivery, and stimulus-reactive delivery, makes them attractive candidates as drug excipients. LR-CD-based nanocarriers can be designed to respond to pH, temperature, enzymes, or redox conditions, enabling the selective activation of drugs in diseased tissues. This system highlights the duality of the LR-CDs: they are not only excipients but also an important part of the architecture that increases stability, circulation time, and targeting precision. LR-CD-based systems have great potential in next-generation precision medicine because they reduce off-target toxicity and enhance the therapeutic payload delivery (Haimhoffer et al., 2019).

Food and nutraceutical industry

LR-CDs are particularly useful in the encapsulation of volatile flavor and aroma molecules because of their larger hydrophobic cavities, which can effectively host larger and more complex molecules or multicomponent molecules that smaller CDs are unable to effectively host. The process of creating stable inclusion complexes can be used to preserve the sensory properties of sensitive aromatic compounds by preventing loss due to evaporation, oxidation, and photochemical or thermal degradation, thereby maintaining their aromatic sensory characteristics during processing, storage, and the product's shelf life. LR-CDs have also been found to possess great potential in suppressing unpleasant or objectionable aromas in food, nutraceutical, and functional beverage formulations, besides their role of stabilizing volatile flavor notes. Their modulation capabilities also enable them to control the release kinetics, which further facilitates controlled or sustained aroma delivery and improves consumer perception and flavor retention. In addition, LR-CD encapsulation enhances the flexibility of formulations, allowing the use of volatile or unstable compounds in powdered, liquid, or heat-processed food systems without significant quality loss. All these advantages demonstrate LR-CDs as effective agents in contemporary food technology and flavor engineering (Cao et al., 2020).

LR-CDs have also been shown to be very effective in stabilizing sensitive vitamins and antioxidant molecules, in particular those that are highly hydrophobic, chemically unstable, or readily oxidized. LR-CDs are able to entrap large bioactive molecules, including vitamin E, carotenoids, polyphenols, and other antioxidant-rich phytochemicals, more efficiently compared to conventional α-, β-, or γ-CDs due to their much larger size. LR-CD inclusion complexes help protect these molecules from adverse environmental factors such as heat, light, pH changes, and oxygen exposure, thereby maintaining their bioactivity throughout processing and storage. The spontaneous and energetically preferred encapsulation of α-tocopherol in CD26 has been demonstrated using molecular dynamics simulations, and it was shown that LR-CD cavities and fat-soluble antioxidants are strongly bound and structurally compatible. Such interactions enhance stability and dispersibility in water systems. Vitamin E acetate complexation with enzymatically produced LR-CDs has been reported to improve water solubility by nearly 800-fold. These properties make LR-CDs extremely appealing as carriers in functional foods, dietary supplements, and nutraceuticals, enabling enhanced bioavailability and controlled release of antioxidant nutrients. LR-CDs show great promise as a modern excipient to formulate health-promoting products of the next generation due to their capacity to maximize absorption, minimize oxidation waste, and facilitate targeted delivery (Kuttiyawong et al., 2015).

Environmental applications

LR-CDs have demonstrated great potential in environmental cleansing, as they possess larger cavity volumes, greater structural plasticity, and the capacity to form robust host–guest complexes with various contaminants. Many of the early studies were conducted on β-CD derivatives. Results have shown that functionalized CD-based systems, including EDTA-modified β-CD-chitosan composites, can effectively extract heavy metals such as Hg^{2+} and simultaneously adsorb organic dyes such as methylene blue and methyl orange, with high uptake capacities of 160–180 mg/g. These findings highlight the potential of cyclic polysaccharides as a new adsorbent for wastewater treatment. LR-CDs are expected to be the most effective at retaining larger pollutants, such as polyaromatic hydrocarbons, endocrine-disrupting chemicals, pesticide residues, and pharmaceutical contaminants, due to their larger hydrophobic cavities. Their biodegradable, nontoxic properties and ability to enter membranes, gels, and polymer matrices also contribute to their use in sustainable, high-performance environmental cleanup approaches (Zhao et al., 2017).

Material science and nanotechnology

LR-CDs are more flexible building blocks towards the creation of complex supramolecular products, due to their larger cavities, enhanced conformational flexibility, and their capability of forming strong but reversible host–guest binding. These properties enable LR-CDs to be assembled into architectures such as molecular necklaces, hydrogels, nanosponges, and polyrotaxanes, with highly tunable mechanical, chemical, and responsive properties. The LR-CD derivatives (which are functionalized), including cholesterol-modified or spermine-substituted, can be self-assembled into stable nanogels that can deliver therapeutic nucleic acids, such as VEGF-targeting siRNA, straight to tumor tissues. Hydrogels prepared via LR-CD self-healing require little drug loading and can be triggered by pH, temperature, or enzymes, enabling fine-tuning of cargo release. LR-CD hydrogels have the potential to find broad biomedical applications, including targeted drug delivery, wound healing, tissue regeneration, and the creation of dynamic, stimuli-responsive biotherapeutic surfaces, owing to their biocompatibility and structural adaptability (Kuttiyawong et al., 2015).

CD nanosponges are three-dimensional networks of cross-linked polymeric structures made of CD and have been of growing interest as versatile drug-delivery vehicles for controlled drug release. Whereas standard nanosponges use small-ring CDs, LR-CDs offer considerable benefits due to their significantly increased number of cavities and greater flexibility. These characteristics enable LR-CD-based nanosponges to load large therapeutic biomolecules, such as proteins, enzymes, peptides, nucleic acids, and growth factors, with greater loading capacity and stability. The porous, sponge-like structure allows them to be released in a sustained, controlled manner, making them particularly useful for delivering therapeutics over an extended period. The LR-CD scaffold promotes tissue regeneration by supporting cell adhesion, facilitating the diffusion of nutrients and oxygen, and delivering essential bioactive agents gradually in tissue engineering. Their tunability, biodegradability, and compatibility with natural polymers such as collagen, chitosan, and hyaluronic acid further increase their potential in regenerative medicine, wound healing, and scaffold-based drug-delivery systems (Sherje et al., 2017).

Functionalization and derivatization of large-ring cyclodextrins

LR-CDs can be derivatized to fit specific therapeutic purposes to increase their application as advanced excipients and nanocarrier ingredients in contemporary drug-delivery systems (Gao et al., 2025). The functionalization enables the attachment of targeting moieties to LR-CDs and improves their accuracy in drug delivery. The derivatization of LR-CDs has the primary goal to enhance their solubility, stability, and functional properties in pharmaceutical formulations, especially in intravenous and oral formulations. Property enhancement can be performed at the native CD hydroxyls. Chemical alterations such as methylation and sulfation increase hydrophilicity, enabling effective incorporation of lipophilic or poorly soluble drugs. Hydroxypropylation (e.g., HP-CD) disrupts hydrogen bonding to increase water solubility, a property typical of pharmaceutical excipients. Hydrophobic guests are also encouraged on Methylated CDs (e.g., MCD). Polyethylene glycol (PEG)ylation and hydroxypropylation also enhancemolecular recognition, reduce intermolecular aggregation, and improve biological compatibility. Such modifications may also enable targeting, controlled release, and longer circulation time.

Although most research focuses on α-CD, hydroxypropylation of rice-based LR-CDs under enzyme catalysis has recently demonstrated improvements in equivalent solubility (Haroon Khalid et al., 2020). Polyrotaxane conjugation involves polymer precipitation of CD onto PEG or polypropylene glycol (PPG) backbones. Polyrotaxanes that have been threaded with siRNA or terminated with peptide stoppers can be dethreaded upon stimulus and can also release their cargo on demand. Camptothecin and paclitaxel (anticancer agents) have been grafted onto α-CD polyrotaxane derivatives, which exhibit sustained-release characteristics and improved in vivo performance. The method can be applied to LR-CDs (Li et al., 2023). PEGylated CD conjugates can also be functionalized with folate, RGD peptides, or other antibodies to promote receptor-mediated uptake, thereby selectively targeting cancerous cells or other diseased tissues. These changes enhance the efficiency of targeting and treatment outcomes (Kost et al., 2020).

Even though LR-CDs are not widely used in commercial preparations relative to α-CD derivatives, there is new evidence to suggest the strong pharmaceutical potential of such materials. Enzymatically hydroxypropylated LR-CDs have shown high solubility in aqueous solutions and can form stable micellar networks that entrap hydrophobic anticancer drugs, such as paclitaxel, achieving an almost 30% drug loading and greatly extended systemic circulation in vivo. Also, PEG-based sulfated LR-CDs have been developed to nanocapsulate anti-MMP-9 siRNA, allowing its released in the tumor microenvironment over seven days and to achieve significant tumor growth inhibition. These functionalized LR-CD systems demonstrate the potential of future biocompatible drug-delivery systems (Arima, 2021).

The versatile characteristics of the LR-CDs are their flexibility of cavities and tunable functional groups, which allow integration into a wide variety of smart, stimulus-responsive delivery systems. LR-CD-polymer hydrogels can be used to release encapsulated therapeutics specifically at the target site in an acidic tumor microenvironment, where further swelling promotes the release of active ingredients. PEGylated LR-CD architectures also increase circulation time, and their thermo-responsive character enables drug release at or slightly above physiological temperatures. These are currently in preclinical testing for anticancer treatments, where they have been shown to have better tumor penetration and less systemic toxicity. Moreover, the use of LR-CDs with redox-sensitive or enzyme-cleavable linkers is also useful for developing multitrigger responsive systems with highly controlled and personalized drug delivery (Rossi et al., 2018).

Challenges, limitations, and regulatory aspects of large-ring cyclodextrins

Even though LR-CDs provide extended hydrophobic spaces and are highly adaptable hosts, their commercialization encounters significant obstacles. Major issues include production and purification costs. Also, there is a lack of toxicological information and ambiguous regulatory pathways for pharmaceutical and food applications, which require stringent safety tests and standard analytical techniques before their use in industry.

There have been considerable difficulties in the synthesis of LR-CDs, both through the enzymatic and chemical routes. Production via CGTase yields mixtures of α-, β-, and γ-CDs, along with multiple LR-CD homologs between DP 9–22, with yields not exceeding 10%. The cost of the biocatalyst, like CGTase, makes the process difficult to scale up. Furthermore, the reaction conditions must be strictly controlled to regulate the distribution of ring sizes. Chemical synthesis, which offers greater structural control, relies on template-directed or multistep coupling strategies, including the use of protecting groups and specialized catalysts. The cost of reagents and the requirement for controlled reaction atmospheres make the process expensive and hard to scale. Purification also makes production more complicated, necessitating preparative-scale HPLC, size-exclusion chromatography, or solvent fractionation methods. Massive amounts of solvents are needed in extraction and purification, which adds to operational costs further. This, in turn, makes the price per gram of LR-CDs far more expensive than α-CD, which can be commercially purchased (~USD 2–3/kg). These methods are time-consuming and cannot be modified to work with industrial scales easily. The characterization of LR-CDs is also challenging due to the close interrelationship between the size of rings, which leads to small mass and structural variations. Sophisticated instruments such as MALDI-TOF MS, 2D NMR, and high-resolution spectroscopy are necessary, but they require skills and equipment that are not always available in typical commercial or industrial facilities (Liu et al., 2023).

Although these technical and economic restrictions pose a significant impediment to large-scale manufacturing, LR-CDs exhibit better performance in terms of solubility, stabilizing properties, and nanocarrier capabilities, particularly for bulky hydrophobic drugs and complex supramolecular systems. Thus they are now justified in terms of commercialization in just high-value applications, such as targeted therapeutics, precision drug-delivery systems, or highly functional materials, where their structural benefits are significant to production expenses (Ngawiset et al., 2023).

Natural CDs are generally safe. Chemically modified CDs (e.g., methyl- or sulfobutyl-α-CD) have been found to cause concerns regarding nephrotoxicity or ototoxicity when taken in high doses. CDs cross-linked and given to rodents at a high dose (2000 mg/kg/day) were extremely nontoxic, indicating that the structural modification itself is not a biological effect. Data on LR-CD-specific toxicity remain scarce and should undergo additional safety profiling. Moreover, in vitro data show that LR-CDs can reduce peptide toxicity, as evidenced by the fact that ultrashort peptide-LR/CD inclusion complexes retained their antibacterial activity despite increased cytotoxicity in L929 fibroblasts (Liu et al., 2023). The in vivo study with the help of reactive oxygen species-reactive LR-CD nanoplatforms showed that mice were repeatedly administered with no organ damage, and their blood markers were found to be stable (Li et al., 2022).

LR-CDs are qualitatively different from regular CDs, which typically undergo new chemical entity evaluation and are considered safe (GRAS), and are allowed in most drugs and foods. Regulatory authorities, including the Food and Drug Administration and the European Medicines Agency, will require complete toxicological information, impurity profiles, stability assays, pharmacokinetics (absorption, distribution, metabolism, excretion), and clinical data. Development of appropriate analytical methods to detect and purify ring-size isoforms by scaling up available MALDI-TOF and advanced NMR techniques is required to bring LR-CDs closer to approval. Ring-specific synthesis is needed to streamline production, and this can be done by scalable green chemical templating or CGTase engineering. Specific economic analysis of the costs and benefits of different applications would permit targeting high-value applications. In addition, comprehensive in vitro and in vivo safety studies focusing on organ-specific toxicities and chronic exposure would be needed (Cao et al., 2020).

Future perspectives and research directions

The LR-CDs with ≥nine degrees of polymerization are still of great interest as macrocycles because of their larger hydrophobic cavities, flexible conformations, and their ability to be chemically modified. Along with the current development of scientific innovation, LR-CDs are being incorporated into next-generation biomedical systems, which consist of gene delivery platforms, targeted nanocarriers, and smart hydrogels. Their biodegradability and tunable structure also position them well for use in green chemistry, environmental remediation, and sustainable material design. Future directions include increasing the specificity of ring-size, scalable synthesis, and potential uses of precision of ring-size, including personalized medicine, biosensing, and personalized multifunctional supramolecular assemblies.

In gene therapy, CD-based supramolecular carriers are gaining popularity. Xu et al. characterized polyrotaxanes, grafted-CD polymers, and supramolecular hydrogels, which were reported to be fruitful carriers of sustained gene delivery with modular release and increased transfection efficiency (Xu et al., 2019). Mendonça et al. (2021) described siRNA-modified CD nanoparticles and blood–brain barrier (BBB)-shuttle peptides that have shown safe brain endothelium-permeable delivery of gene-binding cargo in models of Huntington disease. Although these studies were performed with compact CDs, LR-CDs are preincubated with large nucleic acids or medicinal cargo due to their expanded cavities, which offer an interesting future opportunity. CD coatings in electrochemical and optical sensors are used for selective sensing of pollutants and biomarkers. Li et al. (2025) reviewed CD-functionalized electrodes made of carbon nanostructures, graphene, and metal nanoparticles, which can detect targets with ultra-sensitivity via host–guest chemistry. The high binding capacity of LR-CDs could enable higher detection of larger molecules, such as peptides or protein fragments (Li et al., 2025). Hydrogels based on CD crosslinking are emerging in the field of tissue engineering. The CD-polysaccharide hydrogels are applied to dress wound, reconstruct bone, and myocardial regeneration (Damiri et al., 2023). These hydrogels are self-healing, injectable, and have tunable mechanical performance. The high-quality cavity in the LR-CDs could even encapsulate cells, protect growth factors, and release bioactives so that they will be a future-proof regenerative therapy (Sapsford & Michieletto, 2025).

A smart material is made by chemically grafting LR-CDs onto polymeric backbones. PEG or PPG chains of CD-based polyrotaxanes, which are mechanically threaded onto a PEG or PPG chain surface, have been shown to dethread and release cargo on exposure to stimuli such as pH or enzymatic cleavage. In current systems, most use α-CD, although similar constructions with LR-CDs are under development for novel materials (Xu et al., 2019). Nanocarriers comprising CDs, nanoparticles, and liposomes are becoming more sophisticated. In 2021, it was demonstrated that Huntington's disease was treated with modified CD nanoparticles delivering siRNA via the BBB by targeting the RVG peptide (Mendonça et al., 2021). Hydrogel hybrids undergo changes over time and in space. The release of drug from CD-polymer hybrids is controlled by both pH and temperature. LR-CDs encapsulate larger biomolecules, such as proteins or growth hormones, and provide controllable release rates, making them suitable for use as smart biomedical implants and devices (Omidian et al., 2025).

The LR-CDs have high potential in the personalization of treatment for individual patients. LR-CD carriers can be programmed to achieve specific biodistribution, release characteristics, and cursor dose by controlling ring-size and modulating functional groups, for example, by including targeting ligands or enzymatically cleavable groups. Semisynthetic constructs based on cholesterol-filled, spermine-substituted LR-CD nanogels have been shown to deliver siRNA to renal tumor cells and they show tumor suppression and accurate targeting (Musuc, 2024). CD-based nanostructures are also being considered as contrast agents in theranostics and imaging. Image-guided administration may be made possible with the introduction of CDs in conjugation with imaging probes (spanning MRI, CT, and fluorescent dyes) through lipid nanoparticles. Although no recorded LR-CD imaging systems are available, studies of CD-linked nanoparticles and CD-metal–organic frameworks (MOFs) in imaging suggest that this could be used. LR-CDs have conjugatable functional moieties that can be used to deliver targeted therapy and to serve as modular theranostic platforms containing diagnostic payloads (Gadade & Pekamwar, 2020).

LR-CDs are biodegradable and manufactured out of renewable starch, which predetermines their sustainable application. Recent progress of CD-MOFs based on CD with biocompatible metals offers scalable, green materials for gas storage, environmental pollutant removal, and controlled release systems. These MOFs would provide LR-CDs that increase pore size to enable the sequestration of larger organic compounds, which is ideal for environmental remediation and, therefore sustainable packaging. LR-CDs offer large cavities that are well-suited to encapsulating complex genetic payloads. Liu et al. demonstrated a platform of cyclodextrin-hyaluronic acid nanoparticle platform that incorporated two UV/NIR triggers to deliver siRNA to cancer cells efficiently in hypoxic environments (Liu & Liu, 2022). These breakthroughs are positive news for next-generation nonviral gene therapies, as they suggest a path for LR-CD-based systems that can package larger mRNA constructs or CRISPR elements. More recent developments of CD-functionalized electrodes demonstrate that they are capable of detecting peptides and proteins with high sensitivity using focused host–guest interactions. LR-CDs can detect larger or conformationally insensitive biomolecules, such as cytokines or neuropeptides, by improving capture efficiency and binding and can be used in point-of-care and early-disease settings (Li et al., 2025). CD-based hydrogels have also found use in injectable and self-healing networks in tissue engineering. Roy et al. (2023) outlined the characteristics of CD supramolecular hydrogels, including mechanical and gelation properties that could be varied. Also, increased cavities in LR-CDs would further enrich these systems with better loading of growth factors or peptides and controlled release for possible applications in wound healing, heart repair, or bone regeneration. Theranostic nanocarriers are becoming intelligent. The CDs integrated into a lipid-based nanoparticle platform, together with poly (amidoamine), demonstrated a higher plasmid DNA delivery rate and transfection efficiency (Tian et al., 2022). Alpha-CD/polybutylcyanoacrylate nanoparticles responsive to ultrasound have been created to treat and image plaques in atherosclerosis models (Mehta et al., 2022). These hybrid systems might be used to incorporate LR-CD scaffolds that could significantly enhance the payload capacity, allowing simultaneous loading of therapeutic drugs and imaging dyes to provide accurate diagnostics and responsive treatment (El-Husseiny et al., 2022). CD-derived structures (CD-MOFs) have become popular in green chemistry due to their proven capabilities for pollutant uptake and reaction catalysis. Although alpha-CDs are used in most applications, LR-CD-derived MOFs would capture the bulky colors, pesticide residues, and pharmaceutical contaminants due to enlarged pores. LR-CDs present a major breakthrough in green catalysis and green cleanup of the environment when used in conjunction with biodegradable design (Dummert et al., 2022). Next-generation materials exhibit responsiveness. Examples of pH-, temperature-, and light-sensitive inclusion complexes responsive to poloxamer hydrogel have been demonstrated using CDs. LR-CD-based nanosponges offer opportunities to deliver therapeutic agents specifically and precisely to the site of action by carrying larger therapeutic agents, such as peptides and functional proteins, and by regulating their release after exposure to environmental stimuli (Wang et al., 2023).

Conclusion

LR-CDs represent a powerful new class of macrocyclic hosts that merge biocompatibility with exceptional structural flexibility and functional diversity. Their enlarged cavities enable superior encapsulation of bulky, hydrophobic, and complex molecules, positioning them at the forefront of next-generation drug delivery, food stabilization, environmental remediation, and smart material design. Although current challenges in synthesis, purification, and regulatory approval remain significant, continued advances in enzyme engineering, analytical technologies, and scalable production will rapidly accelerate their translational potential. With their renewable nature and broad application spectrum, LR-CDs are poised to become indispensable tools in precision medicine, green chemistry, and advanced materials—unlocking innovations far beyond the capabilities of classical CDs.

AI disclosure

During the preparation of this work, the author(s) used ChatGPT in order to improve readability. After using this tool/service, the author(s) reviewed and edited the content as needed and take(s) full responsibility for the content of the publication.

References

Arima, H. (2021). Twenty years of research on cyclodextrin conjugates with PAMAM dendrimers. *Pharmaceutics, 13*(5), 697. https://doi.org/10.3390/pharmaceutics13050697.

Cao, C., Xu, L., Xie, P., Hu, J., Qi, J., Zhou, Y., & Cao, L. (2020). The characterization and evaluation of the synthesis of large-ring cyclodextrins (CD_9 –CD_{22}) and α-tocopherol with enhanced thermal stability. *RSC Advances. Royal Society of Chemistry (RSC), 10*(11), G584–G591. https://doi.org/10.1039/c9ra10748g.

Cao, C., Deng, C., Hu, J., & Zhou, Y. (2022). Formation and molecular dynamics simulation of inclusion complex of large-ring cyclodextrin and 4-terpineol. *Journal of Food Science, 87*(10), 4609–4621. https://doi.org/10.1111/1750-3841.16303.

Crini, G., Fourmentin, S., Fenyvesi, É., Torri, G., Fourmentin, M., & Morin-Crini, N. (2018). Cyclodextrins, from molecules to applications. *Environmental Chemistry Letters, 16*(4), 1361–1375. https://doi.org/10.1007/s10311-018-0763-2.

Damiri, F., Fatimi, A., Santos, A. C. P., Varma, R. S., & Berrada, M. (2023). Smart stimuli-responsive polysaccharide nanohydrogels for drug delivery: a review. *Journal of Materials Chemistry B, 11*(44), 10538–10565. https://doi.org/10.1039/d3tb01712e.

Dummert, S. V., Saini, H., Hussain, M. Z., Yadava, K., Jayaramulu, K., Casini, A., & Fischer, R. A. (2022). Cyclodextrin metal–organic frameworks and derivatives: recent developments and applications. *Chemical Society Reviews, 51*(12), 5175–5213. https://doi.org/10.1039/d1cs00550b.

El-Husseiny, H. M., Mady, E. A., Hamabe, L., Abugomaa, A., Shimada, K., Yoshida, T., Tanaka, T., Yokoi, A., Elbadawy, M., & Tanaka, R. (2022). Smart/stimuli-responsive hydrogels: Cutting-edge platforms for tissue engineering and other biomedical applications. *Materials Today Bio, 13*, 100186. https://doi.org/10.1016/j.mtbio.2021.100186.

Endo, T., & Ueda, H. (2004). "Large ring cyclodextrins - Recent progress. *Fabad Journal of Pharmaceutical Sciences. Japan, 29*(1), 27–38.

Endo, T., Zheng, M., & Zimmermann, W. (2002). Enzymatic synthesis and analysis of large-ring cyclodextrins. *Australian Journal of Chemistry, 55*(1–2), 39–48.

Endo, Tomohiro (2011). Large-ring Cyclodextrins. *Trends in Glycoscience and Glycotechnology. Forum: Carbohydrates Coming of Age, 23*(130), 79–92. https://doi.org/10.4052/tigg.23.79.

Esteso, M. A., & Romero, C. M. (2024). Cyclodextrins: Properties and Applications. *International Journal of Molecular Sciences. Spain: Multidisciplinary Digital Publishing Institute (MDPI), 25*(8). https://doi.org/10.3390/ijms25084547.

French, D., Pulley, A. O., Effenberger, J. A., Rougvie, M. A., & Abdullah, M. (1965). Studies on the Schardinger dextrins. *Archives of Biochemistry and Biophysics, 111*(1), 153–160. https://doi.org/10.1016/0003-9861(65)90334-6.

Fujiwara, T., Tanaka, N., & Kobayashi, S. (1990). Structure of δ-Cyclodextrin 13.75 H2O. *Chemistry Letters, 19*(5), 739–742. https://doi.org/10.1246/cl.1990.739.

Gadade, D., & Pekamwar., S. S. (2020). Cyclodextrin Based Nanoparticles for Drug Delivery and Theranostics. *Advanced Pharmaceutical Bulletin, 10*(2), 166–183. https://doi.org/10.34172/apb.2020.022.

Gao, Y., Sun, L., Qiao, C., Zhang, H., Wang, Y., Liu, Y., Feng, R., & Zang, Y. (2025). Cyclodextrin-based delivery systems for chemical and genetic drugs: Current status and future. *Carbohydrate Polymers, 352*, 123174. https://doi.org/10.1016/j.carbpol.2024.123174.

Guedes, A. M., Santos Alves, T. F., Salústio, P. J., Cabral-Marques, H. M., & Ribeiro, M. H. L. (2023). Design of a *Cyclodextrin* bioproduction process using *Bacillus pseudofirmus* and *Paenibacillus macerans*. *Future Pharmacology, 3*(3), 568–584. https://doi.org/10.3390/futurepharmacol3030035.

Haimhoffer, Á., Rusznyák, Á., Réti-Nagy, K., Vasvári, G., Váradi, J., Vecsernyés, M., Bácskay, I., Fehér, P., Ujhelyi, Z., & Fenyvesi, F. (2019). Cyclodextrins in drug delivery systems and their effects on biological barriers. *Scientia Pharmaceutica, 87*(4), 33. https://doi.org/10.3390/scipharm87040033.

Hansen, K. H., Hindsgaul, O., Svensson, B., & Madsen, R. (2025). Enzyme-mediated dynamic combinatorial chemistry enables large-scale synthesis of β-cyclodextrin. *Journal of the American Chemical Society, 147*(16), 13851–13858. https://doi.org/10.1021/jacs.5c02055.

Ismail, A., Kerdpol, K., Rungrotmongkol, T., Tananuwong, K., Ueno, T., Ekasit, S., Muangsin, N., & Krusong, K. (2021). Solubility enhancement of poorly water-soluble domperidone by complexation with the large ring cyclodextrin. *International Journal of Pharmaceutics, 606*, 120909. https://doi.org/10.1016/j.ijpharm.2021.120909.

Ivanov, P. M., & Jaime, C. (2004). Insights into the structure of large-ring cyclodextrins through molecular dynamics simulations in solution. *The Journal of Physical Chemistry. B, 108*(20), 6261–6274. https://doi.org/10.1021/jp037527t.

Ji, X. Y., Wang, Y., Zhang, H., Liu, X., Li, J., Zhao, Y., & Chen, L. (2024). Advances in cyclodextrins and their derivatives in nano-delivery systems. *Pharmaceutics, 16*(8), 1054. https://doi.org/10.3390/pharmaceutics16081054.

Jin, X., Zhu, L., Xue, B., Zhu, X., & Yan, D. (2019). Supramolecular nanoscale drug-delivery system with ordered structure. *National Science Review, 6*(6), 1128–1137. https://doi.org/10.1093/nsr/nwz018.

Kerdpol, K., Rungrotmongkol, T., Nutho, B., Krusong, K., Poo-arpon, R. P., & Hannongbua, S. (2021). Encapsulation of α-tocopherol in large-ring cyclodextrin containing 26 α-D-glucopyranose units: A molecular dynamics study. *Journal of Molecular Liquids, 339*, 116802. https://doi.org/10.1016/j.molliq.2021.116802.

Khalid, S. H., Bashir, M., Asghar, S., Mallhi, T. H., & Khan, I. U. (2020). Effect of cyclodextrin derivatization on solubility and efficacy of drugs. In S. Karakuş (Ed.). *Colloid Science in Pharmaceutical Nanotechnology*IntechOpenhttps://doi.org/10.5772/intechopen.90364.

Kobayashi, M., Mihara, K., & Matsuda, K. (1986). Dextransucrase from *Leuconostoc mesenteroides* NRRL B-512F: Characterization of the enzyme bound to Sephadex gel. *Agricultural and Biological Chemistry, 50*(3), 551–556. https://doi.org/10.1080/00021369.1986.10867236.

Kost, B., Brzeziński, M., Socka, M., Baśko, M., & Biela, T. (2020). Biocompatible Polymers Combined with Cyclodextrins: Fascinating Materials for Drug Delivery Applications. *Molecules (Basel, Switzerland), 25*(15), 3404. https://doi.org/10.3390/molecules25153404.

Krusong, K., Ismail, A., Wangpaiboon, K., & Pomgsawasdi, P. (2022). Production of large-ring cyclodextrins by amylomaltases. *Molecules (Basel, Switzerland), 27*(4), 1446. https://doi.org/10.3390/molecules27041446.

Kuttiyawong, K., Saehu, S., Ito, K., & Pongsawasdi, P. (2015). "Synthesis of large-ring cyclodextrin from tapioca starch by amylo- maltase and complex formation with vitamin E acetate for solubility enhancement" *Process Biochemistry, 50*(12), 21G8–217G8. https://doi.org/10.1016/j.procbio.2015.10.014.

Larsen, D., & Beeren, S. R. (2019). Enzyme-mediated dynamic combinatorial chemistry allows out-of-equilibrium template-directed synthesis of macrocyclic oligosaccharides. *Chemical Science, 10*(43), 9981–9987. https://doi.org/10.1039/c9sc03983j.

Li, X., Fu, L., Lv, Y., Zhang, R., Karimi-Maleh, H., Chen, F., & Zhao, S. (2025). Cyclodextrin-based architectures for electrochemical sensing: From molecular recognition to functional hybrids. *Analytical Methods, 17*(21), 4300–4320. https://doi.org/10.1039/D5AY00612K.

Li, X., Ye, M.,, Gao, Y., Hou, S., Jia, P., Kang, Y., Xue, P., & Xu, Z. (2022). The systematic evaluation of physicochemical and biological properties *in vitro* and *in vivo* for natural silk fibroin nanoparticles. *Advanced Fiber Materials, 4*(5), 1141–1152. https://doi.org/10.1007/s42765-022-00167-2.

Liu, G., Zhu, C., Wang, Y., & Huang, F. (2018). Cyclodextrin-based host–guest supramolecular hydrogel and its application in biomedical fields. *Polymer Chemistry, 9*(25), 3436–3449. https://doi.org/10.1039/C8PY00730F.

Liu, Y. H., & Liu, Y. (2022). Highly effective gene delivery based on cyclodextrin multivalent assembly in target cancer cells. *Journal of Materials Chemistry B. China: Royal Society of Chemistry, 10*(G), 958–9G5. https://doi.org/10.1039/d1tb02585f.

Li, X., Liu, J., & Qiu, N. (2023). "Cyclodextrin-Based Polymeric Drug Delivery Systems for Cancer Therapy," *Polymers. MDPI AG, 15*(G). https://doi.org/10.3390/polym150G1400.

Li, X., Liu, J., & Qiu, N. (2023). Cyclodextrin-based polymeric drug delivery systems for cancer therapy. *Polymers, 15*(G). https://doi.org/10.3390/polym150G1400.

Mehta, S., Bongcaron, V., Nguyen, T. K., Jirwanka, Y., Maluenda, A., Walsh, A. P. G., Palasubramaniam, J., Hulett, M. D., Srivastava, R., Bobik, A., Wang, X., & Peter, K. (2022). An Ultrasound-Responsive Theranostic Cyclodextrin-Loaded Nanoparticle for Multimodal Imaging and Therapy for Atherosclerosis. *Small (Weinheim an der Bergstrasse, Germany), 18*(31), 2200967. https://doi.org/10.1002/smll.202200967.

Mendonça, M. C. P., Cronin, M. F., Cryan, J. F., & O'Driscroll, C. M. (2021). Modified cyclodextrin-based nanoparticles mediated delivery of siRNA for huntingtin gene silencing across an *in vitro* BBB model. *European Journal of Pharmaceutics and Biopharmaceutics, 169*, 309–318. https://doi.org/10.1016/j.ejpb.2021.11.003.

Musuc, A. M. (2024). Cyclodextrins: Advances in Chemistry, Toxicology, and Multifaceted Appli- cations. *Molecules. Romania: Multidisciplinary Digital Publishing Institute (MDPI), 29*(22). https://doi.org/10.3390/molecules29225319.

Ngawiset, S., Ismail, A., Murakami, S., Pongsawasdi, P., Rungrotmongkol, T., & Krusong, K. (2023). Identification of crucial amino acid residues involved in large ring cyclodextrin synthesis by amylomaltase from *Corynebacterium glutamicum. Computational and Structural Biotechnology Journal, 21*, 899–909. https://doi.org/10.1016/j.csbj.2023.01.011.

Omidian, H., Akhzarmehr, A., & Gill, E. J. (2025). Cyclodextrin–Hydrogel Hybrids in Advanced Drug Delivery," *Gels, 11*(3). https://doi.org/10.3390/gels11030177.

Osaki, M., Takashima, Y., Kamitori, S., Yamaguchi, H., & Harada, A. (2019). X-ray crystal structures of α-cyclodextrin–5-hydroxypentanoic acid, β-cyclodextrin–5-hydroxypentanoic acid, β-cyclodextrin–ε-caprolactone, and β-cyclodextrin–ε-caprolactam inclusion complexes. *Journal of Inclusion Phenomena and Macrocyclic Chemistry, 96*(1–2), 93–99. https://doi.org/10.1007/s10847-019-00953-3.

Pardhi, D. S., Rabadiya, K. J., Panchal, R. R., Raval, V. H., Joshi, R. G., & Rajput, K. N. (2023). Cyclodextrin glucanotransferase: fundamentals and biotechnological implications. *Applied Microbiology and Biotechnology, 107*(19), 5899–5907. https://doi.org/10.1007/s00253-023-12708-9.

Poulson, B. G., Alsulami, Q. A., Sharfalddin, A., El Agammy, Emam F., Mouffouk, F., Emwas, A.-H., Jaremko, L., & Jaremko, M. (2021). Cyclodextrins: Structural, Chemical, and Physical Properties, and Applications. *Polysaccharides, 3*(1), 1–31. https://doi.org/10.3390/polysaccharides3010001.

Pulley, A. O., & French, D. (1961). "Studies on the Schardinger dextrins. XI. The isolation of new Schardinger dextrins" *Biochemical and Biophysical Research Communications, 5*(1), 11–15. https://doi.org/10.1016/0006-291X(61)90071-7 undefined.

Rossi, B., Bottari, C., Comez, L., Corezzi, S., Paolantoni, M., Gessini, A., Masciovecchio, C., Mele, A., Punta, C., Melone, L., Fiorati, A., Radulescu, A., Mangiapia, G., & Paciaroni, A. (2018). Structural and molecular response in cyclodextrin-based pH-sensitive hydrogels by the joint use of Brillouin, UV Raman and Small Angle Neutron Scattering techniques. *Journal of Molecular Liquids, 271*, 738–746. https://doi.org/10.1016/j.molliq.2018.08.141.

Roy, A., Ghosh, S., Das, D., & Banerjee, R. (2023). Chemical modification of β-cyclodextrin towards hydrogel formation. *Carbohydrate Polymers, 306*, 120576. https://doi.org/10.1016/j.carbpol.2023.120576.

Saokham, P., Muankaew, C., Jansook, P., & Loftsson, T. (2018). Solubility of Cyclodextrins and Drug/Cyclodextrin Complexes. *Molecules (Basel, Switzerland), 23*(5), 1161. https://doi.org/10.3390/molecules23051161.

Sapsford, E., & Michieletto, D. (2025). "Topologically-crosslinked hydrogels based on γ-cyclodextrins. *Communications Chemistry, 8*(1). https://doi.org/10.1038/s42004-025-014G9-3.

Sherje, A. P., Dravyakar, B. R., Kadam, D., & Jadhav, M. (2017). Cyclodextrin-based nanosponges: A critical review. *Carbohydrate Polymers, 173*, 37–49. https://doi.org/10.1016/j.carbpol.2017.05.086.

Sonnendecker, C., Thürmann, S., Przybylski, C., Zitzmann, F. D., Heinke, N., Krauke, Y., Monks, K., Robitzki, A. A., Belder, D., & Zimmermann, W. (2019). Large-Ring Cyclodextrins as Chiral Selectors for Enantiomeric Pharmaceuticals. *Angewandte Chemie International Edition, 58*(19), 6411–6414. https://doi.org/10.1002/anie.201900911.

Tian, B., Xiao, D., Hei, T.,, Ping, R., Hua, S., & Liu, J. (2020). The application and prospects of cyclodextrin inclusion complexes and polymers in the food industry: A review. *Polymer International, 69*(7), 597–603. https://doi.org/10.1002/pi.5992.

Tian, M., Xin, X., Wu, R., Guan, W., & Zhou, W. (2022). Advances in intelligent-responsive nanocarriers for cancer therapy. *Pharmacological Research, 178*, 106184. https://doi.org/10.1016/j.phrs.2022.106184.

Wang, S., Wei, Y., Wang, Y., & Cheng, Y. (2023). Cyclodextrin regulated natural polysaccharide hydrogels for biomedical applications—a review. *Carbohydrate Polymers, 313*, 120760. https://doi.org/10.1016/j.carbpol.2023.120760.

Xu, C., Wu, Y.-L., Li, Z., & Xian Jun, L. (2018). Cyclodextrin-based sustained gene release systems: a supramolecular solution towards clinical applications. *Materials Chemistry Frontiers, 3*(2), 181–192. https://doi.org/10.1039/c8qm00570b.

Yu, G., & Chen, X. (2019). "Host-Guest Chemistry in Supramolecular Theranostics" *Theranostics, 9*(11), 3041–3074. https://doi.org/10.7150/thno.31G53.

Zhao, F., Repo, E., Dulin, Y., Chen, L., Simo, K., Tang, J., Iakovleva, E., Tam, K. C., & Sillanpää, M. (2017). One-pot synthesis of trifunctional chitosan-EDTA-β-cyclodextrin polymer for simultaneous removal of metals and organic micropollutants. *Scientific Reports, 7*(1), 15811. https://doi.org/10.1038/s41598-017-16222-7.

CHAPTER

Applications of cyclodextrins

14

Drashti Patel[1,2], Rushikesh Joshi[2], Rakeshkumar Panchal[1] and Kiransinh N. Rajput[1]
[1]Department of Microbiology & Biotechnology, University School of Sciences, Gujarat University, Ahmedabad, Gujarat, India,
[2]Department of Biochemistry & Forensic Science, University School of Sciences, Gujarat University, Ahmedabad, Gujarat, India

Introduction

Cyclodextrins (CDs) represent one of the most versatile classes of cyclic oligosaccharides used across scientific, industrial, medical, and environmental disciplines due to their unique molecular architecture and remarkable inclusion complex-forming ability (Szejtli, 1998). These molecules are composed of α-(1,4)-linked D-glucopyranose units arranged in a toroidal or truncated cone-shaped geometry, which creates a hydrophobic internal cavity capable of hosting a wide range of guest molecules while leaving the hydrophilic outer surface exposed to aqueous environments (Loftsson et al., 2019). This amphiphilic property allows CDs to encapsulate hydrophobic compounds without forming covalent bonds, resulting in noncovalent inclusion complexes that alter the physicochemical properties—solubility, stability, volatility, reactivity, and bioavailability of the guest molecules (Del Valle, 2004). The three naturally occurring homologs, α-CD (six glucose units), β-CD (seven glucose units), and γ-CD (eight glucose units), each exhibit cavity sizes suitable for guests of specific molecular dimensions, enabling tailored applications in food, pharmaceutical, environmental, cosmetic, and material industries (Astray et al., 2009).

The scientific importance of CDs arises from their supramolecular behavior, which is governed by host–guest chemistry that modulates the thermodynamic equilibrium between free and bound states of bioactive and industrially valuable compounds (Crini, 2014). This ability to modulate molecular properties without chemically altering the guest molecule has positioned CDs as indispensable functional agents in formulation science, where they are crucial for stabilizing volatile substances, masking undesirable tastes or odors, enhancing solubility, improving dissolution characteristics, and regulating release profiles (Loftsson et al., 2016). Moreover, the functional versatility of CDs has expanded through the development of an extensive array of chemically modified derivatives, such as hydroxypropyl-β-cyclodextrin (HPβCD),

Cyclodextrin Glucanotransferase. DOI: https://doi.org/10.1016/B978-0-443-41694-1.00012-5

sulfobutylether-β-cyclodextrin (SBEβCD), randomly methylated β-cyclodextrin (RAMEB), and carboxymethylated CDs, each with improved aqueous solubility and enhanced binding capacity for specific applications (Khatoon et al., 2025). These modifications have broadened the applicability of CDs to accommodate hydrophobic compounds of varying polarity, molecular weight, and steric complexity, thereby solidifying their role as technologically indispensable excipients and functional macromolecules.

CDs are increasingly central to sustainable development because they enable environmentally friendly processes, reduce the use of organic solvents, support green extraction methods, and enhance the degradation or removal of pollutants in soil and water (Syeda et al., 2022). Their biodegradability, nontoxicity, and compatibility with biological systems make them particularly attractive for eco-friendly industrial applications (Crini & Lichtfouse, 2019). CDs also play a vital role in biomedicine, not only as excipients but also as active pharmaceutical ingredients (APIs) themselves, as illustrated by the development of sugammadex, a modified γ-CD designed to encapsulate and neutralize neuromuscular blocking agents during surgery (Bom et al., 2009).

This chapter offers a comprehensive, application-focused exploration of CDs, detailing their wide-ranging roles in the food, pharmaceutical, cosmetic, agricultural, environmental, chemical, textile, and material science industries. The discussion is structured to illuminate the multifaceted contributions of CDs to product quality, formulation science, sustainability, and emerging nanotechnology and biomedical innovations. Throughout this chapter, emphasis is placed on how the intrinsic supramolecular chemistry of CDs drives these diverse applications, highlighting their ability to function as molecular carriers, stabilizers, solubility enhancers, controlled-release agents, affinity modulators, pollutant traps, and functional modifiers. In doing so, the chapter underscores the expanding importance of CDs as pivotal molecular tools in contemporary science and technology.

Structural and chemical basis of cyclodextrin functionality

CDs derive their functional capabilities from their distinctive toroidal structure, characterized by a relatively hydrophobic cavity formed by the inward-facing C–H groups and glycosidic oxygen atoms and a hydrophilic exterior composed of hydroxyl groups located at the rim of the cone (Szejtli, 1998). This compartmentalized structure enables CDs to encapsulate hydrophobic guests in their cavities while maintaining water solubility due to their hydrophilic surfaces, leading to the formation of stable inclusion complexes stabilized by hydrophobic interactions, van der Waals forces, hydrogen bonding, and the release of high-energy water molecules from the cavity interior (Saokham & Loftsson, 2017). The diameter of the CD cavity determines the specificity and affinity for particular guest molecules, with α-CD being suitable for small aliphatic chains, β-CD for aromatic molecules, and γ-CD for relatively bulky or polycyclic compounds (Del Valle, 2004).

The inclusion complexation process modifies the properties of the guest molecules in a predictable and controllable manner, which is why CDs are used across an exceptionally diverse range of applications (Astray et al., 2009). For instance, encapsulation can increase the aqueous solubility of poorly soluble drugs by several orders of magnitude, reduce the volatility of aromatic compounds used in fragrances, prevent oxidative degradation of sensitive nutrients or bioactive molecules, and mask undesirable tastes or odors in pharmaceuticals and food products (Zuidam & Heinrich, 2009). In some cases, CDs can also alter the reactivity or photostability of the guest molecules, providing further functional benefits in formulation systems (Crini, 2014). Moreover, complexation can reduce the irritation potential of certain compounds by preventing direct contact with biological tissues, a property widely exploited in dermatological and ophthalmic preparations (Uekama, 2004).

Chemical modification of CDs has expanded their solubility and inclusion capabilities, generating derivatives with tailored physicochemical properties better suited for specific technological applications. HPβCD, for instance, possesses superior aqueous solubility and forms more stable complexes with many drugs, making it among the most widely used CD derivatives in pharmaceutical formulations (Loftsson et al., 2016). SBEβCD carries a negative charge, enhancing its solubility and enabling interactions with positively charged guest molecules, which has led to its approval in several injectable medicines (Uekama, 2004). Methylated CDs, such as RAMEB, exhibit significantly increased lipophilicity and binding capacity, making them valuable for solubilizing highly hydrophobic or bulky compounds (Crini, 2014). These structural advances have enabled CDs to address increasingly complex formulation challenges, including the delivery of macromolecules, stabilization of sensitive actives, controlled release of agrochemicals, and removal of persistent organic pollutants.

The supramolecular chemistry of CDs has also paved the way for their integration into advanced materials, such as nanosponges, hydrogels, nanofibers, molecularly imprinted polymers, and metal–organic frameworks (MOFs), which leverage the inherent host–guest capabilities of CDs to create smart systems capable of targeted delivery, stimuli-responsive release, selective adsorption, and catalytic enhancement (Krabicová et al., 2020). These emerging applications highlight the expanding role of CDs in modern technology, particularly in areas requiring precise molecular recognition, environmentally benign synthesis, and high-performance functional materials. Fig. 14.1 provides an overview of the major industrial sectors utilizing CDs, illustrating how their fundamental host—guest inclusion capability translates into sector-specific functional advantages ranging from solubility enhancement and controlled release to pollutant adsorption and chiral recognition, thereby delivering measurable technological, economic, and environmental benefits.

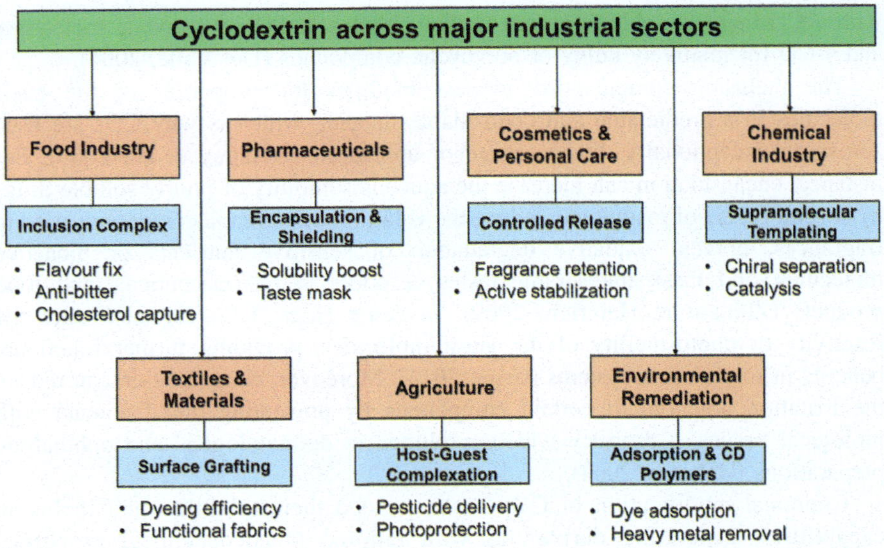

FIGURE 14.1

Key applications of cyclodextrins across major industrial sectors.

Cyclodextrins in the food industry

The food industry represents one of the most extensive and diverse application sectors for CDs due to their capacity to improve food quality, enhance sensory attributes, stabilize sensitive ingredients, reduce undesirable components, and support innovative functional food formulations (Astray et al., 2009). Their use is driven by the ability to modify bioactive or volatile compounds through molecular encapsulation, which protects labile components from degradation, controls release profiles, and reduces volatility or reactivity (Szejtli, 2004). Furthermore, CDs support the creation of healthier food alternatives by enabling the removal of undesirable constituents such as cholesterol, bitter components, and off-flavors (Szente & Szejtli, 2004).

Flavor encapsulation and stabilization

CDs are widely employed for the encapsulation of flavor compounds to enhance stability, reduce volatility, and improve shelf life in various food matrices (Marques, 2010). Many volatile flavor substances, including essential oils, terpenes, aldehydes, and esters, are prone to oxidation, thermal decomposition, and UV-induced degradation, which can significantly diminish product quality during storage (Reineccius, 2007). Encapsulation within CD cavities protects these labile molecules from environmental stressors by shielding them from light, oxygen, and heat, thereby preserving their integrity and extending their functional lifespan

(Gonzalez Pereira et al., 2021). For instance, β-CD complexation has been shown to improve the heat stability of L-menthol by as much as 86-fold during candy production, demonstrating the remarkable capability of CDs to retain volatile aromatic compounds under harsh processing conditions (Reineccius et al., 2004).

Additionally, CDs enhance sensory quality by controlling the release of flavor molecules, ensuring sustained aroma intensity and reducing premature evaporation (Marques, 2010). Encapsulation of essential oils such as rosemary, clove, oregano, or cinnamon not only stabilizes their aromatic profile but also preserves their functional antioxidant and antimicrobial activities, which contribute to improved food safety and extended product shelf life (Zhu et al., 2021). In fruit products, CD inclusion complexes prevent the oxidation of delicate flavor molecules such as citral or limonene, which are prone to degradation upon exposure to air or light (Gonzalez Pereira et al., 2021). This stabilization is particularly valuable in beverages, confectionery, snacks, and packaged foods, where flavor retention is crucial for consumer acceptance.

Taste masking and bitterness reduction

CDs play a vital role in eliminating or reducing undesirable tastes such as bitterness, astringency, or off-flavors that naturally occur in certain food ingredients or develop during processing (Szejtli & Szente, 2005). Bitterness often arises from hydrophobic compounds such as alkaloids, flavonoids, terpenes, peptides, and certain lipids, which can be selectively encapsulated within CD cavities, preventing them from interacting with taste receptors on the tongue (Rahman et al., 2024). This property allows CDs to significantly enhance the palatability of fruit juices, dairy products, protein-rich foods, and plant-based formulations (Xie et al., 2025).

In grapefruit juice, for example, β-CD has been successfully applied to reduce bitterness associated with limonin and naringin, achieving reductions ranging from 30% to 59%, depending on processing conditions (Rahman et al., 2024). Similar improvements have been reported for goat cheese, where CDs help suppress the pungent flavor associated with branched-chain fatty acids (Xie et al., 2025). CDs also mask the bitterness of polyphenols such as (+)-catechin and epicatechin, allowing the inclusion of these health-promoting molecules in functional food products without compromising taste (Szejtli & Szente, 2005). This ability to modulate sensory perception underscores the value of CDs in developing consumer-friendly food formulations enriched with bioactive compounds.

Cholesterol reduction in dairy and egg products

One of the major contributions of CDs to health-oriented food processing lies in their capability to selectively remove cholesterol from high-fat foods such as milk, butter, eggs, and cream (Priyanka & Christophe, 2017). β-CD is particularly effective due to its hydrophobic cavity size, which matches the molecular dimensions of cholesterol, allowing efficient formation of inclusion complexes that can be readily separated from the food matrix (Dai et al., 2022). Studies have demonstrated removal efficiencies ranging from 80% in milk and butter to 98% in cream and soft cheese, providing a technological solution for producing cholesterol-reduced and cholesterol-free food products (Priyanka & Christophe, 2017).

The mechanism involves the solubilization of cholesterol within the hydrophobic CD cavity, followed by centrifugal or filtration-based removal of the CD–cholesterol complex during processing (Dai et al., 2022). This technique is advantageous over traditional extraction methods because it avoids the use of organic solvents, maintains sensory attributes, and preserves nutritional value (Szejtli, 2004). However, regulatory considerations apply, as β-CD has an acceptable daily intake limit of 5 mg/kg bodyweight/day due to historical concerns regarding residual solvent contamination, even though modern production methods eliminate such risks (Park et al., 2011). In contrast, α- and γ-CD are generally recognized as safe by regulatory agencies and have no specified consumption limits (Munro et al., 2004). Despite these considerations, CD-assisted cholesterol removal remains a widely adopted technology for producing healthier dairy and egg products.

Texture improvement and antistaling in baked products

CDs contribute significantly to improving the texture, moisture retention, and shelf-life stability of baked goods, particularly bread, pastries, and cakes, by modifying starch retrogradation behavior (Gujral et al., 2003). Retrogradation, primarily driven by amylopectin recrystallization, leads to bread staling characterized by crumb firming, reduced elasticity, and loss of sensory quality. CDs disrupt this recrystallization process by forming inclusion complexes with starch degradation products and amylose-lipid fragments, thereby reducing intermolecular interactions that promote crystallinity (Yang et al., 2022). Studies demonstrate that the incorporation of β-CD into bread formulations reduces retrogradation enthalpy by more than 60% after 7 days of storage, indicating a marked suppression of starch retrogradation (Gujral et al., 2003). Furthermore, CDs enhance gas retention in dough systems by stabilizing gluten networks and improving viscoelastic properties, resulting in softer crumbs and improved loaf volume (Yang et al., 2022). Their ability to interact with both hydrophilic and hydrophobic components in dough matrices provides a molecular mechanism for maintaining texture quality and prolonging shelf stability in bakery products.

Production of functional oligosaccharides for health-promoting foods

CDs are used to produce maltooligosaccharides and related carbohydrate derivatives that serve as functional dietary fibers and prebiotics, supporting the growth of beneficial gut microbiota (Bláhová et al., 2023). The structural properties of these oligosaccharides, characterized by their degree of polymerization and digestibility profile, enable selective fermentation by probiotics such as *Bifidobacterium* and *Lactobacillus* species, thereby contributing to gut health and metabolic balance (Bláhová et al., 2023). Certain maltooligosaccharides, such as maltotetraose, also exhibit biological activities including antibacterial effects against foodborne and plant pathogenic bacteria, offering potential as natural preservatives in functional foods (Ji et al., 2024). Their incorporation into beverages, dairy products, and nutraceutical formulations contributes to the nutritional and functional enhancement of modern food products targeting health-conscious consumers.

Cyclodextrins in the pharmaceutical and medical sectors

The pharmaceutical industry represents the most scientifically advanced and economically significant domain for CD applications, owing to its ability to dramatically improve the solubility, stability, bioavailability, and therapeutic performance of APIs (Khatoon et al., 2025). Poor aqueous solubility affects more than forty percent of new chemical entities, posing major obstacles to oral absorption and clinical efficacy, a challenge for which CDs provide a highly effective supramolecular solution (Jacob & Nair, 2018).

Enhancement of drug solubility and bioavailability

The most prominent pharmaceutical application of CDs lies in improving the solubility of hydrophobic drugs by incorporating them within the CD cavity, increasing their apparent aqueous solubility and dissolution rate (Del Valle, 2004). Hydrophobic APIs such as itraconazole, naproxen, indomethacin, and remdesivir demonstrate significant increases in solubility when formulated as CD inclusion complexes (Almeida et al., 2023). Enhanced solubility directly correlates with improved absorption, particularly for oral and mucosal routes, resulting in better therapeutic outcomes (Loftsson, 2015). Studies have shown that HPβCD can enhance the solubility of itraconazole up to 20-fold by forming highly stable inclusion complexes, demonstrating the strong potential of CD-based systems for addressing solubility challenges in drug development (Almeida et al., 2023).

Beyond solubility enhancement, CD complexation contributes to improved membrane permeability, reduced drug precipitation upon dilution, and increased chemical stability in physiological environments (Jacob & Nair, 2018). These advantages render CDs essential excipients in oral, parenteral, ophthalmic, nasal, and transdermal drug delivery systems, where they help increase the efficiency of therapeutic molecules and allow reduced drug dosages without compromising clinical effectiveness (Loftsson et al., 2016).

Stabilization of therapeutic agents

CDs protect sensitive drugs from degradation caused by oxidative stress, hydrolysis, heat, light, and pH fluctuations by physically isolating unstable moieties within their hydrophobic cavities (Nicolaescu et al., 2025). Therapeutic molecules such as ascorbic acid, omeprazole, nicardipine, and nitroglycerin show enhanced stability when formulated with CDs, ensuring prolonged shelf life and consistent therapeutic efficacy (Nicolaescu et al., 2025). The stabilization mechanism mainly involves shielding labile bonds from reactive species, preventing decomposition in aqueous or acidic environments (Uekama, 2004). This application is particularly valuable in formulations requiring long-term storage or exposure to fluctuating environmental conditions, such as ophthalmic preparations, topical gels, and protein-based formulations.

Taste masking and odor control for patient compliance

The taste and aroma of drugs significantly influence patient compliance, especially in pediatric and geriatric populations, and CDs offer a nontoxic and highly effective strategy for masking undesirable sensory attributes (Adamkiewicz & Szeleszczuk, 2023). Hydrophobic APIs such as ibuprofen, famotidine, and cetirizine exhibit strong bitterness or unpleasant odors, which can be suppressed by encapsulating these molecules within CD cavities (Kaushik et al., 2024). The supramolecular barrier created by the CD prevents direct interaction between the bitter compounds and taste receptors, allowing formulations to become more palatable without altering pharmacological activity (Adamkiewicz & Szeleszczuk, 2023). In addition, CDs can delay the release of bitter drugs until after swallowing, enhancing patient acceptance in orally disintegrating tablets, syrups, and chewable formulations (Kaushik et al., 2024).

Controlled and targeted drug delivery systems

CDs play a vital role in the design of advanced drug delivery systems, including nanoparticles, nanofibers, nanosponges, MOFs, and polymeric carriers, enabling controlled drug release, targeted delivery, and improved therapeutic profiles (Topuz & Uyar, 2024). CD-based nanosponges, formed through crosslinking reactions that generate porous three-dimensional structures, encapsulate hydrophobic drugs such as paclitaxel, doxorubicin, and curcumin, providing controlled and sustained release kinetics (Swaminathan et al., 2016). These nanosponges enhance stability, reduce systemic toxicity, and allow site-specific delivery when functionalized with targeting ligands such as peptides or antibodies (Krabicová et al., 2020). Similarly, CD-containing MOFs serve as highly tunable drug reservoirs capable of responding to pH, temperature, or enzyme-triggered stimuli, enabling precision release in tumor microenvironments or inflamed tissues (Topuz & Uyar, 2024). Table 14.1 highlights

Table 14.1 Marketed pharmaceutical formulations containing cyclodextrins.

Drug name/ product	Therapeutic class	Cyclodextrin used	Purpose of CD in formulation	Marketed brand
Itraconazole	Antifungal	Hydroxypropyl-β-CD (HPβCD)	Enhances the solubility and oral bioavailability of poorly soluble itraconazole	Sporanox® oral solution
Ziprasidone mesylate	Antipsychotic	Sulfobutylether-β-CD (SBEβCD)	Improves solubility for IV administration	Geodon® injection
Voriconazole	Antifungal	SBEβCD	Stabilizes and solubilizes the drug in an injectable form	Vfend® IV formulation
Aripiprazole lauroxil	Antipsychotic	SBEβCD	Enables nano-suspension and controlled release	Aristada Initio®
Diclofenac sodium	NSAID	HPβCD	Increases solubility, improves topical penetration	Voltaren® gel (CD-derived variants)
Piroxicam	NSAID	β-CD	Enhances dissolution and reduces gastric irritation	Multiple generic CD-based tablets
Nitroglycerin	Vasodilator	β-CD	Stabilizes volatile API and improves shelf life	Used in several sublingual formulations
Dexamethasone	Corticosteroid	HPβCD	Improves solubility for ophthalmic use	Ophthalmic steroid eye drops
Meloxicam	NSAID	γ-CD	Improves solubility and speeds the onset of action	Movalis® fast tablets
Sugammadex	Neuromuscular block reversal	Modified γ-CD	Drug acting as API itself; encapsulates rocuronium	Bridion®, first CD-based API

representative marketed pharmaceutical formulations incorporating CDs, demonstrating how specific CD derivatives enhance drug solubility, stability, bioavailability, safety, and controlled delivery, and in unique cases such as sugammadex, function directly as an API enabling targeted molecular encapsulation.

The controlled delivery capabilities of CDs also extend to hydrophilic peptide and protein therapeutics, where inclusion complexes help minimize aggregation, improve mucosal penetration, and maintain functional stability during storage and administration (Loftsson et al., 2016). These advantages position CDs as essential tools in the development of next-generation drug delivery platforms for personalized and targeted therapies.

Cyclodextrins as active pharmaceutical ingredients

The transformative role of CDs in pharmaceutical science is exemplified by the development of modified CDs that act as therapeutic agents rather than passive excipients. The most notable example is sugammadex, a modified γ-CD designed to encapsulate and neutralize steroidal neuromuscular blocking agents such as rocuronium during surgical anesthesia (Bom et al., 2009). Sugammadex forms a highly stable inclusion complex with rocuronium, enabling rapid reversal of neuromuscular blockade and improving the safety of anesthesia recovery (Pirvu et al., 2025). Another example is the use of HPβCD as an investigational treatment for Niemann–Pick Type C disease, where it helps remove accumulated intracellular cholesterol and reduce neurodegeneration (Liu, 2012). These applications demonstrate the potential of chemically tailored CDs to act directly on biological pathways, expanding their utility beyond formulation science into therapeutic intervention.

Cyclodextrins in the cosmetic industry

The cosmetic and personal care industries extensively utilize CDs for stabilizing active ingredients, improving fragrance performance, enhancing skin delivery, and optimizing product texture (Perinelli et al., 2020).

Fragrance stabilization and controlled release

Volatile fragrance molecules are prone to evaporation and oxidation, leading to diminished scent intensity and shelf stability, challenges that CDs address through inclusion complexation (Szejtli, 2004). Encapsulation protects perfume components such as terpenes, aldehydes, and essential oils from degradation by environmental stressors, ensuring prolonged aromatic release in products such as deodorants, shampoos, soaps, and lotions (Perinelli et al., 2020). The controlled release resulting from CD complexation allows fragrances to be gradually emitted through friction, moisture, or temperature changes, enhancing consumer experience and reducing formulation costs by minimizing fragrance loss (Perinelli et al., 2020).

Delivery and protection of cosmetic active ingredients

CDs improve the solubility, stability, and bioavailability of poorly water-soluble cosmetic actives, including retinol, vitamin E, coenzyme Q10, and certain botanical extracts, by forming inclusion complexes that protect these compounds from oxidation, photodegradation, and hydrolysis (Saffarionpour & Diosady, 2025). Retinol, for instance, is highly sensitive to light and heat, but CD encapsulation markedly improves its stability and reduces skin irritation by controlling release rates (Saffarionpour & Diosady, 2025). CDs are also used to enhance the delivery of antiaging, whitening, and antiinflammatory agents, enabling better skin penetration and reduced destructive interactions with other formulation components (Moon et al., 2018). Table 14.2 summarizes the diverse applications of CDs in cosmetic

Table 14.2 Cyclodextrins in cosmetic and personal care products.

Cosmetic category	Cyclodextrin application	Active compounds	Functional benefit
Fragrances/perfumes	Encapsulation for long-lasting release	Terpenes, essential oils, and aldehydes	Prolonged scent, reduced volatility
Skin-care actives (antiaging)	Protection and solubilization	Retinol, retinoids, coenzyme Q10, peptides	Enhanced stability; reduced irritation
Whitening/brightening agents	CD-assisted synthesis or delivery	α-Arbutin, vitamin C, niacinamide	Controlled release and improved skin penetration
Sunscreens	Stabilization of UV filters	Avobenzone, octinoxate, benzophenones	Increased photostability and reduced degradation
Deodorants and antiperspirants	Odor capture and fragrance retention	Isovaleric acid, perfume blends	Odor neutralization; slow fragrance release
Hair-care products (shampoos, conditioners)	Encapsulation of volatile actives	Menthol, botanical extracts	Cooling effect prolongation; enhanced fragrance
Makeup formulations	Conversion of liquids to powders	Oils, waxes, lipophilic pigments	Improved texture and ease of formulation
Dermatological treatments	Targeted delivery and reduced irritation	Salicylic acid, azelaic acid	Controlled release and better tolerability

and personal care formulations, highlighting their role in encapsulation, stabilization, controlled release, and targeted delivery of active ingredients to enhance product performance, sensory attributes, stability, and skin tolerability.

Texture modification and product optimization

CDs influence the rheology and textural properties of cosmetic formulations by interacting with emulsifiers, lipids, and surfactants. They can convert oily or liquid actives into dry, free-flowing powders suitable for use in compact powders, facial masks, and scrubs (Adeoye et al., 2017). CDs help reduce greasiness and improve the sensory profile of creams and gels by modulating the release of lipophilic components (Adeoye et al., 2017). Furthermore, CDs serve as processing aids for synthesizing specialized cosmetic ingredients such as α-arbutin, a tyrosinase inhibitor used for skin brightening (Moon et al., 2018).

Cyclodextrins in chemical and material sciences

CDs have become indispensable tools in chemical synthesis and materials engineering due to their ability to modulate reactivity, improve selectivity, and provide molecular recognition capabilities within supramolecular systems (Crini, 2014). Their unique host–guest chemistry allows CDs to stabilize reactive intermediates, alter transition-state energies, and influence stereochemical outcomes, making them valuable supramolecular catalysts in organic synthesis. By forming inclusion complexes with reactants, CDs can preorganize molecules into favorable orientations, thereby increasing reaction rates and conferring regio- or stereoselectivity in oxidation, reduction, and substitution reactions (Chen & Jiang, 2011). These advantages make CDs central to green chemistry approaches, offering milder reaction conditions, lower energy consumption, and reduced solvent requirements.

In materials science, CDs are used to functionalize polymers, nanocomposites, hydrogels, and fibers, enabling enhanced mechanical, chemical, and biological properties (Villalonga et al., 2007). CDs can be covalently grafted onto polymer matrices to generate functional materials with selective binding properties, increased hydrophilicity, or improved compatibility with active ingredients. CD-based polymers have been widely explored for controlled release systems, stimuli-responsive materials, and molecularly imprinted polymers used in analytical devices, biosensors, and selective adsorption systems (Krabicová et al., 2020). In advanced applications, CDs are used to construct MOFs, which incorporate CD molecules as organic linkers, creating biocompatible porous structures capable of selective adsorption and catalysis (Topuz & Uyar, 2024). These innovations illustrate how CDs serve as structural and functional building blocks in emerging supramolecular materials.

CDs also play a major role in chiral separation, which is essential for the pharmaceutical and fine-chemical industries, where differentiating enantiomers is critical due to differing biological activities. CDs incorporated into high-performance liquid chromatography stationary phases or capillary electrophoresis buffers provide chiral discrimination through selective host–guest interactions, allowing efficient resolution of chiral molecules that otherwise show similar physicochemical properties (Barman et al., 2025). This molecular-level selectivity has transformed CD-based chromatographic systems into reliable tools for analytical quality control.

Textile industry

The textile industry employs CDs for dyeing efficiency, fiber modification, and functional textile development due to their ability to encapsulate colorants, fragrances, antimicrobial agents, and UV-protective compounds (Ghazal & Beltagy, 2025). CDs enhance dyeing uniformity by encapsulating hydrophobic dye molecules, increasing their solubility in aqueous dye baths, and enabling deeper penetration into fibers. This reduces the need for environmentally harmful organic solvents and synthetic surfactants, promoting greener and more sustainable dyeing processes (Cireli & Yurdakul, 2006). The controlled release afforded by CDs reduces dye aggregation, prevents uneven coloration, and enhances color fastness under washing and light exposure (Tariq et al., 2024).

In functional textiles, CDs are grafted onto cellulose, cotton, or synthetic fibers to create bioactive textiles capable of delivering perfumes, antimicrobial agents, insect repellents, and skin-beneficial compounds. When CDs are covalently bonded to the fiber, they can serve as reservoirs for active molecules that are slowly released through friction, moisture, or thermal activation, enabling long-lasting effects without repeated chemical application. This technology has been successfully applied to cosmetotextiles, where slow-release fragrances or skin-softening agents provide added value to garments (Tariq et al., 2024). CDs also improve the UV protection of fabrics by enhancing the loading and stabilization of UV-absorbing molecules such as cinnamates or benzophenones (Cireli & Yurdakul, 2006). These contributions highlight the role of CDs as multifunctional agents enhancing performance, durability, and user comfort in textile products.

Detergent, pulp, and paper industries

CDs support advanced cleaning formulations by enhancing the solubility, emulsification, and removal of hydrophobic soils in detergents. CDs encapsulate oily dirt particles and odor-causing molecules such as volatile organic compounds and sweat metabolites, facilitating their solubilization and removal during washing processes.

This encapsulation mechanism also helps prevent redeposition of soils onto fabrics, ensuring superior cleaning performance (Zehentgruber et al., 2011). CDs are incorporated into household and industrial detergents to improve the stability of fragrances and cleaning agents, contributing to longer-lasting scent and product shelf life.

In the pulp and paper industry, CDs address the challenge of "stickies"—adhesive contaminants originating from recycled paper, inks, and adhesives that accumulate on machinery and degrade paper quality. CDs encapsulate hydrophobic sticky components, reducing their tackiness and preventing their deposition onto manufacturing equipment (Amiri & Amiri, 2017). This improves operational efficiency, reduces maintenance downtime, and enhances recycled paper quality. CDs also assist in deinking processes by solubilizing ink residues, allowing more effective removal and improving the brightness and purity of recycled paper pulp. These applications demonstrate the value of CDs in promoting sustainable recycling and improving manufacturing efficiency.

Cyclodextrins in agriculture

CDs are increasingly used in agriculture to enhance the performance, stability, and environmental safety of agrochemicals, including insecticides, herbicides, fungicides, and plant growth regulators. CDs form inclusion complexes with active ingredients, protecting them from photodegradation, volatilization, and hydrolysis, thereby prolonging field persistence and reducing application frequency (Liu et al., 2022). This leads to more efficient pest management with lower environmental load and minimized toxic runoff.

Controlled release formulations utilizing CDs help maintain effective concentrations of pesticide actives while reducing nontarget exposure and improving environmental compatibility. For example, complexation of hydrophobic insecticides with β-CD enhances their soil mobility and bioavailability while decreasing toxicity to nontarget organisms (Waris et al., 2021). CDs also improve the solubility and foliar penetration of herbicides, leading to greater efficacy at lower dosages (Liu et al., 2022). These advantages support precision agriculture approaches and sustainable farming practices.

Beyond pesticide delivery, CDs influence plant physiology directly. β-CDs can interact with seed enzymes such as amylases, slowing starch breakdown and delaying germination to achieve coordinated sprouting in certain crops (Duan et al., 2025). Such germination control enhances crop uniformity and yield. CDs can also facilitate the delivery of plant growth regulators such as auxins, gibberellins, or cytokinins by improving their solubility and stability (Waris et al., 2021). In agricultural biotechnology, CDs serve as protective carriers for DNA, RNA, and protein formulations used in gene delivery and plant protection technologies (Liu et al., 2022). These examples indicate a broadening scope for CDs as functional agents in modern agriculture.

Cyclodextrins in environmental remediation

CDs play a crucial role in contemporary environmental remediation processes due to their safety, biodegradability, and exceptional ability to form inclusion complexes with organic and inorganic pollutants (Morin-Crini et al., 2021). Their molecular structure enables them to trap hydrophobic contaminants, improve their solubility, and facilitate biodegradation, making CDs valuable tools for soil and water treatment. Table 14.3 outlines the diverse spectrum of environmental pollutants effectively removed or immobilized using native and modified CDs, emphasizing their mechanisms of chelation, host–guest inclusion, adsorption, and solubility enhancement that collectively enable efficient wastewater treatment, soil bioremediation, air purification, and mitigation of emerging persistent contaminants.

Removal of heavy metals and synthetic dyes

CD-based polymers and nanocomposites are widely applied for removing heavy metals such as cadmium, copper, lead, and nickel from contaminated water via coordination, hydrogen bonding, and electrostatic interactions. Derivatized CDs, including methylated γ-CD and carboxymethylated β-CD, show enhanced metal ion binding capacity due to the presence of additional chelating functional groups (Maphuhla & Oyedeji, 2025). These materials can be incorporated into filtration systems or adsorbent beds, offering high adsorption capacity and reusability.

CD nanosponges and water-insoluble CD polymers (CDPs) efficiently remove synthetic dyes such as methylene blue, azo dyes, and rhodamine B from textile wastewater through host–guest interactions and porous adsorption mechanisms (Syeda et al., 2022). Their selectivity, rapid adsorption, and structural regenerability offer advantages over conventional activated carbon-based systems.

Removal of emerging contaminants

Emerging pollutants, including pharmaceuticals, hormones, pesticides, and polyfluoroalkyl substances, pose significant environmental risks due to their persistence and widespread use. CD-based adsorbents are uniquely suited for capturing these contaminants due to their versatile host–guest chemistry and structural adaptability. CDPs can bind a broad range of contaminants, while maintaining reusability across multiple adsorption–desorption cycles, supporting sustainable water purification (Morin-Crini et al., 2021).

Cyclodextrins in soil remediation

In soil remediation, CDs enhance the extraction and biodegradation of hydrophobic organic pollutants, including polycyclic aromatic hydrocarbons, polychlorinated

Table 14.3 Environmental pollutants removed or immobilized using cyclodextrins.

Pollutant type	Representative examples	Cyclodextrin form/ derivative	Mechanism of removal	Application outcome
Heavy metals	Pb^{2+}, Cd^{2+}, Cu^{2+}, Ni^{2+}	Carboxymethyl-β-CD; β-CD polymer nanocomposites	Chelation, electrostatic binding	Improved removal efficiency in wastewater treatment
Polycyclic aromatic hydrocarbons	Naphthalene, anthracene, fluorene	HPβCD, RAMEB	Increased solubility → enhanced biodegradation	Accelerated soil bioremediation
Polychlorinated biphenyls (PCBs)	PCB-28, PCB-52	HPβCD	Solubilization + bioavailability for microbes	Faster microbial breakdown in contaminated soil
Pesticides and agrochemicals	Atrazine, chlorpyrifos, DDT	β-CD polymers, CD–activated carbon composites	Inclusion complexation, adsorption	Reduced toxicity and improved removal
Synthetic dyes	Methylene blue, Congo red, and azo dyes	Water-insoluble β-CD polymers, nanosponges	Porous adsorption + host–guest binding	Efficient textile wastewater decolorization
Pharmaceutical residues (emerging contaminants)	Carbamazepine, ibuprofen, diclofenac	CD polymers (CDPs), HPβCD	Host–guest trapping, enhanced filtration	High-capacity, regenerable adsorbents
PFASs and persistent pollutants	PFOA, PFOS	Specialized CD-based sorbents	Inclusion + hydrophobic capture	Effective removal of hard-to-treat contaminants
Volatile organic compounds	Benzene, toluene	β-CD aerogels, CD-functionalized filters	Encapsulation and vapor-phase capture	Air purification, odor control

biphenyls, and pesticides that strongly bind to soil particles (Landy et al., 2012). HPβCD and RAMEB significantly increase the aqueous solubility of these contaminants, facilitating their transfer from soil to water phases where microbial degradation is more effective (Landy et al., 2011). Because CDs do not bind strongly to soil particles and are nontoxic to microbial communities, their presence accelerates natural biodegradation pathways and supports bioremediation strategies (Fava & Ciccotosto, 2002).

Advanced and emerging industrial applications of CGTase

CGTase is widely recognized as the principal enzyme responsible for CD synthesis from starch; however, its broader catalytic versatility extends far beyond simple cyclization reactions (Biwer et al., 2002; van der Veen et al., 2000). In addition to cyclization, CGTase exhibits coupling, disproportionation, and hydrolytic activities, collectively positioning it as a powerful carbohydrate-engineering biocatalyst with expanding relevance in food technology, industrial biotechnology, and sustainable bioprocessing (Qi & Zimmermann, 2005). Increasing research attention has therefore shifted toward exploiting CGTase not only for CD production but also for targeted glycosylation, functional carbohydrate synthesis, sweetener modification, and circular bioeconomy strategies (Kelly et al., 2009).

One of the most promising emerging applications of CGTase lies in enzymatic sweetener modification and taste engineering. Many plant-derived sweet compounds, including steviol glycosides and related diterpene sweeteners, exhibit undesirable bitterness or lingering aftertaste that limits their commercial acceptance (Kometani et al., 1994). CGTase-mediated transglycosylation enables the addition of glucose residues to these molecules, thereby modifying their hydrophilicity, receptor interaction profiles, and sensory perception (Qi & Zimmermann, 2005). Enzymatic glycosylation of stevioside has been shown to produce rebaudioside derivatives with reduced bitterness and improved sweetness quality while maintaining low caloric value (Kometani et al., 1994). The regioselective nature of CGTase-catalyzed glycosylation allows precise structural modification under mild aqueous conditions, avoiding harsh chemical reagents and supporting clean-label processing trends (Biwer et al., 2002). As global demand for natural, low-calorie sweeteners continues to increase, CGTase-based glycosylation technologies provide a sustainable and industrially scalable solution for improving organoleptic properties without compromising safety or regulatory standards (Qi & Zimmermann, 2005).

CGTase plays an important role in the production of low-glycemic and functional carbohydrates. Through its disproportionation activity, the enzyme catalyzes controlled restructuring of starch-derived glucans into defined maltooligosaccharides with specific degrees of polymerization (van der Veen et al., 2000). These carbohydrates exhibit slower digestibility and lower glycemic indices compared with conventional mono- and disaccharides, making them valuable in diabetic-friendly foods and metabolic health formulations (Qi & Zimmermann, 2005). Unlike acid hydrolysis, which generates heterogeneous sugar mixtures, CGTase-mediated conversion yields more uniform carbohydrate profiles, including maltotetraose-rich syrups and slowly digestible dextrins (Biwer et al., 2002). Such carbohydrates contribute to sustained energy release, improved glycemic control, and modulation of gut microbiota, supporting their integration into sports nutrition products, infant formulas, and specialized clinical dietary preparations (Qi & Zimmermann, 2005).

The brewing and fermentation industries also benefit from CGTase applications in carbohydrate profile optimization. During starch degradation in brewing, CGTase can modify dextrin composition, thereby influencing fermentable sugar availability, residual body, and mouthfeel characteristics of beer and malt-based beverages (van der Veen et al., 2000). By selectively altering glucan chain distribution, CGTase enables fine control over alcohol yield and sensory properties without interfering with yeast metabolism (Biwer et al., 2002). Its compatibility with moderate temperature and pH conditions allows integration into mashing or postmashing stages, offering brewers an additional enzymatic tool for specialty beverage formulation (Qi & Zimmermann, 2005).

Another expanding area of application involves enzymatic glycosylation of bioactive phytochemicals for nutraceutical and functional food industries. Numerous plant-derived compounds, including flavonoids and phenolic antioxidants, exhibit limited aqueous solubility and chemical instability, restricting their bioavailability (Kometani et al., 1994). CGTase-mediated glycosylation improves solubility and enhances oxidative stability by introducing glucose moieties that alter physicochemical behavior (Qi & Zimmermann, 2005). Glycosylated derivatives of quercetin, hesperetin, and catechins have demonstrated improved storage stability and reduced bitterness while maintaining antioxidant activity (Kometani et al., 1994). Importantly, CGTase-based synthesis proceeds under environmentally benign conditions, supporting green chemistry principles and reducing reliance on multistep chemical glycosylation routes (Biwer et al., 2002).

CGTase also holds significant potential in circular bioeconomy initiatives and agro-industrial waste valorization. Starch-rich agricultural by-products such as broken rice, cassava residues, potato peel starch, and corn-processing waste streams can serve as substrates for CGTase-mediated conversion into high-value CDs, and functional oligosaccharides (Qi & Zimmermann, 2005). This biotransformation approach transforms low-cost biomass into commercially valuable carbohydrate derivatives while reducing environmental waste burdens (Biwer et al., 2002). Integrating CGTase production into biorefinery models enables multiproduct valorization strategies in which starch hydrolysis, oligosaccharide synthesis, and CD production occur within interconnected processing systems (van der Veen et al., 2000).

From a process engineering perspective, enzyme immobilization has greatly enhanced the industrial feasibility of CGTase applications. Immobilized CGTase systems demonstrate improved thermal stability, operational longevity, and reusability across multiple reaction cycles. Continuous reactor configurations incorporating immobilized CGTase allow high-substrate-concentration processing and facilitate product recovery, thereby improving volumetric productivity and reducing downstream purification requirements (Martins & Hatti-Kaul, 2002). Advances in support materials and bioreactor design have expanded options for industrial-scale deployment and process intensification (Qi & Zimmermann, 2005).

Recent developments in protein engineering and synthetic biology have further expanded CGTase utility. Site-directed mutagenesis targeting active-site residues has enabled alteration of product specificity, particularly shifting distribution toward higher-value γ-CD (Kelly et al., 2009). Engineering strategies have also enhanced thermostability, pH tolerance, and catalytic efficiency, enabling application under diverse industrial conditions (van der Veen et al., 2000). Heterologous expression in engineered microbial hosts increases enzyme yield and secretion efficiency, while synthetic biology approaches aim to construct integrated strains capable of direct starch-to-CD conversion (Kelly et al., 2009). Such strategies streamline carbohydrate biomanufacturing, reduce energy inputs, and improve overall process sustainability.

In specialty chemical manufacturing, CGTase-catalyzed transglycosylation reactions are increasingly explored for the synthesis of glycosylated intermediates used in food, cosmetic, and fine-chemical industries (Biwer et al., 2002). Compared with chemical glycosylation, enzymatic modification offers superior regioselectivity, reduced solvent requirements, and milder reaction conditions (Qi & Zimmermann, 2005). This catalytic precision facilitates scalable production of functional glycosides with enhanced solubility and stability profiles (Kometani et al., 1994). As regulatory and environmental pressures continue to favor greener manufacturing routes, CGTase-based catalytic systems are likely to gain further industrial importance (Biwer et al., 2002).

Economically, the expanding CD market remains fundamentally dependent on efficient CGTase production and optimization strategies (Qi & Zimmermann, 2005). Increasing demand for γ-CD and modified derivatives has stimulated research into extremophilic CGTase variants and AI-guided protein engineering approaches (Kelly et al., 2009). Continuous bioreactor systems and cost-reduction strategies aimed at lowering enzyme production expenses per kilogram of CD will further determine future competitiveness (Martins & Hatti-Kaul, 2002). Collectively, these developments demonstrate that CGTase is not merely a CD-producing enzyme but a multifunctional and strategically important biocatalyst central to carbohydrate biotechnology, sustainable industrial processing, and next-generation functional ingredient synthesis (van der Veen et al., 2000).

Nanotechnology and advanced cyclodextrin derivatives

Advances in nanotechnology have significantly expanded the functional scope of CDs by enabling their integration into nanoscale delivery systems, smart carriers,

Table 14.4 Selected cyclodextrin derivatives and their functional advantages.

Cyclodextrin type	Key properties	Major applications
α-Cyclodextrin (α-CD)	Small cavity (4.7–5.3 Å), hydrophilic exterior	Encapsulation of short-chain lipids, dietary fiber, specialty food formulations
β-Cyclodextrin (β-CD)	Medium cavity (6.0–6.5 Å)	Flavor stabilization, drug solubilization, cholesterol removal, cosmetics
γ-Cyclodextrin (γ-CD)	Large cavity (7.5–8.3 Å), highest water solubility	Bulky molecule inclusion, nutraceuticals, and sensitive actives
Hydroxypropyl-β-CD	Highly water-soluble, low toxicity	Injectable drug formulations, protein stabilization, ophthalmics
Sulfobutylether-β-CD	Anionic derivative, excellent parenteral solubility	Injectable APIs (e.g., ziprasidone) have improved safety in IV formulations
Randomly methylated β-CD	High binding strength, lipophilic interior	Solubilization of very hydrophobic drugs, pesticides, and dyes
Carboxymethyl-β-CD	Metal-chelating ability	Heavy metal removal, wastewater treatment
Cyclodextrin polymers/nanosponges	Porous, crosslinked networks, high loading	Controlled drug release, nanocarriers, pollutant adsorption
Large-ring cyclodextrins	>8 glucose units, flexible cavities	Encapsulation of macromolecules, nanomaterials, and advanced drug delivery

and supramolecular assemblies. Advanced CD derivatives such as nanosponges, stimuli-responsive polymers, MOFs, and surface-functionalized nanoparticles offer enhanced loading capacity, targeted delivery, and controlled release of bioactive compounds. These next-generation systems combine the intrinsic host–guest chemistry of CDs with nanostructured engineering, paving the way for precision drug delivery, environmental detoxification, and high-performance material applications.

Table 14.4 presents selected natural and chemically modified CDs, highlighting how variations in cavity size, solubility, charge, and structural architecture confer distinct functional advantages that enable their tailored application across food systems, pharmaceuticals, environmental remediation, and advanced material technologies.

Cyclodextrin-based nanosponges and hydrogels

Nanosponges constructed from crosslinked CDs have emerged as promising nanocarriers for drugs, antioxidants, and environmental pollutants due to their high porosity, stability, and tunable surface properties. These structures encapsulate both hydrophilic and hydrophobic molecules, offering controlled release profiles beneficial in drug delivery, cosmetics, and pollutant remediation (Krabicová et al., 2020). CD-based hydrogels—formed through polymerization reactions involving CDs and polyacrylates, polysaccharides, or synthetic polymers—have been developed for tissue engineering, wound care, and biosensing due to their biocompatibility, mechanical stability, and ability to incorporate therapeutic molecules (Mohamadhoseini & Mohamadnia, 2025).

Stimuli-responsive cyclodextrin systems

Stimuli-responsive CD materials that react to changes in pH, temperature, redox conditions, or enzymatic activity are widely researched for intelligent drug delivery and diagnostic applications. These systems release encapsulated agents only in targeted microenvironments, such as tumor tissues or inflamed regions, improving therapeutic specificity and reducing systemic toxicity (Topuz & Uyar, 2024). The ability to construct multifunctional supramolecular architectures using CDs underscores their central role in precision medicine and smart material design.

Large-ring cyclodextrins

Large-ring cyclodextrins (LR-CDs), consisting of nine or more glucose units, represent an emerging frontier in supramolecular chemistry due to their expanded cavity sizes and unique structural flexibility. These macrocycles can encapsulate bulkier hydrophobic molecules, polymers, and even biomolecules that cannot fit into conventional α-, β-, or γ-CDs (Musuc, 2024).

LR-CDs offer novel inclusion geometries and binding modes, enabling applications in drug delivery, polymer chemistry, and nanostructured material synthesis. Their increased cavity size makes them suitable for stabilizing large pharmaceutical compounds, improving solubility, and facilitating controlled release (Musuc, 2024). The development of efficient synthetic pathways for producing high-purity LR-CDs remains a key research focus, with ongoing advances expected to expand their applicability across multiple industrial sectors.

Table 14.5 summarizes representative CD–guest interactions, illustrating how specific native and modified CDs enable targeted encapsulation of diverse molecules to enhance stability, solubility, controlled release, and environmental remediation, ultimately translating into measurable industrial benefits across food, pharmaceutical, cosmetic, agricultural, and wastewater treatment sectors.

Table 14.5 Representative cyclodextrin–guest interactions and their industrial outcomes.

Guest molecule/class	Cyclodextrin used	Purpose of complexation	Industrial outcome
Essential oils (clove, cinnamon, rosemary)	β-CD, HPβCD	Protect from oxidation and heat	Extended shelf life, sustained aroma release
Bitter compounds (limonin, catechins)	β-CD	Taste masking through encapsulation	Improved palatability of juices and functional foods
Cholesterol	β-CD, methyl-β-CD	Selective inclusion of hydrophobic cholesterol	80%–98% cholesterol removal in dairy/egg products
Hydrophobic drugs (itraconazole, paclitaxel)	HPβCD, RAMEB, SBEβCD	Solubility and bioavailability enhancement	Successful oral/injectable formulations
Volatile fragrances	β-CD, CD polymers	Stabilization and slow release	Long-lasting scent in personal care products
Pesticides (pyrethroids, herbicides)	β-CD, RAMEB	Protection from photolysis; controlled release	Reduced environmental load, fewer applications
PAHs and PCBs	HPβCD, CD polymers	Increased solubility and biodegradation	Accelerated soil bioremediation
Synthetic dyes (methylene blue, azo dyes)	Water-insoluble β-CD polymers	Adsorption via inclusion and surface interactions	Wastewater decolorization and reuse

Future perspectives and conclusion

CDs have emerged as powerful supramolecular tools due to their multifunctional capacity to encapsulate, stabilize, solubilize, and modulate the properties of a wide array of molecules across food, pharmaceutical, cosmetic, agricultural, environmental, textile, chemical, and materials industries. Their biodegradability, safety, and exceptional versatility align with global priorities for sustainable development, green chemistry, and environmentally responsible manufacturing (Crini & Lichtfouse, 2019).

Future innovations in CD science are expected to arise from advanced chemical modifications, nanotechnology integration, stimuli-responsive material design, and the expanding development of LR-CDs with enhanced structural capabilities (Musuc, 2024). CD polymers and nanosponges are poised to transform targeted drug delivery, water purification, tissue engineering, and controlled-release agrochemicals due to their tailored porosity and supramolecular interactions (Krabicová et al., 2020). Improvements in CD-based environmental remediation techniques will continue to support sustainable pollutant management, particularly for persistent organic pollutants and emerging contaminants (Morin-Crini et al., 2021).

Overall, CDs stand as indispensable components of modern industrial biotechnology, formulation science, and environmental engineering. Their intrinsic structural properties, combined with their adaptable chemistry, ensure that CDs will remain central to future scientific and technological advancements, shaping sustainable solutions across multiple sectors.

References

Adamkiewicz, L., & Szeleszczuk, Ł. (2023). Review of applications of cyclodextrins as taste-masking excipients for pharmaceutical purposes. *Molecules (Basel, Switzerland), 28*(19), 6964.

Adeoye, O., Figueiredo, A., & Marques, H. C. (2017). Cyclodextrins and skin disorders: Therapeutic and cosmetic applications. In *Carrier-mediated dermal delivery*. Jenny Stanford Publishing, 463–501.

Almeida, B., Domingues, C., Mascarenhas-Melo, F., Silva, I., Jarak, I., Veiga, F., & Figueiras, A. (2023). The role of cyclodextrins in COVID-19 therapy—A literature review. *International Journal of Molecular Sciences, 24*(3), 2974.

Amiri, S., & Amiri, S. (2017). *Cyclodextrins: Properties and industrial applications*. John Wiley & Sons.

Astray, G., Gonzalez-Barreiro, C., Mejuto, J. C., Rial-Otero, R., & Simal-Gandara, J. (2009). A review on the use of cyclodextrins in foods. *Food Hydrocolloids, 23*(7), 1631–1640.

Barman, K., Islam, M. M., Das, K. S., Singh, N., Priya, S., Kurmi, B. D., & Patel, P. (2025). Recent advances in enantiorecognition and enantioseparation techniques of chiral molecules in the pharmaceutical field. *Biomedical Chromatography, 39*(2), e6073.

Biwer, A., Antranikian, G., & Heinzle, E. (2002). Enzymatic production of cyclodextrins. *Applied Microbiology and Biotechnology, 59*(6), 609–617.

Bláhová, M., Štefuca, V., Hronská, H., & Rosenberg, M. (2023). Maltooligosaccharides: Properties, production and applications. *Molecules (Basel, Switzerland), 28*(7), 3281.

Bom, A., Hope, F., Rutherford, S., & Thomson, K. (2009). Preclinical pharmacology of sugammadex. *Journal of Critical Care, 24*(1), 29–35.

Chen, G., & Jiang, M. (2011). Cyclodextrin-based inclusion complexation bridging supramolecular chemistry and macromolecular self-assembly. *Chemical Society Reviews, 40*(5), 2254–2266.

Cireli, A., & Yurdakul, B. (2006). Application of cyclodextrin to the textile dyeing and washing processes. *Journal of Applied Polymer Science, 100*(1), 208–218.

Crini, G. (2014). A history of cyclodextrins. *Chemical Reviews, 114*(21), 10940–10975.

Crini, G., & Lichtfouse, E. (2019). Advantages and disadvantages of techniques used for wastewater treatment. *Environmental chemistry letters, 17*(1), 145–155.

Dai, Y., Zhong, J., Li, J., Liu, X., Wang, Y., & Qin, X. (2022). Interaction mechanism of cholesterol/β-cyclodextrin complexation by combined experimental and computational approaches. *Food Hydrocolloids, 130*, 107725.

Duan, X., Fan, Y., Liu, Q., & Ding, Y. (2025). An efficient approach for β-cyclodextrin production from raw ginkgo seed powder through high-temperature gelatinization and enzymatic conversion. *Catalysts, 15*(2), 108.

Fava, F., & Ciccotosto, V. (2002). Effects of randomly methylated-β-cyclodextrins (RAMEB) on the bioavailability and aerobic biodegradation of polychlorinated biphenyls in three pristine soils spiked with a transformer oil. *Applied Microbiology and Biotechnology, 58*(3), 393–399.

Ghazal, H., & Beltagy, Z. S. (2025). Sustainability in textile industry using cyclodextrins to improve the properties of textiles. *Journal of Textiles, Coloration and Polymer Science, 22*(1), 15–21.

Gonzalez Pereira, A., Carpena, M., Garcia Oliveira, P., Mejuto, J. C., Prieto, M. A., & Simal Gandara, J. (2021). Main applications of cyclodextrins in the food industry as the compounds of choice to form host–guest complexes. *International Journal of Molecular Sciences, 22*(3), 1339.

Gujral, H. S., Guardiola, I., Carbonell, J. V., & Rosell, C. M. (2003). Effect of cyclodextrinase on dough rheology and bread quality from rice flour. *Journal of Agricultural and Food Chemistry, 51*(13), 3814–3818.

Jacob, S., & Nair, A. B. (2018). Cyclodextrin complexes: Perspective from drug delivery and formulation. *Drug Development Research, 79*(5), 201–217.

Ji, H., Liu, J., McClements, D. J., Bai, Y., Li, Z., Chen, L., & Jin, Z. (2024). Malto-oligosaccharides as critical functional ingredient: A review of their properties, preparation, and versatile applications. *Critical Reviews in Food Science and Nutrition, 64*(12), 3674–3686.

Kaushik, P., Mittal, V., & Kaushik, D. (2024). Unleashing the potential of β–cyclodextrin inclusion complexes in bitter taste abatement: Development, optimization and evaluation of taste masked anti-emetic chewing gum of promethazine hydrochloride. *AAPS PharmSciTech, 25*(6), 169.

Kelly, R. M., Dijkhuizen, L., & Leemhuis, H. (2009). The evolution of cyclodextrin glucanotransferase product specificity. *Applied Microbiology and Biotechnology, 84*(1), 119–133.

Khatoon, H., Faudzi, S. M. M., & Sohajda, T. (2025). Mechanisms and therapeutic applications of β-cyclodextrin in drug solubilisation and delivery systems. *Chemistry & Biodiversity*e00359.

Kometani, T., Terada, Y., Nishimura, T., Takii, H., & Okada, S. (1994). Purification and characterization of cyclodextrin glucanotransferase from an alkalophilic Bacillus species and transglycosylation at alkaline pHs. *Bioscience, Biotechnology, and Biochemistry, 58*(3), 517–520.

Krabicová, I., Appleton, S. L., Tannous, M., Hoti, G., Caldera, F., Rubin Pedrazzo, A., & Trotta, F. (2020). History of cyclodextrin nanosponges. *Polymers, 12*(5), 1122.

Landy, D., Mallard, I., Ponchel, A., Monflier, E., & Fourmentin, S. (2011). Cyclodextrins for remediation technologies. *Environmental Chemistry for a Sustainable World: Nanotechnology and Health Risk, Volume 1*, 47–81.

Landy, D., Mallard, I., Ponchel, A., Monflier, E., & Fourmentin, S. (2012). Remediation technologies using cyclodextrins: An overview. *Environmental chemistry letters, 10*(3), 225–237.

Liu, B. (2012). Therapeutic potential of cyclodextrins in the treatment of Niemann–Pick type C disease. *Clinical lipidology, 7*(3), 289–301.

Liu, Y., Chen, Y., Gao, X., Fu, J., & Hu, L. (2022). Application of cyclodextrin in food industry. *Critical Reviews in Food Science and Nutrition, 62*(10), 2627–2640.

Loftsson, T. (2015). Formulation of drug-cyclodextrin complexes. In *Percutaneous penetration enhancers chemical methods in penetration enhancement: Drug manipulation strategies and vehicle effects*. Springer, 189–205.

Loftsson, T., Moya-Ortega, M. D., Alvarez-Lorenzo, C., & Concheiro, A. (2016). Pharmacokinetics of cyclodextrins and drugs after oral and parenteral administration of drug/cyclodextrin complexes. *Journal of Pharmacy and Pharmacology, 68*(5), 544–555.

Loftsson, T., Saokham, P., & Couto, A. R. S. (2019). Self-association of cyclodextrins and cyclodextrin complexes in aqueous solutions. *International Journal of Pharmaceutics, 560*, 228–234.

Maphuhla, N. G., & Oyedeji, O. O. (2025). Exploring the efficacy of methylated gamma-cyclodextrin (M-γ-CD) in the removal of heavy metals in soil systems. *Applied Sciences, 15*(4), 2076–3417.

Marques, H. M. C. (2010). A review on cyclodextrin encapsulation of essential oils and volatiles. *Flavour and Fragrance Journal, 25*(5), 313–326.

Martins, R. F., & Hatti-Kaul, R. (2002). A new cyclodextrin glycosyltransferase from an alkaliphilic *Bacillus agaradhaerens* isolate: Purification and characterisation. *Enzyme and Microbial Technology, 30*(1), 116–124.

Mohamadhoseini, M., & Mohamadnia, Z. (2025). Advances in β-cyclodextrin-driven self-healing materials: Molecular design and multifunctional applications. *Advanced Materials, 37*(44), e07227.

Moon, S. S., Lee, H. J., Mathiyalagan, R., Kim, Y. J., Yang, D. U., Lee, D. Y., & Yang, D. C. (2018). Synthesis of a novel α-glucosyl ginsenoside F1 by cyclodextrin glucanotransferase and its in vitro cosmetic applications. *Biomolecules, 8*(4), 142.

Morin-Crini, N., Lichtfouse, E., Fourmentin, M., Ribeiro, A. R. L., Noutsopoulos, C., Mapelli, F., & Crini, G. (2021). Remediation of emerging contaminants. In *Emerging contaminants*. In *Remediation, Vol. 2*, Springer International Publishing, 1–106.

Munro, I. C., Newberne, P. M., Young, V. R., & Bär, A. (2004). Safety assessment of γ-cyclodextrin. *Regulatory Toxicology and Pharmacology, 39*, 3–13.

Musuc, A. M. (2024). Cyclodextrins: Advances in chemistry, toxicology, and multifaceted applications. *Molecules (Basel, Switzerland), 29*(22), 5319.

Nicolaescu, O. E., Belu, I., Mocanu, A. G., Manda, V. C., Rău, G., Pîrvu, A. S., & Ciocîlteu, M. V. (2025). Cyclodextrins: Enhancing drug delivery, solubility and bioavailability for modern therapeutics. *Pharmaceutics, 17*(3), 288.

Park, J. H., Choi, K. H., & Kwak, H. S. (2011). Single-and 14-day repeat-dose toxicity of cross-linked β-cyclodextrin in rats. *International Journal of Toxicology, 30*(6), 700–706.

Perinelli, D. R., Palmieri, G. F., Cespi, M., & Bonacucina, G. (2020). Encapsulation of flavours and fragrances into polymeric capsules and cyclodextrins inclusion complexes: An update. *Molecules (Basel, Switzerland), 25*(24), 5878.

Pirvu, A. S., Varut, R. M., Trasca, D. M., Stoica, G. A., Radivojevic, K., Carmen, S., & Popescu, C. (2025). Cyclodextrins as active therapeutic agents: Beyond their role as excipients. *Pharmaceuticals, 18*(10), 1592.

Priyanka, M., & Christophe, B. (2017). Chemical changes in dairy products resulting from cholesterol removal using beta cyclodextrins. *Current Nutrition & Food Science, 13*(4), 309–318.

Qi, Q., & Zimmermann, W. (2005). Cyclodextrin glucanotransferase: From gene to applications. *Applied Microbiology and Biotechnology, 66*(5), 475–485.

Rahman, N., Nasir, M., Bharti, M., & Samdani, M. S. (2024). Fabrication of zero-valent iron nanoparticles impregnated cross-linked chitosan grafted β-cyclodextrin for removal of cloxacillin from aqueous environment. *Journal of Inorganic and Organometallic Polymers and Materials, 34*(4), 1654–1677.

Reineccius, G. (2007). Controlled release of flavour in food products. In *Modifying flavour in food*. Woodhead Publishing, 169–184.

Reineccius, T. A., Reineccius, G. A., & Peppard, T. L. (2004). Utilization of β-cyclodextrin for improved flavor retention in thermally processed foods. *Journal of Food Science, 69*(1), FCT58–FCT62.

Saffarionpour, S., & Diosady, L. L. (2025). Cyclodextrins and their potential applications for delivering vitamins, iron, and iodine for improving micronutrient status. *Drug Delivery and Translational Research, 15*(1), 26–65.

Saokham, P., & Loftsson, T. (2017). γ-Cyclodextrin. *International Journal of Pharmaceutics, 516*(1-2), 278–292.

Swaminathan, S., Cavalli, R., & Trotta, F. (2016). Cyclodextrin-based nanosponges: A versatile platform for cancer nanotherapeutics development. *Wiley Interdisciplinary Reviews: Nanomedicine and Nanobiotechnology, 8*(4), 579–601.

Syeda, S. E. Z., Nowacka, D., Khan, M. S., & Skwierawska, A. M. (2022). Recent advancements in cyclodextrin-based adsorbents for the removal of hazardous pollutants from waters. *Polymers, 14*(12), 2341.

Szejtli, J. (1998). Introduction and general overview of cyclodextrin chemistry. *Chemical Reviews, 98*(5), 1743–1754.

Szejtli, J. (2004). Past, present, and future of cyclodextrin research. *Pure and applied chemistry, 76*(10), 1825–1846.

Szejtli, J., & Szente, L. (2005). Elimination of bitter, disgusting tastes of drugs and foods by cyclodextrins. *European Journal of Pharmaceutics and Biopharmaceutics, 61*(3), 115–125.

Szente, L., & Szejtli, J. (2004). Cyclodextrins as food ingredients. *Trends in Food Science & Technology, 15*(3-4), 137–142.

Topuz, F., & Uyar, T. (2024). Recent advances in cyclodextrin-based nanoscale drug delivery systems. *Wiley Interdisciplinary Reviews: Nanomedicine and Nanobiotechnology, 16*(6), e1995.

Uekama, K. (2004). Design and evaluation of cyclodextrin-based drug formulation. *Chemical and pharmaceutical bulletin, 52*(8), 900–915.

Del Valle, E. M. (2004). Cyclodextrins and their uses: A review. *Process biochemistry, 39*(9), 1033–1046.

van der Veen, B. A., Uitdehaag, J. C., Dijkstra, B. W., & Dijkhuizen, L. (2000). Engineering of cyclodextrin glycosyltransferase reaction and product specificity. *Biochimica et Biophysica Acta (BBA)-Protein Structure and Molecular Enzymology, 1543*(2), 336–360.

Villalonga, R., Cao, R., & Fragoso, A. (2007). Supramolecular chemistry of cyclodextrins in enzyme technology. *Chemical Reviews, 107*(7), 3088–3116.

Waris, K. H., Lee, V. S., & Mohamad, S. (2021). Pesticide remediation with cyclodextrins: A review. *Environmental Science and Pollution Research, 28*(35), 47785–47799.

Xie, W., Ning, K., Chen, Z., Xia, J., Azat, S., Liu, W., & Xu, Q. (2025). Switchable adsorption/desorption bisphenol A and its derivatives by allyl-β-cyclodextrin functionalized intelligent molecularly imprinted polymer membrane. *New Journal of Chemistry, 49*(5), 1814–1826.

Yang, D., Yang, S., Mu, M., Liu, X., Zhao, L., Xu, Z., & Ge, L. (2022). Multifunctional β-cyclodextrin–poly (ethylene glycol)–cholesterol nanomicelle for anticancer drug delivery. *ACS Applied Bio Materials*, *5*(11), 5418–5431.

Zehentgruber, D., Lundemo, P., Svensson, D., & Adlercreutz, P. (2011). Substrate complexation and aggregation influence the cyclodextrin glycosyltransferase (CGTase) catalyzed synthesis of alkyl glycosides. *Journal of Biotechnology, 155*(2), 232–235.

Zhu, Y., Li, C., Cui, H., & Lin, L. (2021). Encapsulation strategies to enhance the antibacterial properties of essential oils in food system. *Food Control, 123*, 107856.

Zuidam, N. J., & Heinrich, E. (2009). Encapsulation of aroma. In *Encapsulation technologies for active food ingredients and food processing*. Springer New York, 127–160.

Index

Note: Page numbers followed by "*f*" and "*t*" refer to figures and tables, respectively.

D

Q

R

S

Printed and bound by CPI Group (UK) Ltd, Croydon, CR0 4YY
22/09/2026
02220473-0016